T0326391

The enforceability of the human right to adequate food

Also available in the 'European Institute for Food Law series':

European Food Law Handbook
Bernd van der Meulen and Menno van der Velde
ISBN 978-90-8686-082-1
www.WageningenAcademic.com/foodlaw

Fed up with the right to food?
The Netherlands' policies and practices regarding the human right to adequate food
edited by: Otto Hospes and Bernd van der Meulen
ISBN 978-90-8686-107-1
www.WageningenAcademic.com/righttofood

Reconciling food law to competitiveness
Report on the regulatory environment of the European food and dairy sector
Bernd van der Meulen
ISBN 978-90-8686-098-2
www.WageningenAcademic.com/reconciling

Governing food security
Law, politics and the right to food
edited by: Otto Hospes and Irene Hadiprayitno
ISBN: 978-90-8686-157-6; e-book ISBN: 978-90-8686-713-4
www.WageningenAcademic.com/EIFL-05

Private food law
Governing food chains through contract law, self-regulation, private standards, audits and certification schemes
edited by: Bernd van der Meulen
ISBN: 978-90-8686-176-7; e-book ISBN: 978-90-8686-730-1
www.WageningenAcademic.com/EIFL-06

Regulating food law
Risk analysis and the precautionary principle as general principles of EU food law
Anna Szajkowska
ISBN: 978-90-8686-194-1; e-book ISBN: 978-90-8686-750-9
www.WageningenAcademic.com/EIFL-07

european food law association
association européenne de droit alimentaire

The enforceability of the human right to adequate food

A comparative study

Bart F.W. Wernaart

Wageningen Academic
P u b l i s h e r s

ISBN: 978-90-8686-239-9
e-ISBN: 978-90-8686-791-2
DOI: 10.3920/978-90-8686-791-2

ISSN 1871-3483

First published, 2013

© Wageningen Academic Publishers
The Netherlands, 2013

Wageningen Academic Publishers
P.O. Box 220
6700 AE Wageningen
The Netherlands
www.WageningenAcademic.com
copyright@WageningenAcademic.com

The content of this publication and any liabilities arising from it remain the responsibility of the author.

Preface

This research marks the end of an extremely intense, inspiring, vivid and turbulent period. I would like to go back 12 years, when this all started as I made the choice to study international law at Tilburg University. At the same time I was studying drums and mallets at the Conservatory in the same city. I gradually found my way through the University curriculum, and ran into the ear-splitting lectures of Mr Frank Vlemminx, PhD. While in this book it will become apparent that I do not always agree with his points of view, he most certainly inspired me to learn more about this wonderful field of expertise that is called human rights. Since we had a thing or two in common, not in the least the fact that Frank was not only a lawyer but also artistically active as a painter, he would later be my thesis supervisor. During an internship at Wageningen University and Research Centre, I became acquainted with Professor Bernd van der Meulen, who introduced me more specifically to one of the most violated rights in the world: the right to adequate food. After graduating, I found a job as lecturer at Fontys University of Applied Sciences. In the meantime I was composing a rock opera to finish my second study at the Conservatory: conducting. My employer gave me the opportunity to start a research project, and I became an external PhD student at the WUR, under the supervision of Bernd. I was now able to do all the things that I admired doing: teaching, playing music, and doing research. A perfect match for which I am ever grateful.

Finalising this book marks the end of a period of approximately five years of conducting intense research. I cannot even begin to imagine how much I have learned from the whole exercise. Writing a PhD thesis is a very lonely experience. However, during this period, I was never truly alone. I therefore wish to thank many people from the bottom of my heart.

First of all, of course Bernd, who never lost his patience in guiding me through the whole process. He has the rare gift of making a person feel encouraged, determined and self-assured about the research on the way back home, while feeling completely lost on arrival.

Next to that all my colleagues at Fontys University, who were always very supportive and interested in my progress, and to the management of my department, which gave me the financial support and time necessary to complete this book.

Furthermore, I would like to thank all my students whom I had the joy of teaching throughout the years. I would dare to say that I've learned at least as much from you as you might have learned from me.

For me, music is my first and primary way of expressing myself. Throughout the year, I have had the honour to play music in my role as a drummer and conductor with many people. A special thank you goes to my percussion group in Valkenswaard and both my choirs in Heusden, for sharing your talents with me. You have no idea what the beat of your drums and the sound of your voices mean to me. A big applause for my rock band the Seasons, for we did some incredible things lately, and to the Ameezing band and vocals: it seems that I've become Bart Wernaart, PhD, after all.

Also, I would like to thank my friends. Some of them I have known for a lifetime, others just recently jumped into my life. I count myself lucky with such a wonderful group of amazing persons who keep surprising me with the humour, wisdom and solid friendship they share with me.

My family in law-to-be, for their continuous support and encouragement.

My wonderful brothers, Peter and Geert. If a comparist would compare us, the conclusion would be that we are so similar in our differences. It is a joy to see the paths they chose so far, and I am extremely proud of them.

My father who, from the very beginning, believed in my scientific capabilities, is one of the most intelligent persons I have ever encountered, and has been simply indispensable during the whole process.

Then, I am blessed with two power-women around me.

My mother, whose passion for teaching the very young amongst us certainly ended up somewhere in my DNA, and who has always been a safe haven when times were a bit rough.

My fiancée, Sylvia, the love of my life, who said 'yes' to my marriage proposal. Being with a man who is from time to time losing himself somewhere between books, reports, documents and Court rulings requires a lot of love and determination. I will not try to put into words how thankful I am for that.

Bart Wernaart

Valkenswaard 2013

Table of contents

Abbreviations

(A selection of the most relevant and frequently used abbreviations was made)

AB	Administratiefrechtelijke Beslissingen (Dutch case law on administrative law)
AU	African Union
BS	Belgische Staatscourant (Belgian Government Gazette)
CA	Constitutional Act
CEDAW	Convention on the Elimination of All Forms of Discrimination Against Women
CESCR	Committee on Economic, Social and Cultural Rights
COA	Centraal Orgaan opvang Asielzoekers (Dutch reception organisation for asylum-seekers)
CSO	Civil Society Organisation
EC	European Community
ECHR	European Convention on Human Rights
ECOSOC rights	Economic, social and cultural rights
ESC	European Social Charter
EU	European Union
FAO	Food and Agriculture Organisation
ICESCR	International Covenant on Economic, Social and Cultural Rights
ICCPR	International Covenant on Civil and Political Rights
ICRC	International Convention on the Rights of the Child
IGWG	Intergovernmental Working Group
ILO	International Labour Organisation
JV	Jurisprudentie Vreemdelingenrecht (Dutch case law on aliens' law)
LJN	Landelijk Jurisprudentie Nummer (Dutch case law numbering)
NATO	North Atlantic Treaty Organisation
NGO	Non-governmental Organisation
NJB	Nederlands Juristenblad (Dutch legal magazine)
OAS	Organisation of American States
OAU	Organisation of African Union
OP-ICESCR	Optional Protocol to the International Convention on Economic, Social and Cultural Rights
UN	United Nations
UPR	Universal Periodic Review
OCMW	Organieke wet betreffende de openbare centra voor maatschappelijk welzijn (Organic Law on Public Centres for Social Welfare)
SK	Sociaalrechtelijke Kronieken (Belgian case law of social law).
TAR	Tijdschrift voor Ambtenarenrecht (Dutch case law on civil servant law)
USZ	Uitspraken Sociale Zekerheid (Dutch case law on social security)
WHO	World Health Organisation

Part 1
Introduction, methodology
and *tertium comparationis*

I. Introduction

1.1 A right for all and a right for each

> It is paradoxical, but hardly surprising, that the right to food has been endorsed more often and with greater unanimity and urgency than most other human rights, while at the same time being violated more comprehensively and systematically than probably any other right.[1]

Although Philip Alston wrote this line back in 1984, it is unfortunately still an accurate reflection of reality.[2] Worldwide, there are too many people who have no access to adequate food. This means that worldwide, there are too many *persons* who have no access to adequate food. The slight difference between the two previous sentences perhaps explains why this thesis was written in the first place, as will be discussed below.

It is not surprising that the right to food is more urgently recalled in the context of developing countries, where ravaging famines but also less obvious undernourishment problems are a painful reality. The right to food may then be discussed in the context of complex global issues, closely related to matters of high politics and the world economy. In this light, many valuable research has been conducted, and inspiring books were written, all unanimously – although from different perspectives – concluding that hunger should be put to an end, is an unnecessary evil, and a shame on humanity: *'hunger is a political condition'*[3], *'hunger is not a question of fate: it is manmade'*[4], *'hunger, disease, the waste of lives that is extreme poverty are an affront to us all'*[5], *'the global community stands indicted for knowing much about how to reduce hunger, but not doing so'*[6], *'theirs is no reason that world hunger should continue'*[7], *'this is a silent holocaust, repeated year after year.'*[8] Of course, those statements and their urgency can only be underlined. While the right to adequate food is often discussed in the context of global issues, involving a large number of people, focusing more on the right to adequate food

---

[1] P. Alston and K. Tomasevski (eds.), The right to food, International Studies in Human Rights, Utrecht, 1984, p. 9.

[2] See for the most accurate date, the reports of the Special Rapporteur on the right to adequate food, available at: www.srfood.org, or the Food and Agricultural Organisation website: www.fao.org.

[3] G. McGovern, *Third Freedom, Ending hunger in our time*, Rowman & Littlefield Publishers Inc: Lanham, 2001.

[4] E/CN.4/2003/54, 10 January 2003, Section 7, report of the Special Rapporteur on the Right to food submitted to the Commission on Human Rights.

[5] Bono, in: J.D. Sachs, *The end of poverty, economic possibilities for our time*, London: Penguin Books, 2006.

[6] C.F. Runge, B. Senauer, P.G. Pardey and M.W. Rosegrant, *Ending hunger in our lifetime, food security and globalization*, Baltimore and London: the Johns Hopkins University Press, 2003.

[7] J. Ziegler, C. Golay, C. Mahon and S.-A. Way, *The fight for the right to food, lessons learned*, Geneva: The graduate institute, 2011.

[8] G. Kent, *Freedom from want: The human rights to Food*, Washington D.C.: Georgetown University Press, 2005.

'to all', this thesis starts from a different perspective, focusing on the right 'to each' individual. The Universal Declaration of Human Rights stipulates that *'all people are born free and equal in dignity and rights.'*[9] This reflects the principle idea of human rights: simply because a person is born, that person, without exception, has human rights, to live a life in dignity. Adequate food is one of those rights, as recognised by many international human rights treaties.[10] Within the framework of the UN human rights system, and elsewhere, the importance of an enforceable international human right to adequate food has often been stressed. A right is after all not a right if it is not enforceable, and individuals are not empowered to claim their rights.[11] However, considering the global issues that relate to hunger and malnutrition, it is not easy to talk about individuals on such a massive scale, let alone the enforceability of such a right in an operating an accessible Court. In such a situation, it is questionable whether an individual will find that the fact that he or she was born, and therefore has a right to adequate food, will make any difference in their case, and matters will probably have to be solved at a global or near-global level. But what about a country in which theoretically the circumstances allow this right to be enjoyed by 'each' individual? A country that has an effective and accessible Judiciary, has rule of law, is democratic, is relatively prosperous, has an extensive system of social security, and has a domestic food production that exceeds the population's need? Since there are always people, even in the most developed countries, which somehow fall through all the safety nets that are in place, would a human rights then make any difference for them? Human rights then naturally may function as an ultimate safety net, with the purpose to catch every single individual that somehow falls through the cracks of the national safety net not being able to have access to her/his most basic needs in order to lead a life in dignity. In that situation, would the right to adequate food be of any help for the most vulnerable in society? In other words: would this right make a difference for that single individual searching for food somewhere in a trash can? It is after all no new information that *'even in functioning democracies, the Courts do not always succeed in enforcing rights especially for the poor, the marginalised, and the unpopular.'*[12]

This might be demonstrated by the fact that during the writing process of my master thesis I ran into two Court rulings in which direct applicability of the right to food was discussed. One ruling came from the Dutch Central Court of Appeal,

---

[9] 217 A (III), 10 December 1948, *The universal declaration on human rights*, Article 1.

[10] See into detail: Chapter 3.

[11] See Chapter 3.

[12] See the contribution to the preface by K. Roth, at the moment of writing Executive Director of Human Rights Watch, in: D. Moeckli, S. Shah, S. Sivakumaran (eds.), D. Harris (cons. ed.), *International human rights law*, Oxford University Press, 2010.

the other from the Belgian Constitutional Court[13], both a Court of Last Instance.[14] The cases show, despite their different national legal context, one important resemblance: an illegally residing foreigner can *inter alia* invoke the right to an adequate standard of living, as stipulated in Article 11 ICESCR. The Dutch Court ruled, freely translated, that Article 11 ICESCR was not generally binding and by its nature not directly applicable, for its content is not sufficiently precise to distil concrete rights that can be invoked in a Court of law.[15] The Belgian Court had a different approach, ruling that it was not unreasonable to assume different obligations regarding the needs of persons lawfully residing on Belgium territory on the one hand, and illegally residing persons on the other hand, implying that only a legally residing person might successfully invoke the right to an adequate standard of living. To put this in other words: in both cases, while both Courts seemed reluctant in applying Article 11 ICESCR, invoked by an illegally residing asylum-seeker, the Dutch Court came to its conclusion using *trias politica* arguments to justify non-application of the norm, while the Belgian Court simply excluded the requesting party from the *ratio personae* of the invoked norm.

At first glance, this legal practice certainly does not correspond with the reports both countries submit to various human rights institutions. In their country reports on the implementation of the ICESCR, both the Netherlands and Belgium consistently report that the countries fulfil their obligations under Article 11 ICESCR. This is basically done by summing up all the initiatives taken during the reporting period, mostly in the field of social security, and assuring that the countries have a food supply that is more than adequate to fill the population's needs.[16] Also, the legal practice as described above does not seem to be widely known in the international arena. For instance, in several researches on the embedding of the right to food in the world's national Constitutions, the FAO concluded that in both the Netherlands and Belgium the right to food was implicitly recognised in the Constitution, by recognising a broader right: in the Netherlands, the right to an adequate standard of living (Article 20 (1) of the Constitution) and in Belgium the right to the means necessary to lead a dignified life (Article 23 of the Constitution).[17] According to the same research, the Dutch Constitution is amongst a small group

---

[13] At that moment under its former name, the Court of Arbitration.

[14] The inspiration for the search for these rulings came after reading a contribution of my former supervisor at Tilburg University for the Dutch Association for Comparative Law, in which he briefly referred to a different approach between Dutch and Belgian Courts regarding direct applicability of certain economic, social and cultural rights, including Article 11 ICESCR: F.M.C. Vlemminx, *De autonome rechtstreekse werking van het EVRM, De Belgische en Nederlandse Rechtspraak over verzekeringsplichten ingevolge het EVRM*, Deventer: Kluwer, 2002a.

[15] Central Court of Appeal, 25 May 2005, LJN AP0561.

[16] See for the Netherlands for example: E/C.12/NLD/4-5, 17 July 2009 Sections 219-234; see for Belgium for example: E/C.12/BEL/3, 21 September 2006, Sections 418-558 (especially 541-558). See for more details: Chapters 9 and 13.

[17] L. Knuth and M. Vidar, *Constitutional and legal protection of the right to food around the world*, Right to food Studies, Rome: FAO, 2011.

of four countries in which the Constitution clearly states that treaties have an equal or higher status than national law (reference was made to Article 94 of the Constitution). Furthermore, '*several sources, including case law, laws on treaties and reports to the human rights bodies*' indicated that Belgium is amongst another group of countries in which there is supremacy of international law over national law. The research suggests that in the Netherlands and Belgium, the right to food is indirectly embedded in the Constitution, and the Constitutional mechanisms of both countries ensure that the right to food as embedded in the international treaties signed by the Netherlands and Belgium is directly applicable in the national legal order of these countries.[18]

In addition, the legal practice appears not to be in line with the UN and other human rights institutions purpose to realise human rights for each individual to deny its effect in Court cases. Although it is certainly not undisputed in the global arena, as will be demonstrated in Chapter 3, the obligations under Article 11 ICESCR include – according to various UN organs and scholars specialised in the matter – that the right should be enforceable through the national Courts.

1.2 Research objective and questions

It is the main objective of this research to contribute to the improvement of the effective realisation of the right to food through enforceability. Therefore, the main research objectives are:
1. to gain knowledge about the enforceability of the right to food as embedded in UN human rights instruments in the Netherlands and Belgium through law comparison; and
2. where necessary critically evaluate both approaches in view of the UN human rights doctrine regarding the enforceability of the right to food.

It is expected that comparing the Dutch practice with the Belgian leads to more insight regarding the implementation of the right to food in two countries in which nothing of significance stands in the way of an enforceable right to food. That legal practice is translated into periodic reports under the reporting obligations recognised by the Member States of the UN human rights treaties. This is the basic source of information for the UN treaty bodies, which greatly influences their view on the effect of the human rights standards in the domestic legal order of the reporting countries. This means that some data might get 'lost in translation', so to speak, as the above example suggests. To put this in other words: the Judiciary may respond differently to the question whether or not the right to adequate food has direct effect than the Governments through their reporting. Also, as discussed above, the response of the Judiciaries and Governments of the Netherlands and

---

[18] L. Knuth and M. Vidar, *Constitutional and legal protection of the right to food around the world*, Right to food Studies, Rome: FAO, 2011; see in particular the conclusion, Chapter 6.

Belgium may be different compared to the expected legal practice according to the United Nations human rights system. Therefore, the main research question of this thesis is:

> What are the legal factors that explain the differences and similarities regarding the response of the Dutch and Belgium Judiciaries and Governments to the enforceability of the right to adequate food in view of the UN human rights system?

This research thus implies a triple comparison.
I.   a comparison between the legal practice in Belgium and the Netherlands (what the countries really do);
II. a comparison with those legal practices and the reporting behaviour of both countries (what the countries say they do); and
III. a comparison between the legal practice and the interpretation on the enforceability of the right to food within the UN human rights system (what the countries should do).

It is the comparing process, as discussed in Chapter 2, that leads to in-depth insight, and provides multiple dimensions in the evaluating process. To be able to compare, it is necessary to first collect and analyse the relevant data. Because of that, three sub-questions need to be answered:
1. To what extent is the right to adequate food perceived to be an enforceable right within the UN human rights system?
2. A. What is the response of the national Judiciaries of the Netherlands and Belgium when the right to food as stipulated in the UN human rights system is invoked by individuals?
   B. And how can this response be explained?
3. What do the Governments of the Netherlands and Belgium communicate in their reports in the United Nations arena regarding the enforceability of the right to food in their domestic legal order?

This data then will be compared and critically evaluated in view of the UN human rights system, and where possible, recommendations to all actors involved will be made for improving the realisation of the right to food through enforceability.

1.3 Demarcations, terminology and references

As will be demonstrated in Chapter 3, the field of human rights is closely interrelated to complex global and national issues. In this light it is not easy to address one single issue, without addressing another, correlated problem. Therefore, it is important to be very clear on the boundaries of this research, to avoid a superficial or a too general approach, or to avoid writing a book that is overly extensive and unreadable. Also, there is a need to clarify some terminology and the use of references in this book.

1. This thesis is about the enforceability of the right to adequate food as embedded in UN human rights treaties. It does not concern (if there is any) the enforceability of a national right to food, nor does it concern national standards that are adopted with the purpose of realising the international right to food in the national legal order. Although the implementation on the national level through legislation of the right to food is one important obligation of the Member States of the UN human rights treaties, it is a different obligation than ensuring that the right to food is directly applicable in the national Courts. The last obligation is important in view of the purpose of a human rights in a developed country: to offer a safety net when all other measures fail to work. This is especially important considering the fact that both the Netherlands and Belgium are of the opinion that they generously fulfil all the obligations implied by the right to food. If that were true, no one would have to invoke the Article in front of a Court.

2. During the research process, I found out that within the context of certain research questions the right to food as such was not an issue in itself. Therefore, instead, it was frequently necessary to explore a broader range of human rights (Economic, Social and Cultural Rights in general), that were treated and considered in a similar way, as one category of rights. So, where possible, my research focuses on the human right to adequate food, and where necessary the focus was widened to ECOSOC rights in general.

3. As will be demonstrated in Chapter 3, the right to food is embedded in the international human rights system in many different ways. In this research, there will be a focus on the UN Provisions that most directly stipulate the right to adequate food. Those include Article 11 ICESCR, Articles 24 and 27 ICRC, and Article 12 CEDAW. Occasionally however, especially in the Chapters on the reporting cycles (Chapters 9 and 13) there is reason to also discuss other Provisions, due to the fact that the Netherlands and Belgium may assume a strong inter-relationship between the right to food and the other Provisions, or chose to report relevant matters on the implementation of another Provision. The latter may also be caused by the UN reporting guidelines and formats. This will then be discussed in the relevant Chapters.

   The ESC is mainly excluded from this research, as will be discussed in Section 2.4, due to the fact that the ESC does not stipulate the right to food, most individuals who invoke the right to food do not fall under the *ratio personae* of the ESC, and including the ESC would hinder the establishment of a coherent *tertium comparationis*. As it will appear however, for other reasons an analysis of the Dutch case law on the direct applicability of Article 6 ESC and the corresponding Parliamentary History will be necessary to eventually determine the criterion used amongst the Dutch Courts to establish whether or not an international provision is directly applicable.

4. This thesis explores the enforceability of the international right to adequate food by individuals. 'Individuals' does not mean the same as citizens. This

thesis is about the enforceability of the international right to adequate food by each individual, regardless the residential status of the person.

5. The Netherlands are a country that are part of the Kingdom of the Netherlands, which also included Aruba and the Netherlands Antilles until the reforms of 2010, and since then includes Curaçao, St. Maarten and Aruba as sovereign States, and Bonaire, St. Eustatius and Saba as special municipalities. Since it is the Kingdom of the Netherlands that signs and ratifies the UN treaties, they also apply to the overseas parts of the Kingdom, and thus it is the Kingdom that is responsible for submitting the reports on the implementation of human rights in all parts of the Kingdom. Furthermore, the Supreme Court is also Court of Last Instance for the overseas territories. No evidence could be found that in the overseas territories a profoundly different legal doctrine exists on the enforceability of the right to food. However, considering the research questions and objective of this thesis, the focus of interest should be on the European part of the Kingdom. Therefore, but only when of significant relevance, the case law and reports of the overseas territories are included in this thesis. This is especially discussed in more detail in Section 9.1.2.

6. In both literature and legal practice, a variety of terms is used to describe the effect of international law in domestic law. The Dutch Courts use the term 'directly applicable' as an equivalent to 'binding for all persons' (the latter standard is stipulated in Articles 93 and 94 CA), while the terms 'direct effect' and 'self-executing' can be found as well. All terms are also used here to describe a similar concept.[19] The Belgian Courts prefer to use the wording 'direct effect.' In the international arena, mostly terms as 'justiciability', and 'self-executing' are being used. The Committee on Economic, Social and Cultural Rights seems to consider a standard as 'self-executing' when it can be applied without the interference of a national Legislator, while 'justiciability' in this context covers a broader category of standards that may have some effect in case law. 'Self-executing' standards in the jargon of the UN are more or less similar to the national concepts of standards that are 'binding for all persons' or have 'direct effect.' To avoid national or international interpretations in the problem statements or research questions, a meta-standard is used in this research as a generic term, which is 'enforceability.'

7. When referring to sessions with a treaty body on a report on the implementation of a UN treaty, reference is made to the body as one entity, while during those meetings usually one body Member asks questions. Since the body, outside the scope of those meetings, does act as one entity, this reference is continued when referring to the sessions. However, to be more accurate, in the footnote reference is made to the exact Committee Member. For reasons of consistency, the same is done when referring to a Delegate of the Netherlands or Belgium.

8. The reference of the sources is as much as possible in accordance with the generally accepted rules of academic legal writing style. Furthermore, especially

---

[19] C.A.J.M. Kortmann, *Constitutioneel recht* (6ᵉ druk), Deventer: Kluwer, 2008, pp. 183-185.

in case of references to specialised sources such as Parliamentary History, case law, UN documents or other international sources of a non-literary nature, the references are similar to the way of referring of those bodies that create the sources or are mostly in use in the field. In the case law analyses, especially concerning the more recent cases, I frequently chose to refer to the well-organised online databases instead of using the more traditional approaches in referring to publications of selected legal case law.

9. While I studied sources in four languages (English, Dutch, French, and occasionally German), the language used in this thesis is English only. Therefore, to avoid misunderstandings on languages, quotes from different languages were translated into English for the main text, and the quote in its original language is included in a footnote.

10. Legal practice is an ever changing and lively field of expertise. During the writing process of this thesis, the text was rewritten several times to catch up with the latest developments in the field, making sure that this book is as accurate as possible. However, one needs to draw a line somewhere. This thesis is up to date through the end of 2012. While there are no limitations back in time regarding the case law analyses, there is a strong focus on the period 2000-2012, since this would give the most accurate and up-to-date impression on the legal practices.

1.4 Book structure

One might expect a section on the structure of a book in the first Chapter. However, since the organisation of this book is greatly determined by the methodology used, it makes more sense to discuss this in the Chapter on methods. Therefore, an explanation on the book structure will be included in Section 2.7.2.

2. Methods

2.1 Introduction

As demonstrated in Chapter 1, the purpose of this research is '*to gain knowledge about the enforceability of the right to food as embedded in international human rights instruments in the Netherlands and Belgium, and where necessary critically evaluate both approaches in view of the UN human rights doctrine regarding the enforceability of the right to food.*' It is especially the comparison of two countries in view of the international human rights system that may lead to more insight, since the Courts in both countries appear to reason differently regarding the implementation of the same right in a similar case, and the international communication of the countries seems not to correspond with the actual legal practice.

The methods used in this study therefore naturally include legal comparative methods. However, there are two reasons to show some restraint in classifying this research as a comparative legal study. Firstly, as the scientific quality of comparative law is not uncontested and its methods are not (yet) very coherent, I need to address the question to what extent comparative law as such is a toolbox from which tools can be taken legitimately to conduct scientific legal research, and where comparative methods might be complemented with other legal methods, such as legal theory on legal reasoning. Secondly, this research focusses on an international standard that has been adopted for a large part in the UN human rights framework. This system is based on certain ideas on law, including the idea of a top-down approach towards human rights, with the ultimate goal of successful enforcement by the national Courts, including enforceability for those who invoke the right in a trial. In fact, domestic enforcement is the core focus of interest of this research. Starting from these rather defined presumptions on how international law should work, may however oppose to the more flexible nature of most methods on comparative law: as will appear below, most comparists prefer to start from more neutral perspectives.

This Chapter explores what methodological tools are most suitable for this research. To this end, an impression will be given of the state of the art of comparative law in Section 2.2. As it will appear, in comparative law literature, certain issues are almost unanimously recurring topics, and from a methodological perspective deserve further elaboration in view of this research. Therefore firstly, in Section 2.3, the purpose of the law comparison will be discussed. Secondly, in Section 2.4, the choice of country will be further elaborated. Thirdly, comparative methods in itself will be discussed, starting from the traditional functional method as proposed by Zweigert and Kötz (Section 2.5.1), criticism on this method (Section 2.5.2), and alternatives to this method (Section 2.5.3). Then it is time to consider what specific needs this research has in relation to a comparative method (Section 2.5.4), and based on that, a choice will be made for a particular understanding of a functional method, inspired by Ralf Michaels (Section 2.5.5). Subsequently, in Section 2.6, I will discuss the research direction that follows from the constructive move as proposed by Michaels. In Section 2.6.1, the research direction of the first sub-research question will be discussed. In Section 2.6.2, the direction of the second sub-research question will be explored, which involves further clarification on legal reasoning methods. In Section 2.6.3, the research direction of the third sub-research question will be discussed, and in Section 2.6.4 the approach concerning the three comparisons and evaluation will be explained. In Section 2.7, some concluding remarks will be made, and the structural layout of the book will be explained.

2.2 Comparative law methodology

Comparative law as an applied method is not a recent phenomenon.[20] It has mostly been used for practical purposes such as transferring law to other legal systems (e.g. in a colonisation context), borrowing foreign law (e.g. in domestic law making processes), and unifying law (e.g. in the creation of European Union legislation). As a science however, often perceived to have emerged during the 1900 Congress for Comparative Law in Paris, comparative law has always been regarded with some suspicion, lacking a clear methodology,[21] and operating somewhere on the edge of law, sociology, anthropology, political science, history and other fields of science, which may suggest that comparative law has a multi-disciplinary or even interdisciplinary flavour.[22] For that matter, some authors prefer to classify comparative law only as a method, or consider only macro-comparison as a true science.[23] If there has been an established scientific method for comparative law, this must have been the functional method, proposed by Zweigert and Kötz in a very short Chapter on methodology.[24] While appearing to have had a sort of a monopoly position on a method for comparative law for many years, their functional method has been criticised by various authors, while it has not been replaced by a coherent theory on methodology yet. It can be said that comparative law as a science is going through a very interesting period, on the one hand lacking a clear method and therefore apparently an incomplete field of science, while on the other hand, as a result, very vivid and creative texts are published, with an open minded, positive and fresh approach to the future of comparative law.

---

[20] C. Donahue, *Introduction, comparative law before the Code Napoléon*, in: M. Reimann and R. Zimmermann (eds.), *The Oxford handbook of comparative law*, Oxford: Oxford University Press, 2006; K. Zweigert and H. Kötz, *An introduction to comparative law*, Oxford University Press, 1998, Chapter 34.

[21] E. Örücü, *Chapter 2, Developing comparative law*, in: E. Örücü and D. Nelken (eds.), *Comparative law, a handbook*, Portland: Hart Publishing, 2007.

[22] D. Nelken, *Chapter 1, Comparative law and comparative legal studies*, in: E. Örücü and D. Nelken (eds.), *Comparative law, a handbook*, Portland: Hart Publishing, 2007.

[23] F. Gorlé, G. Bourgeois, H. Bocken, F. Reyntjens, W. de Bondt and K. Lemmens, *Rechtsvergelijking*, Mechelen: Kluwer, 2007, Section 24.

[24] K. Zweigert and H. Kötz, *An introduction to comparative law*, Oxford University Press, 1998, Chapter 3.

Regardless of the apparent lack of a coherent methodology for comparative law, some methodological issues are commonly discussed in literature, and need some further elaboration here as well, for determining in this research to which extent these discussions may contribute to the formulation of a suitable comparative legal method.

2.3 Purposes of comparative law

In most major works on comparative law, the need is felt to dedicate some special thoughts to possible purposes of comparative law. Comparative law can be used for many purposes, mostly depending on the aim or profession of the comparist. Zweigert and Kötz originally introduced five functions of comparative law: knowledge, an evaluative function, an interpretational function, an educational function and a function to unify law. This classification is still widely used amongst comparative lawyers.

Of course, firstly, knowledge is the primary aim of law comparison whereby the research aim is to gain insight in different (legal) solutions to a social problem. That would not be possible by ('normal') legal research that would not cross boundaries.[25] Although 'knowledge' as a purpose for a science is rather plain, some interesting thoughts can be dedicated to the matter, as will be demonstrated below when discussing Michaels' view on functionalism. In this research, it is expected that the in-depth knowledge on the enforceability of the right to food in the Netherlands and Belgium will offer valuable insights in the relation between an international human rights standard and domestic legal influences on its enforceability in the national Courts, that would normally not be available for institutions in the global arena that strive for the realisation of this right, and contributes to the national debates on the enforceability of human rights. This objective of comparative law seems to be the most scientific one (or the least practical one if you like), and can be distinguished from most other functions that are generally introduced. These functions namely have a more practical background, and are in most literature described from the perspective of the Legislature (evaluative function), a judge (interpretational function), or international institutions (unifying law function).

A second function distinguished by Zweiger and Kötz is the evaluative function, mostly referred to as to determine the better law, often for legislative purposes: in the legislative process, a Legislator could choose to compare different legal solutions to a social problem, and use the superior solutions he/she finds in other legal systems as a guide for its own design for a legal solution.[26] Zweigert and

---

[25] K. Zweigert and H. Kötz, *An introduction to comparative law*, Oxford University Press, 1998, Chapter 2, Section I.

[26] K. Zweigert and H. Kötz, *An introduction to comparative law*, Oxford University Press, 1998, Chapter 2, Section II.

Kötz even consider the evaluation of what has been discovered as a duty of the comparist. They consider that – despite the fact that evaluating is a subjective exercise – if the comparist is not doing it, probably no one will.[27] Some authors consider such a purpose invalid. Glenn for instance, while referring to the two Latin words that are united in 'comparc' (*cum* and *pare*), argues that literally, comparing means the bringing and keeping together of equals. In this light, he underlined that comparing should not be about determining the better law, but finding the similarities and differences in peacefully co-existing legal traditions.[28] Michaels calls for restraint in determining the better law as well, arguing that if one wants to evaluate different legal solutions to a problem, the evaluation criteria should at least be different from the criteria the comparison is based on.[29] In this research however, it is certainly one of the aims to critically evaluate the data that is found in both countries in view of the UN human rights system. As demonstrated in Chapter 1, this research was conducted from a strong belief that it is of the utmost importance to strive for better realisation of the human right to adequate food. Apparently, domestic enforcement of this right in the Netherlands and Belgium, both countries in which the realisation of this right should not constitute too big a problem, is hard to achieve. Only answering the question why this is the case would prove to be an interesting, but also an incomplete exercise: it is the ambition of this research to also critically evaluate the legal practice in both countries concerning the enforceability of the right to food in view of the UN human rights system, and make recommendations for the countries compared as well as the UN concerning the matter. Of course it will never be the intention to determine *in abstracto* which legal system is the better one,[30] and as much as possible both compared countries will be considered to be equal. In that sense I will take heed of Glenn's warning. But it is the belief of the author, in line with Zweiger and Kötz, that the job of a (comparative) lawyer is not only to make sense of legal data, but also to have an opinion on that legal data, in order to determine what both countries might learn from one another, and to contribute to the improvement of the effective realisation of the right to food through enforceability, by critical evaluation and recommendations.

A third function is the use of law comparison for the construction of Court rulings.[31] In this, the dialogic method discussed below is commonly used. Inspiration from rulings abroad may contribute to a broader view on the domestic interpretation

---

[27] K. Zweigert and H. Kötz, *An introduction to comparative law*, Oxford University Press, 1998, Chapter 3, Section VII.

[28] H.P. Glenn, *Chapter 4, Comparing*, in: E. Örücü and D. Nelken (eds.), *Comparative law, a handbook*, Portland: Hart Publishing, 2007.

[29] R. Michaels, *Chapter 10, The functional method of comparative law*, in: M. Reimann and R. Zimmermann (eds.), *The Oxford handbook of comparative law*, Oxford: Oxford University Press, 2006.

[30] See also: R. Michaels, *Chapter 10, The functional method of comparative law*, in: M. Reimann and R. Zimmermann (eds.), *The Oxford handbook of comparative law*, Oxford: Oxford University Press, 2006.

[31] K. Zweigert and H. Kötz, *An introduction to comparative law*, Oxford University Press, 1998, Chapter 2, Section III.

of legal rules. Also, concerning the interpretation of international law, it might lead to a more coherent application of these standards, or can be argued to result in '*ius commune*' regarding the implementation of, for instance, human rights.[32] In this research, as will appear from Chapters 4 and 12, the legal reasoning of the Courts was hardly based on a dialogic method.

Fourth, Zweigert and Kötz argue that comparative law should play an important role in legal education, to broaden the student's mind.[33] May this research contribute to that goal!

Finally, law unification as a purpose of comparative law can be traced back to 19th century ideas on natural law, where comparists used to set the aim to find universal or natural law through law comparison, a goal set for instance during the 1900 Congress for Comparative Law in Paris. Since the Second World War, such ideas on the existence of natural law were abandoned, most likely due to the formation of socialist countries and the decolonisation process. Instead, during the post-war period attempts were made to harmonise law in different international arenas concerning a variety of legal issues, in particular for maintaining peace and security.[34] A highly developed example is the harmonisation of European economic law. The European Legislator frequently compares the national laws of the EU Member States preceding the adoption of a supranational standard, in a process of harmonising (economic) legislation amongst the Member States. Of course, also the worldwide adoption of several human rights instruments may be seen as a form of harmonising/unifying law, although the high moral content of these standards amongst human rights lawyers often leads to an opposite starting point in which the position is defended that instead of a process of unifying different legal standards, there is already a universal standard every human being should be able to invoke, simply because she/he was born. From that point of view all human rights, as if they were natural law, are the same throughout the world. Perhaps not surprisingly, there is some criticism on such universal approach of human rights amongst comparative lawyers, whose job it naturally is to determine differences and similarities in law: '*Natural law thinking failed to appreciate that law is rooted less in a universal condition than in the specific conditions of different cultures.*'[35] And: '*Similar legal concepts can mean different things in different contexts. The lesson for human rights lawyers is that they ignore the different institutions contexts in which interpretation takes place and the different power relations in these jurisdictions*

---

[32] C. McCrudden, *Chapter 16, Judicial comparativism and human rights*, in: E. Örücü and D. Nelken (eds.), *Comparative law, a handbook*, Portland: Hart Publishing, 2007.

[33] K. Zweigert and H. Kötz, *An introduction to comparative law*, Oxford University Press, 1998, Chapter 2, Section IV.

[34] F. Gorlé, G. Bourgeois, H. Bocken, F. Reyntjens, W. de Bondt and K. Lemmens, *Rechtsvergelijking*, Mechelen: Kluwer, 2007, Section 23.

[35] R. Cotterrell, *Chapter 6, Is it so bad to be different?* in: E. Örücü and D. Nelken (eds.), *Comparative law, a handbook*, Portland: Hart Publishing, 2007.

at their peril.'[36] According to McCrudden, using law comparison to unify law in this context may lead to 'persuasive comparison', providing for arguments that justify a single undisputed interpretation of human rights,[37] while despite the fact that most human rights instruments will sum up almost the same list of human rights in general terms, their specific meaning is definitely not understood in one undisputed way, and subject to local cultural particularities.[38] This kind of criticism on a universal approach towards human rights offers a valuable explanation on why human rights might not work as universal as originally intended on the one hand, and gives some food for thought on the usefulness to strive for a similar understanding of this right. On the other hand, in the context of this research, it does little more than advising to be cautious not to err into persuasive comparison when trying to find out what specific domestic influences cause differences or similarities in the understanding of human rights amongst Courts. While the standard under investigation may be the result of an attempt to unify law, it is not necessarily the purpose of this research.

To conclude, in this research, as discussed in Chapter 1, it is the ambition:
* to gain knowledge about the enforceability of the right to food as embedded in international human rights instruments in the Netherlands and Belgium through law comparison; and
* where necessary to critically evaluate both approaches in view of the UN human rights doctrine regarding the enforceability of the right to food.

In comparative law terms the objective of this research could be classified as the function of knowledge and the evaluative function.

2.4 Choice of country

Another recurring topic in comparative legal literature is the choice of country. In this research, three entities are involved: a global institution in which global human rights are developed and its realisation is facilitated – the United Nations-, and two countries/legal systems that both agreed to take the responsibilities that come with the ratification of these human rights. One specific element of the responsibility to implement the right to food – enforceability in the Courts – in both countries will be compared with the authoritative interpretation on the matter shared in this global institution (see Chapter 3), and as a result, the enforceability of the right to food in both countries with one another. The choice to include the UN as a global institution in the comparison is obvious, for it is in the context of this organ that

---

[36] D. Nelken, *Chapter 1, Comparative law and comparative legal studies*, in: E. Örücü and D. Nelken (eds.), *Comparative law, a handbook*, Portland: Hart Publishing, 2007.
[37] C. McCrudden, *Chapter 16, Judicial comparativism and human rights*, in: E. Örücü and D. Nelken (eds.), *Comparative law, a handbook*, Portland: Hart Publishing, 2007.
[38] C. McCrudden, *Chapter 16, Judicial comparativism and human rights*, in: E. Örücü and D. Nelken (eds.), *Comparative law, a handbook,* Portland: Hart Publishing, 2007.

the standard under discussion was created. Although there are – as will appear from Chapter 3 – regional conventions and charters that stipulate human rights as well, they are in principle excluded from this research. The regional human rights treaty ratified by the Netherlands and Belgium that is closest related to the right to food must be the European Social Charter.[39] There are three reasons to exclude the ESC from this thesis. Firstly, although this Charter does recognise some related rights such as the right to social security and the right to benefit from social welfare services,[40] the Charter does not stipulate the right to food. Secondly, since it will appear that in both the Netherlands and in Belgium the right to food is mostly invoked by non-European foreigners, the *ratione personae* of the ESC might be considered not to include them.[41] Thirdly, including the ESC would also mean including the Council of Europe as a global institution, next to the United Nations, which would seriously complicate the establishment of a coherent *tertium comparationis*. The choice for the Netherlands and Belgium as legal systems in which the enforceability of this standard will be discussed needs some more explanation however.

When comparing law, the choice that is made for the compared legal systems must of course be well-founded, but also here, no coherent methodological clues can be found in literature on comparative law. Some authors claim that one will have to compare legal systems that show some resemblance,[42] for instance legal entities that belong to the same family, culture, or tradition, while others challenge comparists instead to include differing legal systems to maximise the learning outcome.[43] Another advice might be to include in the comparison at least legal systems that could be called a 'parent system', most clearly representing one legal family.[44] But to be honest, mostly the advice comes down to using your common

---

[39] However, it will appear to be necessary to analyse Dutch case law and Parliamentary History on Article 6 ESC, to clarify the criterion used by the Dutch Courts to determine whether or not an international Provision has direct effect in view of Articles 93 and 94 CA (see especially Chapter 8).

[40] European Social Charter (revised), 3 May 1996, Strassbourg, in: European Treaty Series 163, Articles 12 and 14. See also more in detail Section 3.3.3.

[41] See the preamble of the ESC, that stipulates *'considering that the aim of the Council of Europe is the achievement of greater unity between its Members for the purpose of safeguarding and realising the ideals and principles which are their common heritage and of facilitating their economic and social progress, in particular by the maintenance and further realisation of human rights and fundamental freedoms.'* Furthermore, consideration 19 states that: *'Migrant workers who are nationals of a Party and their families have the right to protection and assistance in the territory of any other Party.'* A good example is the Belgian ruling of the Constitutional Court, 51/94, 29 June 1994, consideration B.5.6. An illegally residing foreigner would not fall under the protection of the ESC, for the different treatment *in casu* was not based on nationality but on residence status.

[42] D. Nelken, *Chapter 1, Comparative law and comparative legal studies*, in: E. Örücü and D. Nelken (eds.), *Comparative law, a handbook,* Portland: Hart Publishing, 2007.

[43] G. Dannemann, *Chapter 11, Comparative law: study of similarities or differences?* in: M. Reimann and R. Zimmermann (eds.), *The Oxford handbook of comparative law*, Oxford: Oxford University Press, 2006.

[44] K. Zweigert and H. Kötz, *An introduction to comparative law*, Oxford University Press, 1998, Chapter 3, Section IV; W. Pintens, *Inleiding tot de rechtsvergelijking*, Leuven: Universitaire Pers Leuven, 1998, Section 98.

sense, and explaining the choices you make.[45] Not surprisingly, from a practical point of view, it appears safe to say that the comparist must have some familiarity with the system that is under comparison, regarding issues as language, legal sources and culture; legal sources for instance can best be studied in its original form and language.[46]

In this research, the choice was made to include the Netherlands and Belgium for several reasons. Firstly, the inspiration for starting this research came from the comparison between two Court cases, one Dutch and one Belgian, in which the underlying facts and issues were rather similar, but the outcome was remarkably different. The alleged similarity of the facts at first glance, and the apparent difference between the Dutch and Belgium Courts' reasoning regarding the enforceability of the right to food therefore needed to be questioned. Secondly, the research is specifically about the enforceability of the right to food. Enforceability implies that the right to food is not only a general right for all, discussed in the context of pressing global issues. It is a right for each, which makes the element of enforceable rights for individuals so important. Belgium and the Netherlands roughly show similarities on variables that are not directly part of this research, which enable a thorough investigation on enforceable rights for each. As demonstrated in Chapter 1, both countries have a relatively effective and accessible Judiciary, have rule of law, are democratic, are relatively prosperous, and have a domestic food production that exceeds the population's need. In addition, there is no war, and there are no famines, or recent State-threatening disasters. These are all conditions that should theoretically not stand in the way of enforceability of the right to food in both countries for each individual. This would obviously be different when a country was included in the comparison that for instance would have no functioning Judiciary, would be very poor, and would be devastated by a famine. Also, as considered in Chapter 1, both countries report to the UN treaty bodies that they comply with the international obligations coming forth from recognising the right to adequate food, and are considered by the FAO studies to be countries in which this right is directly applicable. Thirdly, the author has the Dutch nationality, is familiar with both countries regarding their cultures and legal system, and speaks the languages used in the Belgian Judiciary (French, German and Dutch), which is – as mentioned above – a practical prerequisite to perform the research, and enables him to study all legal sources in its original language.

Within this context, in comparative law literature the question is frequently raised whether the legal entities that are compared should be determined by

---

[45] F. Gorlé, G. Bourgeois, H. Bocken, F. Reyntjens, W. de Bondt and K. Lemmens, *Rechtsvergelijking*, Mechelen: Kluwer, 2007, Section 44.
[46] F. Gorlé, G. Bourgeois, H. Bocken, F. Reyntjens, W. de Bondt and K. Lemmens, *Rechtsvergelijking*, Mechelen: Kluwer, 2007, Section 58.

the boundaries of a nation State.[47] On the one hand, one might argue that state borders at least determine to a certain extent what jurisdiction, political reality and language(s) is/are common. In addition, this particular comparison is done in the context of the UN international human rights system, in which it is assumed that it are nation States that agree on the implementation of the human rights enshrined in the treaties they ratify. Therefore, using a State as a starting point for law comparison seems to be a reasonable approach. On the other hand, it is obvious that globalisation has a strong influence on the sovereignty of these States,[48] especially in Europe, where law and policy are not only made by national Governments, but also by international institutions, such as the European Union, the WTO, UN institutions, and to a certain extent by international human rights bodies. At a decentralised (or 'sub-national') level, as we will see especially in the case of Belgium (the regions and communities), legislative or political decisions and its effect must not be overlooked either.[49] For these reasons, some authors choose to use terms like 'legal culture' or 'legal tradition' to avoid a geographic limitation determined by the boundaries of a State. In this research however, the term 'State' will be used, for the reasons mentioned above. In the case of Belgium, the Regions, the Communities and the Federal State, have equal legislative powers. Therefore, throughout this research the State entity/entities that is/are relevant will of course be taken into consideration.

2.5 The functional method

2.5.1 The functional method proposed by Zweigert and Kötz

The functional method as proposed by Zweigert and Kötz was intended to be the methodological principle of comparative law, from which all other rules of comparative law would follow. The most important point made by Zweigert and Kötz is that when comparing legal rules, it only makes sense to compare (legal) rules that fulfil the same function. This approach helps the comparist to find the right rules to compare, and prevents him/her from looking for legal constructions in an alien legal system on places that would – from the perspective of his or her own legal system – be a logical place to look for them, thereby jumping to the wrong conclusions: a classic mistake made in comparative law. Zweigert and Kötz underlined that this *'proposition rests on what every comparist learns, namely that the legal system of every society faces essentially the same problems, and solves these problems by quite different means though very often with similar results.'* And: *'The question to which any comparative study is devoted must be posed in purely functional*

---

[47] W. Twining, *Chapter 3, Globalization and comparative law,* and R. Cotterrell, *Chapter 6, Is it so bad to be different?* In: E. Örücü and D. Nelken (eds.), *Comparative law, a handbook,* Portland: Hart Publishing, 2007.
[48] D. Nelken, *Chapter 5, Defining and using the concept of legal culture,* in: E. Örücü and D. Nelken (eds.), *Comparative law, a handbook,* Portland: Hart Publishing, 2007.
[49] D. Nelken, *Chapter 5, Defining and using the concept of legal culture,* in: E. Örücü and D. Nelken (eds.), *Comparative law, a handbook,* Portland: Hart Publishing, 2007.

terms; the problem must be stated without any reference to the concepts of one's own legal system.[50] These propositions imply that firstly, the comparist needs to be capable of placing a legal construction outside the national circumstances and considering the construction purely in its neutral functional context – to prevent her/him to look at the construction without the constraints of looking from the perspective of one's own (or another) legal system. Secondly, Zweigert and Kötz introduced the *praesumptio similitudinis*, meaning that: *'different legal systems give the same or very similar solutions, even as to detail, to the same problems of life, despite the great differences in their historical development, conceptual structure, and style of operation.'*[51] In short: whereas the legal construction under comparison may differ significantly – although it has the same function in society – the solution to the social problem is mostly similar.

2.5.2 Criticism on the functional method

As stated above, the functional method as proposed by Zweigert and Kötz seemed to have had a monopoly position on methods in law comparison for a few decades, and is still proposed (perhaps slightly altered) by authors of handbooks on legal comparison.[52] However, the method is increasingly contested amongst comparists, with a variety of credible arguments, but – unfortunately – without replacing it with one or more coherent methodological alternatives. I cannot do justice to all the creative work in which this method is respectfully criticised without having to write my own handbook on legal comparison. Here, I will concentrate on four common points of criticism on the functional method, as voiced by different authors. Firstly, the question can be raised whether it is a good idea to separate the functionally equivalent legal construction from its national background. Secondly, it can be questioned whether a functional approach enables the comparist to regard the compared function into sufficient detail and in its proper relation with society. Thirdly and fourthly, the presumption of universal problems that are mostly solved in equal manners (*praesumptio similitudinis*) can be criticised.

Firstly, Zweigert and Kötz advise to separate a legal solution to a social problem from its cultural background by comparing it without any reference to concepts of its own legal system, thereby striving for strict neutrality of the comparist. The tendency to strive for neutrality could possibly be explained by the fear of the influence of a political agenda in law comparison, such as the search for universal law before and after the Second World War. Some authors consider such an approach to be impossible and even undesirable. This can be argued from a subjective and an objective point of view. Subjective, because a comparist cannot

---

[50] K. Zweigert and H. Kötz, *An introduction to comparative law*, Oxford University Press, 1998, p. 34.

[51] K. Zweigert and H. Kötz, *An introduction to comparative law*, Oxford University Press, 1998, p. 39.

[52] F. Gorlé, G. Bourgeois, H. Bocken, F. Reyntjens, W. de Bondt and K. Lemmens, *Rechtsvergelijking*, Mechelen: Kluwer, 2007; W. Pintens, *Inleiding tot de rechtsvergelijking*, Leuven: Universitaire Pers Leuven, 1998.

neutrally consider foreign legal systems due to the simple fact that he or she is preconditioned by its own familiar legal system. Also, the choice of the comparist to compare certain legal constructions with certain criteria in certain countries is often a personal choice, depending on the interest, agenda and competences of the comparist, and thus never a neutral choice.[53] Separating a legal construction from its culturally determined background will thus hardly lead to the desired result of more neutrality. Rather, it is better to put some effort in thoroughly explaining the choices made, instead of claiming to be neutral.[54] In addition, from an objective point of view, it can be argued whether it makes sense to consider a legal construction without reference to a historical, political, cultural, traditional context, for those elements naturally influence law and legal practice,[55] and allowing those elements in the research may offer valuable insights into the element that is under comparison. This last argument then often leads to extensive discussions on how to study or compare elements of legal cultures.[56]

Secondly, the abstraction of legal concepts that results from translating them into merely functionally equivalent solutions to the same problem can be considered to be an unanalysed assumption rather than a given fact,[57] and could lead to an oversimplified (or even vague) impression of the compared legal construction. This approach may lack the concreteness necessary to be of use in scientific research. Also, the approach may easily ignore that it is not necessarily so that all societies choose to use law as a solution to a social problem, or even may have chosen a legal response that has no practical use in that society. Other elements, mostly culturally determined, may also respond to social problems, and should therefore be included in the research rather than being separated from it. Zweigert and Kötz already underlined that comparists sometimes will have to search for functionally equivalent solutions outside the law.[58] In addition, the approach presupposes that there was a social problem first, and as a response to that came

---

[53] N. Jansen, *Chapter 9, Comparative law and comparative knowledge*, in: M. Reimann and R. Zimmermann (eds.), *The Oxford handbook of comparative law*, Oxford: Oxford University Press, 2006; W. Devroe, *Rechtsvergelijking in een context van europeanisering en globalisering*, Leuven: Uitgeverij Acco, 2010.

[54] N. Jansen, *Chapter 9, Comparative law and comparative knowledge*, in: M. Reimann and R. Zimmermann (eds.), *The Oxford handbook of comparative law*, Oxford: Oxford University Press, 2006. See also: D. Nelken, *Chapter 1, Comparative law and comparative legal studies*, in: E. Örücü and D. Nelken (eds.), *Comparative law, a handbook*, Portland: Hart Publishing, 2007.

[55] R. Cotterrell, *Chapter 6, Is it so bad to be different?* In: E. Örücü and D. Nelken (eds.), *Comparative law, a handbook*, Portland: Hart Publishing, 2007; N. Jansen, *Chapter 9, Comparative law and comparative knowledge*, H. Muir Watt, *Chapter 17, Globalization and comparative law*, in: M. Reimann and R. Zimmermann (eds.), *The Oxford handbook of comparative law*, Oxford: Oxford University Press, 2006; C.J.P. van Laer, *The applicability of comparative concepts*, Electronic Journal of Comparative Law Vol. 2.2, August 1998.

[56] R. Cotterrell, *Chapter 6, Is it so bad to be different?* In: E. Örücü and D. Nelken (eds.), *Comparative law, a handbook*, Portland: Hart Publishing, 2007.

[57] R. Cotterrell, *Chapter 6, Is it so bad to be different?* In: E. Örücü and D. Nelken (eds.), *Comparative law, a handbook*, Portland: Hart Publishing, 2007.

[58] K. Zweigert and H. Kötz, *An introduction to comparative law*, Oxford University Press, 1998, p. 39.

a functionally equivalent (legal) concept that intends to solve the problem. This merely confirms the existence and comparability of the different (legal) solutions to a given social problem, but it reveals very little about the effect of the chosen (legal) solution. A solution may have unintended side-effects that have a certain effect on society or may even cause new social problems. Also, the solution might not be as effective as intended when adopted, or may be even completely ineffective. Therefore, this interconnection between society and legal solutions must not be overlooked too easily, and where relevant included in the research.

Thirdly, Zweigert and Kötz seem to suggest that all societies somehow face the same problems. The presumption that all legal systems need to respond to the same problems tends to imply that social problems are universal, an assumption that shows characteristics of legal naturalism. Also here, it can be argued that it is pointless not to consider a social problem in its proper cultural circumstances, since one needs to *'bear in mind the extent to which 'social problems' are culturally constructed rather than given.'*[59]

Fourthly, the *praesumptio similitudinis* is criticised because such an approach would favour or presuppose similarities over dissimilarities in law. Also here, the approach might be accused of showing characteristics of legal naturalism. A *praesumptio similitudinis* that is understood as presupposing, or even favouring similarities over dissimilarities is criticised by many authors for a variety of reasons.[60] Some authors advocate a neutral attitude towards differences or similarities, and some even call to 'celebrate dissimilarities.' David Nelken argues that *'instead of taking a position a priori in favor of similarity or difference, it may be more productive to ask why we expect to find one or the other.'*[61]

2.5.3 Alternatives to functionalism?

Indeed, some alternate approaches were proposed, although without establishing a coherent and generally accepted method on comparative law. In this section, some examples will be discussed: the formation of ideal type solutions, a dialogic method, a factual method, a dogmatic approach, and a contrastive approach.

The formation of immanent concepts, or ideal type solutions in law comparison is an approach through which comparists formulate typical legal solutions in more abstract words, using these concepts as tools for comparison without the

---

[59] D. Nelken, *Chapter 1, Comparative law and comparative legal studies*, in: E. Örücü and D. Nelken (eds.), *Comparative law, a handbook*, Portland: Hart Publishing, 2007.
[60] In Section 2.5.5 I will discuss Michaels' understanding of a *praesumptio similtudinis*, which is different.
[61] D. Nelken, *Chapter 1, Comparative law and comparative legal studies*, in: E. Örücü and D. Nelken (eds.), *Comparative law, a handbook*, Portland: Hart Publishing, 2007.

need to cut the concept loose from its national circumstances.[62] The approach indeed meets the criticism on the functional method that it is an unrealistic goal to cut a legal concept loose from its national background, and focuses on the comparability of legal concepts itself, without a reference to functional equivalence. Although I am inclined to favor an approach in which a legal system is studied and compared while using the same language as a lawyer originating from the compared country only would do, explaining legal concepts in a neutral language as such – so called second order language – may be a helpful approach to consider and explain foreign legal systems in a more descriptive way, and may facilitate a more empirical comparison.[63] However, this method requires some similarities regarding the compared national concepts, for without those such an abstract immanent concept has limited use, and one may even propose to fall back on the functional method.[64] In addition, for this research I see limited added value in abstracting a legal concept, using it almost as a *tertium comparationis,* while simultaneously the same concept should be regarded in its national legal context and be compared (almost) deliberately in a non-functional context, perhaps in the urge to find means to empirically compare these legal concepts. Since that comparison would only reveal the similarities and dissimilarities between legal systems in relation to an immanent concept, and not lead to a decent comparison in which it is explained how and why a certain legal concept is used.

In the context of human rights case law it is interesting to mention a dialogic approach. Also in this approach towards comparison, there is no need to separate a legal construction from its domestic context. This approach is based on an international dialogue between legal experts, and is in general increasingly used amongst judges worldwide: the method '*often involves judges considering what occurs in other jurisdictions as well as their own in order to appreciate dimensions of the issue that might not otherwise have been as apparent.'*[65] Judges may use examples of other jurisdictions to reach conclusions in their own verdicts. This method allows judges to compare and learn from legal arguments and rulings, and to use elements, thoughts and ideas of foreign, domestic and international verdicts or legislation in their own legal culture if desired.[66] According to McCrudden, '*it is in the development of this dialogic method applied to the problem of incompletely theorized agreements in*

---

[62] See for instance: C.J.P. van Laer, *The applicability of comparative concepts,* Electronic Journal of Comparative Law Vol. 2.2, August 1998.

[63] N. Jansen, *Chapter 9, Comparative law and comparative knowledge,* in: M. Reimann and R. Zimmermann (eds.), *The Oxford handbook of comparative law,* Oxford: Oxford University Press, 2006.

[64] C.J.P. van Laer, *The applicability of comparative concepts,* Electronic Journal of Comparative Law Vol. 2.2, August 1998.

[65] C. McCrudden, *Chapter 16, Judicial comparativism and human rights,* in: E. Örücü and D. Nelken (eds.), *Comparative law, a handbook,* Portland: Hart Publishing, 2007.

[66] In the context of constitutional law, see: D.S. Law, *Generic Constitutional Law,* Minnesota Law Review, 2005, volume 89:652; V.C. Jackson, *Constitutional comparisons: convergence, resistance, engagement,* Harvard Law Review, 2005, volume 119:109.

human rights that the most fruitful role for judicial comparativism may lie.'[67] In his view, it is the national Courts that contribute to a more coherent understanding of human rights through this method of comparative law.[68] This method thus not necessarily starts from normative assumptions, but rather leaves room for cultural diversity in law, which is in contrast with a universalistic approach regarding human rights, or with the aforementioned 'persuasive comparison.' While this research starts from a normative assumption – which is: the right to food should be enforceable – this understanding of a comparative method applied on human rights challenges a too rigid top-down approach concerning the realisation of human rights. For further methodological purposes a dialogic method seems not usable in this research.

Another approach may be a more factual method, in which the idea of a functional relation is abandoned, and components in two or more different legal systems are merely compared in terms of similarity and dissimilarity. This approach broadens the possibility for comparison, for now literally everything can be compared, even the *'number nine with Chicago.'*[69] This may have certain advantages, for there is no need for complex presumptions before starting the comparison, and may to a certain extent lead to knowledge that may be overlooked in the first place when using a method based on functionality. On the other hand, a factual method would be very casuistic, and strips the compared concepts partly from their meaning, for they are not put in a functional context: it remains unclear to what end a concept operates in society, and what its (cultural) background is.[70]

A dogmatic approach is usually mentioned in contrast to a functional approach. Where a dogmatic approach only considers formal legal sources, or 'the law in the books', the functional approach also includes 'the law in action', which differs greatly compared to law in formal sources. Therefore, the dogmatic approach is often portrayed as obsolete, leading to false results. Still, for a first impression of a legal system, a brief survey through merely formal legal sources may be helpful.[71]

In some handbooks, the contrastive approach is mentioned, an approach that presupposes such insurmountable differences between the compared systems, that it will only make sense to use a comparison to contrast the systems, to gain

---

[67] C. McCrudden, *Chapter 16, Judicial comparativism and human rights*, in: E. Örücü and D. Nelken (eds.), *Comparative law, a handbook*, Portland: Hart Publishing, 2007.
[68] C. McCrudden, *Chapter 16, Judicial comparativism and human rights*, in: E. Örücü and D. Nelken (eds.), *Comparative law, a handbook*, Portland: Hart Publishing, 2007.
[69] H.P. Glenn, *Chapter 4, Com-paring*, in: E. Örücü and D. Nelken (eds.), *Comparative law, a handbook*, Portland: Hart Publishing, 2007.
[70] R. Michaels, *Chapter 10, The functional method of comparative law*, in: M. Reimann and R. Zimmermann (eds.), *The Oxford handbook of comparative law*, Oxford: Oxford University Press, 2006.
[71] W. Pintens, *Inleiding tot de rechtsvergelijking*, Leuven: Universitaire Pers Leuven, 1998, Section 91.

more perspective. This is a method that was especially used to compare socialist legal systems with Western legal systems.[72]

All in all the above responses are interesting alternatives to a functional method, although in an attempt to meet the criticism on elements of the functional method as proposed by Zweigert and Kötz, the authors' solutions also raise methodological questions that are not easily answered. While the current debates on methods in comparative law are very refreshing and extremely interesting, comparists seem to struggle to develop methodological tools that on the one hand do not exclude any relevant influences on the development of law, while on the other hand as a result do not become too vague or undefined for practical use.

2.5.4 Demarcations on functionalism in this research

To compare the legal reality of the Netherlands and Belgium concerning the enforceability of the right to food in the domestic Courts in view of the global human rights system, I will of course need a suitable comparative method. As demonstrated above, a method of functionalism proposed by Zweigert and Kötz is now increasingly criticised, while on the other hand hardly any coherent and usable methods are developed to replace this functional method – especially none suitable for this research. However, as will be discussed in this section, the research questions offer a demarcation in the research method in itself, and as set out below in Section 2.5.5, in that context a modernised understanding of functionalism seems to offer suitable methodological tools to perform the comparison in this research.

In this particular research, following from the research questions, important elements in the comparison are already predetermined, in line with the UN human rights system. If I would put this in terms of functionalism, both the social problem and responding institution are – at least formally and to a certain extent – predetermined, due to the fact that both countries signed and ratified the UN treaties in which the right to food is recognised, and thus have agreed formally that everyone should have the right to food. It is assumed that ratifying the right to food as a consequence means that the domestic Courts accept enforceability of this right. In a functional approach then, the social problem is that individuals must be able to enforce their right to food. The institutions that should respond to this are also, to a certain extent, predetermined by the UN human rights system: firstly it is the domestic Courts that should respond to this social problem by making the right enforceable. In addition, to ensure this enforceability in the global arena, a reporting system has been implemented, in which Governments

---

[72] W. Pintens, *Inleiding tot de rechtsvergelijking*, Leuven: Universitaire Pers Leuven, 1998, Section 94; W. Devroe, *Rechtsvergelijking in een context van europeanisering en globalisering*, Leuven: Uitgeverij Acco, 2010, Section 63.

of Member States have a duty to report on the implementation of the right to food, of which the enforceability is an element. Another institution therefore is the Governments that should respond to the social problem by reporting on the legal practice regarding the enforceability of the right to food.

In this research then, functionalism as a method will be useful only when applied in a more narrow way. Instead of using a method of functionalism in order to determine from a blank position which institutions in society respond to a certain problem in which particular manner, functionalism will rather be used to determine how and why the predetermined institutions respond to the shared social problem, in view of an international human rights system. This demarcation is most likely in contradiction with comparative methods, for – as demonstrated above – predetermining the institutions that are part of the research does not fit in with most approaches to comparative law, and especially not with functionalism. The determination of the institutions that respond to a social problem is in particular a fundamental step in the comparative process, which is now more or less being excluded. Nonetheless, there are some good reasons for that in this research. Firstly, this research is constructed from the perspective of an existing human rights system, and the countries compared are discussed in view of this system. As demonstrated above, this system obliges Member States to respond in a particular manner to a social problem using a particular institution. The focus of attention therefore is not necessarily on how particular institutions respond to the problem, although of course (especially in the context of Belgium) it needs to be clarified exactly which Courts or tribunals respond to the social problem. Instead, this research focuses on in which manner the institutes respond to the problem. Furthermore, the social response needs to be clarified. Since the reports are a translation in the global arena, provided for by the Governments, of the legal practice in the domestic legal order, the focus will be in this thesis to explain this response of the Courts. Of course, explaining the response of the Governments would be another interesting research focus, although it remains to be seen whether a legal thesis is a suitable place to do so. A method on comparative law in this research therefore should function firstly as a search tool for relevant data concerning how Courts and Governments respond to the enforceability of the right to food in view of the UN human rights system. In other words: to find out what countries do, what they say they do, and what they should do (sub-questions 1, 2a and 3). Secondly, the method will be used to clarify the response of the social institutions (the Courts) to the social problem (sub-question 2b); and thirdly, as a tool for performing the comparisons (I, II, and III). As mentioned before, it is also the ambition of this research to make recommendations to both the compared countries and the UN regarding the enforceability of the right to food. A comparative method is thus required to structurally confront the compared entities, and learn from that process.

But also from a methodological point of view the demarcation is interesting. The fact that such a demarcation in comparative law contradicts the most forward ideas on comparative law, mostly due to the fact that such an approach would overlook important domestic particularities, portrays but also leads to questions about the strict top-down approach of the UN system. Especially when within these predetermined elements of the comparison there appear to be profound differences in approach between the two compared countries. It is then the differences within the (given) similarities that are the most exciting to discuss in view of the evaluative purpose of this research.

2.5.5 Ralf Michaels' reflections on the functional method

In his contribution to the Oxford Handbook of Comparative Law,[73] Ralf Michaels critically reviews the comparative method as proposed by Zweigert and Kötz, and introduces an interesting modern understanding of the method: a functional method based on equivalence functionalism and on an epistemology of constructive functionalism seems to offer adequate methodological tools in this research, as will be discussed in this section.

Before discussing Michaels' view on functionalism, first some historic contextual remarks on functionalism in comparative law should be made. The functional method was introduced during a time in which comparative law basically focussed on private legal issues, usually comparing legal constructions from Western legal systems, while nowadays comparative lawyers increasingly focus on also non-Western countries and other branches of law than private law. Perhaps, most of the criticism discussed above on the functional approach as proposed by Zweigert and Kötz can be traced back to this rather narrow approach towards comparative law. For instance, at first glance private legal solutions may not be as culturally determined as non-private legal issues, and therefore more easily distilled from their legal background.[74] This is however a statement that can easily be contested, for also private legal solutions are of course culturally determined, albeit perhaps not always at first sight. Furthermore, it may be more convincing to conclude that only Western (European) countries face the same social problems and therefore ultimately look for similar (or equivalent) solutions than applying such theories of a *praesumptio similitudinis* on a worldwide scale. Also, shortly after a period of war over differences, the urge to unify and therefore stress similarities is understandable.[75] Whatever the circumstances may have been for Zweigert and Kötz to define the functional method as they did, it is quite clear that the method

---

[73] R. Michaels, *Chapter 10, The functional method of comparative law*, in: M. Reimann and R. Zimmermann (eds.), *The Oxford handbook of comparative law*, Oxford: Oxford University Press, 2006.
[74] G. Dannemann, *Chapter 11, Comparative law: study of similarities or differences?* in: M. Reimann and R. Zimmermann (eds.), *The Oxford handbook of comparative law*, Oxford: Oxford University Press, 2006.
[75] R. Michaels, *Chapter 10, The functional method of comparative law*, in: M. Reimann and R. Zimmermann (eds.), *The Oxford handbook of comparative law*, Oxford: Oxford University Press, 2006.

as originally proposed is in need for a renewed understanding. This does not necessarily mean however that the general ideas behind such a functional method are invalid for further use in comparative law.

In the first half of his Chapter,[76] Michaels underlines that the term 'functionalism' in itself needs some further elaboration, for the word is borrowed from other sciences, and may be understood in quite different meanings. Defining seven concepts of functionalism (finalism, adaptionism, classical functionalism, instrumentalism, refined functionalism, epistemological functionalism, and equivalence functionalism), Michaels argues that legal functionalists almost randomly choose parts of all these concepts and use them *'regardless of their incompatibility.'* This may be caused by the fact that *'the founders of the functional method were more pragmatically than methodologically interested.'* Michaels argues that equivalence functionalism seems to be the most fruitful concept for legal functionalism *'both because it is the most robust concept in sociology and because it represents the central element of functionalist comparative law as developed by Rabel and Zweigert'* and explains its meaning: *'functional equivalence means that similar problems may lead to different solutions; the solutions are similar only in their relationship to the specific function under which they are regarded.'* Crucial to Michaels' understanding of this concept is firstly that by acknowledging that solutions to a problem are merely functionally equivalent, the classic understanding of the idea of a *praesumptio similitudinis,* and the urge to search for ideal types of solutions, or similar solutions to problems, is abandoned. Secondly and as a consequence, Michaels underlines that instead of merely describing causal relations (as demonstrated above, the criticism is that this might lead to oversimplified and unspecific perceptions of legal reality) equivalence functionalism recognises the uniqueness of a solution to a problem in a given society – *'its decision for one against all other (functionally equivalent) solutions'* – which implies that the researcher needs to fully understand the background of this solution in society, instead of separating the solution from its background.

In the second half of the chapter,[77] Michaels explains what in his opinion the purpose (he refers to this as a 'function') should be of this equivalence functionalism. He thoroughly discusses seven possible functions of functionalism, roughly based on the purposes and aims of comparative law as proposed by Zweigert and Kötz:[78] *'(1) the epistemological function of understanding law; (2) the comparative function of achieving comparability; (3) the praesumptive function of emphasizing similarity; (4) the formalizing function of system building; (5) the evaluative function of determining the better law; (6) the universalizing function of preparing legal unification; and (7) the*

---

[76] R. Michaels, *Chapter 10, The functional method of comparative law,* in: M. Reimann and R. Zimmermann (eds.), *The Oxford handbook of comparative law,* Oxford: Oxford University Press, 2006, Section II.

[77] R. Michaels, *Chapter 10, The functional method of comparative law,* in: M. Reimann and R. Zimmermann (eds.), *The Oxford handbook of comparative law,* Oxford: Oxford University Press, 2006, Section III.

[78] K. Zweigert and H. Kötz, *An introduction to comparative law,* Oxford University Press, 1998, Chapter 2.

critical function of providing tools for the critique of law.' He concludes that regarding the last four functions one needs to be cautious, for they can be criticised from a methodological point of view. He proposes a focus on using comparative law as an explanatory tool, and therefore advocates the epistemological function, possibly complemented by a comparative and a presumptive function.

Michaels argues that the first function of functionalism is to *'make sense of the data we find'*, but contrary to for instance a factual method (as discussed above), epistemology in a functional method focuses on specific relations, that is the understanding of the relationship between a social problem and the solution provided for by this society in law (both dogmatic/law in the books and law in action) and outside law, including its cultural background. Indeed, Michaels here meets the criticism that a functional approach would not take into account the cultural background of a (legal) solution to a social problem. He explains that the interest of a functionalist in culture only differs from for instance a sociologist, in the sense that a functionalist will have the tendency to translate culture (or a part of it), among other things, into a functional relation, a process in which the perception of an 'insider' in that culture is not necessarily taken as a starting point: the perception of the functionalist plays an important role as well, depending on the researcher.

A second function of functionality is, according to Michaels, a comparative function, an approach that is certainly not unanimously supported.[79] To be able to compare, one needs an invariant element to compare with, in comparative law often referred to as *tertium comparationis*. This suggests a certain universality (or at least a shared common feature) of either the response to a problem or the problem itself. In this light, as demonstrated above, some authors prefer to distinguish idealised solutions to problems, or a commonly shared vocabulary to more neutrally distinguish solutions to a problem which can even be used as *tertium comparationis*.[80] Michaels thinks otherwise, since responses to a problem can hardly be considered to be of a universal nature; in his view, it is more likely to consider the shared social problem as a *tertium comparationis*. Also here, Michaels meets the criticism to the assumption of universal social problems. In the first place he argues that indeed some social problems, when they are formulated in abstract general terms could be considered to be universal problems, such as the need to survive. The more specific a problem is defined, the less universal it will be. Starting from a rather general common problem, results in obtaining a very complex, but also very rich comparison. Interestingly, Michaels adds that in

---

[79] For instance: H.P. Glenn: '*Similar legal concepts can mean different things in different contexts. The lesson for human rights lawyers is that they ignore the different institutions contexts in which interpretation takes places and the different power relations in these jurisdictions at their peril.*' H.P. Glenn, *Chapter 4, Comparing*, in: E. Örücü and D. Nelken (eds.), *Comparative law, a handbook*, Portland: Hart Publishing, 2007.
[80] For instance: K. Zweigert and H. Kötz, *An introduction to comparative law*, Oxford University Press, 1998, Chapter 3, Section VI.

functional epistemology, it is not even necessary to start from shared universal problems, and argues that '*once the formulation of a problem is understood as a constructive move rather than an empirical one, the universality of problems is likewise a constructive move rather than a mere representation of reality.*' He argues that a social problem not necessarily causes a social solution in that order, implying that it is therefore not always easy to find such a response. For that reason, he underlines that a solution needs to be identified from other possible solutions. In this lies the added value of comparison: functionally equivalent solutions may be revealed when compared to other legal systems. In comparing, one could also reveal the underlying social problem as a consequence: '*this reasoning is of course circular – it goes from problems to functions and from functions to problems, and it appears justified for constructivist comparative law as interpretation because it mirrors the hermeneutic circle between the comparist and the legal systems observed that is characteristic of comparative law.*' The activity of functionalist comparison in itself is then a constructive move, ultimately leading to a better understanding of both the chosen solutions to a social problem and the construction of the underlying social problem, without the need to presume one of these elements.

The third function Michaels describes is the presumptive function. This function can be traced back to the short section mentioned above, as written by Zweigert and Kotz, and has been heavily criticised, basically due to the fact that such an approach would favour similarities over differences, and it implies universality of both solutions and the problem. According to Michaels, the concept of a *praesumptio similitudinis* as introduced by Zweigert and Kötz is not as presumptive as it seems, and often misunderstood. Regarding the presumed similarity of the solutions to a problem, Michaels argues that it only makes sense to compare functionally equivalent solutions to a problem (he refers to this as institutions), and therefore the compared institutions must be the same as regarding to their function. This however does not imply that the institutions are similar or dissimilar, and consequently, a *praesumptio similtidunis* does not favour similarity over dissimilarity: '*functionalism leads to comparability of institutions that can thereby maintain their difference even in the comparison. It neither presumes, nor does it lead to, similarity.*' In this light, Michaels discusses the assumed universality of social problems: '*if only functionally equivalent institutions are comparable, then by definition these institutions must be similar in the sense that they respond to the same problem.*' Again, Michaels underlines that functionalism should be considered to be a constructive step rather than an empirical one: '*The claim of universality of a problem is a first interpretative step that can be challenged, but this is a fruitful way of making sense of one legal system in relation to another.*' According to Michaels, a universal problem may thus offer a starting point in doing comparative law research, but the researcher may find reason, along the way, to change this first assumption, depending on what he or she encounters in the compared legal systems. The process of comparing then leads to better understanding (epistemological function) of the compared legal systems, and the compared solutions may lead to a reformulation of the underlying problem, and vice versa. So, also here, Michaels considers the activity of a constructive move as a hermeneutic circle that ultimately leads to knowledge. As demonstrated above, in this research, not only the social problem is predetermined, but also the institutions that respond to the social problem. The focus of attention is therefore not the search for in which manner institutions respond to a social problem, but how and why.

The comparative scheme as introduced by Michaels is a very flexible one with a continuous open end. In this research however, most elements of this scheme are predetermined. Despite this, a functional method based on equivalence functionalism and on an epistemology of constructive functionalism is a suitable methodological approach for several reasons. Firstly, it gives direction to the data finding process. Michaels proposes to start the research from a shared universal problem, but is cautious in holding on to the same universal problem throughout the entire research, which explains his hermeneutic approach. In this research, there is no need for reassessing the shared universal problem, for it is a given fact that both compared countries signed and ratified the same treaty Provisions. At most, the response to the given social problem (A) might instantaneously be a response to other related (or unrelated) social problems (B, C, etc.). Mapping this can offer valuable explanations of the precise background of the countries' responses to the social problem (A), which may be overlooked when discussing the relation between a solution and one social problem only. In addition, while the institutions that respond to the social problem (the Courts through case law and the Governments through fulfilling *inter alia* its reporting duties), are given beforehand, it is the way how and why these institutions respond to the problem that is of interest here. To determine this, the constructive move, as thus labeled by Michaels, can be applied, albeit in a slightly different manner: not necessarily as a continuing hermeneutic circle, but as a way to find and explain facts within a given context. The constructive move has thus a rigid core that already puts the research into a certain direction. Therefore, it is not the question if it is the Courts or Governments that respond to the social problem, but rather how and why, which represents the non-rigid part of the constructive move. For instance, it will be beyond dispute that in both compared countries Courts respond in a way to the social problem that the right to food needs to be enforced. On the other hand, in a Court case, there is a complexity of legal facts that need to be judged upon, with different legal questions and interests that are at stake. Therefore, it is unlikely that a Court responses to one social problem only when the right to food is invoked. Following the constructive move, the legal reasoning patterns of the Courts may thus determine the subsequent research that is necessary to explain how and why the Courts respond the way they do (as will be discussed in Section 2.6.2), and may also lead to the discovery of related (or unrelated) social problems the Courts respond to simultaneously. In this way, there is still a moving back and forth between social problem and social response in a hermeneutic background, while the rigid core that is the normative starting point of this research remains rigid.

Second, the method offers a scheme in which the legal practice of both compared countries, the Governmental reporting on this practice and the *tertium comparationis* can be structurally contrasted and compared.

In a simple diagram, a simplified methodological overview looks like given in Figure 2.1.

While Michaels calls for cautiousness regarding a possible evaluative function of comparative law, he is not necessarily opposing such an approach as long as the criteria of evaluation are different from the criteria of comparability. He considers that a comparison is a more neutral and scientific activity, while evaluation is rather a *'policy decision, a practical judgment, under conditions of partial uncertainty.'*[81] In this research, enforceability of the right to food is supported, and evaluative considerations and recommendations will be aimed at the improvement of realising the right to food.

2.6 The constructive move into more detail

A functional method based on equivalence functionalism and on an epistemology of constructive functionalism can thus, as discussed above, be used as a constructive move to find and clarify data – the sources that must be explored in order to finally perform the comparison – albeit in this research with a rigid core, determined by the UN human rights system. The research question encompasses the social problem (the right to food is an enforceable right), the institutions that respond to

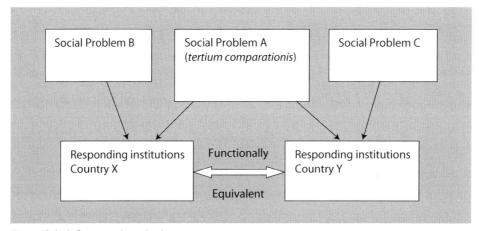

Figure 2.1. A functional method.

---

[81] R. Michaels, *Chapter 10, The functional method of comparative law*, in: M. Reimann and R. Zimmermann (eds.), *The Oxford handbook of comparative law*, Oxford: Oxford University Press, 2006, Section III.5.

that problem in the Netherlands and Belgium (the Legislature and the Government), and the *tertium comparationis* (the right to food in the UN human rights system):

> What legal factors explain the differences and similarities regarding the response of the Dutch and Belgium Judiciary and Government to the enforceability of the right to adequate food in view of the UN human rights system?

To answer this question, three sub-questions need to be answered:

1. To what extent is the right to adequate food perceived to be an enforceable right within the UN human rights system? (tertium comparationis)
2. What is the response of the national Judiciary of the Netherlands and Belgium when the right to food as stipulated in the UN human rights system is invoked by individuals, and how can this response be explained?(social institution responding to the social problem)
3. What do the Governments of the Netherlands and Belgium communicate in their reports in the United Nations arena regarding the enforceability of the right to food in their domestic legal order? (social institution responding to the social problem)

In Sections 2.6.1-2.6.3, the research direction that follows from the constructive move in relation to data-finding in order to answer the sub-questions will be discussed. So far, mainly in view of the first research purpose: *to gain knowledge about the enforceability of the right to food as embedded in UN human rights instruments in the Netherlands and Belgium through law comparison.* In Section 2.6.4, the subsequent comparing and evaluating will be further discussed, mainly in view of the second purpose: *where necessary critically evaluate both approaches in view of the UN human rights doctrine regarding the enforceability of the right to food.*

2.6.1 The research direction of sub-question 1

It is the purpose of the first sub-question to specifically determine the constructed shared social problem, or the *tertium comparationis* in this research. An answer must be provided to the question: '*To what extent is the right to food perceived to be an enforceable right within the UN human rights system?*' In Chapter 3, this sub-question will be answered. Firstly, by making an inventory of the international, regional and national Provisions that stipulate the right to food, or are related to the right to food (and thereby are of importance to distract its exact meaning), as well as the points of view of the international institutions that respond to the need to further clarify the meaning of that right. These are basically UN or UN related bodies: (1) the treaty-based bodies that oversee *inter alia* the various reporting procedures; (2) the FAO that stimulates the effective realisation of the right to food by hosting *inter alia* intergovernmental summits and gatherings and by supporting extensive research on the matter; and (3) the thematic mandates concerning the right to food: the Special Rapporteurs. The authoritative interpretation of these institutions will be used to analyse further. Secondly, by understanding the meaning of the phrase 'adequate food', especially in the context of the question whether the right to food is sufficiently precise to be an enforceable right in a domestic Court. Thirdly, by comprehending the exact obligations for Member States which are the result of ratifying the right to food. It is in that context that the matter of enforceability will be clarified.

2.6.2 The research direction of sub-question 2

The second sub-question is: '*What is the response of the national Judiciary of the Netherlands and Belgium when the right to food as stipulated in the UN human rights system is invoked by individuals, and how can this response be explained?*'

To this end, a thorough analysis of case law will be made in Chapter 4 (the Netherlands) and Chapter 12 (Belgium), to determine through what legal reasoning pattern the Courts respond to the enforceability of the right to food.

2.6.2.1 On legal reasoning

While there has scarcely been conducted thorough research to methods of legal reasoning,[82] it is helpful to put the legal reasoning of the Courts in a more theoretical context to give further direction to the research process, especially by selecting relevant sources to examine with the purpose to explain these legal reasoning patterns applied by the Courts.

Peter Wahlgren – with a background in Scandinavian legal theory – is one of the few who constructed a methodological model on legal reasoning in which the legal reasoning of a Court regarding the application of a standard is structured and categorised. Scandinavian law is, like Dutch and Belgian law, in its core driven by Statutes, and not by case law. In that light, methodological essays on legal theory on the methods of legal reasoning with an Anglo/American background are presumably less relevant here, and therefore Wahlgren's work is referred to as an example in this particular research. As demonstrated below, the methodological subdivision of the legal reasoning process as proposed by Wahlgren helps us further in the more detailed determination of the research direction concerning sub-question 2.

In essence, Wahlgren subdivides the legal reasoning process into six interrelated steps: (1) identification; (2) law-search; (3) interpretation; (4) evaluation; (5) formulation; and (6) learning. In short, it is assumed that a Court needs to '*define*

---

[82] P. Wahlgren, *Legal reasoning, a jurisprudential model*, Scandinavian Studies in Law, volume 40, 2000, pp. 199-282.

a situation in a legally relevant way', which is the process of identification. The identification stage then continues simultaneously with the law or standard search, since both stages are interdependent: while recognising relevant juridical facts, the standard search will develop and vice versa. Wahlgren explains that such a standard search is mostly determined by some kind of hierarchy of legal sources. He underlines that in continental jurisprudence mostly *'prevailing legislation is almost always the most important. Thereafter, legislative preparatory materials may have to be inspected. In addition, case reports may supply examples on how a rule is to be applied in practice. Important legal knowledge may also be found in works of jurisprudence.'* Wahlgren underlines that the importance of sources may differ per country or area of law, making it impossible to formulate universal principles on the search for law based on the hierarchy of legal sources. Wahlgren argues that in modern-day law however, there will be hardly any rule that can be applied without further interpretation by the Courts, mainly because normally a rule does not equally apply to a specific case. The basic challenge then is how to deal with a certain degree of vagueness in written standards. Wahlgren distinguishes three approaches in jurisprudence concerning interpretation of written standards. Firstly, a contextual interpretation, in which a Court attempts to clarify a written standard in relation to a specific case by *'finding explicit indicators from the legal system.'* This could be done by the search for definitional support elsewhere in law, in which through specification and generalisation through law in different hierarchical stages a better understanding is generated of the relevant standard (for instance the search for clarification of a Constitutional standard in substantive domestic law). Another manner is to look for intentional support, in which *'it is sometimes possible to find explicit statements concerning intentions behind legal propositions within the legal system.'* Wahlgren suggests that the analysis of legislative preparatory material may lead to such knowledge. Also, the search for methodological support may result in a better contextual interpretation, in which the Court will comply with legal rules that offer guidance in how to interpret a standard. A second approach towards interpretation is classified by Wahlgren as supplementing interpretation: a method mostly used in case the legal system does not provide for clear indicators on how to interpret a standard. A Court then tries to settle a case by transforming legal constructions, for instance by comparing the underlying situation to previous ones, and when the previous solution applies analogously to the current case (analogous reasoning or reasoning *a contrario*). Another way of supplementing interpretation is the extensive or restrictive interpretation of a legal standard. A third approach distinguished by Wahlgren is the interpretation of priority: in case rules collide, Courts use rules of priority (such as *lex superior legi inferiori derogate*) to solve the matter, and decide what standard will be applied in what specific case. Wahlgren argues that the specific choice in approach and method made by the Court is in some cases rather obvious, especially when clear intentional or methodological support can be found. In other cases however, *'interpretation is tied up with the process of evaluation and with the purpose of the legal system.'* In that case, the Court will weigh up the effect of alternate interpretation methods, and choose an interpretation, led by evaluation goals. This already demonstrates the interrelation between the phase of interpretation and evaluation. A fourth step in legal reasoning is distinguished by Wahlrgen as rule of application, a step that *'is performed when an individual case has been subsumed under a general description.'* But before the intended application of the rule is formulated, the Court will consider whether the consequences of the application are acceptable in law. The stage of rule of application then serves as a checkpoint whether the previous steps have been performed correctly. When application of a rule seems reasonable, the process will gradually move towards the next step of evaluation. When application of a rule leads to absurd outcomes, *'another round of identification, interpretation and law-search may have to be initiated.'* According to Wahlgren, the process of rule of application is determined by accuracy and by formal rules. Accuracy then means that the standard application should be exact and correct, that follows from two meta-standards or fundamental principles in law: the principle of legality, meaning that all verdicts must be in accordance with the law, and the principle of justice or equality, meaning that *'similar cases should be treated a similar way.'* Furthermore, formal rules are rules that *'define when and where rules of substantive law may be applied'*, (promulgation rules and rules concerning jurisdiction) and *'the appropriate legal forum and the necessary competence of legal decision makers'* (rules concerning competence). The close link between the phase of standard application with both the previous steps and the following evaluation phase shows that *'rule application can thus be described as a recurring process in which a lawyer is able to specify more and more adequate rule-applications through the succeeding and recurring instances of identification, law-search, interpretation and evaluation. Rule application described in this way stresses the fact that many phases of the legal reasoning process depend on each other and are conducted in a more or less parallel way.'* Then, Wahlgren describes the evaluation phase, in which the effects of the legal decisions are evaluated and where needed be adjusted to the purpose of the law. This is also referred to as teleology. Although there are many resemblances between the interpretative phase and the evaluative phase, and therefore both stages cannot always be distinguished from one another in the legal reasoning process, the basic distinction is that in the interpretative stage *'the objective is to relate a legal proposition to a legal rule'* through interpretation, while in the evaluative stage the entire legal decision is evaluated. There are different views on the legality of the effect of evaluation on the actual outcome of a decision. Wahlgren distinguishes between an objective teleological method and a subjective teleological method. In the first approach, the objective purpose of law needs to be discovered, which means that the Courts' investigation goes beyond the intention of the Legislature only. The Courts may also *'adjust the hypothesized original purpose to the succeeding social changes.'* In the latter approach, the *'lawyer must try to find out about (and rely on) the intentions of the lawmaker such as the way they were when the law was formulated.'* An approach that shows much resemblance with the interpretation discussed before through intentional support. Ultimately, the final step in the legal reasoning process is the phase of formulation, in which the Court's verdict is *'performed in speech or in writing, or even by means of combining the two.'*

2.6.2.2 The research direction

It would appear that the Dutch Judiciary highly values the opinion of the Legislature regarding the enforceability of the right to food. The Dutch Courts struggle with a Constitutional construction that determines the relationship between the Legislature and the Courts regarding the decision on whether an international Provision has direct effect. Since the reasoning of the Court appears to be influenced by the Constitutional system in Articles 93 and 94 CA, that regulates the effect of international law in the domestic legal order. In Chapter 5, the background of this Constitutional construction will be explored in literature, since its exact meaning remains unclear. Remarkably, most authors who clarify the functioning of this Constitutional construction refer to the Parliamentary History of the Constitutional reforms in which this construction was introduced and altered, but appear to draw very different conclusions. Therefore, an analysis of this Parliamentary History will be conducted in order to relate the Courts' response to the enforceability of the right to food to this Constitutional construction. Furthermore, in the analysed case law, more than once a reference is made to the Parliamentary History of the ratification of the relevant treaties, to underline the verdict of the Court. The Judiciary clearly is unwilling to oppose the views of the Legislature, and is very cautious not to exceed its legal function concerning the interpretation of treaty Provisions. The legal reasoning of the Dutch Courts is almost unanimous, and shows resemblances with the intentional support approach mentioned above, or subjective teleological method, depending on which stage one addresses the exact considerations of the Court. Therefore, the Parliamentary History of the adoption of human rights instruments (Chapters 6 and 7) will be explored in order to find data on why the Courts conclude that the law is not directly applicable.

In Belgium, the Courts reason differently: not once a reference was made to Parliamentary documents. The Courts technically do not deny the direct effect of the right to food, but instead seem to limit the *ratione personae* of the right. The case law is not as unanimous, due to the differences in function of the Courts (including the Courts of last instance): the Belgian Courts and Legislature are part of a very complex division of powers. Therefore, before analysing case law, a thorough analysis will be made of the Belgian Judiciary (Chapter 11). As it appears, the Courts respond with its case law-each from the perspective of its own function – to changes in legislation. In these verdicts, the right to food is not uncommonly invoked. To understand the Belgian case law on enforceability of the right to food, it is therefore necessary to consider this in the light of the actions of the Legislature, which will be dealt with altogether in Chapter 12. The verdicts of the Belgian Judiciary appear to show some resemblances with the aforementioned objective evaluative method, where the social desirability of a verdict is often the core of the considerations of the Courts. In addition, at least the Constitutional Court, operating within the *objective litigation*, seems to also focus on the question whether a case is similar to previous cases, which shows resemblances with analogous (or *e-contrario*) interpretation, and the principle of justice (equality). This is caused by several developments concerning Constitutional reforms that increasingly assigned reviewing powers to the Constitutional Court. Therefore, in Chapter 11, a thorough analysis of the historical development of the Constitutional Court in the context of the Belgian Constitutional system is included to fully understand its legal reasoning.

2.6.3 The research direction of sub-question 3

The third sub-question is: '*What do the Governments of the Netherlands and Belgium communicate in their reports in the United Nations arena regarding the enforceability of the right to food in their domestic legal order?*' As discussed above, one part of the implementation of international human rights treaties is the duty of Governments to report on the progress made so far. These reporting procedures then were designed to support the progressive realisation of human rights. Also here, the institution that responds to the social problem is predetermined (Governments), and also the form in which the response is communicated is predetermined into detail, as will be discussed in Chapter 3 (Section 3.5.8). Since the right to food is most clearly recognised in the ICESCR, the ICRC, and the CEDAW, an analysis of the reporting history of both countries and the subsequent discussions between the Governments, third parties (such as NGOs) and the UN bodies installed to this means, will focus on those three treaties, in particular on the enforceability of ECOSOC rights and the right to food. Also, the Universal Periodic Review reports will be analysed, since they generally discuss the human rights implementation of both countries. This will be done in Chapters 9 (the Netherlands) and 13 (Belgium).

2.6.4 Comparing and evaluating law

As stated in Chapter 1, next to answering the sub-questions in order to find and explain the relevant data, three comparisons are necessary to answer the main research question:
I. a comparison between the legal practice in Belgium and the Netherlands (what the countries really do);
II. a comparison with those legal practices and the reporting behaviour of both countries (what the countries say they do); and
III. a comparison between the legal practice and the interpretation on the enforceability of the right to food within the UN human rights system (what the countries should do).

Comparison II will be dealt with in conclusion of the country Section itself, that is Chapter 10 (the Netherlands) and Chapter 14 (Belgium), for in those Sections the national legal practice is compared with the reporting behaviour of the same country, and this suits better in the part of the book dedicated to that country. Also, this comparison is of a different nature than the other two, since two responses of institutions to the same social problem are compared at that, whereas in comparison I the functionally equivalent social responses of both countries are compared, and in comparison III those social responses are compared to the *tertium comparationis* are. Therefore, comparisons I and III are interrelated, and need to be discussed together in one coherent Chapter. Comparison I and III will be performed after all the data has been found and evaluated in Chapter 15.

Furthermore, a second purpose of this research is to: *2) where necessary critically evaluate both approaches in view of the UN human rights doctrine regarding the enforceability of the right to food.* Therefore, in Chapters 10, 14, and 15, an evaluative part will be included, in which the data that is found, as well as the comparison, will be summarised and evaluated.

Based on that, recommendations will subsequently be made to both the UN and the countries compared in order to improve the effective realisation of the right to food through enforceability in Chapter 16.

2.7 Conclusion and structure of the book

2.7.1 Concluding remarks

The purpose of this research is thus twofold. Firstly, to gain knowledge through a law comparison about the implementation of the right to food as embedded in international human rights instruments in the Netherlands and Belgium. Secondly, recommendations on the enforceability of the right to food will be made to the UN and the two countries compared. The methodological approach of the comparison will be a modern interpretation of functionalism – introduced by Ralf Michaels – that can be qualified as a functional method based on equivalence functionalism and on an epistemology of constructive functionalism. The constructive move that comes from this approach to functionalism will be used to determine the research direction regarding the sources that need to be examined and compared concerning the threefold question how Member States should respond, do respond and say they respond to the need of enforceability of the right to food. This constructive move is not as flexible as Michaels proposed, but has a rigid core that is determined by the UN human rights system. The countries' social responses will be compared after that. The level of comparison will be on both micro and macro levels, as it is traditionally required in a (functionalistic) law comparison, with – naturally – a focus on Constitutional and administrative law.

2.7.2 Structure of the book

Therefore, as discussed above, the structure of this book will be as follows (Figure 2.2). This thesis consists of four parts.

Part one consists of Chapters 1-3. Chapter 1 introduces the research objectives and research questions, and explains the demarcations of the research, alongside with some remarks on terminology and the use of references. In Chapter 2, the methods used in this thesis are introduced and clarified. In Chapter 3, the first sub-question will be answered, determining the *tertium comparationis* needed for the law comparison.

Then, part two consists of Chapters 4-10, and discusses the Dutch part of this research. In Chapter 4, all relevant case law of the Dutch Judiciary will be analysed, answering sub-question 2A. Two main conclusions lead to further research. As it appears, firstly, the Courts struggle with the Constitutional structure as embedded in Articles 93 and 94 CA. Therefore, the effective functioning of those Articles will be discussed in Chapter 5. Secondly, the Courts seem to greatly value the opinion of the Legislature in deciding on the effect of the right to food in the domestic legal order. The opinion of the Legislature is analysed in Chapters 6 and 7, while discussing the right to food and the enforceability of human rights

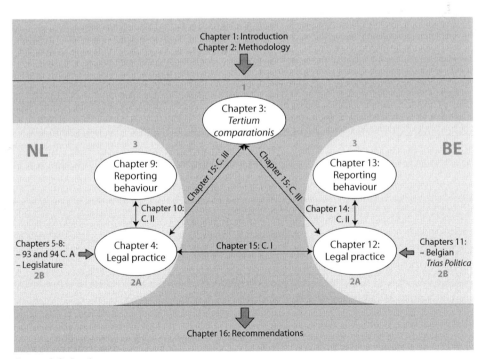

Figure 2.2. Book structure.

separately. It is still unclear however, what criteria are exactly used amongst the Courts to determine whether or not an international standard is directly applicable. Therefore, Chapter 8 addresses this issue into more detail, discussing the apparent sole exception to the rule that ECOSOC rights are not directly applicable: the right to strike as embedded in Article 6 ESC. Chapters 5-8 thus answer sub-question 2B. In Chapter 9, the Dutch reports on the implementation of the right to food will be analysed, answering sub-question 3. In Chapter 10, conclusions will be drawn, evaluating and comparing the legal practice and the reporting behaviour (comparison II).

Part three consists of Chapters 11-14, and discusses the Belgian part of this research. In Chapter 11, the complex Constitutional structure of Belgium will be explored, in which the Belgian Judiciary operates. The analysis of the Belgian *trias politica* is necessary to fully understand the Belgian case law, and will turn out to mainly answer sub-question 2b. Therefore, this Chapter precedes Chapter 12, in which an analysis of relevant case law of the Belgian Courts will be made, with a focus on the case law of the Constitutional Court, answering sub-question 2a. In Chapter 13, the Belgian reports on the implementation of the right to food will be analysed, answering sub-question 3. In Chapter 14, conclusions will be drawn, evaluating and comparing the legal practice and the reporting behaviour (comparison II).

Part four consists of Chapters 15 and 16. In Chapter 15, the legal practice of the Netherlands will be compared to the legal practice of Belgium (comparison I), and both will be compared to the interpretation on the enforceability of the right to food within the UN human rights system (comparison III). In Chapter 16, recommendations will be made.

3. The enforceability of the international human right to adequate food

3.1 Introduction

The right to adequate food is intensely embedded in the international human rights system that evolved in the period after the Second World War, but also strongly interrelated with complex matters – both from a global practical perspective and from a legal theoretical point of view – that are not easily solved. Therefore, it appeared to be not easy to give meaning to the content of the right to food, and to help Member States of the United Nations to implement this right in a suitable way in their national legal systems. It is no surprise then, that a large web of international institutions, each functioning within their own competences and mandates, are involved in the process of further developing the right to food. To establish a *tertium comparationis* for this research, it is necessary to discover what the right to food as a global right encompasses, especially on the matter of enforceability. To this end, this Chapter firstly provides a global historical overview of the most important steps in the legal development of the right to food since World War II (Section 3.2). Secondly, the right to food as it is embedded in the international, regional and domestic human rights systems will be discussed (Section 3.3). Thirdly, the full meaning of 'adequate food' will be explored (Section 3.4), and fourthly, the state obligations implied by the human right to adequate food will be critically reviewed (Section 3.5).

3.2 The legal development of the concept of 'right to food' over time

After the Second World War, the United Nations were founded with as main objective to maintain peace and security. The UN Charter may be regarded as its Constitution,[83] establishing the primary bodies of the UN – The General Assembly, the Security Council and The Economic and Social Council – and providing for procedures that enable these primary bodies to install secondary bodies. One of those secondary bodies was the Commission on Human Rights, established at the first meeting of the Economic and Social Council to help the Council fulfil their task.[84] The Commission spent their first three years of existence writing the Universal Declaration of Human Rights, which was adopted by the General Assembly on 10 December 1948.[85] Article 25 UDHR stipulates the human rights to

---

[83] Charter of the United Nations, 1945.

[84] Economic and Social Council Resolution 5 (I), 16 February 1946. The decision was Based on Article 68 UN Charter: *'The Economic and Social Council shall set up commissions in economic and social fields and for the promotion of human rights, and such other commissions as may be required for the performance of its functions.'*

[85] 217 A (III), 10 December 1948, *The universal declaration on human rights.*

an adequate standard of living, including adequate food. Originally, the intention was to create a worldwide human rights system, consisting of three parts: a declaration, a binding treaty and a monitoring/accountability mechanism. However, due to different opinions between Eastern and Western countries on the one hand and developing and developed countries on the other,[86] the choice was made to draw two Covenants instead of one: one Covenant containing civil and political rights (ICCPR), and the other economic, social and cultural rights (ICESCR). Article 11 ICESCR stipulates the human rights with regard to an adequate standard of living, including food. It took some time, until 1966, before the Covenants were adopted by the General Assembly and opened for signature,[87] and it took another decade before they were ratified by the required 35 States to enter into force.[88] The third step of the international human rights system was integrated in these two Covenants: the State Parties of the ICESCR have the obligation to submit reports on the measures which they have adopted and the progress made in achieving the observance of the ICESCR rights.[89] The Committee on Economic, Social and Cultural Rights was established in 1985,[90] in order to monitor the implementation of the ICESCR. This UN body receives and considers the submitted country reports. To stimulate the further implementation of the ICESCR in the national legal systems of its Members, the Committee regularly writes General Comments to the treaty, clarifying in more detail the content of its Provisions and advising on how to implement the Covenant. In 2008, the General Assembly adopted an Optional Protocol to the ICESCR to strengthen this monitoring system,[91] establishing a procedure for individual communications, inter-state communications and an inquiry procedure. The Committee on Economic, Social and Cultural rights is also the competent body to receive those communications.

In 1974, The Universal Declaration on the Eradication of Hunger and Malnutrition was adopted at the World Food Conference.[92] Increasing food production and sharing resources more adequately were considered to be key elements in order to combat hunger, a common responsibility to the international community. Due to the failure of reaching the goals of the World Food Conference, the FAO organised

---

[86] Manisuli Ssenyonjo, *Economic, social and cultural rights in international law*, Portland: Hart Publishing, 2009, Chapter one, Section II, Section 1.13 and 1.16, Section IV, Section 1.48-1.49.

[87] Both adopted in the same resolution: A/RES/21/2200, 16 December 1966, *International Covenant on Economic, Social and Cultural Rights, International Covenant on Civil And Political Rights And Optional Protocol To The International Covenant On Civil And Political Rights.*

[88] The ICESCR entered into force on 3 January 1976 and the ICCPR entered into force on 23 March 1976.

[89] A/RES/21/2200, 16 December 1966, *International Covenant on Economic, Social and Cultural Rights,* part IV.

[90] Economic and Social Council Resolution 1985/17, 28 May 1985.

[91] A/RES/63/117, 10 December 2008, *Optional Protocol to the International Covenant on Economic, Social And Cultural Rights,* adopted by the General Assembly.

[92] Universal declaration on the eradication of hunger and malnutrition, adopted on 16 November 1974 by the World Food Conference convened under General Assembly resolution A/RES/3180 (XXVIII) of 17 December 1973; endorsed by General Assembly resolution A/RES/3348 (XXIX) of 17 December 1974.

the World Food Summit in 1996, bringing together close to 10,000 representatives on the highest level from 185 countries, who reaffirmed '*the right of everyone to have access to safe and nutritious food, consistent with the right to adequate food and the fundamental right of everyone to be free from hunger.*'[93] Without doubt, the most important ambition of the World Food Summit was to reduce the number of undernourished people to halve their level no later than by 2015.[94] Furthermore, the participating States of the World Food Summit committed themselves '*to clarify the content of the right to adequate food and the fundamental right of everyone to be free from hunger, (...) and to give particular attention to implementation and full and progressive realisation of this right as a means of achieving food security for all.*' In response, the Committee on Economic, Social and Cultural Rights adopted General Comment 12 on the right to food. In this General Comment, the Committee defines the term 'adequate food' more precisely, and points out the different types of obligations of Member States resulting from the right to food: the duty to respect, protect and fulfil.[95] In addition, with the same purpose, the High Commissioner for Human Rights convened expert consultations on the right to food.[96]

Once again on invitation of the FAO, a World Food Summit was held in 2001. The Summit was originally not foreseen, but had '*been prompted by the concern that the target set in the 1996 Rome Declaration may not be achieved.*'[97] The summit resulted in another declaration: the declaration of the World Food Summit, five years later. It must be noted however that the developed countries were mostly not represented on the highest level, in contrast to most developing countries. The right to food was again reaffirmed, but not after difficult negotiations: some participating countries preferred the idea of 'food security' over a right-based approach.[98] An important outcome was Section 10 of the declaration, that mandated the FAO to '*establish an International Working Group (...) to draw a set of voluntary guidelines to support the Member States' efforts to achieve the progressive realisation of the right to adequate food in the context of national food security*', in order to strengthen the respect for all human rights and fundamental freedoms.[99]

---

[93] Rome declaration on world food security, 13 November 1996.

[94] Rome declaration on world food security, 13 November 1996. Later, in the Millennium Goals, the goal was set to reduce by half the proportion of people who suffer from hunger by 2015. See the Millennium Goals, adopted by the UN General Assembly in 2000; A/55/L.2, 18 September 2000, General Assembly Resolution, Section 19.

[95] E/C.12/1999/5, 12 May 1999, CESCR, *General Comment 12, Right to Adequate Food.*

[96] Reports of the High Commissioner for Human Rights: E/CN.4/1998/21, 15 January 1998; E/CN.4/1999/45, 20 January 1999; E/CN.4/2001/148, 30 March 2001.

[97] H.E. Carlo Azeglio Ciampi, President of the Italian Republic, 10 June 2002, inaugural ceremony of the World Food Summit, five years later, available at: http://www.fao.org/DOCREP/MEETING/005/Y7106E/Y7106E02.htm.

[98] A/57/356, 27 August 2002, J. Ziegler, *Report of the Special Rapporteur on the right to food to the General Assembly*, Sections 7-21. See also: J. Ziegler, C. Golay, C. Mahon and S.-A. Way, *The fight for the right to food, lessons learned*, Geneva: The graduate institute, 2011, Chapter 1.

[99] Declaration of the World Food Summit: five years later, 10-13 June 2002, Section 10.

During their 123rd session, the FAO installed the Intergovernmental Working Group (IGWG) as requested. In the period March 2003 – September 2004, the IGWG held four sessions and an intersessional meeting. After harsh negotiations,[100] the IGWG adopted the final version of *'the voluntary guidelines to support the progressive realisation of the right to adequate food in the context of national food security.'* The guidelines give practical guidance on 19 topics, based upon three underlying dimensions: adequacy, availability and accessibility.[101]

A third World Food Summit was held in November 2009: the World Summit on Food Security.[102] In the final declaration, the goal that was set during the first World Food Summit to halve the number of people who suffer from hunger or malnutrition by 2015 was reaffirmed,[103] and the declaration contained commitments and actions, that would lead to food security. A worrying decline in interest seemed to emerge: the participating representatives did not meet the amount and high level of representation compared to the first two Summits. The Declaration does little more than to generally reaffirm already existing targets and commitments.

Meanwhile, the Commission on Human Rights had mandated several working groups to investigate human rights situations. In this light, Jean Ziegler has functioned as Special Rapporteur on the right to food in the period 2000-2008.[104] Since 2008, his position was replaced by Olivier de Schutter,[105] and another Special Rapporteur, Catarina de Albuquerque, was appointed to examine the right to safe drinking water and sanitation.[106] The Sub-Commission on the Promotion and Protection of Human Rights[107] also installed mandates on the right to food and the right to water. Mr Asbjørn Eide[108] has functioned as Special Rapporteur on the right to food, and Mr El Hadji Guissé as Special Rapporteur on the right to drinking

---

[100] A/59/385, 27 September 2004, J. Ziegler, *Report of the Special Rapporteur on the right to food to the General Assembly*, Chapter III, in particular Section 27. See also A. Oshaug, Chapter 6, *The Netherlands and the making of the voluntary guidelines on the right to food*, in: O. Hospes and B. van der Meulen (eds.), *Fed up with the right to food*, Wageningen: Wageningen Academic Publishers, 2009.

[101] *The voluntary guidelines to support the progressive realisation of the right to adequate food in the context of national food security*, adopted by the 127th session of the FAO Council, November 2004.

[102] FAO Doc. WSFS 2009/2, Rome, 16-18 November 2009, *Declaration of the world food summit on food security.*

[103] FAO Doc. WSFS 2009/2, Rome, 16-18 November 2009, *Declaration of the world food summit on food security*, objective 7.1.

[104] Commission on Human Rights Resolution 2000/10, 17 April 2000, Section 10.

[105] Mr O. de Schutter was appointed Special Rapporteur on the right to food on March 26, 2008, by the Human Rights Council, his mandate was extended for another three years in: A/HRC/RES/13/4, 14 April 2010, Human Rights Council Resolution.

[106] A/HRC/RES/7/22, 28 March 2008, Human Rights Council Resolution. The mandate was extended for another three years in: A/HRC/RES/16/2, 8 April 2011, Human Rights Council Resolution.

[107] Before also named: *'Sub-Commission on the Prevention of Discrimination and Protection of Minorities.'*

[108] E/CN.4/Sub.2/1999/12, 28 June 1999, *Updated study on the right to food, submitted by Mr A. Eide in accordance with Sub-Commission decision 1998/106.*

water.[109] In their reports and activities, the Special Rapporteurs contributed to the development and realisation of the right to food.

The Commission on Human Rights was replaced by the Human Rights Council in 2006,[110] and the Sub-Commission on the Promotion and Protection of Human Rights was replaced by the Advisory Council in 2007.[111] All previous resolutions adopted by the UN organs concerning the right to food were reaffirmed,[112] including the mandate of the Special Rapporteur on the right to food.[113]

To conclude, it appears that – very roughly – three pillars can be distinguished within UN context that contribute to the further development of the right to food since World War II (Figure 3.1). The first pillar seems to be the general human rights treaties and the ICESCR related Committee on Economic, Social and Cultural

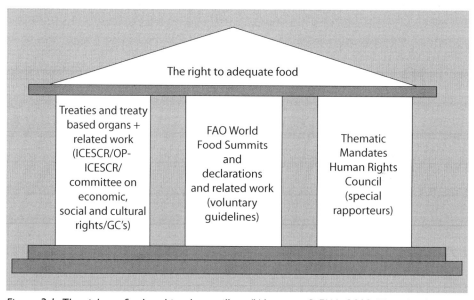

Figure 3.1. The right to food and its three pillars (Wernaart, B.F.W., 2010. The plural wells of the right to food. In: Hospes, O. and Hadiprayitno, I. (eds.) Governing food security, law, policies and the right to food. Wageningen Academic Publishers, Wageningen, the Netherlands).

---

[109] Reports submitted by Mr El Hadji Guissé: E/CN.4/Sub.2/1998/7, 10 June 1998; E/CN.4/Sub2/2004/20, 14 July 2004; E/CN.4/Sub.2/2005/25, 11 July 2005.

[110] A/res/60/251, 3 April 2006, 3 April 2006, *Human Rights Council*, Section 6.

[111] A/HRC/RES/5/1, Human Rights Council Resolution, 18 June 2007, and A/HRC/RES/6/102, 27 September 2007, Human Rights Council Resolution.

[112] A/HRC/RES/7/14, 27 March 2008, Human Rights Council Resolution.

[113] A/HRC/RES/6/2, 27 September 2007, Human Rights Council Resolution., Section 2, *Mandate of the Special Rapporteur on the right to food*. For a previous description of the mandate as written by the Human Rights Council, see Commission on Human Rights Resolutions 2000/10, 17 April 2000, Sections 8-12, and 2001/25, 20 April 2001, Section 11.

rights. The second pillar that could be distinguished is the work done within FAO context: the adoption of the World Food Summit declarations and the work that resulted from these summits, such as the adoption of the voluntary guidelines. The third pillar would be the work done by the Special Rapporteurs. Needless to say, that the delineation between these three pillars is somewhat arbitrary as the proposed pillars are closely interrelated with each other, but distinguishing them may help to better understand the way the right to food has been developed so far. Also, it needs to be noted here that the specified treaties for the protection of particular groups are not included in this overview.

3.3 The right to food in international, regional and domestic human rights systems

The right to food is mostly discussed in the context of Article 11 ICESCR. Limiting the right to food to the interpretation of this Article only would however do no justice to the full meaning of the right to food. A survey through international human rights instruments shows that the right to food is also recognised directly in other documents that mostly aim at the protection of a particular group of individuals (Section 3.3.1), and is furthermore inextricably linked to other human rights or human rights-related issues (Section 3.3.2). Also, the right to food is recognised outside the UN context on a regional level (Section 3.3.3). Finally, it is the purpose of international human rights to be implemented in the domestic legal order of states. To this end, referring to FAO research on its domestic implementation worldwide,[114] the right to food might be recognised directly or indirectly in Constitutions or used as a directive principle (Section 3.3.4).

3.3.1 The right to food stipulated as independent right

The freedom from want is mentioned in the preamble of some international human rights documents, being one of the considerations that lead to the adoption. The Universal Declaration on Human Rights, the International Covenant on Civil and Political Rights and the International Covenant on Economic, Social and Cultural rights are such documents. Also, the right to food is recognised as an independent right in documents that aim at the protection of a particular group of individuals. In this light, reference can be made to Article 27 of the International Covenant for the Rights of the Child,[115] Article 28 of the Convention on Persons with Disabilities,[116] and in the context of healthcare and pregnancy, Article 12 (2) of The Convention on the Elimination of All Forms of Discrimination against

---

[114] L. Knuth and M. Vidar, *Constitutional and legal protection of the right to food around the world*, Right to food Studies, Rome: FAO, 2011.

[115] A/RES/44/25, 20 November 1989, *Convention on the Rights of the Child*, Article 27.

[116] A/RES/61/106, 24 January 2007, *The Convention on the Rights of Persons with Disabilities*, Article 28.

Women,[117] and Article 24 (2) of the International Covenant for the Rights of the Child.[118] Furthermore, the right to food is stipulated in several Articles of the Geneva Conventions and protocols[119] on humanitarian law that stipulate the right to food in the specific context of the beneficiaries of the treaties.[120]

3.3.2 The right to food in relation to other human rights

The Vienna declaration underlines that '*all human rights are universal, indivisible and interdependent and interrelated.*'[121] Indeed, the right to food is inextricably linked to other human rights[122] and is therefore frequently mentioned in their context, by stressing the importance of adequate food in realising the other human rights and vice versa.

First and foremost, non-discrimination Provisions are often at the core of human rights instruments and usually not restricted by exception clauses. Therefore, the Committee on Economic, Social and Cultural Rights pointed out with regard to the ICESC that '*discrimination in access to food...constitutes a violation of the Covenant.*'[123] The right to food may be an important issue in case of persistent discrimination of a particular group of people, and is therefore also recognised in

---

[117] A/RES/34/180, 18 December 1979, *Convention on the Elimination of All Forms of Discrimination Against Women*, Article 12(2).

[118] A/RES/44/25, 20 November 1989, *Convention on the Rights of the Child*, Article 27.

[119] Geneva Conventions and Protocols that include references to the right to food are: First Geneva Convention for the Amelioration of the Condition of the Wounded and Sick in Armed Forces in the Field, 12 August 1949, Articles 32 (2) jo Article 27; Third Geneva Convention Relative to the Treatment of Prisoners of War, 12 August 1949, Articles 20, 26, 28, 46, 51, 72; Fourth Geneva Convention relative to the Protection of Civilian Persons in Time of War, 12 August 1949, Articles 15, 23, 49, 50, 55, 59, 76, 87, 89, 100, 108, 127; Protocol Additional to the Geneva Conventions of 12 August 1949, and relating to the Protection of Victims of International Armed Conflicts, 8 June 1977, Articles 54, 69, 70; Protocol Additional to the Geneva Conventions of 12 August 1949, and relating to the Protection of Victims of Non-International Armed Conflicts, 8 June 1977, Articles 5, 14, 18.

[120] In the four Geneva Conventions and their three protocols the human right to adequate food is recognised for the following groups of persons: medical personnel of a neutral country assisting one of the parties to a conflict, prisoners of war in general, prisoners of war who are being evacuated or transferred, civilians, detained civilians, and persons whose liberty is restricted. The starvation of civilians as means of pressure is forbidden in national and international armed conflicts, as well as the deliberate destruction of foodstuffs and drinking water. Forced displacements of civilians leading to starvation are prohibited. There are also international rules concerning the protection of humanitarian assistance in occupied territories and during non-international armed conflicts. Also shipment/delivery of means of existence – including food – for prisoners of war or detained civilians should be allowed. In case of the establishment of a neutralised zone, the delivery of food supplies for (among others) the wounded and sick combatants or non-combatants and civilians should be agreed upon amongst the conflicting parties.

[121] A/CONF.157/23, 12 July 1993, the world conference on human rights, *Vienna declaration and programme of action.*

[122] See also A. Eide, Chapter 11, *Adequate standard of living*, Section 5, in: D. Moeckli, S. Shah, S. Sivakumaran (eds.), D. Harris (cons. ed.), *International human rights law*, Oxford University Press, 2010.

[123] E/C.12/1999/5, 12 May 1999, CESCR, *General Comment 12, Right to Adequate Food*, Sections 18 and 19. See also: E/C.12/GC/20, 10 June 2009, CESCR, *General Comment 20, Non-Discrimination in Economic, Social and Cultural Rights (Art. 2, Section 2)*, in particular Sections 6, 23, and 30.

human rights treaties for specific groups, as demonstrated above. The principle of non-discrimination thus strengthens the right to food.

There is furthermore a strong interconnection between the right to self determination and the realisation of human rights in general: '*the right to self determination is of particular importance because its realisation is an essential condition for the effective guarantee and observance of individual human rights and for the promotion and strengthening of those rights. It is for that reason that States set forth the right of self-determination in a Provision of positive law in both Covenants and placed this Provision as Article 1 apart from and before all of the other rights in the two Covenants.*[124] In order to realise the right to food, State Parties must comply with the Provisions recognising the right to self-determination. The right is most often addressed in relation to the accessibility of food, and in this regard frequently mentioned in a problem-specific context, such as land access in rural areas,[125] or access to resources in poor fishing communities.[126] Access to resources is reported as particularly problematic for indigenous people[127] and woman.[128] They often experience difficulties in gaining access to resources that are necessary for food production, mostly as a result of discriminatory circumstances, which also demonstrates the interrelationship with non-discrimination Provisions.

In addition, there is a clear link between the right to health or healthcare and the right to food. Unhealthy eating habits or malnutrition lead to bad health, and bad health may prevent an individual from consuming adequately.[129] The right to healthcare is of particular importance for weaker groups in a society. For these groups, access to adequate nutrition is of vital importance for the realisation of

---

[124] The Human Rights Committee, 13 March 1984, *General Comment 12, The right to self-determination of peoples* (Article 1), Section 1.

[125] A/57/356, 27 August 2002, J. Ziegler, *Report of the Special Rapporteur on the right to food to the General Assembly*, Chapter III.

[126] A/59/385, 27 September 2004, J. Ziegler, *Report of the Special Rapporteur on the right to food to the General Assembly*, Chapter IV.

[127] For instance: A/RES/61/295, 2 October 2007, *United Nations Declaration on the Rights of Indigenous Peoples;* A/HRC/RES/7/14, 27 March 2008, Human Rights Council Resolution, Section 12; A/RES/62/164, 13 March 2008, General Assembly Resolution, Section 12; A/CONF.151/26/Rev.1 (Vol.1.), Rio de Janeiro, 3-14 June 1992, *Agenda 21*, Chapter 26: *Recognising and strengthening the role of indigenous people and their communities; The voluntary guidelines to support the progressive realisation of the right to adequate food in the context of national food security*, adopted by the 127th session of the FAO Council, November 2004, preamble, Section 8.1; A/60/2005, 12 September 2005, J. Ziegler, *Report of the Special Rapporteur on the right to food to the General Assembly*, Chapter III. See also L. Knuth, *The right to food and indigenous people, how can the right to food help indigenous people?* Rome: FAO, 2009, especially Section 1.3.1.

[128] For instance: A/58/330, 28 August 2003, J. Ziegler, *Report of the Special Rapporteur on the right to food to the General Assembly*, especially Section 22; A/CONF.177/20, Beijing, China, 4-15 September 1995, *Report of the fourth world conference of women*; A/HRC/RES/7/14, 27 March 2008, Human Rights Council Resolution, Sections 4-5.

[129] See for instance: Office of the United Nations High Commissioner for Human Rights and the World Health Organisation, factsheet no. 31, *The right to health,* Geneva: UN, 2008.

their right to health or healthcare, especially with regard to women and young children.[130] In this context the right to breast-feeding is of utter importance.[131] The Committee on Economic, Social and Cultural Rights explained that *'the right to health embraces a wide range of socio-economic factors that promote conditions in which people can lead a healthy life, and extends to the underlying determinants of health, such as food and nutrition, access to safe and potable water...'*[132] In their view, the obligations of States include at least: *'to ensure access to the minimum essential food which is nutritionally adequate and safe, to ensure freedom from hunger to everyone.'*[133]

In some circumstances, the right to food may be considered to be a prerequisite to fulfil the right to life. Although the ICCPR is not too often viewed in the context of economic, social and cultural rights, Article 6 ICCPR implies, according to the Human Rights Committee, a duty for Member States to *'take all possible measures to reduce infant mortality and to increase life expectancy, especially in adopting measures to eliminate malnutrition and epidemics.'*[134] Furthermore, to withhold (access to) food with the purpose to destroy life is a clear violation of Article II (c) of the Convention on the Prevention and Punishment of the Crime of Genocide.[135]

It will be no surprise that there is a close relationship between the right to food and the right to social security. This can clearly be demonstrated when one considers the meaning of Articles 26 and 27 of the Convention on the Rights of the Child. Article 26 ICRC stipulates that the State Parties *'shall recognise for every child the right to benefit from social security (...) the benefits should (...) be granted, taking into account the resources and the circumstances of the child and persons having responsibility for the maintenance of the child...'*[136] Article 27 ICRC states that *'parents or other responsible for the child have the primary responsibility to secure, within their abilities and financial capacities, the conditions of living necessary for the child's development.'* The State Parties *'shall take appropriate measures to assist parents and others responsible for the child to implement this right and shall in case*

---

[130] For instance: A/RES/34/180, 18 December 1979, *Convention on the Elimination of All Forms of Discrimination Against Women*, Article 12; A/RES/44/25, 20 November 1989, *Convention on the Rights of the Child*, Article 24 (c); A/CONF.177/20, Beijing, China, 4-15 September 1995, *Report of the fourth world conference on women*, especially Chapter IV C.

[131] For instance: European Social Charter (revised), Strasbourg, 3.V.1996, Article 8; A/RES/44/25, 20 November 1989, *Convention on the Rights of the Child*, Article 24 (e). See also: Arun Gupta, Chapter 5, *International Obligations for Infants' Right to food*, in: G. Kent (ed.), *Global obligations for the right to food*, Lanham: Rowman and Littlefield Publishers Inc, 2008.

[132] E/C.12/2000/4, 11 August 2000, CESCR, *General Comment 14, The right to the highest attainable standard of health*, Section 4.

[133] E/C.12/2000/4, 11 August 2000, CESCR, *General Comment 14, The right to the highest attainable standard of health*, Section 43 (b).

[134] UN Human Rights Committee, 30 April 1982, *General Comment 6: Article 6, Right to Life*, Section 5.

[135] A/RES/260 (III) (A), 9 December 1948, *International Convention on the Prevention and Punishment of the Crime of Genocide.*

[136] A/RES/44/25, 20 November 1989, *Convention on the Rights of the Child*, Article 26.

of need provide material assistance and support programmes, particularly with regard to nutrition, clothing and housing.'[137] The realisation of the right to food is thus a primary responsibility for parents or others responsible for the child. The State Party has the obligation to assist those who are primarily responsible for the child when they fail to meet with their responsibilities. It is a small step to see this in the context of social security, when social security should be granted taking the resources and circumstances of the persons who are responsible for the child into account. On the other hand, when we take a closer look at the country reports submitted by the Member states of the ICESCR, it appears that while discussing Article 11 of this Covenant, the level of social security is sometimes used as a way to express to what extent the Member state complies with the realisation of the right to an adequate standard of living, including the right to food.[138] The validity of such arguments is not self-evident: the right to social security is recognised in Article 9 of the same Covenant, and both rights should not be lumped together too easily. Although there is an obvious link between the right to food and the right to social security, access to the benefits of a social security system does not automatically lead to access to food. Therefore, the Committee on Economic, Social and Cultural Rights held that the right to social security must at least provide the benefited with access to a minimum life standard, in accordance with Article 11 ICESCR.[139] In this light, it is interesting to note that the right to social security is, more often than the right to food, recognised in international non-discrimination treaties, including the International Convention on the Elimination of All Forms of Racial Discrimination,[140] the Convention on the Elimination of All Forms of Discrimination against Women,[141] the Convention relating to the Status of Refugees,[142] the International Convention on the Protection of the Rights of All Migrant Workers and Members of Their Families,[143] and, as already mentioned above, the Convention on the Rights of the Child.[144]

Another interesting relationship lies between the right to food and the right to education. According to the CESCR, education should fulfil the requirements of availability, accessibility, acceptability and adaptability. The Committee underlined

---

[137] A/RES/44/25, 20 November 1989, *Convention on the Rights of the Child*, Article 27.

[138] For instance, the Dutch Government states in its third periodic report while discussing Article 11 that: *'the Netherlands have a comprehensive system of social benefits guaranteeing its citizens an adequate minimum income.'* See: E/1994/104/add.30, 23 August 2005, *Third periodic reports submitted by State Parties under Articles 16 and 17 of the Covenant, Addendum, the Netherlands*, Section 329.

[139] E/C.12/GC/19, 4 February 2008, CESCR, *General Comment 19, The Right to Social Security (Art. 9)*. See in particular Sections 18, 22, 28, and 59.

[140] A/RES/2106 (XX), 21 December 1965, *The International Convention on the Elimination of All Forms of Racial Discrimination*, Article 5 (e) (iv).

[141] A/RES/34/180, 18 December 1979, *Convention on the Elimination of All Forms of Discrimination Against Women*, Article 11 (e).

[142] A/RES/429 (IV), 14 December 1950, *Draft Convention relating to the Status of Refugees*, Article 24 (1) (b).

[143] A/RES/45/158, 18 December 1990, *The International Convention on the Protection of the Rights of All Migrant Workers and Members of Their Families*, Article 27.

[144] A/RES/44/25, 20 November 1989, *Convention on the Rights of the Child*, Article 26.

that the availability at least includes, amongst others, sanitation for both sexes and clean drinking water.[145] It is often argued that the right to food and the right to education go hand in hand when it comes to the development of a country.[146] On the one hand, hunger may affect the learning abilities of the student, especially on a young age when the child is in a crucial stage of development, which has a devastating long term effect on both the individual, but also on a larger scale of the economic possibilities of a region. On the other hand, school meals may stimulate class attendances and therefore generally increase the level of education of a region, which may have positive long term effects on the economic achievements of this region.[147]

Also, the right to food is often interrelated with a specific matter or a problem that involves a combination of human rights. For example, the CESCR underlined the importance of the right to food in the specific context of the relationship between economic sanctions and respect for economic social and cultural rights.[148] The Special Rapporteur on the right to food usually gives a sound overview in his reports on the most topical issues with which the right to food and possibly other human rights are interrelated.[149]

Noteworthy here, in the context of emergency food aid, is the Food Assistance Convention, adopted in London on 12 April 2012 which will enter into force on 1 January 2013.[150] Although it does not stipulate the right to food, and the instrument is far from perfect, the Convention aims to *'save lives, reduce hunger, improve food security, and improve the nutritional status of the most vulnerable populations.'*[151]

3.3.3 The right to food in regional documents

Since the creation of a global human rights system, the need was felt in many parts of the world to also adopt a regional human rights system parallel to the global system, which would take the more specific needs, historical context, values and culture of a specific region into account. In these regional human rights systems, the right to adequate food is also embedded.

---

[145] E/C.12/1999/10, 8 December 1999, CESCR, *General Comment 13, The Right to Education (Art.13)*, Section 6.

[146] See for instance: S. Vivek, Chapter 8, *Global Support for School Feeding*, in: George Kent (ed.), *Global obligations for the right to food*, Lanham: Rowman and Littlefield Publishers Inc, 2008.

[147] The benefits of school meals for development and peace are fiercely underlined by G. McGovern in: *Third freedom, ending hunger in our time*, Lanham: Rowman & Littlefield Publishers Inc, 2001.

[148] E/C.12/1997/8, 12 December 1997, CESCR, *General Comment 8, The relationship between economic sanctions and respect for economic, social and cultural rights*, Section 3.

[149] The reports of O. de Schutter are available at: www.srfood.org; the reports of J. Ziegler are available at www.righttofood.org.

[150] The Food Assistance Convention, adopted on 12 April 2012, London.

[151] See especially the commentary of the International Red Cross, at: www.ifrc.org.

3.3.3.1 The African Charter on Human and Peoples' Rights

The African Charter on Human and Peoples' Rights was adopted on the initiative of the Organisation of African Unity (OAU) – since 26 May 2001 the African Union (AU) – in 1981 and entered into force in 1986.[152] The right to adequate food is not specifically recognised in the Charter, although Article 21 stipulates that '*all peoples shall freely dispose of their wealth and natural resources.*'

A protocol to the Charter on Human Rights concerning the rights of women does however explicitly recognise the right to food security for women.[153] Furthermore, in 1990, the OAU adopted the African Charter on the Rights and Welfare of the Child[154] that stipulates the right to adequate food in Articles 14 and 20. Also, in 2009, the AU adopted the African Union Convention for the Protection and Assistance of Internally Displaced Persons in Africa (Kampala Convention).[155] In this convention the States Parties pledge themselves to provide internally displaced persons with adequate humanitarian assistance, including food and water.[156] Also, Members of armed groups '*shall be prohibited from denying internally displaced persons the right to live in satisfactory conditions of dignity, security, sanitation, food, water, health and shelter...*'[157]

The Charter established the African Commission on Human and Peoples' Rights to protect and promote human rights and advise on the interpretation of the Charter.[158] Article 62 establishes a reporting procedure, obliging the Member States to submit a report on the implementation of the rights enshrined in the Charter every two years, which are made public and (on invitation) discussed in public meetings with the Commission.[159] In 1998 however, the monitoring mechanisms of the African human rights system was significantly strengthened by the adoption of an additional protocol that installed the African Court on Human and Peoples' Rights, a Court that is mandated to make legally binding

---

[152] OAU Doc. CAB/LEG/67/3 rev. 5, 21 I.L.M. 58, June 1981, African Charter on Human and Peoples' Rights.
[153] Additional Protocol to the African Charter on Human and Peoples' Rights and the Rights of Women in Africa, 11 July 2003, Article 15.
[154] OAU Doc. CAB/LEG/24.9/49, July 1990, African Charter on the Rights and Welfare of the Child.
[155] African Union Convention for the Protection and Assistance of Internally Displaced Persons in Africa (Kampala Convention), 22 October 2009.
[156] African Union Convention for the Protection and Assistance of Internally Displaced Persons in Africa (Kampala Convention), 22 October 2009, Article 9 (2) (b).
[157] African Union Convention for the Protection and Assistance of Internally Displaced Persons in Africa (Kampala Convention). 22 October 2009, Article 7 (5) (c).
[158] OAU Doc. CAB/LEG/67/3 rev. 5, 21 I.L.M. 58, June 1981, African Charter on Human and Peoples' Rights, part II.
[159] See: www.achpr.org for the website of the African Commission on Human and Peoples' Rights. On this website, all periodic country reports are published.

decisions[160] and is competent in considering all cases that concern the human rights recognised in the African Charter and any other relevant human rights instrument ratified by the States concerned.[161] The Commission, State Parties and African Intergovernmental Organisations may submit a case to the Court.[162] In addition, Member States may accept the competence of the Court to receive cases submitted by NGOs or individuals.[163] The Court's jurisdiction thus includes economic, social and cultural rights, and possibly also the right to food, a development which was praised by the Special Rapporteur on the right to food.[164] However, it remains to be seen whether the Court will produce effective case law,[165] since so far the Court only ruled in a limited number of cases as of 2008. Also, it often rules that it has no jurisdiction in the matter.[166]

3.3.3.2 The Asian Human Rights Charter

Asia is the only region in the world without a regional treaty on human rights broadly adopted by its Governments. However, several non-governmental initiatives were taken to stimulate regional recognition of human rights. For instance, a large group of NGOs and individuals created the Asian Human Rights Charter, a document with the intention to '*deepen the Asian debate on human rights, to present the peoples' views on human rights as against those of some Asian leaders who claim that human rights are alien to Asia and to promote political, social and legal reforms for ensuring human rights in the countries of the region.*'[167] The Charter recognises the right to food in Article 7.1, and underlines in Article 14.2 that: '*arbitrary arrests, detention, imprisonment, ill-treatment, torture, cruel and inhuman punishment are common occurrences in many parts of Asia. Detainees and prisoners are often forced to live in unhygienic conditions, are denied adequate food and healthcare and are prevented from having communication with, and support from, their families.*'

---

[160] Additional Protocol to the African Charter on Human and Peoples' Rights on the establishment of an African Court on Human and Peoples' Rights, 10 June 1998, Article 28-30.

[161] Additional Protocol to the African Charter on Human and Peoples' Rights on the establishment of an African Court on Human and Peoples' Rights, June 1998, Article 3.

[162] Additional Protocol to the African Charter on Human and Peoples' Rights on the establishment of an African Court on Human and Peoples' Rights, June 1998, Article 3.

[163] Additional Protocol to the African Charter on Human and Peoples' Rights on the establishment of an African Court on Human and Peoples' Rights, June 1998, Article 34 (6) jo. 3.

[164] E/CN.4/2002/58, 10 January 2002, J. Ziegler, *Report of the Special Rapporteur on the right to food*, Section 64.

[165] G. Mukundi Wachira, *African court on human and peoples' rights: ten years on and still no justice*, Minority Rights Group International, 2008.

[166] See the official website of the Court: www.african-court.org.

[167] Asian Human Rights Charter, a peoples' charter, declared in Kwangju, South Korea, 17 May 1998, preface.

3.3.3.3 *The Arab Charter on Human Rights*

In the Arab world, several human rights instruments have been adopted throughout the past decades. The Cairo Declaration on Human Rights in Islam,[168] adopted in 1990, can be seen as a reaction to the UN human rights mechanisms that were based on the UDHR. A group of Muslim countries felt insufficiently involved in the creation of the UDHR, and therefore adopted their own declaration on human rights. The right to adequate food is recognised in case of armed conflict, for children, and for every individual in Articles 3, 7, and 17. The Declaration had a predecessor that also recognised the right to adequate food.[169] Both declarations are heavily criticised for the fact that they affirm that Shari'ah law is the only legal source.[170] In 2004, the League of Arab States adopted a stronger legal instrument, the Arab Charter on Human Rights,[171] in which both the foremost UN human rights treaties and the Cairo declaration are reaffirmed.[172] The right to food is recognised in Article 38, and Article 39 recognises the right to health, which includes the obligation for Member States to take measures that include the *'Provision of the basic nutrition and safe drinking water for all.'* Furthermore, the right to self-determination is underlined in the preamble, the equality between men and women is recognised (especially Article 3 (3)), and special protection is embedded for children (especially Article 34 (3)) and disabled persons (Article 40). The implementation of the Charter is monitored by the Arab Human Rights Committee through a reporting system. Every Member State has the obligation to submit an initial report within one year from the date on which the charter entered into force, and from then on a periodic report every three years.[173] Since the Charter entered into force in 2008, its effect on the realisation of human rights has to be awaited.[174]

3.3.3.4 *The Charter of the Organisation of American States*

In 1948, the Organisation of American States (OAS) adopted the Charter of the Organisation of American States. Article 34 of the Charter states that proper

---

[168] Cairo Declaration on Human Rights in Islam, 5 August 1990, adopted at the Nineteenth Islamic Conference of Foreign Ministers.

[169] Universal Islamic Declaration of Human Rights, 19 September 1981, adopted by the Islamic Council, Article XVIII.

[170] See for instance Articles 24-25 of the Cairo Declaration on Human Rights in Islam, 5 August 1990, adopted at the Nineteenth Islamic Conference of Foreign Ministers.

[171] Arab Charter on Human Rights, May 22, 2004, League of Arab States. See also: M. Amin Al-Midani, M. Cabanettes (translation) and S.M. Akram (revision), *Arab Charter on Human Rights 2004*, Boston University International Law Journal, Volume 24, Fall 2006, Number 2, pp. 147-164.

[172] Arab Charter on Human Rights, May 22, 2004, League of Arab States, preamble.

[173] See Articles 45-47 for the installment procedure of the Committee; see Articles 48-49 for the reporting procedure.

[174] See for the most recently updated information on the Arab Human Rights Committee: www.arableagueonline.org.

nutrition and modernisation of rural life and land reforms are basic goals in order to support *inter alia* equality of opportunity, elimination of poverty and equal distribution of wealth and income.

In 1969, the American Convention on Human Rights was adopted by the OAS, and entered into force on 1978.[175] The Convention has a strong focus on civil and political rights, and goes no further than obliging States to guarantee a progressive development of the economic, social and cultural rights (Article 26). The Inter American Court of Human Rights supervises a State complaint procedure (Chapter VIII), while the Commission may receive individual petitions or communications of alleged human rights violation by a Member State, and supervises a corresponding friendly settlement procedure (Chapter VII). The Committee furthermore promotes the respect for human rights through *inter alia* make recommendations and prepare reports and studies, and *'to request the Governments of the Member states to supply it with information on the measures adopted by them in matters of human rights.'*[176] There is thus no formal reporting system through which the implementation of the rights is supervised.

The San Salvador Additional Protocol to the American Convention on Human Rights, adopted in 1988, does however stipulate economic, social and cultural rights, and recognises the right to adequate food in Article 12. Furthermore, the protocol has a non-discrimination Provision (Article 3), recognises the right to just, equitable, and satisfactory conditions of work (Article 7), the right to social security (Article 9), and the right to health (Article 10). Special protection is underlined for the family (Article 15), children (Article 16), the elderly (Article 17) and persons with a handicap (Article 18). The protocol includes a monitoring procedure that involves the obligation of Member States to submit periodic reports on the implementation of the Protocol, and a very limited procedure for complaints regarding trade union rights and the right to education, both supervised by the Inter-American Commission on Human Rights.[177] The Member States have an obligation to adopt measures *'to the extent allowed by their available resources, and taking into account their degree of development, for the purpose of achieving progressively and pursuant to their internal legislations, the full observance of the rights recognised in this Protocol'*[178], and to enact domestic legislation if the rights are not already guaranteed by domestic law *'in accordance with their Constitutional processes and*

---

[175] The American Convention on Human Rights, 22 November 1969, adopted at the Inter-American Specialised Conference on Human Rights.

[176] The American Convention on Human Rights, 22 November 1969, adopted at the Inter-American Specialised Conference on Human Rights, Article 41.

[177] The San Salvador Additional Protocol to the American Convention on Human Rights, November 1988, adopted by the General Assembly of the Organisation of American States, Article 19.

[178] The San Salvador Additional Protocol to the American Convention on Human Rights, November 1988, adopted by the General Assembly of the Organisation of American States, Article 1.

the Provisions of this Protocol, such legislative or other measures as may be necessary for making those rights a reality.'[179]

3.3.3.5 The European Social Charter

The European Social Charter was adopted in 1961,[180] and revised in 1991.[181] The ESC is the counterpart of the European Convention on Human Rights, is strongly based on ILO legislation, and has therefore a focus on labour-law. Perhaps due to this focus, the Charter contains no Provision that specifically recognises the right to food, although in Article 8, the right for employed woman to have sufficient time to nurse (including breast-feeding) their infant is recognised. Nevertheless, some Provisions may contribute to the realisation of the right to food, such as the right to safe and healthy working conditions (Article 3), the right to a fair remuneration sufficient for a decent standard of living for themselves and their families (Article 4), the right to protection of health (Article 11), the right to social security (Article 12), the right to social and medical assistance (Article 13) and the right to benefit from social welfare services (Article 14). Also, the rights of children (Article 7), women (Article 8), the family (Article 16), and migrant workers (Article 19) receive additional protection in the Charter.[182] The European Committee of Social Rights (ECSR) supervises the monitoring of the implementation of the legal obligations of the Charter through a reporting system[183] and, under an additional protocol, through a collective complaint procedure.[184]

3.3.4 The right to food in national Constitutions

In several countries, the right to food is recognised in the national Constitutions. According to an FAO right to food study in 2011,[185] the right to food is recognised explicitly in the Constitution of 23 Countries. In addition, the right is recognised implicitly, for instance by means of a broader right or by a directive principle, in the Constitutions of 33 countries. Due to the direct effect of international Provisions, the right to food has effect in at least another 51 countries. Overall, the right to food is thus legally applicable in 107 countries. As already mentioned

---

[179] The San Salvador Additional Protocol to the American Convention on Human Rights, November 1988, adopted by the General Assembly of the Organisation of American States, Article 2.

[180] European Social Charter, 18 October 1961, Turin, in: European Treaty Series 35.

[181] European Social Charter (revised), 3 May 1996, Strassbourg, in: European Treaty Series 163.

[182] The Articles referred to are numbered in this manner in the 1991 Charter. It must be noted however that some countries only ratified the 1961 edition, which is organised differently.

[183] Originally embedded in part IV of the 1961 Charter, later altered in the Amending Protocol of 1991 reforming the supervisory mechanism, 21 October 1991, Turin, in: European Treaty Series 142, and currently referred to in Section IV, Article C of the 1991 Charter.

[184] Additional Protocol to the European Social Charter Providing for a System of Collective Complaints, 9 November 1995, Strassbourg, in: European Treaty Series 158.

[185] Lidija Knuth and Margret Vidar, *Constitutional and Legal Protection of the Right to food around the World*, Right to food Studies, Rome: FAO, 2011.

in Chapter 1, the Netherlands and Belgium are amongst those Countries. The research however seems to be based on a dogmatic law comparison, and certainly not on a functional one, which is understandable, considering the enormous scope of the research. The actual impact of the human right to food however can only be assessed by more intense analysis of national case law in which human rights Provisions are invoked, in relation to the Constitutional structure in which the national Judiciary operates.

3.4 The meaning of 'adequate food'

When exploring the meaning of the right to adequate food, and the obligations for States that follow from ratifying the right, it appears that this is most often discussed in the context of Article 11 ICESCR. Most likely because in this Provision the right was stipulated as a global legal right for the first time, and not specifically focussed on a particular vulnerable group. Thus it has the broadest scope. Therefore, in this section and in Section 3.5, the right is primarily discussed in view of Article 11 ICESCR.

One argument to reject the enforceability of the right to food is that the formulation of the right in especially Article 11 ICESCR is too imprecise to be applied directly by domestic Courts without the adoption of further domestic legislation in which the right is specified.[186] To this end, various attempts were made in the international arena to further clarify the content of the right to food. In this process, the meaning of 'adequate food' as well as the content of the State Duties implied by the right to food was elaborated. The first will be discussed in this section, the latter in Section 3.5.

Article 11 of the ICESCR was written in cooperation with the FAO, which proposed sub Section (b), to provide for a legal basis for their 'freedom from hunger campaign' that was set up in 1960. It explains why sub (a) refers to '*the right to adequate food*', and sub (b) refers to '*freedom from hunger.*'[187] During the negotiations that led to the final draft of the ICESCR in the early nineteen sixties, the content of Article 11 was debated heavily. For instance, the Dutch Delegation stated that the Article '*was too detailed, covered many matters which went beyond the competence of the Third Committee, was not consistent with the bald statements relating to the rights of housing and clothing, was more appropriate as a declaration than a legally-binding instrument.*'[188] Together with a group of other Member States, this Government believed that food-related problems were so diverse that different approaches were required. They considered that the formulation of the Provisions in Article

---

[186] E/CN.4/2002/58, 10 January 2002, J. Ziegler, *Report of the Special Rapporteur on the right to food*, Section 35.
[187] See also the contribution from P. Alston, in: P. Alston and K. Tomasevski (eds.), *The right to food, International Studies in Human Rights,* Utrecht: SIM, 1984.
[188] A/C.3/SR.1266, 1963, Section 57-63.

11 was too specific to be compatible with that requirement. Article 11 ICESCR is therefore – as well as many other ICESCR Provisions – a compromise between countries who favoured a strong wording including clear obligations and countries that preferred a larger margin of discretion for Member States to implement the right in a way that suits best considering the particularities of the specific country.

In its final form, Article 11 ICESCR stipulates that:
1. The States Parties to the present Covenant recognise the right of everyone to an adequate standard of living for himself and his family, including adequate food, clothing and housing, and to the continuous improvement of living conditions. The States Parties will take appropriate steps to ensure the realisation of this right, recognising to this effect the essential importance of international co-operation based on free consent.
2. The States Parties to the present Covenant, recognising the fundamental right of everyone to be free from hunger, shall take, individually and through international co-operation, the measures, including specific programmes, which are needed:
   a. To improve methods of production, conservation and distribution of food by making full use of technical and scientific knowledge, by disseminating knowledge of the principles of nutrition and by developing or reforming agrarian systems in such a way as to achieve the most efficient development and utilisation of natural resources.
   b. Taking into account the problems of both food-importing and food-exporting countries, to ensure an equitable distribution of world food supplies in relation to need.

In Section 2, the Provision specifies the right to food on several aspects. In the first place, the right to adequate food means that people should be free from hunger. It is widely understood that the right to food implies both the absence of hunger and the absence of malnutrition. Hunger (also referred to as undernourishment or undernutrition) means that a person lacks a diet of an adequate quantity, meaning that the person has no access to a sufficient amount of calories. Malnutrition means that a person lacks a diet of an adequate quality, meaning that the person has no access to a sufficiently varying diet that includes all necessary substances to lead a healthy life.[189] These concepts have been further defined by institutions such as the FAO, WFP, WHO and UNICEF.[190] In the second place, Section 2 mentions that the right to food should be realised by States individually and through international co-operation, and further explains how this should be done.

---

[189] See for instance: E/CN.4/2001/53, 7 February 2001, J. Ziegler, *Report of the Special Rapporteur on the right to food*, Section 16; J. Ziegler, C. Golay, C. Mahon and S.-A. Way, *The fight for the right to food, lessons learned*, Geneva: The graduate institute, 2011, Chapter 1; G. Kent, *Freedom from want: The human rights to food*, Washington D.C.: Georgetown University Press, 2005, Chapter 1.
[190] See for further information the websites of these organisations, for instance: www.wfp.org/hunger.

In Section 1, the right to adequate food is recognised as part of an adequate standard of living. The CESCR underlined in General Comment 12 that: *'the right to adequate food is realised when every man, woman and child, alone or in community with others, has physical and economic access at all times to adequate food or means for its procurement.'*[191] According to the Committee, the meaning of 'adequate food' encompasses the concept of 'availability' and 'accessibility', and thus the right to food implies: *'The availability of food in a quantity and quality sufficient to satisfy the dietary needs of individuals, free from adverse substances, and acceptable within a given culture.'* And also: *'The accessibility of such food in ways that are sustainable and that do not interfere with the enjoyment of other human rights.'*[192] The CESCR argued that the right to food implied that food should be 'adequate' which is *'to a large extent determined by prevailing social, economic, cultural, climatic, ecological and other conditions'* and *'sustainable'*, which *'incorporates the notion of long-term availability and accessibility.'*[193]

It is generally accepted that the right to adequate food includes the right to water.[194] The CESCR underlined that *'the right to water clearly falls within the category of guarantees essential for securing an adequate standard of living, particularly since it is one of the most fundamental conditions for survival.'*[195] According to the Committee, the meaning of adequate water may vary in different situations, but in all circumstances, the factors of availability, quality and accessibility are applicable.[196]

Furthermore, as already discussed in Section 3.3.2, the right to food implies the right to breast-feeding, as stipulated in Article 24 (2) ICRC.

It is obvious that it is not possible to exactly define what adequate food is for each individual in each specific situation. On the other hand, the work done in the international arena to further clarify the concept of the right to food provides for some general sub-elements of the right to food, that makes it easier to apply the right to food in specific situations. Besides it seems reasonable to at least recognise

---

[191] E/C.12/1999/5, 12 May 1999, CESCR, *General Comment 12, Right to Adequate Food*, Section 6.

[192] E/C.12/1999/5, 12 May 1999, CESCR, *General Comment 12, Right to Adequate Food*, Section 8.

[193] E/C.12/1999/5, 12 May 1999, CESCR, *General Comment 12, Right to Adequate Food*, Section 7.

[194] The Commission on Human Rights requested the Special Rapporteur on the right to food in 2001 *'to pay attention to the issue of drinking water, taking into account the interdependence of this issue and the right to food'*. See: Commission on Human Rights Resolution 2001/25, 20 April 2001. See furthermore: A/56/210, 23 July 2001, J. Ziegler, *Report of the Special Rapporteur on the right to food to the General Assembly*, Chapter IV; E/CN.4/2003/54, 10 January 2003, J. Ziegler, *Report of the Special Rapporteur on the right to food*, Chapter II. See also: E/C.12/2002/11, 20 January 2003, CESCR, *General Comment 15, The Right to Water*. Furthermore, see the works of the Special Rapporteurs on the right to water: Mr El Adji Guissé for the sub-commission on human rights, and Mrs C. de Albuquerque for the Human Rights Commission.

[195] E/C.12/2002/11, 20 January 2003, CESCR, *General Comment 15, The Right to Water*, Section 3; see also (earlier) E/1996/22, annex IV at 97 (1995), 24 November 1995, CESCR, *General Comment 6, The Economic, Social and Cultural Rights of Older Persons*, Section 32.

[196] E/C.12/2002/11, 20 January 2003, CESCR, *General Comment 15, The Right to Water*, Section 12.

a core content of the right to food that implies specific State obligations, as will be discussed below. As Vlemminx pointed out, it seems rather odd to claim that Article 11 ICESCR is too vague to apply in a particular case when it is quite clear that in the underlying case, the individual who invokes the Provision has no access to housing, clothes and food at all.[197]

3.5 State obligations regarding the right to food

Rights imply duties, but it has never been easy to reach universal consensus on what State duties are recognised in the global Human Rights system, especially regarding ECOSOC rights. This section will therefore discuss the difficulties in determining what duties are implied by economic, social and cultural rights in general, and the right to food in particular. Section 3.5.1 will focus on the question why States are responsible for the realisation of human rights, and Section 3.5.2 will discuss the matter of universality of human rights, and the extent to which states are bound to realise these human rights. Section 3.5.3 will focus on the interdependence and indivisibility of all human rights, and will in that context discuss the classical division of human rights in civil and political rights on the one hand, and economic, social and cultural rights on the other hand. Furthermore, a frequently proposed typology of State duties that can be applied on all human rights will be introduced. In this light, Article 2 ICESCR will be discussed in Section 3.5.4, mostly explaining why this Article cannot be understood as an unlimited and 'duty free' margin of discretion for States to realise the rights enshrined in the ICESCR. In Section 3.5.5, the domestic enforceability of ECOSOC rights in general will be discussed, and the domestic enforceability of the right to food will be explored in Section 3.5.6. Furthermore, Section 3.5.7 will discuss the matter of State reservations that can be made to human rights instruments. Finally, Sections 3.5.8 and 3.5.9 will introduce and discuss the different monitoring systems (reporting procedures and complaint mechanisms) that are embedded in or later added to the different international human rights treaties.

3.5.1 Human rights and the State

Legal instruments in which human rights are embedded are usually aimed at States, which then have certain responsibilities towards the realisation of these rights. Naturally, the realisation of human rights is not the sole responsibility of a Government: all actors in society should contribute to the realisation of human

---

[197] F.M.C. Vlemminx, *Een nieuw profiel van de grondrechten, een analyse van de prestatieplichten ingevolge klassieke en sociale grondrechten*, Den Haag: Boom Juridische uitgevers, 2002b, Chapter 5, Section 9, referring to: District Court of 's-Gravenhage, 6 September 2000, Rawb 2001, p. 55.

rights,[198] starting with the individual who seeks to enjoy human rights.[199] However, the role of non-governmental actors is not so often discussed in the context of legal obligations. Besides some soft legislation such as non-legally binding declarations,[200] the role of non-governmental actors is sometimes referred to in preambles of human rights treaties,[201] or occasionally specifically addressed, such as Article 27 (2) ICRC that stipulates that it is the primary responsibility of the parents or others who are responsible for the child '*to secure, within their abilities and financial capacities, the conditions of living necessary for the child's development.*' Mostly however, international human rights documents are addressed to its Member States: '*At its core, a human right is a claim against the Government, a claim the Government must do or desist from doing specific things to further human dignity.*'[202]

---

[198] G. Kent, *Freedom from want: the human rights to food*, Washington D.C.: Georgetown University Press, 2005, Chapter 6. In this Chapter, Kent introduces the '*rings of responsibility*', involving all relevant actors in society.

[199] A. Eide, Chapter 11, *Adequate standard of living*, in: D. Moeckli, S. Shah, S. Sivakumaran (eds.), D. Harris (cons. ed.), *International human rights law*, Oxford University Press, 2010.

[200] For instance: OECD, *OECD Guidelines for multinational enterprises*, OECD Publishing, 2011, available at http://dx.doi.org/10.1787/9789264115415-en, especially part I, Section IV; E/CN.4/Sub.2/2003/12, 26 August 2003: *Draft standards on the responsibilities of transnational corporations and other business enterprises with regard to human rights;* Tripartite Declaration of Principles concerning Multinational Enterprises and Social Policy, adopted by the Governing Body of the International Labour Office at its 204th Session (Geneva, November 1977) as amended at its 279th (November 2000) and 295th Session (March 2006); and in more than one occasion in: *The voluntary guidelines to support the progressive realisation of the right to adequate food in the context of national food security*, adopted by the 127th session of the FAO Council, November 2004. See also: General Assembly, resolution 42/115, 7 December 1987, *The impact of property on the enjoyment of human rights and fundamental freedoms;* UN Commission on Human Rights, Resolution 1987/18, 10 March 1987, *On the impact of property on the economic and social development of Member States;* E/C.12/1999/5, 12 May 1999, CESCR, *General Comment 12, Right to Adequate Food*, Section 20; E/CN.4/2004/10, 9 February 2004, J. Ziegler, *Report of the Special Rapporteur on the right to food*, especially Chapter III; A/58/330, 28 August 2003, J. Ziegler, *Report of the Special Rapporteur on the right to food to the General Assembly*, especially Chapter III; E/CN.4/2006/44, 16 March 2006, J. Ziegler, *Report of the Special Rapporteur on the right to food*, Chapter III, Section D; A/HRC/7/5, 10 January 2008, 10 January 2008, J. Ziegler, *Report of the Special Rapporteur on the right to food to the Human Rights Council*, Chapter III, Section D. See also M. Brady, Chapter 4, *Holding Corporations Accountable for the Right to food*, in: G. Kent (editor), *Global obligations for the right to food*, Rowman and Littlefield Publishers Inc, 2008.

[201] For instance, the preambles of both the ICCPR and the ICESCR mention that the State Parties agree upon the Articles, '*Realising that the individual, having duties to other individuals and to the community to which he belongs, is under a responsibility to strive for the promotion and observance of the rights recognised in the present Covenant...*'

[202] See for instance the contribution of E. Bates, Chapter 1, *History*, in: D. Moeckli, S. Shah, S. Sivakumaran (eds.), D. Harris (cons. ed.), *International human rights law*, Oxford: Oxford University Press, 2010. See furthermore: G. Kent, *Freedom from want: the human rights to food*, Washington D.C.: Georgetown University Press, 2005, p. 80.

3.5.2 Universality of human rights and State consent to be bound by them

An interesting question in this context is whether a State has to ratify a human rights treaty first before it is bound to it. The usual approach in international law is to respect the State's sovereignty and free will, and thus a State cannot be bound to a treaty without its consent, which reflects a positivistic legal point of view. On the other hand, one could argue that all States have human rights obligations, even without ratifying the treaties in which these rights are recognised, due to the universality of human rights, in line with the naturalism principle that people have human rights simply because they were born. There seems to be tension between defining and recognising human rights in declarations and treaties under the wings of international positive law on the one hand, and presenting it as universal or even natural law on the other hand. It is no secret that the UDHR was not universally accepted at the time of its declaration, and the UN human rights treaties are not unanimously ratified. Also, States frequently made reservations to human rights treaties, as will be discussed below. In addition, without willing to generalise too easily, the fact that the need was felt to regionally adopt human rights treaties with an equivalent content to the global treaties for more suitable and better accepted implementation, at least suggests that there is no (entirely) universally accepted understanding of human rights. Would this mean that human rights are thus not universal and only effective in the countries that ratified the treaties, to the extent of the content to which they are willing to commit to? Or in the words of Christopher McCrudden: '*This is the issue of whether the obligations that human rights impose depend on the State for their existence or exist irrespective of State recognition.*' This resembles the classic discussion between legal positivism and natural law: '*Are human rights legal rights because they are incorporated into positive law, or are they legal rights irrespective of whether they have been incorporated into any particular legal system, because they are already included in what we consider foundational to any legal system?*'[203]

Frédéric Mégret argues that human rights treaties are not normal international treaties, due to the fact that such a treaty is not primarily about commitments between states. Instead, she considers that human rights are about a commitment from a State towards individuals, and may in certain aspects even be about a commitment *ergo omnes* towards the international community. Also, the content of the treaties are of a high moral value that stands '*above and beyond States' consent to be bound by them.*' Therefore, balancing between naturalism and legal positivism, Mégret argues that '*human rights obligations have a life of their own that takes over as soon as States have manifested their initial commitment to be bound. (…) States are solemnly committing to something which they were already, at least*

---

[203] C. McCrudden, *Chapter 16, Judicial comparativism and human rights*, in: E. Örücü and D. Nelken (eds.), *Comparative law, a handbook*, Portland: Hart Publishing, 2007.

morally or philosophically, obliged to recognise.'[204] As appealing as this may sound, it does not alter the fact that the form in which human rights are formulated is that of international treaties which need to be ratified before they enter into force between States, whereas the content concerns the primary prerequisites for an individual to lead a dignified life. This 'high moral value' transcends State borders from a lot of perspectives and makes it a global issue. But the legal mechanisms that exist in the international arena seem to be – for other valid reasons such as the respect for cultural diversity, state sovereignty and the principle of legality – incapable of giving human rights a true universal meaning. Therefore, in the previous Chapter, I already argued that instead of discussing the possibilities or impossibilities of universal human rights, it may be more fruitful to explain similarities and differences on the implementation of these human rights between legal cultures. This knowledge may be more helpful in building a more coherent global human rights system, with a more realistic attitude towards similarities and differences, than continuing this debate on the universality of human rights.

3.5.3 The interdependence and indivisibility of human rights and State Duties

Although it is often stressed that all human rights are interdependent and indivisible, history shows that on a global level civil and political rights were developed differently compared to economic, social and cultural rights. As stated above,[205] due to differences of opinion amongst the UN Member States, the civil and political rights on the one hand, and the economic, social and cultural rights as declared in the UDHR on the other hand, were introduced in two separate treaties: the ICCPR and the ICESCR. While Western countries emphasised the civil and political rights, advocating freedom from the State, the socialist countries favored the economic, social and cultural rights, considering that freedom is to be ensured by an active State providing for basic needs.[206] This mindset has led to different perceptions of State obligations concerning the two types of rights, or even to classifications as 'generations of rights', in which the differences, or perhaps even a hierarchy between civil and political rights (first generation), economic, social and cultural rights (second generation) and sometimes also solidarity rights (third generation) is emphasised.[207] In this context, civil and political rights are traditionally linked to negative State duties to refrain from intervention, and

---

[204] F. Mégret, Chapter 6, *The Nature of state obligations*, in: D. Moeckli, S. Shah, S. Sivakumaran (eds.), D. Harris (cons. ed.), *International human rights law*, Oxford University Press, 2010.

[205] Especially in Section 3.2.

[206] A. Eide, in his contribution to: O. Hospes and I. Hadiprayitno (eds.), *Governing food security, law, policies and the right to food*, Wageningen: Wageningen Academic Publishers, 2010, Chapter 4; M. Ssenyonjo, *Economic, social and cultural rights in international law*, Portland: Hart Publishing, 2009, Chapter one, Section II, Section 1.13 and 1.16, Section IV, Section 1.48-1.49.

[207] For the first time introduced by K. Vasak, *Human rights: a thirty-year struggle: the sustained efforts to give force of law to the universal declaration of human right*, UNESCO Courier 30:11, Paris: United Nations Educational, Scientific, and Cultural Organisation 1977.

economic, social and cultural rights are linked to positive State obligations, in which States have a duty to progressively realise the rights concerned. This has led to a somewhat undifferentiated view amongst some Member States. They assume that civil and political rights require immediate implementation and entail enforceable rights, while economic, social and cultural rights leave Member States a margin of discretion to progressively realise the rights through policy over time. They consider that the latter are not as clearly formulated as civil and political rights, and therefore are of a non enforceable nature.[208]

This view can be questioned not only from a substantive point of view (as will be done below), but also based on other historical developments on human rights law. In the first place, in UN treaties recognising human rights for specific groups of beneficiaries, both civil and political rights as well as economic, social and cultural rights are present. In the second place, on a regional level, only in Europe and America the choice was made to adopt separate human rights treaties equivalent to the ICCPR and the ICESCR. The African Charter on Human and Peoples' Rights, the Asian Human Rights Charter, and the Cairo Declaration on Human Rights in Islam recognise civil and political rights alongside with economic, social and cultural rights.[209] So, historically the division between rights is not necessarily a globally accepted practice, but it clearly has had extensive consequences. As Ssenyonjo puts it: '*Although the division reflected essentially the contrasting interests of the Cold War division between the West and the East, it continues to impact negatively on the realisation of ESC rights. This means that any right not considered a 'first generation' civil or political right is not given the same value by the international community, and, as such, violations of 'lesser' valued 'second or third generation' human rights are more tolerated.*'[210]

For decades, the subordination of economic, social and cultural rights to civil and political rights is opposed by many scholars[211] and UN institutions[212] whose field of interest contains ECOSOC rights. Through studies and interpretative declarations, attempts were made to further clarify the meaning of ECOSOC

---

[208] In literature, see for instance: P.W.C. Akkermans, C. Bax and L. Verhey, *Grondrechten, grondrechten en grondrechtsbescherming in Nederland,* Heerlen: Kluwer, 2005, especially Sections 3.4 and 3.9.

[209] See Section 3.3.3.

[210] M. Ssenyonjo, *Economic, social and cultural rights in international law,* Portland: Hart Publishing, 2009, Section 1.13.

[211] For instance: P. Alston and K. Tomasevski (eds.), *The right to food,* International Studies in Human Rights, Utrecht, 1984; G. Kent, *Freedom from want: the human rights to food,* Washington D.C.: Georgetown University Press, 2005, especially Chapters 4-7; M. Ssenyonjo, *Economic, social and cultural rights in international law,* Portland: Hart Publishing, 2009, especially part I; F. Vlemminx, *Een nieuw profiel van de grondrechten, een analyse van de prestatieplichten ingevolge klassieke en sociale grondrechten,* Den Haag: Boom Juridische uitgevers, 2002b; G. Maes, *De afdwingbaarheid van sociale grondrechten,* Antwerp: Intersentia 2003.

[212] See for instance the many reports of the Special Rapporteurs on the right to food and the General Observations of the Committee on Economic, Social and Cultural Rights to the Country Reports.

rights, and to differentiate between the typology of State obligations, in order to make the rights 'more solid law' and ultimately to facilitate national applicability and enforceability as much as possible.[213]

Rather than civil and political rights implying negative obligations and ECOSOC rights implying positive obligations, most UN institutions prefer to apply three different types of State Obligations on all sorts of human rights: the duty to respect, the duty to protect and the duty to fulfil. A typology that was introduced in the early eighties by various authors,[214] and applied to the right to food by Asbjørn Eide when he introduced this typology in the UN system in his capacity as Special Rapporteur.[215] The typology is now being used by most UN institutions that deal with ECOSOC rights, such as the Committee on Economic, Social and Cultural Rights in its General Comments, and the Special Rapporteurs in their reports.

The Committee on Economic, Social and Cultural Rights summarised the State duties and applied it for the first time to an ECOSOC right in their General Comment 12, regarding the right to food: *'The obligation to respect existing access to adequate food requires State's parties not to take any measures that result in preventing such access. The obligation to protect requires measures by the State to ensure that enterprises or individuals do not deprive individuals of their access to adequate food. The obligation to fulfil (facilitate) means the State must pro-actively engage in activities intended to strengthen people's access to and utilisation of resources and means to ensure their livelihood, including food security. Finally, whenever an individual or group is unable, for reasons beyond their control, to enjoy the right to adequate food by the means at their disposal, States have the obligation to fulfil (provide) that right directly. This obligation also applies for persons who are victims of natural or other disasters.'*[216] The duty to fulfil is subdivided, as already suggested by Asbjørn Eide,[217] in a duty to facilitate and a duty to provide. In General Comment 15 on the right to water, the Committee added a third element to the duty to fulfil – the duty to promote – that obliges *'the State Party to take steps to ensure that there is appropriate education concerning hygienic use of water, protection of water sources and methods to minimise water wastage.'*[218] In literature, additional duties are suggested, but not widely used in practice, for example the duty to assist, and its counterpart,

---

[213] G. Kent, *Freedom from want: the human rights to food*, Washington D.C.: Georgetown University Press, 2005, Chapter 5.

[214] The earliest reference to such a typology found by the author was in various contributions to: P. Alston and K. Tomasevski (Eds.), *The right to food*, International Studies in Human Rights, Utrecht, 1984; see especially the contributions of H. Shue: 'The interdependence of duties,' and G.J.H. van Hoof: 'The legal nature of economic, social and cultural rights: a rebuttal of some traditional views.'

[215] E/CN.4/Sub.2/1999/12, 28 June 1999, *Updated study on the right to food, submitted by Mr A. Eide in accordance with Sub-Commission decision 1998/106.*

[216] E/C.12/1999/5, 12 May 1999, CESCR, *General Comment 12, Right to Adequate Food,* Section 15.

[217] E/CN.4/Sub.2/1999/12, 28 June 1999, *Updated study on the right to food, submitted by Mr. Asbjørn Eide in accordance with Sub-Commission decision 1998/106,* Section 53.

[218] E/C.12/2002/11, 20 January 2003, CESCR, *General Comment 15, The Right to Water,* Section 25.

the duty to seek and use effectively international assistance in the context of international development aid.[219]

In a traditional explanation, one might conclude that ICCPR rights basically imply obligations to respect, and ICESCR rights obligations to fulfil, portraying the difference between freedom from the State, and freedom through State actions. This typology of duties however opens the possibility to further nuance the State obligations when they are applied to the right in both the ICCPR and the ICESCR.

For example, Article 7 ICCPR stipulates that *'no one shall be subjected to torture or to cruel, inhuman or degrading treatment or punishment. In particular, no one shall be subjected without his free consent to medical or scientific experimentation.'* To fully realise this right, States should indeed refrain from torturing, but should also protect right holders from being tortured by private parties. In addition, the State should act to shape the conditions of a torture-free society through the adoption and execution of law and policy. In comparison, the Committee demonstrated in General Comment 12 that all three obligations are of importance to fully realise the right to food.[220]

This approach can even be further differentiated, as Vlemminx shows us,[221] by applying positive and negative obligations to all types of State Duties. For instance, refraining from torturing is only an effective right if there is a clear law that prohibits torturing, and victims from torturing can defend their rights through effective juridical remedies. States thus have a duty to provide for this legislation and judicial system in order to be able to respect the right. Granting a right holder social benefits in order to enable him – or herself to obtain adequate food is from a right-based point of view meaningless, if the State can withdraw this benefit at any time. The benefits would resemble charity rather than being a right. The State should thus abstain from withdrawing the social benefits arbitrarily.

Nuancing this typology of duties may lead to two conclusions. In the first place, the fear of many States that right holders could claim that the State has to provide them with these rights endlessly – thus creating a disproportional financial burden for the State – seems to be unnecessary. The duty to provide is only required when all other duties seem to be ineffective, and would normally only be necessary as a 'safety net'.[222] This would be perfectly in line with the purpose of a human right, as considered in Chapter 1. In the second place, the margin of discretion that is implied by progressively realising rights does not exclude the existence of

---

[219] M. Ssenyonjo, *Economic, social and cultural rights in international law,* Portland: Hart Publishing, 2009, Sections 2.56-2.66.
[220] E/C.12/1999/5, 12 May 1999, CESCR, *General Comment 12, Right to Adequate Food,* Section 15.
[221] F.M.C. Vlemminx, *Een nieuw profiel van de grondrechten, een analyse van de prestatieplichten ingevolge klassieke en sociale grondrechten,* Den Haag: Boom Juridische uitgevers, 2002b.
[222] E/C.12/1999/5, 12 May 1999, CESCR, *General Comment 12, Right to Adequate Food,* Section 15.

minimum core obligations that need to be realised immediately, to not deprive the treaty Provisions of its 'raison d'être.'[223] Not all duties described above require a realisation through policy in the longer term. Some duties imply immediate action and specific entitlements, such as the duty to provide in case of emergency.

3.5.4 The meaning of Article 2 (I) ICESCR

It is often understood by States that from Article 2 ICESCR, it follows that the substantive rights stipulated in the ICESCR merely imply the duty to progressively realise these rights, leaving a wide margin of discretion for the States to choose a suitable way to do so, taking into account the specific needs of a country, and as a result implying no enforceable rights. The exact wording of Article 2 (1) is: *'Each State Party to the present Covenant undertakes to take steps, individually and through international assistance and co-operation, especially economic and technical, to the maximum of its available resources, with a view to achieving progressively the full realisation of the rights recognised in the present Covenant by all appropriate means, including particularly the adoption of legislative measures.'* The meaning of this Article has been further clarified by the authoritative interpretation of the CESCR in their General Comment 3.[224]

In General, the CESCR observed that too often the differences between Articles 2 ICCPR[225] and 2 ICESCR were emphasised, while there were also similarities between the Covenants. Referring to the work of the International Law Commission,[226] the CESCR argued that the rights embedded in the ICESCR *'include both what may be termed...obligations of conduct and obligations of result.'*[227] In this understanding, the obligation to conduct means *'to take steps'* to work towards *'achieving progressively the full realisation of the rights'*, which might be understood as the obligations of result. The Committee argued that the meaning of the phrase 'to take steps', especially considering its French and Spanish translation – *'s'engage à agir'* (to act) and *'a adoptar medidas'* (to adopt measures) – at least implies that *'within a reasonably*

---

[223] E/1991/23, annex III at 86 (1991), 14 December 1990, CESCR, *General Comment 3: The Nature of States Parties' Obligations*, especially Section 10.

[224] E/1991/23, annex III at 86 (1991), 14 December 1990, CESCR, *General Comment 3: The Nature of States Parties' Obligations.*

[225] Article 2 (1) ICCPR stipulates that: *'Each State Party to the present Covenant undertakes to respect and to ensure to all individuals within its territory and subject to its jurisdiction the rights recognised in the present Covenant, without distinction of any kind, such as race, colour, sex, language, religion, political or other opinion, national or social origin, property, birth or other status.'*

[226] The International Law Committee adopted both *The Limburg Principles on the Implementation of the International Covenant on Economic, Social and Cultural Rights*, published in: E/CN.4/1987/17, Annex and Human Rights Quarterly, Vol.9 (1989), pp 122-135; and *The Maastricht Guidelines on Violations of Economic, Social and Cultural Rights*, Maastricht, January 22-26, 1997. Especially in Section 7 of the Maastricht Guidelines, the obligations of conduct and result are discussed.

[227] E/1991/23, annex III at 86 (1991), 14 December 1990, CESCR, *General Comment 3, The Nature of States Parties' Obligations*, Section 1; see also: *The Maastricht Guidelines on Violations of Economic, Social and Cultural Rights*, Maastricht, January 22-26, 1997, Section 7.

short time after the Covenant's entry into force' measures should be taken by the State to realise the full implementation that is '*deliberate, concrete and targeted as clearly as possible towards meeting the obligations recognised in the Covenant.*'[228] In other words: doing nothing is a violation of the Covenant. Furthermore, the Committee indeed recognises that there is a difference between Article 2 ICCPR, requiring immediate recognition of all rights embedded in the ICCPR, and Article 2 ICESCR, requiring a progressive realisation, in the sense that ICESCR rights '*will generally not be able to be achieved in a short period of time.*' On the other hand, it can be argued that also elements of the rights embedded in the ICCPR need to be progressively realised. As Ssenyonio explains, the guarantee of humane treatment in detention, as recognised in Article 10 ICCPR, may require '*the construction of a sufficient number of detention Centres and the development of alternative measures to imprisonment...*'[229] The other way around, it can be defended that the rights embedded in the ICESCR do imply obligations that are of a more immediate nature. The CESCR argued that the obligation to progressive realisation should '*not be misinterpreted as depriving the obligation of all meaningful content.*' This means that the States have the duty to '*move as expeditiously and effectively as possible*' towards full realisation of the ICESCR rights, and '*any deliberate retrogressive measures in that regard would require the most careful consideration and would need to be fully justified by reference to the totality of the rights provided for in the Covenant and in the context of the full use of the maximum available resources.*'[230] The Committee goes even further by stating that after careful consideration of the State reports of over a decade, '*it is of the opinion that a minimum core obligation to ensure the satisfaction of, at the very least, minimum essential levels of each of the rights is incumbent upon every State Party.*'[231] Failing to achieve this would be a violation of the Covenant. According to the Committee, a lack of available resources is no excuse to not guarantee these core obligations, unless the State has proven to have undertaken every effort, '*as a matter of priority*' to guarantee these minima. Regarding the right to food, the Committee underlined that '*a State claiming that it is unable to carry out its obligations for reasons beyond its control therefore has the burden of proving that this is the case and that it has unsuccessfully sought to obtain international support to ensure the availability and accessibility of the necessary food.*'[232] Furthermore, the right to food is violated when there exists '*discrimination in the access to food, as well as to means and entitlements for its procurement...with the purpose or effect of nullifying or impairing the equal enjoyment or exercise of economic, social and*

---

[228] E/1991/23, annex III at 86 (1991), 14 December 1990, CESCR, *General Comment 3, The Nature of States Parties' Obligations,* Section 2.

[229] M. Ssenyonjo, *Economic, social and cultural rights in international law,* Portland: Hart Publishing, 2009, Section 2.23.

[230] E/1991/23, annex III at 86 (1991), 14 December 1990, CESCR, *General Comment 3, The Nature of States Parties' Obligations,* Section 9.

[231] E/1991/23, annex III at 86 (1991), 14 December 1990, CESCR, *General Comment 3, The Nature of States Parties' Obligations,* especially Section 10.

[232] E/C.12/1999/5, 12 May 1999, CESCR, *General Comment 12, Right to Adequate Food,* Section 17.

cultural rights.'[233] Also, the CESCR argued that violations can occur through the direct actions of the State, but also through the action of *'other entities insufficiently regulated by States.'*[234]

Summarising, it appears to be the case that the CESCR understands Article 2 (1) ICESCR in a way that it implies immediate core obligations, obligations of conduct that are of an immediate nature in the sense that the State has to start immediately with moving towards the final goal, which is the full realisation of the Human Right (Figure 3.2).

3.5.5 ECOSOC rights and domestic enforceability

Much has been written about the enforceability of ECOSOC rights in the domestic legal orders of the Member States of the treaties that recognise these rights. The CESCR discussed the matter frequently, basically by giving their interpretation of the phrase *'by all appropriate means'* as stipulated in Article 2 (1) ICESCR. In this context three issues will be discussed. Firstly, the rights must be implemented in the domestic legal order. According to the CESCR, this implies secondly that effective legal remedies must be ensured to those whose rights are violated; and thirdly, the rights must have some juridical effect in the Courts, which is often referred to as 'justiciability'. In their General Comments, the CESCR frequently addressed the issue.

With regard to the implementation in the domestic legal order, the CESCR stressed that although the ICESCR did not specify how the rights of the Covenant should be implemented and that this is up to the States to decide, *'the means used should be appropriate in the sense of producing results which are consistent with the full discharge*

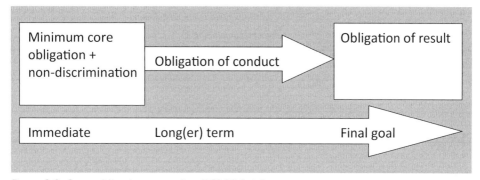

Figure 3.2. State obligations regarding ECOSOC rights.

---

of its obligations by the State Party.'[235] The CESCR observed that the ICESCR Member States use various approaches in their implementation of the CESCR. According to an analysis of the CESCR, some States did not do anything to implement the Covenant, other States adjusted their domestic legislation to implement the Covenant, but did not invoke the specific treaty Provisions, and yet again other States did incorporate the ICESCR in their domestic law by *'giving formal validity in the national legal order'* to the Covenant.[236] The last two approaches pretty much resemble a dualistic and a monistic approach towards the effect of international law in the domestic legal order. Although one might ask whether the CESCR should form an authoritative legal interpretation of the Constitutional form of its Member States, the CESCR expressed their preference for a monistic approach, because *'direct incorporation avoids problems that might arise in the translation of treaty obligations into national law, and provides a basis for the direct invocation of the Covenant rights by individuals in national Courts.'*[237] Furthermore, *'whatever the preferred methodology'*, the CESCR argued that the means used by the State to implement the Covenant must be *'adequate to ensure the obligations under the Covenant'*. The CESCR underlined that *'The need to ensure justiciability...is relevant when determining the best way to give domestic legal effect to the Covenant rights.'*[238] Furthermore, the CESCR argued that it was the responsibility of the State to use the means *'which have proved to be most effective in the country concerned in ensuring the protection of other human rights.'*[239] If a country would choose to use significant different methods to ensure the ICESCR rights compared to other human rights instruments, the country should justify these differences.[240]

Concerning the need for effective juridical remedies, the CESCR underlined that Article 8 UDHR stipulates that *'everyone has the right to an effective remedy by the competent national tribunals for acts violating the fundamental rights granted him by the Constitution or by law.'* The Committee argued that those countries who were also Member State to the ICCPR already obliged to ensure an effective remedy for persons whose rights or freedoms are violated under *inter alia* Article 2 (3) (b) ICCPR, and that despite the fact that the ICESCR had no counterpart to this Provision, the phrase *'all appropriate means'* in Article 2 (1) ICESCR would imply *'the Provision of judicial remedies with respect to rights which may, in accordance with*

---

[235] E/C.12/1998/24, 3 December 1998, CESCR, *General Comment 9, The Domestic Application of the Covenant*, Section 5.

[236] E/C.12/1998/24, 3 December 1998, CESCR, *General Comment 9, The Domestic Application of the Covenant*, Section 6.

[237] E/C.12/1998/24, 3 December 1998, CESCR, *General Comment 9, The Domestic Application of the Covenant*, Section 8.

[238] E/C.12/1998/24, 3 December 1998, CESCR, *General Comment 9, The Domestic Application of the Covenant*, Section 7.

[239] E/C.12/1998/24, 3 December 1998, CESCR, *General Comment 9, The Domestic Application of the Covenant*, Section 7.

[240] E/C.12/1998/24, 3 December 1998, CESCR, *General Comment 9, The Domestic Application of the Covenant*, Section 7.

the national legal system, be considered justiciable.' The CESCR warned that a State that failed to provide these legal remedies, would have to demonstrate that these remedies were not *'appropriate means'* or *'unnecessary'*, and added that *'it will be difficult to show this and the Committee considers that, in many cases, the other "means" used could be rendered ineffective if they are not reinforced or complemented by judicial means.'*[241] According to the CESCR, an effective remedy does not necessarily need to be a judicial remedy *per se*: administrative remedies for instance might also be adequately effective, as long as they are *'accessible, affordable, timely and effective'*, although the possibility to finally appeal to a Court is favoured by the Committee.[242]

Having the right to juridical remedies is not very useful if the rights under dispute cannot be invoked. Therefore, the CESCR stressed the importance of justiciability of ICESCR rights. The Committee held that *'In relation to civil and political rights, it is generally taken for granted that judicial remedies for violations are essential. Regrettably, the contrary assumption is too often made in relation to economic, social and cultural rights. This discrepancy is not warranted either by the nature of the rights or by the relevant Covenant Provisions.'* In this light, the CESCR underlined that *'there is no Covenant right which could not, in the great majority of systems, be considered to possess at least some significant justiciable dimensions.'* The Committee distinguished justiciability, meaning that a matter can be properly resolved by a Court, from self-executing, meaning that a Provision can be applied by a Court *'without further elaboration.'*[243] Regarding the latter, the Committee considered that: *'The Covenant does not negate the possibility that the rights it contains may be considered self-executing in systems where that option is provided for. Indeed, when it was being drafted, attempts to include a specific Provision in the Covenant to the effect that it be considered "non-self-executing" were strongly rejected. In most States, the determination of whether or not a treaty Provision is self-executing will be a matter for the Courts, not the executive or the Legislature. In order to perform that function effectively, the relevant Courts and tribunals must be made aware of the nature and implications of the Covenant and of the important role of judicial remedies in its implementation. Thus, for example, when Governments are involved in Court proceedings, they should promote interpretations of domestic laws which give effect to their Covenant obligations. Similarly, judicial training should take full account of the justiciability of the Covenant. It is especially important to avoid any* a priori *assumption that the standards should be considered to be non-*

---

[241] E/C.12/1998/24, 3 December 1998, CESCR, *General Comment 9, The Domestic Application of the Covenant*, Section 3.

[242] E/C.12/1998/24, 3 December 1998, CESCR, *General Comment 9, The Domestic Application of the Covenant*, Section 9.

[243] E/C.12/1998/24, 3 December 1998, CESCR, *General Comment 9, The Domestic Application of the Covenant*, Section 10. The exact phrase was: *'It is important in this regard to distinguish between justiciability (which refers to those matters which are appropriately resolved by the Courts) and standards which are self-executing (capable of being applied by Courts without further elaboration). While the general approach of each legal system needs to be taken into account, there is no Covenant right which could not, in the great majority of systems, be considered to possess at least some significant justiciable dimensions.'*

self-executing. In fact, many of them are stated in terms which are at least as clear and specific as those in other human rights treaties, the Provisions of which are regularly deemed by Courts to be self-executing.' It is unclear to what end the Committee made the distinction between justiciable and self-executing rights. However, it is my understanding that the CESCR here expresses that some ICESCR Provisions, or elements of the rights they recognised, can have an direct effect when they are invoked because the right is self-executing. A possibility that is greatly depending on the Constitutional structure of a State, and the considerations of the national Courts. Justiciability on the other hand, can have a broader meaning, in the sense that ICESCR Provisions when invoked in Court can somehow be used by the Courts as a guideline or a direction for their verdict. It is however clear that the CESCR considers at least some Provisions to be self-executing, considering that *'many of them are stated in terms which are at least as clear and specific as those in other human rights treaties, the Provisions of which are regularly deemed by Courts to be self-executing.'*[244] The Committee listed *'by way of example'*,[245] and thus not exhaustively, some ICESCR Provisions that *'would seem to be capable of immediate application by judicial and other organs in many national legal systems.'*[246] These Provisions are: 3, 7 (a) (i), 8, 10 (3), 13 (2) (a), (3) and (4), and 15 (3) ICESCR. The Committee underlined that it deemed arguments that would indicate that these Provisions were not self-executing *'difficult to sustain.'*[247]

In addition, as will be demonstrated in Chapters 9 and 13, the CESCR, but also the Committee on the Rights of the Child and The Committee on the Elimination of Discrimination against Women, frequently address the issue of enforceability when assessing the submitted country reports, stating that the treaties should be enforceable in the domestic Courts. During the reporting cycles however, the enforceability of the treaty standards is discussed in general, and mostly not per Article.

3.5.6 The right to food and domestic enforceability

In General Comment 12, the CESCR expressed its authoritative legal interpretation of the right to food, as recognised in Article 11 ICESCR. According to the Committee, in implementing the right *'every State will have a margin of discretion, in choosing its own approaches, but the Covenant clearly requires that each State Party will take whatever steps are necessary to ensure that everyone is free from hunger and as soon*

---

[244] E/C.12/1998/24, 3 December 1998, CESCR, *General Comment 9, The Domestic Application of the Covenant,* Section 11.
[245] E/C.12/1998/24, 3 December 1998, CESCR, *General Comment 9, The Domestic Application of the Covenant,* Section 10.
[246] E/1991/23, annex III at 86 (1991), 14 December 1990, CESCR, *General Comment 3, The Nature of States Parties' Obligations,* Section 5.
[247] E/1991/23, annex III at 86 (1991), 14 December 1990, CESCR, *General Comment 3, The Nature of States Parties' Obligations,* Section 5.

as possible can enjoy the right to adequate food.'[248] Therefore, '*a national strategy to ensure food and nutrition security for all*', is required.[249] Also, the CESCR stressed that States should set '*verifiable benchmarks for subsequent national and international monitoring*', and, in this connection, '*States should consider the adoption of a framework law as a major instrument in the implementation of the national strategy concerning the right to food.*'[250] In addition, States '*shall develop and maintain mechanisms to monitor progress towards the realisation of the right to adequate food for all...*'.[251]

Article 11 ICESCR was not amongst the Articles listed by the CESCR in General Comment 3,[252] and later quoted in General Comment 9,[253] that were considered to be self-executing. Since this list was provided '*by way of example*', this does not automatically mean that Article 11, or some of its elements, cannot be self-executing or justiciable. On this, the CESCR is not so specific. The Committee reaffirms in the specific context of the right to food the requirement of '*effective judicial or other appropriate remedies*', and stresses that victims of a violation of the right should be entitled to '*adequate reparation.*'[254] The CESCR also argued that domestic incorporation of international instruments recognising the right to food '*can significantly enhance the scope and effectiveness of remedial measures...*'. And: '*Courts would then be empowered to adjudicate violations of the core content of the right to food by direct reference to obligations under the Covenant.*'[255] As it seems, the CESCR is somewhat cautious to hold that the right to food is self-executing. On the one hand, the Committee argues that victims of a violation of the right to food should have effective juridical means and should be compensated, which suggests that the right to food does play some juridical role in such situations. On the other hand, the CESCR encourages States to further implement the right, so that Courts will be empowered to directly refer to the ICESCR Provisions in their verdicts, which might suggest that the right is self-executing without the need for further implementation. As it appears, the CESCR tries to balance between stressing that the right to food as such is a justiciable right in the sense that it can play some role in Court cases – respecting the State's margin of discretion – and urging States to implement the right so that it can become self-executing.

The former Special Rapporteur on the right to food, Jean Ziegler, in his reports strongly supported the content of General Comment 12, and even considered

---

[248] E/C.12/1999/5, 12 May 1999, CESCR, *General Comment 12, Right to Adequate Food,* Section 21.

[249] The CESCR's view on such a strategy is further elaborated in Sections 21-28, E/C.12/1999/5, 12 May 1999, CESCR, *General Comment 12, Right to Adequate Food.*

[250] E/C.12/1999/5, 12 May 1999, CESCR, *General Comment 12, Right to Adequate Food,* Section 29.

[251] E/C.12/1999/5, 12 May 1999, CESCR, *General Comment 12, Right to Adequate Food,* Section 31.

[252] E/1991/23, annex III at 86 (1991), 14 December 1990, CESCR, *General Comment 3, The Nature of States Parties' Obligations,* Section 5.

[253] E/C.12/1998/24, 3 December 1998, CESCR, *General Comment 9, The Domestic Application of the Covenant,* Section 10.

[254] E/C.12/1999/5, 12 May 1999, CESCR, *General Comment 12, Right to Adequate Food,* Section 32.

[255] E/C.12/1999/5, 12 May 1999, CESCR, *General Comment 12, Right to Adequate Food,* Section 33.

that the principles set out in this General Comment should be adopted in binding legal standards.[256] He was of the opinion that *'achieving justiciability of the right to food is a prime objective of the Special Rapporteur.'*[257] Therefore, he defended the position that the right to food is a justiciable right in somewhat stronger words, emphasising the equality between Civil and Political Rights on the one hand, and Economic, Social and Cultural Rights in nature and justiciability on the other hand. In this light, the enforcement mechanisms of ECOSOC Rights should equal those of Civil and Political Rights.[258] Ziegler strongly opposed the view of some Western countries that ECOSOC rights would not be justiciable because of its very nature. Ziegler summarised this view, developed somewhere during the Cold War period, by portraying four arguments to deny justiciability of these rights: *'firstly, the right to food was imprecise; secondly, the right to food was subject to the limit of progressive realisation; thirdly, the right to food required resources to be provided; and fourthly, that, in the absence of precise national legislation on the right to food, it was difficult for the Judiciary to fill the gap that properly belonged to the legislative branch of the State.'*[259] Ziegler opposed these arguments by stating that the right to food was actually quite precise, considering the wording of Article 11 ICESCR.[260] Furthermore, he reminded that there are limits to the application of the concept of progressive realisation as demonstrated by the CESCR in its General Comment 12,[261] and that minimum core obligations existed, as demonstrated in General Comment 3.[262] He argued furthermore that the obligation of non-discrimination is not subject to the limitation of progressive realisation.[263] Ziegler opposed to the view that the realisation of the right to food only required resources to be provided, by supporting the more differentiated typology of State duties, widely used in the UN system.[264] Finally, Ziegler explained that there were examples in a number of countries in which ECOSOC rights could be invoked in Courts, quoting the case of *'Government of the Republic of South Africa v. Irene Grootboom and*

---

[256] E/CN.4/2001/53, 7 February 2001, J. Ziegler, *Report of the Special Rapporteur on the right to food*, Section 91.
[257] E/CN.4/2002/58, 10 January 2002, J. Ziegler, *Report of the Special Rapporteur on the right to food*, Section 30.
[258] E/CN.4/2002/58, 10 January 2002, J. Ziegler, *Report of the Special Rapporteur on the right to food*, Sections 32-34.
[259] E/CN.4/2002/58, 10 January 2002, J. Ziegler, *Report of the Special Rapporteur on the right to food*, Section 35.
[260] E/CN.4/2002/58, 10 January 2002, J. Ziegler, *Report of the Special Rapporteur on the right to food*, Section 38.
[261] E/CN.4/2002/58, 10 January 2002, J. Ziegler, *Report of the Special Rapporteur on the right to food*, Section 39.
[262] E/CN.4/2002/58, 10 January 2002, J. Ziegler, *Report of the Special Rapporteur on the right to food*, Sections 40, 42-43.
[263] E/CN.4/2002/58, 10 January 2002, J. Ziegler, *Report of the Special Rapporteur on the right to food*, Section 41.
[264] E/CN.4/2002/58, 10 January 2002, J. Ziegler, *Report of the Special Rapporteur on the right to food*, Sections 44-45.

others'.[265] In this light, he referred to General Comment 9,[266] in which the CESCR argued that putting ECOSOC rights beyond the reach of the Courts is *'arbitrary and incomparable with the principle that the two sets of human rights are indivisible and interdependent.'*[267] Taking this all into consideration, he concluded that *'the right to food can be considered as justiciable by its very nature, and is therefore equal to civil and political rights.'*[268] Zieglers successor, Olivier de Schutter, also strongly expressed the view that the right to food should be a justiciable right, and argued that states should adopt a framework legislation *'ensuring that the right to food is justiciable before national Courts or that other forms of redress are available...'* The obligations implied by the right to food were then more precisely formulated, which would encourage *'Courts or other monitoring mechanisms, such as the human rights institutions...to contribute to ensure compliance with the right to food.'*[269]

It must however be noted that in other UN related institutions in which a more intergovernmental decision-making procedure is maintained, and mostly decisions are based on unanimous consent between the participating countries, there is a more cautious approach towards the enforceability of the right to food. These decisions are often made in the above introduced 'second pillar' in which the FAO hosted the World Food summits, and the resulting Voluntary Guidelines. In no WFS declaration, the enforceability of the right to food is mentioned, or the nature of the right in terms of self-executing or justiciability is recognised. The voluntary Guidelines do reaffirm the typology of duties in which a Country has to respect, protect and fulfil the right to food,[270] but concerning the implementation of the right rather cautiously *'invite'* states *'to consider...whether to include Provisions in their domestic law (...) that facilitates the progressive realisation on the right to food in the context of national food security'*[271], or *'to include Provisions...to directly implement the progressive realisation of the right to food.'*[272] It is obvious that 'inviting to consider to include Provisions to' is not the same as 'having an obligation to'[273]

---

[265] E/CN.4/2002/58, 10 January 2002, J. Ziegler, *Report of the Special Rapporteur on the right to food,* Sections 47-48.

[266] E/CN.4/2002/58, 10 January 2002, J. Ziegler, *Report of the Special Rapporteur on the right to food,* Section 37.

[267] E/C.12/1998/24, 3 December 1998, CESCR, *General Comment 9, The Domestic Application of the Covenant,* Section 10.

[268] E/CN.4/2002/58, 10 January 2002, J. Ziegler, *Report of the Special Rapporteur on the right to food,* Section 49.

[269] A/HRC/9/23, 8 September 2008, O. de Schutter, *Report of the Special Rapporteur on the right to food to the Human Rights Council,* Section 18.

[270] *The voluntary guidelines to support the progressive realisation of the right to adequate food in the context of national food security,* adopted by the 127th session of the FAO Council, November 2004, Section 17.

[271] *The voluntary guidelines to support the progressive realisation of the right to adequate food in the context of national food security,* adopted by the 127th session of the FAO Council, November 2004, guideline 7.1.

[272] *The voluntary guidelines to support the progressive realisation of the right to adequate food in the context of national food security,* adopted by the 127th session of the FAO Council, November 2004, guideline 7.2.

[273] As demonstrated above, a position defended by the Committee on Economic, Social and Cultural Rights and the Special Rapporteurs on the Right to food.

implement the right to food. Regarding the enforceability of the right to food through juridical remedies, the guidelines stipulate that '*administrative, quasi-juridical and judicial mechanisms to provide adequate, effective and prompt remedies accessible, in particular, to Members of vulnerable groups may be envisaged.*'[274] And: '*States that have established a right to adequate food under their legal system should inform the general public of all available rights and remedies to which they are entitled.*'[275] Also here, it is clear that '*remedies may be envisaged*' and '*inform the public of remedies to which they are entitled*' is not the same as '*the right to food is justiciable before national Courts*'[276] or '*effective judicial or other appropriate remedies are required.*'[277] Of course one must not forget that the purpose of the Voluntary Guidelines is to give practical guidance to countries on how to implement the right to food in its domestic legal order, and not an authoritative legal interpretation of the right to food. Especially different opinions on the matter of clear State obligations and accountability mechanisms were cause for stiff and harsh negotiations, even ending up in a negotiation breakdown. The Special Rapporteur, emphasising a right-based approach regarding the Voluntary Guidelines earlier,[278] strongly voiced his disappointment about '*the strong resistance of some other Governments, from developed and the developing world, which worked hard to water down the language of the text on the political and legal obligations implied by the right to adequate food. Many phrases are rendered almost meaningless by the number of caveats introduced by non-committal language, especially in relation to accountability.*'[279] Later however, the Special Rapporteur showed more positivism, perhaps due to the agreement of better objectives elsewhere in the final text, or perhaps due to a realisation that there would be no better outcome considering the difficult negotiation process. The Special Rapporteur on the Right to food hailed the final outcome of the '*ground breaking*' Voluntary Guidelines due to the fact that *inter alia* the right to food was now defined by one worldwide definition, included the obligations to respect, protect and fulfil, the right to food was addressed in an international dimension, the responsibility of Non-State actors was underlined, and practical guidance was given on the implementation and accountability aspects of the right.[280]

---

[274] *The voluntary guidelines to support the progressive realisation of the right to adequate food in the context of national food security*, adopted by the 127th session of the FAO Council, November 2004, guideline 7.2.
[275] *The voluntary guidelines to support the progressive realisation of the right to adequate food in the context of national food security*, adopted by the 127th session of the FAO Council, November 2004, guideline 7.3.
[276] In the line of reasoning of the Special Rapporteur on the right to food Mr J. Ziegler.
[277] In the line of reasoning of the Committee on Economic, Social and Cultural Rights in its General Comment 12.
[278] E/CN.4/2003/54, 10 January 2003, J. Ziegler, *Report of the Special Rapporteur on the right to food*, Chapter I, especially Section 24.
[279] A/59/385, 27 September 2004, J. Ziegler, *Report of the Special Rapporteur on the right to food to the General Assembly*, Chapter III.
[280] E/CN.4/2005/47, 24 January 2005, J. Ziegler, *Report of the Special Rapporteur on the right to food*, Chapter IV.

3.5.7 Reservations

Reservations are allowed under international law,[281] although there is tension between the idea of human rights as natural law with a universal working, and the very rich use by States of reservations to human rights treaties. It could also very well be that the goal of striving for the highest number of Member States to a human rights treaty as possible in the urge to promote human rights in the widest possible way, could lead to easier acceptance of reservations.[282] In this light, the International Court of Justice considered, contrary to the usual approach regarding an international treaty, that to a certain extent a reservation to a human rights treaty, even if other States Countries object to this reservation, cannot exclude the Party from that treaty.[283] In addition, it is questionable whether State Parties are willing to object on one another's reservations,[284] due to the fact that the reservation of one State has no or limited effect on the other[285] or, perhaps even more likely, out of political motives.[286] This is clearly demonstrated by the reluctance of States to make use of existing State communications procedures under human rights treaties.[287] It is also broadly accepted that a new international human rights treaty will enter into force if its meets the requirement of the agreed number of ratifications, even if the ratifying countries have adopted reservations to the treaty.[288] Of course the aim to spread human rights as widely as possible through the world is an admirable one, but it must be noted that the approach mentioned above does not stimulate Member States to ratify international human rights treaties without reservations, or to withdraw existing reservations.

Considering the practice sketched above and the human rights nature of the treaties, the Member States will not be the ones to judge whether reservations are in conformity with the human rights treaty in question. The treaty bodies,

---

[281] See Articles 19-23 of the Vienna Covenant on the Law of Treaties, adopted on 23 May 1966, entered into force on 27 January 1980, United Nations, Treaty Series, vol. 1155, p. 331.

[282] F. Mégret, Chapter 6, *The Nature of State Obligations*, in: D. Moeckli, S. Shah, S. Sivakumaran (eds.), D. Harris (cons. ed.), *International human rights law*, Oxford: Oxford University Press, 2010.

[283] International Court of Justice, advisory opinion of 28 May 1951, Reservations to the Convention on the Prevention and Punishment of the Crime of Genocide.

[284] A possibility that is recognised in Article 20 (5) of the Vienna Covenant on the Law of Treaties, adopted on 23 May 1966, entered into force on 27 January 1980, United Nations, Treaty Series, vol. 1155, p. 331: '*a reservation is considered to have been accepted by a State if it shall have raised no objection to the reservation or by the end of a period of twelve months after it was notified of the reservation or by the date on which it expressed its consent to be bound by the treaty, whichever is later.*'

[285] F. Mégret, Chapter 6, *The nature of state obligations*, in: D. Moeckli, S. Shah, S. Sivakumaran (eds.), D. Harris (cons. ed.), *International human rights law*, Oxford: Oxford University Press, 2010.

[286] M. Ssenyonjo, *Economic, social and cultural rights in international law*, Portland: Hart Publishing, 2009, Section 1.65.

[287] M. Ssenyonjo, *Economic, social and cultural rights in international law*, Portland: Hart Publishing, 2009, Section 1.65.

[288] F. Mégret, Chapter 6, *The nature of state obligations*, in: D. Moeckli, S. Shah, S. Sivakumaran (eds.), D. Harris (cons. ed.), *International human rights law*, Oxford: Oxford University Press, 2010.

which – among others things – receive and comment to the State reports, seem to be the ones most capable to do so. Perhaps from a 'normal' international law point of view the treaty bodies' authority to do so can be discussed, since it would normally be up to the other State Parties of an international treaty to value a State's reservation.[289] However, the treaty bodies indeed started to discuss the reservations made by the Member States during the reporting procedure. As it appears, the treaty bodies usually choose to diplomatically and carefully address the issue of Reservations.[290] The soft approach might be explained by the rather limited accounting mechanisms that exist on human rights law, or perhaps the fear that by condemning ratifications in a too harsh way, the State might withdraw from the entire treaty.[291] There is little coordination amongst the treaty bodies on how to respond to reservations. Regarding the ICESCR, the question of reservations has not been dealt with into detail by the CESCR, and therefore, Ssenyonjo rightfully urges the CESCR to adopt a General Comment on the matter.[292] But with or without a coordinated and clear approach towards reservations, it may come to no surprise that it is the general view of the treaty bodies that Member States should strive for the ratification of human rights treaties without any reservations.

There are various motivations for a State to make a reservation to a human rights treaty. As Frédéric Mégret observes '*A typical reservation is one whereby a state purports to interpret an internationally protected right only in accordance with its domestic, often Constitutional or religious, law.*'[293] As will be demonstrated in especially Chapters 9 and 13, both the Netherlands and Belgium made this kind of reservations to international human rights treaties containing ECOSOC rights.

However, Article 19 (c) of the Vienna Covenant on the Law of Treaties prohibits as a general rule reservations that are not in compliance with the object and the purpose of the Treaty.[294] Similar Provisions are enshrined in most human rights treaties, except for, remarkably, the ICESCR.[295] The prohibition of a reservation that is incompatible with the object and purpose of the treaty seems a difficult to interpret rule due to its general wording, although it suggests at least that a reservation should not affect the *raison d'être* of a treaty.

---

[289] See for a more detailed explanation: M. Ssenyonjo, *Economic, social and cultural rights in international law*, Portland: Hart Publishing, 2009, Sections 5.15-5.23.
[290] As will be considered into more detail in Chapters 9 and 13.
[291] M. Ssenyonjo, *Economic, social and cultural rights in international law*, Portland: Hart Publishing, 2009, Section 5.21.
[292] M. Ssenyonjo, *Economic, social and cultural rights in international law*, Portland: Hart Publishing, 2009, Section 5.05-5.05.
[293] F. Mégret, Chapter 6, *The nature of state obligations*, in: D. Moeckli, S. Shah, S. Sivakumaran (eds.), D. Harris (cons. ed.), *International human rights law*, Oxford: Oxford University Press, 2010.
[294] Article 19 (c), Vienna Covenant on the Law of Treaties, adopted on 23 May 1966, entered into force on 27 January 1980, United Nations, Treaty Series, vol. 1155, p. 331.
[295] M. Ssenyonjo, *Economic, social and cultural rights in international law*, Portland: Hart Publishing, 2009, Section 5.08-5.13.

3.5.8 Reporting procedures

Usually, a body of experts is installed for each UN human rights treaty in order to monitor the implementation of the treaty in the domestic legal orders of its Member States. Generally, the competences of these bodies are *inter alia* to receive and consider the periodic reports submitted by the Member States on the progress that was made on the implementation of the right enshrined in the treaty, and to recommend on the interpretation of the Provisions.

All States have the obligation to submit a general 'Core Document' that should contain '*information of a general and factual nature relating to the implementation of the treaties to which the reporting State is party and which may be of relevance to all or several treaty bodies.*'[296] The States also submit treaty-specific reports, starting with an initial report, which should mostly be submitted within one year after ratification of the treaty, and from then on periodic reports, in which the State reports '*on the measures, including legislative, judicial, administrative or other measures, which they have adopted in order to achieve the enjoyment of the rights recognised in the treaty.*'[297]

The usual report cycle starts with the submission of the State Report. The report is then considered by the competent Committee that usually sends a 'List of Issues' to the Member State, containing questions about the report, or requests for additional data. The Member State concerned will then answer to these written questions and requests by submitting formal '*replies to List of Issues*'. After this, the Member State sends a Delegation to the Committee, usually seated in Geneva once a year, and discusses the report with the Committee in a '*constructive dialogue*' during a couple of sessions. Summaries of these sessions are usually published under the name '*Summary Record*'. Finally, although this was not a general practice during the earlier reporting cycles,[298] the Committee submits a final document on the report containing Concluding Observations, in which the Committee usually highlights the '*positive aspects*' of the reporting cycle, names the '*factors and difficulties impeding the implementation of the Covenant*', and expresses its '*principal subjects of concern and suggestions and recommendations*'. These Concluding Observations may then serve as a guideline for the further implementation of the rights enshrined in the treaty in the upcoming reporting period. Increasingly, there is a role for NGOs during this process, which can sometimes formally submit so-called 'shadow-reports' to the treaty bodies, in which the NGOs report on the implementation of the human rights that are discussed from their point of view. Also, NGO representatives are sometimes present during or even participate in the dialogue between the State

---

[296] HRI/GEN/2/Rev.6, 3 June 2009, *Compilation of guidelines on the form and content of reports to be submitted by State Parties to the International Human Rights Treaties*, Section 27.
[297] HRI/GEN/2/Rev.6, 3 June 2009, *Compilation of guidelines on the form and content of reports to be submitted by State Parties to the International Human Rights Treaties*, Section 2.
[298] For instance, the CESCR started to submit Concluding Observations to State Reports since 1992.

Delegation and the Treaty Body. Since 2000, increasingly, committees started to appoint one or two of their Members as a country rapporteur, who prepare(s) the List of Issues, receive(s) information from NGOs, and lead(s) the sessions on behalf of the Committee.[299]

In general, States that ratified most UN human rights treaties have a very extensive reporting duty, which often leads to delay in the submission of the reports.[300] To facilitate timely submission of periodic reports, the UN Treaty Bodies attempted to harmonise the treaty-specific reporting guidelines.[301]

Under Articles 16 and 17 ICESCR, Member States have the obligation to submit periodic reports *'on the measures which they have adopted and the progress made'* in achieving the ICESCR rights. Article 16 ICESCR authorises the ECOSOC to consider these reports, and thus no specific treaty body was installed. However, the ECOSOC did install the CESCR to assist in their task,[302] and in practice, the CESCR receives and considers these periodic reports submitted by the Member States. In their General Comment 1, the CESCR further clarified the reporting objectives.[303] Furthermore, they adopted reporting guidelines to *'advise States Parties on the form and content of their reports, so as to facilitate the preparation of reports and ensure that reports are comprehensive and presented in a uniform manner by States Parties.'*[304] The States are requested to submit an initial report within two years after ratification, and from then on a periodic report every five years.[305]

More or less comparable reporting procedures exist under the ICRC and the CEDAW. The Committee on the Rights of the Child, installed directly by Article 43 ICRC, overviews the reporting procedure on the implementation of the ICRC. The Member States are obliged to submit an initial report within two years after ratification and from then on a periodic report every five years (Art. 44 ICRC). The Committee on the Elimination of Discrimination against Women, installed directly by Article 17 CEDAW, overviews the reporting procedure on the implementation of the CEDAW. The Member States are obliged to submit an initial report within

---

[299] See for instance regarding the CESCR: E/C.12/2000/6, 7 July 2000, *Substantive issues arising in the implementation of the International Covenant on Economic, Social and Cultural Rights*, Section II C. See also the working methods of the Committee on the Rights of the Child, available via: http://www2.ohchr. org/english/bodies/crc/workingmethods.htm.

[300] This is certainly the case with the Netherlands and Belgium, as discussed in Chapters 9 and 13.

[301] HRI/GEN/2/Rev.6, 3 June 2009, *Compilation of guidelines on the form and content of reports to be submitted by State Parties to the International Human Rights Treaties.*

[302] Economic and Social Council Resolution 1985/17, 28 May 1985.

[303] E/1989/22, annex III at 87 (1989), 24 February 1989, CESCR, *General Comment 1, Reporting by States parties.*

[304] E/C.12/2008/2, 24 March 2009, *Guidelines on treaty-specific documents to be submitted under Articles 16 and 17 of the International Covenant on Economic, Social and Cultural Rights.*

[305] Economic and Social Council Resolution 1988/4, 24 May 1988, Section 6.

one year after ratification and from then on a periodic report every four years (Art. 18 CEDAW).

The effect of these reporting procedures ha been debated more than once. One might argue that through reporting and the subsequent debate with the various UN Committees and the more recent involvement of NGOs, human rights violations might be exposed, and through naming and shaming a contribution can be made to improve the implementation of human rights in a country.[306] On the other hand, one might consider the reporting systems to be weak. Firstly, the supervising bodies lack the power to make real changes, since these bodies are fully depending on the willingness of the Member States to cooperate, whereas their General Comments are hardly binding. Secondly, there are hardly any possibilities to complain about violations of human rights of an individual or a group through an international legal remedy in a monitoring system.

3.5.9 Complaint mechanisms

Part of a monitoring/accountability mechanism, the originally foreseen third step in the establishment of a global human rights mechanism, could be the installation of complaint procedures. In human rights treaties, a monitoring procedure is usually embedded, but a complaint mechanism is not always integrated,[307] and is often installed much later through an Optional Protocol, usually preceded by long and difficult negotiations. For instance, the ICESCR entered into force in 1976, but only since 1990, attempts were made by the Committee to formulate an Optional Protocol to the ICESCR.[308] These attempts did not lead to the desired outcome, mostly due to a lack of response and cooperation of States.[309] Therefore, on the advice of the independent expert Professor Hatem Kotrane,[310] the Commission on Human Rights installed the Open-ended Working Group on the OP-ICESCR some years later in 2002,[311] whose mandate was extended in 2006.[312] Their first draft of

---

[306] See for instance the contribution to the preface by Kenneth Roth, at the moment of writing Executive Director of Human Rights Watch, in: D. Moeckli, S. Shah, S. Sivakumaran (eds.), D. Harris (cons. ed.), *International human rights law*, Oxford: Oxford University Press, 2010.

[307] Although in for instance Article 41-43 ICCPR a (voluntary) State complaints procedure is embedded.

[308] See for instance E/CN.4/1997/105, 18 December 1996, Draft Optional Protocol to the International Covenant on Economic, Social and Cultural Rights, adopted by the Committee on Economic, Social and Cultural Rights.

[309] M. Ssenyonjo, *Economic, social and cultural rights in international law*, Portland: Hart Publishing, 2009, especially Sections 1.57-1.58.

[310] Commission on Human Rights Resolution 2001/30, 20 April 2002.

[311] Commission on Human Rights Resolution 2002/24, 22 April 2002.

[312] A/HRC/1/L.10, 29 June 2006, Human Rights Council Resolution 2006/3.

an Optional Protocol was finished in 2007.[313] The General Assembly adopted the finalized Optional Protocol to the ICESCR in 2008,[314] containing three procedures. Firstly, *'communications…by or on behalf of individuals or groups of individuals, under the jurisdiction of a State Party, claiming to be a victim of a violation of any of the…rights set forth in the Covenant by that State Party'* may be submitted.[315] Secondly, a procedure for 'communications' from Member States *'that claim that another State Party is not fulfiling its obligations under the Covenant'* is embedded in the Protocol.[316] Thirdly, an inquiry procedure was adopted, authorising the Committee to examine alleged violations of the ICESCR *'if the Committee receives reliable information indicating grave or systematic violations by a State Party of any of the economic, social and cultural rights set forth in the Covenant.'*[317] At the time of writing, the protocol had 42 signatories of whom 10 Member States actually ratified,[318] which is exactly the number of ratifications necessary to enter into force, which took place on 5 May 2013.[319] The structure of an individual communications procedure, an inter-State communications procedure and an inquiry procedure is widely used for Optional Protocols to UN human rights treaties, for instance in the Optional Protocols to the CEDAW and the ICRC. The Committee on the Elimination of Discrimination against Women is the competent body to receive complaints under the Optional Protocol to the CEDAW,[320] and the Committee on the Rights of the Child will be the competent body to receive complaints under the Optional Protocol to the ICRC on a communications procedure when 10 States have ratified the Protocol.[321] It is for this research relevant to know that at the time of writing, the Netherlands and Belgium signed the Optional Protocol to

---

[313] A/HRC/6/WG.4/2, 23 April 2007, draft Optional Protocol to the International Covenant on Economic, Social and Cultural Rights; A/HRC/8/WG.4/2, 24 December 2007, revised draft Optional Protocol to the International Covenant on Economic, Social and Cultural Rights; A/HRC/8/WG.4/3, 25 March 2008, (second) revised draft Optional Protocol to the International Covenant on Economic, Social and Cultural Rights.

[314] A/RES/63/117, 10 December 2008, *Optional Protocol to the International Covenant on Economic, Social and Cultural Rights,* adopted by the General Assembly.

[315] A/RES/63/117, 10 December 2008, *Optional Protocol to the International Covenant on Economic, Social and Cultural Rights,* adopted by the General Assembly, Articles 2-9.

[316] A/RES/63/117, 10 December 2008, *Optional Protocol to the International Covenant on Economic, Social and Cultural Rights,* adopted by the General Assembly, Articles 10-14.

[317] A/RES/63/117, 10 December 2008, *Optional Protocol to the International Covenant on Economic, Social and Cultural Rights,* adopted by the General Assembly, Articles 11-12.

[318] See for an official overview of signatories, accessions and ratifications of all UN human rights treaties: http:/treaties.un.org.

[319] A/RES/63/117, 10 December 2008, *Optional Protocol to the International Covenant on Economic, Social and Cultural Rights,* adopted by the General Assembly, Article 18.

[320] A/RES/54/4, 15 October 1999, *Optional Protocol to the Convention on the Elimination of All Forms of Discrimination against Women.*

[321] A/RES/66/138, 27 January 2012, *Optional Protocol to the Convention on the Rights of the Child on a communications procedure,* Article 19. See the preparatory work: A/HRC/17/36, 16 May 2011, Report of the Open-ended Working Group on an Optional Protocol to the Convention on the Rights of the Child to provide a communications procedure, annex (Draft Optional Protocol to the Convention on the Rights of the Child to provide a communications procedure). See Article 19 for the entry into force Provision. At the moment of writing, 37 States signed, while only 6 had ratified the Protocol.

the ICESCR, signed and ratified the Optional Protocol to the CEDAW, and only Belgium signed the Optional Protocol to the ICRC.

Considering the legal practice, there is little experience in the effect of individual complaints concerning ECOSOC rights so far, and there is little enthusiasm amongst States to ratify these procedures. Inter-State procedures are rarely used, which could be explained by several reasons. Firstly, the nature of human rights may play a role: basically, human rights establish a legal relationship between an individual and a State, and not primarily between States. This may explain the reluctance to complain about another State's behaviour when the individuals whose rights are violated do not fall within the State's own jurisdiction.[322] Another explanation could be of a more political nature. When one State complains about the behaviour of another State, the complaining State might fear retaliation complaints about its own behaviour.[323]

3.6 Conclusion

This section shows that since World War II, considerable but also difficult progress was made to develop the right to food in the international human rights arena, which involved many actors who, from their different perspectives and out of different expertise contributed to this. Three pillars can be distinguished to categorise the work that has been done to further clarify the right to food, especially in view of Article 11 ICESCR: the first pillar is the work done by the treaty bodies, such as the CESCR. The second pillar is the work done within FAO context, which includes the World Food Summits and the adoption of the Voluntary Guidelines. The third pillar is the work done in the context of the Human Rights Council, basically involving the works of the Special Rapporteurs.

The right to food is not an isolated human right, but inextricably linked to other human rights and therefore broadly embedded in the international and regional human rights instruments. Regarding the status of enforceability of the right to food, there is a need to thoroughly analyse the respective case law per country, in their proper Constitutional context, as will be done in this research regarding the Netherlands and Belgium.

As it appears, the content of the right to food is often discussed in the context of developing countries. The relevant articles and explanatory documents written in UN context seem to offer enough guidance to at least come to a minimum specification of what 'adequate food' means as a substantive right.

---

[322] F. Mégret, Chapter 6, Section 7, *The nature of obligations*, in: D. Moeckli, S. Shah, S. Sivakumaran (eds.), D. Harris (cons. ed.), *International human rights law*, Oxford: Oxford University Press, 2010.
[323] M. Ssenyonjo, *Economic, social and cultural rights in international law,* Portland: Hart Publishing, 2009, Section 1.65.

In this light, notwithstanding the fact that the realisation of human rights is a responsibility for all, the international human rights system addresses States to realise human rights. There is tension between the concept of universality of human rights on the one hand, and State sovereignty implying a free choice to be bound by treaties, and to how the treaties should be implemented on the other hand.

The traditional distinction made between civil and political rights on the one hand, implying negative state obligations, and economic, social and cultural rights on the other hand, implying positive state obligations has been criticised for decades now. There are many valid arguments to oppose to such a distinction both from a practical point of view and from a legal theoretical perspective. Instead, a typology of duties applied to all human rights seems to do more justice to the meaning of human rights. Such a typology consists of a duty to respect, protect and fulfil. It demonstrates that ECOSOC rights do not leave States an undefined margin of discretion in the realisation of these rights.

According to the CESCR, Article 2 ICESCR implies that there is a minimum core obligation regarding ECOSOC rights, which requires immediate realisation. In addition, there are immediate obligations of conduct to move as expeditiously and effectively as possible towards the full realization of the right. Ultimately, this is an obligation of result. This suggests thus that States that do nothing or little to implement the rights also violate the Convention. Also, the CESCR underlined that discrimination in the access to food constitutes a direct violation of the ICESCR. Furthermore, the phrasing *'by all appropriate means'* implies the Provision of judicial remedies with respect to rights which may, in accordance with the national legal system, be considered justiciable.

It is difficult to establish whether in the UN context the right to food is unanimously considered to be an enforceable right, and if so, what it exactly means, for the UN can hardly be considered to be one single entity speaking with one voice. In general, no evidence could be found that immediate obligations of conduct are enforceable. The discussion on enforceability focuses more on the States' core obligations. While the CESCR did not put Article 11 ICESCR on their list of Provisions that would by nature be self-executing rights in their General Comments 3 and 9, it seems to consider in General Comment 12, with some more caution, that the implementation of the right to food implies the requirement of effective judicial or other appropriate means. Furthermore, the Committee had stressed in general that putting ECOSOC rights beyond the reach of Courts is arbitrary and incomparable with the principle that human rights are indivisible and interdependent, and seems to emphasise this consistently in their assessment of submitted country reports. The Special Rapporteurs on the right to food however, are much more specific in their viewpoints. For instance, Ziegler argued that a core content of the right to food is certainly justiciable, and, in line with the CESCR, underlines that non-discrimination does not fall under the scope of progressive realisation, suggesting that discrimination could be dealt with directly by national Courts. On the other hand, in the UN fora in which intergovernmental decision-making procedures are used, the enforceability seems to be hardly supported.

The latter is also reflected in the lack of sincerity of Countries ratifying human rights treaties, considering the rich use of the possibilities to make reservations to human rights treaties. It is hard for treaty bodies to oppose to these ratifications, due to a lack of binding power. The same can be said of the various reporting procedures that are part of the monitoring systems for the human rights treaties. Although the treaty bodies usually recommend on the reports, and more generally adopt General Comments on the interpretation of the rights enshrined in these treaties, the bodies lack the power to seriously change a State's behaviour if the State is not cooperative. To this end, complaint procedures have been established to the various human rights instruments. However, there is reluctance amongst States to ratify these procedures, and actually make use of them. The system through which these human rights are developed thus strongly depends on the willingness of countries to be successful, and countries are not easily prepared to give up sovereignty. This tension causes a problem that is difficult to solve regarding the realisation of international human rights, and leaves the hope for effective and enforceable human rights in the hands of the domestic Courts.

To conclude, and come to a *tertium comparationis,* it seems thus that the specialised bodies of the UN, basically the CESCR and the Special Rapporteurs, consider with sound arguments that '*the right to food and ECOSOC rights in general can be understood to imply immediate state obligations that should be enforceable through domestic Courts. There seems to be at least a core content, consisting of a minimum substance without whom the right would be stripped of its raison d'être, and a non-discrimination principle, that should be justiciable, or even self-executing*'. However, it is questionable whether this view is largely shared amongst the Member States of the UN, considering their behaviour in the intergovernmental decision-making processes, and towards the treaties in view of their sovereignty.

Part 2
The Netherlands

4. Dutch case law on the enforceability of the right to adequate food

4.1 Introduction

In general, it must be concluded that no evidence could be found that there are differences of any significance between the case law of the Courts on the one hand, and the administrative tribunals on the other hand, regarding the direct applicability of the right to food, or other ECOSOC standards. Furthermore, in Dutch literature, in describing case law, it is not unusual to refer to both the case law of the Courts and the tribunals, in all layers combined. Therefore, in this Chapter, no specific distinction is made between the Courts types, but rather between the types of invoked standards. Where necessary, differences in approach between the types of Courts are of course discussed.

In this chapter, firstly the organisation of the Dutch Judiciary will be discussed (Section 4.2). Then, an analysis will be presented of the case law concerning the direct applicability of international Provisions stipulating or related to the right to food, that is, Article 11 ICESCR (Section 4.3), Article 24, 26, and 27 ICRC (Sections 4.4, 4.5 and 4.6, respectively) and Article 12 CEDAW (Section 4.7). Finally, some conclusions will be drawn (Section 4.8).

4.2 The Dutch Judiciary

The competences of the Dutch Courts and tribunals are regulated in Chapter 6 of the Constitution: in short, Article 112 sub 1 CA entrusts competence to the Judiciary on matters concerning civil rights and obligations and Article 113 CA on matters of criminal law. Article 112 (2) CA stipulates that by statutory regulation the competence to judge in cases regarding administrative law may be entrusted to either the Judiciary, or organs that are not part of the Judiciary (tribunals). In addition, Article 115 CA stipulates the possibility (by statutory regulation or delegated law) to open appeal procedures against administrative decisions before administrative appeal bodies, a possibility that is now rarely used. The organisational structure of the Judiciary is rather different compared to the organisational structure of administrative tribunals, although most safeguards for an independent and impartial Judiciary also apply to those tribunals. In practice however, the term Judiciary is used to cover also the tribunals and appellate bodies that rule in matters of administrative law, especially since the Administrative Court of First Instance is now a Section of the District Courts.[324]

---

[324] See for instance the official website of the Dutch Judiciary: www.rechtspraak.nl.

Since 2008, the Judiciary is organised in three layers of Courts. Mostly, the Court of First Instance is one of the nineteen District Courts. These Courts consist of three different sections, that administer justice in private law, criminal law, and administrative law, and a section that previously was a separate Court of law: the subdistrict section. In the latter, cases concerning labour conflicts, debts, traffic fines, tenancy law and fines concerning offences in criminal law that concern no more than 25,000 Euros are dealt with. Then, there are five Courts of Appeal that are competent to rule in all the matters that are open for appeal judged by the District Courts, except for administrative legal issues. Finally, one could turn to a Court of Last Instance, the Supreme Court that serves as the Court of Cassation. Since 1 January 2013, the judicial organisation was reformed once again, mainly reducing the amount of Courts to eleven District Courts and four Courts of Appeal. Since this thesis only concerns the period to and including 2012, the latest judicial reform will not be of any relevance in this Chapter.[325]

The administrative tribunals are organised differently. Firstly, it needs to be noted that also the Judiciary might be competent in ruling on matters of administrative law, as long as no tribunal has specifically been attributed the competence to rule on the matter (Article 112 (2) CA). These procedures then mostly start with the section on private law matters of the District Courts. Secondly, as demonstrated above, the Administrative Courts of First Instance have been incorporated in the structure of the District Courts. From there on, the appeal procedures are different. Depending on the case, the Central Court of Appeal, the Administrative judicial Division of the Council of State, and the Trade and Industry Appeals Tribunal are the competent appeal bodies for Administrative issues. All three bodies are, with some minor exceptions,[326] also the Court of Last Instance.[327]

As it will appear below, most cases in which the right to food are invoked concern asylum-seekers and other foreigners, or matters related to social security and social benefits. Competent in those cases are the Administrative Courts of first instance, and the appellate body is the Central Court of Appeal.

---

[325] See for instance the official website of the Dutch Judiciary: www.rechtspraak.nl.

[326] For instance, in certain cases involving tax regulations, it is possible to appeal against a decision of the Central Court of Appeal, or the Trade and Industry Appeals Tribunal before the Supreme Court. See for further details: Regulation of Internal Service of the Supreme Court of the Netherlands pursuant to Article 75 Section 4 of the Law on the judicial Organisation. Original title in Dutch: Reglement van Inwendige Dienst van de Hoge Raad der Nederlanden ex Artikel 75 lid 4 Wet op de Rechterlijke Organisatie. Available on: www.rechtspraak.nl.

[327] In this Section, we gratefully made use of the terminology and explanation offered in: L. Prakke and C. Kortmann (edt.), *Constitutional law of 15 EU Member States,* Deventer: Kluwer, 2004, especially p. 632-636.

4.3 Article 11 ICESCR

4.3.1 Rejecting direct applicability

Apart from some summary trial procedures, in which the Courts decided that the question of direct applicability of Article 11 ICESCR is 'too complex' to be answered in such a procedure,[328] or did not use the invoked Provision in their considerations,[329] the Dutch Courts unanimously seem to reject the direct applicability of Article 11 ICESCR. To this end the Courts merely seem to consider that the Provision:

> From the wording of Article 11 ICESR follows that it does not contain standards the Courts can apply directly as yardstick to judge decisions of administrative authorities, because this Provision is insufficiently concrete for such use. Therefore it first needs to be elaborated in national legislation.[330]

The Central Court of Appeal, Court of Last Instance in many aliens' procedures, ruled that:

> The Court considers that this appeal is ineffective, for these Provisions cannot be regarded as binding to all persons, as stipulated in Articles 93 and 94 of the Constitutional Act. Given the wording and scope of both Provisions, it rather concerns generally formulated social objectives, with regard to which the State Parties have committed themselves to pursue and realise those by regulations, rather than a right recognised by those State Parties, that can be invoked by citizens. In this context, reference must be made to the Explanatory Memorandum to the ratification Bill on the ICESCR (Parliamentary Documents, II 1975-1976 13 932, No. 3) it is stated that the Provisions of this Convention "in general" will have no direct effect.[331]

---

[328] District Court of Haarlem, 11 July 2007 LJN BB0998 (summary trial procedure).

[329] See for instance: District Court of Haarlem, 21 June 2007, LJN BA9025 (summary trial procedure); 29 July 2008 LJN BE9491 (summary trial procedure); 29 March 2012, LJN BW2431; Central Court of Appeal, 18 June 2009, LJN NI9928; Council of State, 22 December 2010, AB 2011, 169.

[330] Council of State of 19 April 2007, LJN BA4289.

[331] Central Court of Appeal, 25 May 2004, USZ 2004, 241. Original text in Dutch: '*is de Raad van oordeel dat dit beroep geen doel treft nu deze bepalingen niet kunnen worden aangemerkt als een ieder verbindende bepaling als bedoeld in de artikelen 93 en 94 van de Grondwet. Gelet op de bewoordingen en strekking van bedoelde bepalingen is daarin veeleer sprake van algemeen geformuleerde sociale doelstellingen, tot het nastreven en verwezenlijken waarvan in hun regelgeving de verdragsstaten zich hebben verbonden, dan van een door die verdragsstaten erkend recht, waarop burgers zich in hun nationale rechtsorde zonder meer kunnen beroepen. In dit verband verdient vermelding dat in de Memorie van Toelichting bij de wet tot goedkeuring van het IVESCR (Bijl. Hand. II 1975-1976 13 932, no. 3) is opgemerkt dat de bepalingen van dit verdrag "in het algemeen" geen rechtstreekse werking zullen hebben.*' See for a similar consideration: Central Court of Appeal, 1 November 2005, LJN AU5600; 26 January 2010, LJN BL1686; 2 November 2010, LJN BO3025; 14 March 2011, NJB

Summarising, it seems that the Courts find their arguments to deny direct effect of Article 11 ICESCR in the fact that:
- the Article is not binding on all persons, as stipulated in Articles 93 and 94 CA;
- the Article is not precise enough for concrete use, and therefore further national legislation is required;
- the Legislature, in their ratification Bill, already pointed out that ICESCR rights in general would have no direct effect.

It is remarkable that mostly the Courts deal quickly with the matter of direct applicability, and use one of the aforementioned standard considerations without discussing the particularities of the case. Furthermore, in some cases, the Courts do not even refer to the specific Articles invoked by the claimant, but simply state that the complete ICESCR has no direct effect.[332]

4.3.2 The ICESCR cases

Not surprisingly, most cases in which Article 11 ICESCR is invoked, concern people on low incomes, prisoners, elderly, disabled persons and asylum-seekers. Generally, Article 11 ICESCR is invoked to support demands for social benefits, providing the claimant with minimum means of subsistence. As it appears, Article 11 ICESCR cases come in two categories. The first consist of the cases in which the level of social benefits is under discussion. In these cases, the claimant enjoys social benefits, but s/he considers these benefits to be inadequate and thus not guaranteeing minimum means of subsistence.[333] Here, the right to food is seldom explicitly an issue: the focus is on Dutch legislation and procedures concerning social benefits. The second category consists of cases in which the claimant has no access to social benefits at all, and, consequently, has no means of subsistence that are provided for by the Government. The claimants are mostly asylum-seekers residing in the Netherlands.[334] A majority of the Article 11 ICESCR cases concern this last category.

2011, 755; District Court of Arnhem, 25 May 2007, LJN BA6562. In Central Court of Appeal, 18 June 2004, JB 2004, 303, also a reference was made to the Explanatory Memorandum on the ratification Bill of the ICESCR, when denying direct effect to Article 9 ICESCR.

[332] For instance: 21 November 2007, LJN BB9625; District Court of Amsterdam, 3 December 2008, LJN BG7017, consideration 2.3.4.

[333] For instance: Central Court of Appeal, 1 November 2005, LJN AU5600, 9 May 2006, LJN AX2177, 1 October 2008, LJN BF4589.

[334] For instance: Central Court of Appeal, 25 May 2004, USZ 2004, 241; 8 July 2005, LJN AT910211; 11 October 2007, LJN BB5687; 21 November 2007, LJN BB9625; District Court of 's-Gravenhage, 30 August 2000, LJN AA6959; 23 January 2006, LJN AV0548; District Court of Arnhem, 25 May 2007, LJN BA6562; District Court of Rotterdam, 19 September 2007, LJN BB5715; 24 December 2007, LJN BC0852; District Court Haarlem 8 April 2008 LJN BD3399 (summary trial procedure); District Court of Amsterdam: 4 August 1999, LJN AA4043; 13 March 2001, LJN AB0942.

Once again, two situations can be distinguished in which an asylum-seeker invokes Article 11 ICESCR. In the first situation, the asylum-seeker stays rightfully in the Netherlands (Article 8 Aliens Act), while awaiting a final decision concerning a residence permit (Article 8f-g Aliens Act 2000), or concerning certain administrative procedures (Article 8h Aliens Act 2000). In this situation, the asylum-seekers '*are housed in one of a number of reception centres scattered throughout the country*'[335], but have generally no further rights concerning income support or other social benefits.[336] In the second situation, the asylum-seeker stays illegally in the Netherlands, and has generally no access to social benefits,[337] or possibility to stay in a reception centre.[338] The above is based on the so-called 'Linkage Act'[339], an Act that partly excludes asylum-seekers without residence permit from entitlements to general social benefits in the Netherlands, with the main objective to restrict immigration. The Courts generally rule that this Act does not conflict with international law, and is not disproportionate (it is a suitable means to reach its legitimate purposes).[340] Therefore, in general, this means that an asylum-seeker, unlawfully residing in the Netherlands, or lawfully residing but not having a residence permit, cannot make a successful claim to most general social benefits, and thus also cannot make a successful appeal to a right to food.[341]

4.3.3 Article 11 ICESCR and the authoritative interpretation of the Committee on Economic, Social and Cultural Rights

The Committee on Economic, Social and Cultural Rights in their various Comments and Observations on the implementation of the ICESCR defends the position that some ICESCR Provisions, including Article 11, should be justiciable or even self-executing.[342] This was also brought before the Central Court of Appeal to support claims based on *inter alia* the right to an adequate standard of living.

In a case in 2007, an asylum-seeker, awaiting a decision on his application for a residence permit, appealed to the decision of the local Public Centre for Social Welfare not to grant certain social benefits that would enable him to provide for himself and his younger brother. One of the arguments he put forward was that the State has the duty to provide at least some care, because he was residing

---

[335] Quote from: E/1994/104/Add.30; 23 August 2005, Section 373.

[336] For instance, District Court of Amsterdam, 13 March 2001 LJN AB0942.

[337] For instance, Central Court of Appeal, 21 November 2007 LJN BB9625.

[338] District Court of Amsterdam, 13 March 2001, LJN AB0942.

[339] In Dutch: Koppelingswet.

[340] Central Court of Appeal, 26 June 2001, LJN AB2324.

[341] For instance: Central Court of Appeal, 25 May 2004, USZ 2004, 241; 8 July 2005, LJN AT9102; 11 October 2007, LJN BB5687; 21 November 2007, LJN BB9625; District Court of 's-Gravenhage, 30 August 2000, LJN AA6959; 23 January 2006 LJN AV0548; District Court of Arnhem, 25 May 2007, LJN BA6562; District Court of Rotterdam, 24 December 2007, LJN BC0852; District Court of Haarlem 8 April 2008, LJN BD3399 (summary trial procedure); District Court of Amsterdam: 4 August 1999, LJN AA4043.

[342] See Sections 3.5.5. and 3.5.6.

lawfully in the Netherlands. His Attorney, anticipating the standard considerations with regard to the invoked Articles 9, 11, and 13 ICESCR, argued that in 1986 the Central Court of Appeal had considered with regard to Article 7 ICESCR that it would be incorrect to assume that direct applicability would never be possible.[343] In response, the Court seems to consider Article 7 ICESCR on equal treatment of women and man an exceptional Provision in the Covenant,[344] and argued in the first place that Article 7 ICESCR is quite concrete, unlike Articles 9, 11, and 13 ICESCR. In the second place, the Court stated that the Committee on Economic, Social and Cultural Rights of the United Nations stated in one of their General Comments that some of the Provisions of the ICESCR, including Article 7 ICESCR, are suitable for direct applicability. Apparently, the Court is referring to General Comment 3. In Section 5, the Committee on Economic, Social and Cultural Rights lists 6 ICESCR Provisions that *'would seem to be capable of immediate application by judicial and other organs in many national legal systems.'*[345] This reference by the Court to General Comment 3 is hardly convincing. Firstly, the list in this General Comment not only mentions Article 7, but also includes Article 13 (2) (a) (3), one of the Provisions that were invoked in this case.[346] Secondly, if the General Comments are to carry weight, it is at the least remarkable that General Comment 12, concerning the right to adequate food,[347] was not taken into consideration. In this General Comment, the Committee clarifies *inter alia* that *'any person or group who is a victim of a violation of the right to adequate food should have access to effective judicial or other appropriate remedies at both national and international levels. All victims of such violations are entitled to adequate reparation, which may take the form of restitution, compensation, satisfaction or guarantees of non-repetition. National Ombudsmen and human rights commissions should address violations of the right to food.'*[348]

---

[343] Central Court of Appeal, 3 July 1986, TAR 1986, 215. It must be remarked, however, that the Court had already withdrawn somewhat from this position in a later ruling in which it ruled that direct applicability of one of the provisions of the ICESCR would be 'a total exception from the Covenant's general character.' As will be demonstrated in 8.1., the Provision is later used as an interpretative norm, or simply denied direct effect, except for one incidental case from 1984, in which the Provision was granted direct effect.

[344] Article 7 ICESCR contains *inter alia* the principle guaranteeing women conditions not inferior to those enjoyed by men, with equal remuneration for equal work.

[345] E/1991/23, annex III at 86 (1991), 14 December 1990, CESCR, *General Comment 3, The Nature of States Parties' Obligations,* Section 5.

[346] Along with Articles 3, 7(a)(i), 8, 10, and 15(3) ICESCR.

[347] In General Comment 12, the Committee on Economic, Social and Cultural Rights, responds to Objective 7.4. of the World Food Summit, *'To clarify the content of the right to adequate food and the fundamental right of everyone to be free from hunger, and as stated in the International Covenant on Economic, Social and Cultural Rights and other relevant international and regional instruments, and to give particular attention to implementation and full and progressive realisation of this right as a means of achieving food security for all.'*

[348] E/C.12/1999/5, 12 May 1999, CESCR, *General Comment 12, Right to Adequate Food,* Section 32.

In another case, in 2008, an appellant invoked Article 13 ESC and Articles 11 and 12 ICESCR.[349] He argued that the rights embedded in those Articles could be directly invoked by an individual. With regard to the direct applicability of the ICESCR Provisions, this interpretation was not only based on the Committee on Economic, Social and Cultural rights' General Comments, but also on their Concluding Observations concerning the second and third periodic report on the implementation of the ICESCR. In these observations, the Committee expressed their concern with regard to the fact that the Courts deny the direct applicability of the ICESCR Provisions.[350] For instance, *'It urges the State Party to ensure that the Provisions of the Covenant are given direct effect by its domestic Courts, as defined in the Committee's General Comment 3, and that it promotes the use of the Covenant as a domestic source of law. It invites the State Party to include, in its fourth periodic report, information on case-law concerning the rights recognised in the Covenant.'*[351] The Central Court of Appeal however rejected direct applicability using a standard consideration, and considered that the documents referred to by the appellant did not provide for sufficient basis to deviate from this reasoning.[352]

4.3.4 A substantial right to food?

Despite the rejection of direct applicability of the right to food, there are examples in which the Courts show some compassion for the requesting party, albeit not based on a legal standard. For example, one ruling was given in a dispute between on the one hand a refugee from Somalia with her 23-months' old child and on the other hand the COA.[353] COA is the Dutch agency charged with care for and housing of asylum-seekers. To this end they operate so-called reception centres. In this case the woman appealed to a decision taken by the COA to deny her access to such a reception centre. The reason was that her request for a residence permit had been denied, and it was the COA's policy only to house people awaiting a first decision. However, she was awaiting the decision on a second request. In her argument she took recourse to *inter alia* Article 11 ICESCR. She argued that denying her and her child shelter in a reception centre, in combination with the circumstance that she was not allowed to work in the Netherlands, left her without (adequate) means of subsistence and fully dependent on charity-help.[354] The

---

[349] Central Court of Appeal, 22 December 2008, LJN BG8789.

[350] Central Court of Appeal, 22 December 2008, LJN BG8789, Section 3.2.

[351] E/C.12/NLD/CO/3, 24 November 2006, Section 19.

[352] Central Court of Appeal, 22 December 2008, LJN BG8789, Section 4.2.

[353] Freely translated an abbreviation for: Central Organ for Reception Service for Asylum-seekers.

[354] Remarkably, in another case, the provision of charity aid was held against the claimant: the fact that an asylum-seeker received shelter in a care facility owned and operated by a charity organisation, was one of the arguments based on which the summary trial Court of Haarlem judged that the municipality of Haarlem rightfully rejected an application for housing. See: District Court of Haarlem, 29 July 2008, LJN BE9491 (summary trial procedure). In contrast, in the discussed case ruled by the District Court of Amsterdam, the dependence on charity aid, is taken into account to establish the severe circumstances

case was decided by the District Court in Amsterdam.[355] The Court first rejected direct applicability of Article 11 ICESCR using wordings along the lines quoted above. This did not, however, keep the Court from ruling in favour of the claimant. The applicable policy guidelines required COA to take account of distressing humanitarian circumstances. In the light of the circumstances of the case, among which the fact that the claimant did not have sufficient means of subsistence – including food – to provide for herself and her child, the reasoning given by COA did not convince the Court that COA had sufficiently taken account of distressing humanitarian circumstances. The District Court of Amsterdam quashed the COA's decision, and ruled that it had to reassess the case. As it seems, while the direct applicability of Article 11 ICESCR has been denied, and not considered in the verdict, the absence of an adequate standard of living was clearly a reason for the Court to rule in favour of the applicant, while this argument was not explicitly substantiated with a legal Provision of any kind.

4.3.5 Further developments in the denial of direct applicability

On the 3rd of March 2008, then Dutch Minister of Foreign Affairs Maxime Verhagen held a speech at the Human Rights Council.[356] Verhagen proudly announced *'here today that the Netherlands will join the group of countries who have recognised the right to water as a human right.'* Shortly after this proud announcement and explicitly referring to it, the local Court of Heerlen recognised direct applicability of Articles 11 and 12 of the ICESCR. *In casu*, a water provider (practically having a monopoly in the province) shut down the water supply to a customer, whose payments were overdue. The Court ruled that in doing so the water provider violated the customer's right to water and health as codified in Articles 11 and 12 of the ICESCR. Article 11 ICESCR was thus directly applied. This was done on the initiative of the Court itself (no Article 11 ICESCR claim was made by the customer).[357] However, the water provider appealed against the judgment, and the Court of Appeal overruled the ruling of the local Court of Heerlen on 2 February 2010, based on arguments mainly concerning national legislation.[358] With regard to Articles 11 and 12, the Court of Appeal considered that in line with the reasoning of the Committee on Economic, Social and Cultural rights in its General Comment 15, the right to water did not automatically mean a right to free water, but rather

---

the claimant is facing (District Court of Amsterdam, 13 March 2001, LJN AB0942). In yet another ruling, the dependence on charity aid was considered to be of no influence on existing lawful claims on social entitlements. See: District Court of Haarlem, LJN BC6101 (Article 27-ICRC case).

[355] District Court of Amsterdam, 13 March 2001, LJN AB0942.

[356] See: http://www.rijksoverheid.nl/regering/documenten-en-publicaties/toespraken/2010/02/10/statement-by-maxime-verhagen-at-the-7th-session-of-the-human-rights-council-geneva-3-march-2008.html.

[357] District Court of Maastricht (Section Heerlen), 25 June 2008, LJN BD5759.

[358] Court of Appeal, 's-Hertogenbosch, 2 February 2010, LJN BL6583.

the right to affordable water.[359] A remarkable consideration, since it was obvious that the consumer could not afford the water. The Court ruled furthermore that no successful claim could be based on the invoked ICESCR Articles and that therefore there was no need to determine whether those Articles were directly applicable.[360] Since the ruling of the Court of Appeal, the Dutch Courts have not shown any sign of changing their minds, and keep denying direct applicability of the right to water, and most other ICESCR Provisions.[361]

4.4 Article 24 ICRC

In general, it must be concluded that Article 24 has no direct effect.[362] Some Courts decided not to take the invoked Article into consideration,[363]or ruled that the Provision was clearly not violated.[364] Occasionally, the Article is used as an interpretative standard.[365] In Dutch case law, this standard seems to play hardly any role of significance.

---

[359] The Committee on Economic, Social and Cultural Rights stated in General Comment 15 that '*water, and water facilities must be affordable for all. The direct and indirect costs and charges associated with securing water must be affordable, and must not compromise or threaten the Covenant rights.*' See: E/C.12/2002/11, 20 January 2003, CESCR, *General Comment 15, The Right to Water*, Section 12 (c)(ii).

[360] Court of Appeal, 's-Hertogenbosch, 2 February 2010, LJN BL6583, Section 4.10-4.11.

[361] Central Court of Appeal, 1 October 2008, LJN BF4589; 22 December 2008, LJN BG8776; 22 December 2008, LJN BG8789; 21 May 2009, LJN BI8400; 11 June 2009, LJN BI9325; 30 March 2010, USZ 2010, 166; 14 April 2010, LJN BM3583; 19 April 2010, LJN BM0956; 11 May 2010, LJN BM6748; 28 September 2010, LJN BN9571; 2 November 2010, LJN BO3025; 14 March 2011, NJB 2011, 755; Council of State, 29 June 2011, AB 2011, 327; District Court of Amsterdam, 3 December 2008, LJN BG7017; 12 December 2008, LJN BG6963; 12 December 2008, LJN BG6965; 2 June 2009, LJN BJ3914; 3 May 2011, LJN BQ9532; District Court of Haarlem, 6 May 2009, LJN BI3326 (summary trial procedure); 14 June 2010, LJN BM9368 (summary trial procedure); 29 March 2012, LJN BW2431; District Court of Leeuwarden, 05 July 2010, LJN BN0391 (summary trial procedure).

[362] See for instance: Council of State, 12 April 2007, LJN BA3394; Central Court of Appeal, 23 May 2012, www.rechtspraak.nl, 30 May 2012; District Court of Alkmaar, 20 July 2005, LJN AT9598; District Court of 's-Gravenhage, 24 July 2008, LJN BF0906; 2 March 2010, LJN BM2383; District Court of Zwolle-Lelystad, 19 April 2011, LJN BQ3967; 9 June 2011, LJN BR3569 (summary trial procedure).

[363] See for instance: Council of State, 26 February 2003, JV 2003, 164; 27 April 2007, LJN BA4654; District Court of 's-Gravenhage, 25 March 2004, LJN AO6655; District Court of Groningen, 14 July 2010, LJN BN2935 (summary trial procedure).

[364] See for instance: District Court of Haarlem, 29 August 2007, LJN BB3043 (summary trial procedure).

[365] See for instance: District Court of Amsterdam, 19 December 2005, AWB 04/19508.

4.5 Article 26 ICRC

In general, it must be concluded that due to the Dutch reservation to Article 26 ICRC,[366] minors have no direct entitlement to social security.[367] Some Courts considered instead that, due to its general wording, Article 26 ICRC has no direct effect.[368] Other Courts ruled that the Provision was not violated,[369] or the invocation inadequately substantiated.[370]

4.6 Article 27 ICRC

Where case law on Article 11 ICESCR could not be clearer and more consistent on the rejection of direct effect, a very different approach can be found on the effect of Article 27 ICRC.[371] The case mentioned in Section 4.3.4 revealed that the needs of children present a special circumstance that justifies a more kind approach, compared to the standard legal practice. While in some cases in which Article 27 ICRC was invoked, the Courts chose to not further include the Provision in their

---

[366] The Dutch reservation to Article 26 ICRC: '*The Kingdom of the Netherlands accepts the Provisions of Article 26 of the Convention with the reservation that these Provisions shall not imply an independent entitlement of children to social security, including social insurance.*' See: http://treaties.un.org.

[367] See for instance: Central Court of Appeal, 29 March 2005, LJN AT3468; 8 April 2005, LJN AT4112; 24 January 2006, JB 2006, 66; 7 April 2008, LJN BD0221; 10 July 2008, LJN BD8630; 23 July 2010, LJN BN2492; 5 August 2011, LJN BR4268; Council of State, 13 June 2007, LJN BA7088. See also: District Court Zwolle-Lelystad, 21 April 2011, LJN BQ9140; District Court of Haarlem, 17 May 2005, LJN AT6534 (summary trial procedure); District Court of Dordrecht 23 April 2009, LJN BI8643 (summary trial procedure); 27 May 2011, LJN BR5744 District Court of Leeuwarden 28 November 2005, LJN AU7449; District Court of Zwolle-Lelystad, 21 April 2011, LJN BQ9140; District Court of Arnhem, 13 November 2012, www.rechtspraak.nl, 22 November 2012.

[368] District Court of 's-Gravenhage, 6 September 2000, JV 2000, 224, 2 October 2006, LJN AY9546 (summary trial procedure); Council of State, 13 June 2007, LJN BA7088; 9 April 2008, LJN BC9087; District Court of Zutphen, 12 December 2008, LJN BJ1349. See also: Supreme Court, 23 November 2012, JV 2013, 115, in which the Court ruled that *inter alia* from Article 26 ICRC no individual right to child allowances could be distilled.

[369] Central Court of Appeal, 21 June 2011, LJN BR0385.

[370] District Court of Amsterdam, 18 February 2011, LJN BQ5256.

[371] It must be noted that in writing this Chapter, although it overlapped for the major part with our own data, we thankfully made use of the data collected by J.H. de Graaf, M.M.C. Limbeek, N.N. Bahadur and N. van der Mey, who analysed all Court rulings on the direct applicability of ICRC Provisions in the period 2002-2011. See: J.H. de Graaf,. M.M.C. Limbeek, N.N. Bahadur and N. van der Mey, *De toepassing van het Internationaal Verdrag inzake de Rechten van het Kind in de Nederlandse Rechtspraak*, Nijmegen: Ars Aequi Libri, 2012.

ruling,[372] considered that the invocation was not adequately substantiated,[373] or obviously not violated,[374] there had been uncertainty concerning the direct applicability of this Provision,[375] since the Dutch Government had not mentioned this Provision in their list of ICRC Provisions they considered directly applicable in its explanatory memorandum.[376] Also, the question had not explicitly been dealt with by the Central Court of Appeal.[377] Therefore, there was little coherence in case law amongst the lower Courts. For instance, the District Court of 's-Gravenhage ruled in a case in 2000 that *inter alia* Article 11 ICESCR *and* 27 ICRC were not directly applicable, due to the general wording of both Articles. The Court considered that the Provisions could not be applied without further national legislation, and that it was not the competence of the Court to further specify the invoked international Provisions. In this case thus, direct applicability of Article 11 ICESCR and 27 ICRC was denied using the same consideration.[378] The District Court of Almelo however, had ruled that Article 27 had direct effect, and could be invoked on behalf of an illegally residing child, resulting in the granting of social benefits to cover the costs of maintenance for the child.[379] Some District Courts did not explicitly rule on the direct applicability of Article 27 ICRC, but considered that it stipulated an

---

[372] Council of State, 11 June 2012, www.rechtspraak.nl, 18 June 2012; District Court of Zwolle-Lelystad, 29 April 2009, LJN BJ5171; 3 May 2011, LJN BQ5114; District Court of Haarlem, 23 November 2006, LJN AZ4222 (summary trial procedure); District Court of 's-Gravenhage, 08 November 2007, LJN BB8838 (although Article 27 ICRC was not explicitly invoked, the claimant stated that a child needs to eat, in view of the ICRC. The Court considered that Article 2 ICRC was not directly applicable, and did not further address the issue of food); Central Court of Appeal 20 October 2010, JLN BO3581; 26 April 2011, LJN BQ3795; 13 March 2012, USZ 2012, 101. The fact that the District Court of Amsterdam did not motivate a rejection of certain child allowances while the applicant had invoked the first protocol to the ECHR and Article 27 ICRC, was reason for the Central Court of Appeal to quash the verdict, and redirect the case to the District Court for a new ruling. See: Central Court of Appeal, 6 September 2007, LJN BB6188.

[373] Supreme Court, 3 June 2005, LJN AT3445; District Court of Haarlem, 21 June 2007, LJN BA9025 (summary trial procedure); 11 July 2007, LJN BB0998 (summary trial procedure); 18 February 2011, LJN BQ5256; Court of Appeal, 's-Gravenhage, 16 August 2011, LJN BR6656; District Court of Dordrecht, 23 April 2009, LJN BI8643 (summary trial procedure); Central Court of Appeal, 18 June 2009, LJN NI9928.

[374] District Court of 's-Gravenhage, 1 August 2002, AWB 02/54360 and 02/54358; District Court of 's-Gravenhage, 1 August 2002, AWB 02/54362 and 02/54361; District Court of 's-Gravenhage, 21 March 2008, AWB 07/28996; District Court of Haarlem, 29 August 2007, LJN BB3043 (summary trial procedure); Central Court of Appeal 25 August 2005, LJN AU1850; 21 June 2011, LJN BR0385.

[375] See: C.H. Slingenberg, *Illegale kinderen en recht op bijstand in het licht van het IVK*, in: Migratierecht 2006-02, p. 54-57.

[376] Parlementaire Geschiedenis, II 1992-1993, 22855 (R1451), no. 3, Chapter I, Section 6A. Also in Dutch literature, this 'list' of the Government was underlined. See for instance: G.C.A.M. Ruitenberg, *Het internationaal kinderrechtenverdrag in de Nederlandse rechtspraak*, IVRK-reeks 1, Amsterdam: Uitgeverij SWP, 2003, p.35-37; Margrite Kalverboer and Elianne Zijlstra, *Kinderen uit asielzoekersgezinnen en het recht op ontwikkeling, het belang van het kind in het Vreemdelingenrecht*, Amsterdam: Uitgeverij SWP, 2006 p.16. See in this light also District Court of 's-Gravenhage, 14 August 2003, LJN AM3133, also referring to this list in deciding on the direct effect of Article 3 ICRC. *In casu*, the Court decided that Article 3 should be taken into serious consideration as an interpretative standard.

[377] Central Court of Appeal, 24 January 2006, LJN AV0197.

[378] District Court of 's-Gravenhage, 6 September 2000, JV 2000, 224.

[379] District Court of Almelo, 28 November 2005, LJN AU7003.

important standard that needed to be taken into account seriously, and therefore used the Provision as an interpretative standard in their rulings.[380]

Since 2005, the Central Court of Appeal was occasionally confronted with appeals on Article 16 (1) of the Work and Social Assistance Act, stipulating that: '*to a person not entitled to assistance, the Mayor and Municipal Executive may, taking into consideration all circumstances, notwithstanding this Section, provide assistance if so required due to very urgent reasons.*'[381] The Court ruled that *inter alia* Article 27 ICRC needs to be taken into account in assessing whether or not there was a very urgent reason *in casu* in view of this Provision. Mostly, the Court considered a situation to be urgent when it was obvious that there were inadequate resources to cover the costs of maintenance of the child. First, this was ruled in cases concerning Dutch children,[382] and later also regarding legally residing foreign children.[383] Regarding the latter category, it is interesting to notice that on 8 August 2005, the Court came to this conclusion after considering '*the text of Articles 2, first and second Section, 3, first and second Section, and 27, third Section, of the ICRC and the comments on these Articles of the Committee on the Rights of the Child. The in the treaty stipulated English wording "without discrimination of any kind, irrespective of the child's or his or her parents' or legal guardians'…status" in Article 2, first Section, of the Convention, in conjunction with the other aforementioned Provisions, indicate that the Linkage principle does not constitute a decent justification to completely exclude the possibility of granting benefits solely on behalf of the minor children in a situation in which their parents who are not entitled to social benefits but request for such assistance are unable to cover the costs of food, clothing and other essential costs, necessary for the minor children.*'[384] Similar reasoning can be found in the case of

---

[380] For instance: District Court of 's-Gravenhage, 29 August 2002, LJN AF2534; District Court of Zwolle, 17 February 2003, LJN AF4890, and 19 March 2003, LJN AF6351 and AF6354.

[381] Original text in Dutch: '*Aan een persoon die geen recht op bijstand heeft, kan het college, gelet op alle omstandigheden, in afwijking van deze paragraaf, bijstand verlenen indien zeer dringende redenen daartoe noodzaken.*'

[382] Central Court of Appeal, 29 March 2005, LJN AT3468 (summary trial procedure); 14 June 2005, LJN AT8038; 5 July 2005, LJN AT9963; 13 February 2007, LJN AZ8596; 29 May 2007, LJN BA6523; 2 May 2012, USZ 2012, 158 (although the social benefits were rejected *in casu*). See for the lower Courts: District Court of Dordrecht, 25 September 2008, LJN BG3517 (summary trial procedure); 21 December 2009, LJN BL9388 (summary trial procedure); District Court of 's-Gravenhage, 23 December 2009, LJN BL2473.

[383] Central Court of Appeal, 8 August 2005, LJN AU0687; 24 January 2006, LJN AV0197; 6 October 2009, LJN BK0734; 20 July 2010, LJN BN3318. See for the lower Courts for instance: District Court of Amsterdam, 8 November 2007, LJN BF 1926; 9 February 2010, LJN BL6113; District Court 's-Hertogenbosch, 22 January 2008, LJN BC4003; District Court of Haarlem, 28 February 2008, LJN BC6101; District Court of 's-Gravenhage, 20 April 2011, LJN BQ3199 (summary trial procedure).

[384] Central Court of Appeal, 8 August 2005, LJN AU0687 (summary trial procedure). Original text in Dutch: '*De tekst van de artikelen 2, eerste en tweede lid, 3, eerste en tweede lid, en 27, derde lid, van het IVRK en op de op deze artikelen verschenen commentaren van het Comité voor de rechten van het kind. De in de Engelse verdragstekst voorkomende woorden "without discrimination of any kind, irrespective of the child's or his or her parent's or legal guardians' (…) status" in artikel 2, eerste lid, van het IVRK, bezien in samenhang met de andere zojuist genoemde bepalingen, wijzen er op dat het koppelingsbeginsel geen voldoende rechtvaardiging kan vormen voor het geheel uitsluiten van de mogelijkheid om uitsluitend ten behoeve van de*

24 January 2006, in which the Court ruled that despite the fact that according to Article 27 (2) ICRC the parents or others responsible for the child have the primary responsibility to secure the conditions of living necessary for the child's development, all Government institutions should act in accordance with Article 3 jo 27 (3) ICRC. This means that when decisions must be made that somehow affect the conditions of living of the child – *in casu* the rejection of a claim to social benefits – the best interests of the child should be the first consideration for these institutions.[385] However, Article 16 (2) of the Work and Social Assistance Act excludes illegally residing aliens from the exception to provide for social benefits in case of urgent circumstances. The Central Court of Appeal considered that also in case of illegally residing children, this exclusion was a proportionate means to achieve the goal as established in the 'Linkage-Act' to restrict immigration, and merely ruled that Article 27 had no direct effect.[386] Only occasionally, the Council seems to consider whether the illegaly residing foreigner was in the absolute impossibility to leave Dutch territory, suggesting that this might be an exception to the general exclusion of social benefits.[387]

The view of the Central Court of Appeal to use ICRC Provisions as an interpretative standard was understood by the Minister of Justice as a recognition of the direct applicability of the entire ICRC. This prompted the Minister to broaden the scope of persons entitled to certain social benefits: also children without a residence permit, but legally remaining in the Netherlands had entitlements to income support.[388] In late 2007, the District Court of Rotterdam was led by this assumption of the Minister. In this case, after rejecting direct applicability of Article 11 ICESCR, the Court considered that '*after all, Article 27, third Section, of the ICRC is a directly effective Provision.*'[389] The District Court of Rotterdam ruled that in a situation in which the parents of a child, who are awaiting a decision on their application for a residence permit, and who are receiving a monthly allowance from the

---

minderjarige kinderen bijstand te verlenen in een situatie dat hun om deze bijstand vragende niet-rechthebbende ouders zelf niet in staat zijn de kosten van voeding, kleding en andere essentiële, voor de minderjarige kinderen noodzakelijke kosten te betalen.'

[385] Central Court of Appeal, 24 January 2006, LJN AV0197. See also the annotation to this verdict of C.H. Slingenberg, *Illegale kinderen en recht op bijstand in het licht van het IVK*, in: Migratierecht 2006-02, p. 54-57; see also the annotation of G. Vonk, *Ongewenste kinderen, opmerkingen naar aanleiding van CRvB 24 januari 2006*, Sociaal Maandblad Arbeid, 2006, 131-134.

[386] See for instance: Central Court of Appeal, 9 October 2006, LJN AY9940; 7 April 2008, LJN BD0221; 14 July 2010, LJN BN1274. See in this light also: Central Court of Appeal, 22 December 2008, LJN BG8776 (although Article 27 was not evaluated by the Court, it denied direct effect of *inter alia* Article 11 ICESCR).

[387] See for instance, Central Court of Appeal, 9 October 2006, LJN AY9940.

[388] Adaption on the regulation on the provision of certain categories of aliens; Regulation of the Minister of Justice of 22 December 2006 no. 5458886/06/DVB. Original text in Dutch: Wijziging Regeling verstrekking bepaalde categorieën vreemdelingen; Regeling van de Minister van Justitie van 22 December 2006, no. 5458886/06/DVB. See also: District Court of Haarlem, 28-02-2008, LJN BC6101 and District Court of 's-Gravenhage, 10-04-2008, LJN BC9445. See in particular Article 4 (e).

[389] Original text in Dutch: '*Immers, artikel 27, derde lid, van het IVRK is een rechtstreeks werkende bepaling.*'

Government, can prove that it does not have sufficient money to provide for himself (*in casu* food and housing), the amount of the monthly allowance should be reviewed.[390]

Some Courts did follow the lead of the Central Court of Appeal's ruling, using Article 27 (3) ICRC (sometimes combined with Article 2 or 3 ICRC) as an interpretative standard.[391] In this light, an interesting development must be noted, starting with a verdict of the Court of Appeal of 's-Gravenhage in a summary trial procedure, where three illegally residing children invoked Articles 2, 3, 27, and 37 ICRC, when they were removed from the facilities of a COA. The Court considered *'that the State, as a result of the ratification of the treaties to which these Provisions are part, have the legal duty, to the extent that such Provisions have direct effect, to respect this working, and, to the extent those Provisions only imply instructional standards, through regulations, administrative decisions and measures, and actions, to create a legal and factual situation in which the rights and interests of children who are in the territory of the State are protected and safeguarded in accordance with these Provisions'*[392] Therefore, the COA was not allowed to remove the minors from their facilities. This was confirmed by the Supreme Court in cassation, although the Court added that, especially due to the fact that the mother of the minors did not want to cooperate in leaving the territory, this would not constitute a right to social support that is exactly similar compared to the benefits received before.[393] An interesting verdict, for here also the Supreme Court seems to accept that the invoked ICRC Articles need to be taken into consideration as an interpretative standard. However, in another case, the Court considered that Articles 2, 3, 26,

---

[390] District Court of Rotterdam, 19 September 2007 LJN BB5715.

[391] For instance: District Court of 's-Gravenhage, 22 May 2006, LJN AX4451; 8 November 2007, LJN BB9819; 7 December 2007, LJN BC2933; 10 April 2008, LJN BC9445; 23 December 2009, LJN BL2473; 20 April 2011, LJN BQ3199; Court of Appeal, 's-Gravenhage, 27 July 2010, LJN BN2164; District Court of Amsterdam, 19 April 2007, LJN BA6900; 8 November 2007, LJN BF1926; 9 February 2010, LJN BL6113; District Court of Haarlem, 28 February 2008, LJN BC6101; District Court of Dordrecht, 25 September 2008, LJN BG3517; 6 October 2009, LJN BK0734; 21 December 2009, LJN BL9388; District Court of 's-Hertogenbosch, 22 January 2008, LJN BC4003; District Court of Assen, 10 April 2012, www.rechtspraak.nl, 25 May 2012. In District Court of Haarlem, 21 November 2008, LJN BG6130, the Court considered that Article 27 ICRC was, among other things, adequately taken into account when reducing certain social benefits to a family as a penalty for the refusal to work.

[392] Court of Appeal, 's-Gravenhage, 27 July 2010, LJN BN2164 (summary trial procedure), consideration 3.6. Original text in Dutch: *'dat op de Staat, als gevolg van de ratificatie van de verdragen waarvan deze bepalingen deel uitmaken, de rechtsplicht rust om, voor zover deze bepalingen rechtstreekse werking hebben, die werking te eerbiedigen, alsook om, voor zover deze bepalingen slechts instructienormen bevatten, door middel van regelgeving, bestuurlijke beslissingen en maatregelen, en door feitelijke handelingen een zodanige juridische en feitelijke toestand te creëren dat de rechten en belangen van kinderen die zich op het grondgebied van de Staat bevinden overeenkomstig deze bepalingen worden beschermd en geborgd.'*

[393] Supreme Court, 21 September 2012, NJ 2013, 22. M.M.C. Limbeek understands that in another case, the Supreme Court, on 12 February 2010, LJN BI9729, indirectly rejected the direct effect of *inter alia* Article 27 ICRC by deciding that all other legal remedies could not lead to cassation. However, this verdict rather seems to resort under the category of rulings in which a Court simply does not decide on the matter.

and 27 would not lead to any individual rights to child allowances in case of aliens awaiting a decision on their application for a residence permit.[394]

Other Courts however followed a different path,[395] inspired by the permanent case law of the Council of State, Section administrative law, who consequently rejected direct applicability of Article 27 ICRC. The Council usually offers not much further clarification than that due to the nature and wording of the Article direct applicability was not possible.[396] Remarkably, in some cases the Central Court of Appeal seems to come back from their previous approach, and merely considered that Article 27 ICRC had no direct effect without using the Provision as an interpretative standard. The cases concerned besides the above mentioned denial of social benefits to illegally residing children,[397] disputes on the amount of social benefits granted,[398] the refusal to grant child allowances,[399] the withdrawal of social benefits due to work in illegality,[400] the refusal to grant orphan's benefits,[401] and the export of child allowances in case of a child living abroad.[402] In this light, a remarkable consideration can be found in a case of 23 July 2010,[403] in which the Central Court first ruled that Article 27 ICRC had no direct effect. Furthermore, on request of the plaintiff, the Court considered that the earlier case of 24 January 2006,[404] in which Article 27 ICRC was used as an interpretative standard, concerned an Act that regulated certain social benefits which would provide for a basic minimum. *In casu* however, the case concerned an Act that provided for general social benefits, and was therefore not comparable with the case in 2006, and thus the principles of that case could not be applied here. This case reflects a certain

---

[394] Supreme Court, 23 November 2012, JV 2013, 115.

[395] District Court of Zwolle/Lelystad, 7 June 2006, LJN AY8861; 16 April 2009, LJNBI1369; 21 April 2011, LJN BQ9140; District Court of 's-Gravenhage, 2 October 2006, LJN AY9546; 22 August 2007, LJN BC0745; 20 December 2007, LJN BC1047; 24 January 2008, LJN BC2955; 25 September 2009, LJN BK7090; 7 October 2009, LJN BK3052; 2 March 2010, LJN BM2383; District Court of Amsterdam, 23 April 2009, LJN BJ1021; 30 January 2009, LJN BH4457 (summary trial procedure); 1 June 2012, www.rechtspraak.nl, 15 June 2012; District Court of Arnhem, 29 March 2012, www.rechtspraak.nl, 13 April 2012; 13 November 2012, www.rechtspraak.nl, 22 November 2012.

[396] See for instance: Council of State, 1 March 2005, JV 2005/176; 13 September 2005, JV 2005, 409; 15 February 2007, LJN AZ9524; 13 June 2007, www.rechtspraak.nl, 13 June 2007; 26 November 2007, www.rechtspraak.nl, 2 January 2008; 08 October 2010, LJN BO0685; 13 October 2010, LJN BO0794; 22 February 2012, JV 2012, 200.

[397] Central Court of Appeal, 9 October 2006, LJN AY9940; 7 April 2008, LJN BD0221; 14 July 2010, LJN BN1274.

[398] Central Court of Appeal, 12 June 2007, LJN BA7026; 26 January 2010, LJN BL1686; 30 March 2010, USZ 2010, 166.

[399] Central Court of Appeal, 10 July 2008, LJN BD8630. Although Article 27 was not invoked, see in this light also: Central Court of Appeal, 15 July 2011, JV 2011, 393.

[400] Central Court of Appeal, 2 November 2010, LJN BO3025.

[401] Central Court of Appeal, 13 April 2012, USZ 2012, 161.

[402] Central Court of Appeal, 5 August 2011, LJN BR4248; LJN BR4268 and LJN BR4785.

[403] Central Court of Appeal, 23 July 2010, LJN BN2492. See for a similar reasoning: District Court of Zwolle-Lelystad, 21 April 2011, LJN BQ9140.

[404] Central Court of Appeal, 24 January 2006, LJN AV0197.

---

unspoken basic principle that seems to emerge in Dutch case law, that foreign children should have access to at least their basic needs, mostly through social benefits granted to their parents or legal guardian. However, this principle seems not to apply to illegally residing children as can be demonstrated by the case law of the Central Court of Appeal. Furthermore, considering the above case law, it is difficult to assess whether this principle is uniformly adopted in a coherent way, for there are different approaches amongst the Courts regarding the granting of social benefits that are applied in a seemingly random way. A clear indication on what factors could be decisive in choosing to deny direct applicability of Article 27 ICRC, or (in)directly allowing it to have some weight in the verdicts cannot be distinguished. The Judiciary seems to muster a casuistic approach, which leads to rather contradicting verdicts.

4.7 Article 12 CEDAW

No cases could be found in which the direct applicability of Article 12 CEDAW was considered. Regarding the CEDAW, the Courts did not yet frequently rule on direct applicability of the CEDAW rights. Only recently, several Courts, including the Supreme Court, granted direct effect to Article 7 CEDAW (on the discrimination against women in the political and public life).[405] Also, one example was found in which the Supreme Court reviewed against Articles 1 and 2 CEDAW, concerning non-discrimination in general, albeit amongst many other international Provisions.[406] Both the Supreme Court and the Central Court of Appeal rejected the possibility of direct effect of Article 11 (2) CEDAW (discrimination against women on the grounds of marriage or maternity).[407] It is noteworthy, that the Supreme Court considered that it does not appear from the Explanatory Memorandum on the ratification Bill of the CEDAW whether or not Article 11 CEDAW was not directly applicable. No verdicts in which economic, social and cultural rights as embedded in the CADEW were explicitly invoked could be found.

4.8 Concluding observations

At this point, some observations can be made. Firstly, it seems obvious that in case of the right to food, or more generally in case of economic, social and cultural rights, the role of the Legislature should not be underestimated. The Central Court of Appeal, as a Court of Last Instance, more than once referred to the Parliamentary History on the ratification Bill on the ICESCR, when rejecting

---

[405] Court of Appeal, 's-Gravenhage, 7 September 2005, NJ 2005, 473; Court of Appeal, 's-Gravenhage, 20 December 2007, NJ 2008, 133; Council of State, 5 December 2007, AB 2008, 35; Supreme Court, 9 April 2010, LJN BK4549; all concerning cases against the State regarding the position of the Dutch Political Reformed Party (SGP) on women.

[406] Court of Appeal, 's-Gravenhage, 30 October 2007, NJF 2007, 532.

[407] Supreme Court, 30 May 1986, LJN AC9402; 1 April 2011, JB 2011, 115; 21 September 2012, NJ 2013, 22; Central Court of Appeal, 30 December 2011, USZ 2012, 58.

direct applicability of Article 11 ICESCR, and the statement of a Minister of Foreign Affairs in the international arena led to a remarkable verdict of one of the lower Courts, accepting direct applicability of the right to water. Apparently thus, the opinion of the Legislature seems to be highly valued. Also, but to a lesser extent, the list of the Government included in the Explanatory Memorandum to the ratification Bill of the ICRC seems to play a role in the assessment of direct effect of ICRC Provisions. This will be discussed in-depth in Chapters 6 and 7.

Secondly, the Constitutional structure of Articles 93 and 94 CA seems to play a role of significance. In deciding on right to food cases, often a reference is made to these Constitutional Provisions, stating that the invoked international standard does not match the criterion of being binding on all persons. However, in determining when a Provision is binding on all persons, it seems that the assessment on direct applicability is not based on the same criteria each time. Especially regarding the effect of Article 27 ICRC, it is completely unclear what the criteria are. It seems thus that the Courts are struggling with the Constitutional Structure as embedded in Articles 93 and 94 CA. This matter is further discussed in Chapter 5.

Thirdly, it appears that the Judiciary is under circumstances willing to take a somewhat milder approach when Article 27 ICRC is invoked, compared to Article 11 ICESCR. While direct effect is denied, Article 27, alongside with other ECOSOC rights enshrined in the Treaty, is used as an interpretative standard by the Courts, in a balancing of interests between the urgent situation of the individual and the Linkage principle. In these cases, the international Provisions invoked seem to have a significant effect on the verdict, on top of the domestic legislation. However, this milder approach does not apply to illegally residing children.

Fourthly, it appears from one case that even when formally the direct applicability of the international right to adequate food is rejected, substantially the right may still play a role in the final decision of the Court, although it is not necessarily portrayed as a right or international standard. However, this makes it an obscure non-formal legal standard that is impossible to invoke by a claimant, and can only be voluntarily applied by a Court.

Fifthly, in general it can be concluded that in cases in which international Provisions stipulating the right to food are invoked, the Courts prefer to solve the matter by applying domestic law, and are hesitant in considering legal opinions expressed in the international arena on the application of international standards.

5. Dutch monism and the Constitutional reforms of Article 93 and 94 CA

5.1 Introduction

As concluded in Chapter 4, the Dutch Judiciary seems to struggle with the interpretation of the system that is stipulated in Articles 93 and 94 of the Constitutional Act (CA), in relation to the effect of the internationally embedded right to food in the domestic legal order. It appears that the Parliamentary History of the ratification Acts of those treaties is an important source for the Courts to reach a verdict. As discussed in Chapter 2, in terms of legal reasoning, this could be referred to as the use of intentional support approach, or a subjective teleological method.[408] The constructive move will therefore be naturally leading us to perform a full analysis of this Constitutional system in this Chapter, as well as an analysis of the Parliamentary History of treaties that relate to the right to food in Chapter 6.

In this Chapter I will first study the relevant literature on this matter. It will appear that the Dutch Constitutional system in which the applicability of international standards is regulated can be characterised as a qualified monistic system. However, the exact meaning of this is unclear (Section 5.2), and therefore has led to a rather technical discussion amongst Dutch legal scholars (Section 5.3). This discussion results into several interesting questions with regard to the functioning of this system (Section 5.4). The leading authors on Constitutional law refer unanimously to the Parliamentary History of three Constitutional reforms in defending their position regarding the meaning of Articles 93 and 94 CA. Therefore, after shortly introducing the procedures in Parliament regarding the adoption of law (Section 5.5), I will analyse the Parliamentary Documents on those reforms in Sections 5.6-5.8. I will then as a conclusion discuss what was originally foreseen by the Constitutional Legislator when the system of Articles 93 and 94 Constitutional Act was adopted (Section 5.9).

5.2 Qualified monism

In Constitutional legal studies, one traditionally distinguishes between two different Constitutional concepts through which an international standard can have effect in the domestic legal order: monism and dualism. In monistic Constitutions, the international standard is automatically applicable in the domestic legal order, without the need of a domestic transformation of the international standard. In dualistic Constitutions, a transformation of the international standard into a

---

[408] P. Wahlgren, *Legal reasoning, a jurisprudential model*, Scandinavian Studies in Law, volume 40, 2000, pp. 199-282.

domestic standard is required before the international standard will have any effect in the domestic legal system.

However, it seems unlikely that a Constitutional system will ever be fully monistic or dualistic, and it would do more justice to legal reality to consider monism and dualism as two extremes on one axis. Although it is unanimously accepted that the Netherlands have a monistic Constitution, the monism is often considered to be limited, or moderate. The Dutch Government used the phrase 'qualified monism' in its first Universal Periodic Report.[409] The exact meaning of this Dutch monism however, is not clear and greatly depends on the interpretation given to Articles 93 and 94 of the Dutch Constitutional Act.

In Article 93 Dutch Constitutional Act is elaborated that: *'Provisions of treaties and of resolutions by international institutions which may be binding on all persons by virtue of their contents shall become binding after they have been published.'*[410]

In Article 94 Dutch Constitutional Act is stated that: *'Statutory regulations in force within the Kingdom shall not be applicable if such application is in conflict with Provisions of treaties or of resolutions by international institutions that are binding on all persons.'*[411]

The debate on the exact meaning of the system laid down in both Articles almost exclusively focuses on the meaning of the phrase 'binding on all persons'. This phrase is often considered to be a limitation to monism, and thus may have some serious consequences for the application of international treaty Provisions in the domestic legal order. Also, as a consequence, the relationship between the two Articles (containing both the same disputed phrase) is under dispute.

5.3 Dutch scholars on the meaning of Articles 93 and 94 Constitutional Act

Amongst Dutch scholars there is certainly no agreement on the exact meaning of the system laid down in Articles 93 and 94 Constitutional Act. In this section, an analysis is made of the interpretation of the foremost authors on the matter.

---

[409] A/HRC/WG.6/1/NLD/1, 7 March 2008, The Netherlands, first Universal Periodic Report, Sections 17-19.

[410] Original text in Dutch: *'Bepalingen van verdragen en besluiten van volkenrechtelijke organisaties, die naar haar inhoud een ieder kunnen verbinden, hebben verbindende kracht nadat ze zijn bekendgemaakt.'* Translation of the Dutch text quoted from: http://www.denederlandsegrondwet.nl, a website related to the Dutch Ministry of the Interior and Kingdom Relations.

[411] Original text in Dutch: *'Binnen het Koninkrijk geldende wettelijke voorschriften vinden geen toepassing, indien deze toepassing niet verenigbaar is met een ieder verbindende bepalingen van verdragen en van besluiten van volkenrechtelijke organisaties.'* Translation of the Dutch text quoted from: http://www. denederlandsegrondwet.nl, a website related to the Dutch Ministry of the Interior and Kingdom Relations.

5.3.1 Bellekom, Heringa, Van der Velde and Verhey[412]

In the tenth edition of the 'Compendium van het Staatsrecht', the authors argue that Article 93 Constitutional Act determines what international standards have direct effect in the domestic legal order – those which are binding on all persons – and that those Provisions can be applied by the domestic Courts. Article 94 Constitutional Act then determines basically the hierarchical relation between international standards that are binding on all persons (and thus directly applicable) and national standards: international standards that are binding on all persons are superior to domestic law. This imllies that domestic Courts may review national standards against these superior international standards.

According to the authors, it is solely for the Courts, and not for the Legislature, to decide whether an international standard is binding on all persons. This would follow from the famous railway-strike ruling, where the Supreme Court, contrary to the view expressed by the Legislature, held that the right to strike, as recognised in Article 6 (4) ESC, was directly applicable.[413] The authors argue that the Judiciary mostly uses two approaches in determining whether an international Provision is directly applicable. Firstly, the Court may focus on the question whether the Provision literally recognises a right for the individual. Secondly, the Court may focus more on the content of the Provision. The Court, inspired by the Van Gend en Loos ruling of the European Court of Justice,[414] determines whether the invoked Provision is by its nature capable to bind all persons. In other words: is the Provision specific enough to be applied by a Court without the intervention of the Legislature?[415]

As a consequence, the Judiciary may not review national law against international Provisions that are not binding on all persons. According to the authors of the Compendium, the Constitutional Act remains silent about the (hierarchical) relation between international customary law and domestic standards, but argue that it appears from case law that the Courts are competent to apply international customary law, as long as it does not review national law against these standards.[416]

---

[412] Th.L. Bellekom, A.W. Heringa, J. van der Velde en L.F.M. Verhey, *Compendium van het staatsrecht* (10ᵉ druk), Deventer: Kluwer, 2007.
[413] Th.L. Bellekom, A.W. Heringa, J. van der Velde en L.F.M. Verhey, *Compendium van het staatsrecht* (10ᵉ druk), Deventer: Kluwer, 2007, no. 29. Supreme Court, 30 May 1986, NJ 1986, 668 (railway-strike ruling).
[414] Court of Justice of the European Union, 05-02-1963, 26/62 (Van Gend en Loos Ruling).
[415] Th.L. Bellekom, A.W. Heringa, J. van der Velde en L.F.M. Verhey, *Compendium van het staatsrecht* (10ᵉ druk), Deventer: Kluwer, 2007, no. 29.
[416] The authors refer to the Nyugat ruling, in which the Supreme Court held that it was not competent to review a national standard against international customary law: Supreme Court, 6 March 1959, NJ 1962, 2.

5.3.2 Burkens, Kummeling, Vermeulen and Widdershoven[417]

A slightly different view on the matter is expressed by the authors of 'Beginselen van de democratische rechtstaat'. They argue that it appears from customary Constitutional law that the Netherlands have a monistic system regarding the applicability of international standards. They refer to the 'grenstractaat Aken' ruling, in which the Supreme Court considered that, without reference to a specific (Constitutional) Provision, international law has direct effect in the domestic legal order without the need for transformation into domestic law first.[418] The later adopted Articles 93 and 94 Constitutional Act are therefore not the recognition of a monistic system, but rather a limitation to already existing monism. In the first place, because Article 93 Constitutional Act states that it is required to publish an international standard first before it can be directly applicable. In the second place, because Article 94 Constitutional Act stipulates that the Judiciary (and also the administrative power), may only review national standards against international Provisions that are binding on all persons.

According to the authors, the system embedded in Articles 93 and 94 Constitutional Act only concerns a specific part of international law, and mainly instructs the Judiciary and administrative powers on how to treat that particular part of international law. This part only concerns the international standards that have been published and are binding on all persons. Therefore, international standards that are not binding on all persons have direct effect in the Netherlands, based on the aforementioned customary law, but may not be reviewed against. In this understanding, Article 94 Constitutional Act does not authorise the Judiciary to review against certain international standards, but rather restricts an already existing right in customary law to review against international standards in general to a specific group of international standards. Evidence that the international standards that may be reviewed against is indeed limited was found by the authors in the harmonisation Act ruling[419] in which the Supreme Court held that it was not authorised to review national standards against international standards that were not binding on all persons, and in the Nyugat ruling, in which the Supreme Court ruled that it was not authorised to review against international customary law.

The authors argue that it is for the Courts to decide whether an international standard is binding on all persons, although this decision is not always easy to make. The authors consider that while there might be several approaches for the Courts in determining whether an international standard is binding on all persons, in reality the Courts will only look at the nature and content of the

---

[417] M.C. Burkens, H.R.B.M. Kummeling, B.P. Vermeulen and R.J.G.M. Widdershoven, *Beginselen van de Democratische rechtstaat, inleiding tot de grondslagen van het Nederlandse staats- en bestuursrecht* (5e druk), Deventer: Tjeenk Willink, 2001.

[418] Supreme Court, 3 March 1919, NJ 1919, p. 371 ('Grenstractaat Aken'-ruling).

[419] Supreme Court 14 April 1989, AB 1989, 207/NJ 1989, 469 (harmonisation Act ruling).

invoked standard: '*a rule of international law is binding on all persons if the Provision is that precise and concrete that further elaboration in national law is unnecessary for domestic application*'[420] The authors underline that there is hardly any use for Courts to consider the intentions of the contracting parties regarding the direct applicability of the Provisions, for in most treaties both countries with dualistic and monistic Constitutional systems are involved. This makes it unlikely that matters of direct applicability – something that is hardly relevant for countries with a dualistic Constitutional system – are discussed in the negotiating process preceding the adoption of the treaty, or are mentioned in the text of a treaty. As an example, the authors refer to a famous passage of the railway-strike ruling: '*Whether or not the contracting States intended to grant direct effect to Article 6 ESC is not relevant, whereas neither from the text, nor the history of the formation of the treaty follows that they have agreed that Article 6 ESC is directly applicable. In that situation, according to Dutch law, only the content of the Provision is decisive (...).*'[421] The authors then conclude that the intention of the contracting States is of almost no importance for the Dutch Courts in determining whether a treaty Provision is binding on all persons. Therefore, the Courts are bound to base their decision on the matter solely on the nature and content of the invoked treaty Provision. It is remarkable however, that in another passage, the authors recognise that the Courts quite often are led by the results of the Parliamentary discussion preceding the ratification of the treaty involved on the direct applicability of the international standard. The opinion of the Legislature is not decisive, but is taken into consideration by the Courts.

The basic criterion when determining whether an international standard is directly applicable is thus that a Court must be capable of applying a Provision without the need for intervention of the Legislature, which has to adopt further national legislation to elaborate the international standard. As a rule of thumb, the authors suggest that the Provisions recognising the so-called classic human rights are directly applicable, and the Provisions recognising social human rights are not. Main reason for this is that social rights often imply Government action, including the adoption of further national legislation, which would make direct applicability unlikely. According to the authors, this does not mean that Provisions containing social human rights would have no effect at all in the domestic legal order. The authors argue that Provisions containing social rights might have a 'standstill'-effect, meaning that in case a State causes a situation that deprives people of

---

[420] Original text in Dutch: '*Een bepaling van internationaal recht is een ieder verbindend indien de bepaling dusdanig precies en concreet is dat uitwerking in nationale regelgeving niet nodig is om haar rechtstreeks in het nationale recht te kunnen toepassen.*'

[421] Supreme Court, 30 May 1986, NJ 1986, 668 (railway-strike ruling), consideration 3.2. Original text in Dutch: '*Of verdragsluitende Staten al dan niet hebben beoogd aan artikel 6 ESH directe werking toe te kennen, is niet van belang nu noch uit de tekst, noch uit de geschiedenis van de totstandkoming van het verdrag valt af te leiden dat zij zijn overeengekomen dat aan artikel 6, vierde lid, die werking niet mag worden toegekend. Bij deze stand van zaken is naar Nederlands recht enkel de inhoud van de bepaling beslissend (...).*'

previous benefits that are embedded in a social right, the standstill effect of this right might be reason for a Court to still apply the social human right, even if it would normally be considered to be a Provision that is not directly applicable. Also, the authors refer to Article 7 (a) (i) ICESCR, that despite the fact that it recognises a social human right, was used by the Supreme Court to give a certain interpretation to Article 26 ICCPR.

5.3.3 Van der Pot (adapted by Elzinga and De Lange)[422]

The authors of 'Van der Pot, Handboek van het Nederlandse Staatsrecht' also start from the point of view that the Dutch Constitutional system is of a monistic nature. In their view, evidence of this can be found in the Supreme Court's verdict 'Grenstractaat Aken'[423], an approach that later, after some Parliamentary discussion, was reaffirmed in Article 93 Constitutional Act. During the Constitutional reforms of 1953, it was unclear whether treaty Provisions that were ratified before the adoption of Article 93 Constitutional Act were also superior to national Provisions. Originally, it was the intention of the Government not to decide on this during the Constitutional reforms, but rather to leave the matter to the Judiciary (who until then also did not specifically decide on the matter). Despite this, Article 94 was added to the Constitution by the amendment Serrarens and co[424] to give a more detailed description of the Courts' obligation to review national law against international standards. Consequently, according to the authors, on the one hand the Courts now had an obligation to also review national legislation against international Provisions that were ratified before the adoption of Article 93 Constitutional Act. The Courts are thus obliged to review national legislation against all international standards that are binding on all persons. On the other hand the Supreme Court understood Article 94 Constitutional Act as a prohibition to review national standards against legal standards that would fall outside the scope of Article 94 Constitutional Act, including international customary law and standards that are not binding on all persons. Evidence for this was found in the aforementioned Nyugat ruling.[425]

---

[422] C.W. Van der Pot (adapted by D.J. Elzinga and R. De Lange), *Handboek van het Nederlandse Staatsrecht* (15^e druk), Deventer: Kluwer, 2006.

[423] Supreme Court, 3 March 1919, NJ 1919, p. 371 ('Grenstractaat Aken' ruling).

[424] Parliamentary Documents, II 1951-1952, 2374, no. 32.

[425] Supreme Court, 6 March 1959, NJ 1962, 2. The authors quoted: '*dat blijkens de geschiedenis van artikel 66 met dit artikel bepaaldelijk bedoeld is den strijd te beslechten over de vraag, in hoeverre de Nederlandse rechter het Nederlandse recht op strijd met het internationale recht mag toetsen, en deze toetsing uitdrukkelijk heeft willen beperken tot de gevallen van de in dat artikel vermelde zelfwerkende bepalingen van overeenkomsten.*' Freely translated: '*that, according to the history of Article 66, it was intended to definitely settle the question to what extent the Dutch Court can review Dutch law against international law by this Article, and this review has explicitly been limited to the cases of Articles that are self-operating Provisions.*'

The authors are of the opinion, following from the railway-strike ruling,[426] that the Courts determine whether an international Provision is binding on all persons purely based on the interpretation of the invoked treaty Provision, and mostly not based on other criteria, such as the intention of the contracting States. According to the authors, this approach resulted in permanent case law, in which as a rule of thumb Provisions of the ECHR and ICCPR are considered to be binding on all persons, and Provisions of the ESC and ICESCR are not. However, the authors claim that at this, it is not the particular treaty in which the Provision is embedded that determines whether the Provision is binding on all persons or not, but the content of the Provision itself. To demonstrate this, they underline that there appear to be exceptions to this rule of thumb. As an example, they refer to the fact that the Dutch Courts generally deny direct effect to Articles 14 ECHR and 3, 15 (5), and 23 (4) ICCPR, and grant direct effect to Articles 6 (4) ESC and (under certain conditions) Article 7 (a) (i) ICESCR. In this context, the authors furthermore argue that Provisions of the Convention of the Elimination of All Forms of Racial Discrimination and the Convention Against Torture will probably not be binding on all persons, and direct effect of certain Provisions of the CEDAW and ICRC is not unthinkable.

5.3.4 Kortmann[427]

In his work 'Constitutioneel Recht', Kortmann argues that the Dutch Constitutional system is of a monistic nature. Evidence for this cannot be found in the Constitution itself, but in case law, especially in the 'Grenstractaat Aken' ruling.[428] In this verdict, the Supreme Court decided that an international standard has a double effect. Firstly, on State level, it binds the Dutch State in relation to other countries, and secondly, in domestic law, the Dutch citizen may invoke the standard. According to Kortmann, the function of Article 93 Constitutional Act is not to embed monism in the Dutch Constitution, but only to oblige all public offices (not only the Courts) to apply all international Provisions that are binding on all persons, and only if the international standard has been published. Due to this publication requirement, Kortmann considers that Article 93 Constitutional Act introduces an element of dualism in the Constitution. Since this prerequisite only applies to directly applicable standards, other international standards still have an automatic direct effect according to case law, although they cannot be applied by public offices in their relation to citizens. Therefore, the Dutch Constitutional system is of a monistic nature, but not fully concerning standards that are binding on all persons.

According to Kortmann, it is in the end the Judiciary that decides whether an international standard is binding on all persons, using the wording, nature, purpose

---

[426] Supreme Court, 30 May 1986, NJ 1986, 668 (Railway strike ruling).
[427] C.A.J.M. Kortmann, *Constitutioneel Recht* (6ᵉ druk), Deventer: Kluwer, 2008.
[428] Supreme Court, 3 March 1919, NJ 1919, p. 371 ('Grenstractaat Aken'-ruling).

and legislative history of the standard as leading criteria. In rare occasions, the Court may pay some attention to the national legal context in which the standard should operate. Also, occasionally, the views expressed in Parliamentary History may influence the decision on the matter, but are certainly not binding to the Court.

Article 94 Constitutional Act then stipulates that the Courts are bound not to apply a national standard that is in contradiction with an international standard that is binding on all persons. Furthermore, Kortmann makes some additional remarks on the exact meaning of the Constitutional Provision. Firstly, he considers that Article 94 Constitutional Act not only entails an obligation to review national standards against international standards that are binding on all persons, but also a prohibition to review national standards against international standards that are not binding on all persons or international customary law. Secondly, Kortmann considers the relation between international standards and domestic law. He argues that strictly speaking Article 94 Constitutional Act has no effect on the relation between international standards and domestic law, for the Courts should only review in a specific situation the application of a national standard *in concreto* against an international standard that is binding on all persons, and not against the standard *in abstracto*, although he observes that, in some cases, the Courts seem to review the standard in general. If the Court rules that application of a domestic standard is not in accordance with the disputed international standard binding on all persons, the Court has the obligation to disregard the domestic standard. However, it may also occur that a Court decides that the Court finds that the application of a national standard is in contradiction to an international standard, but does not want to rule on the matter, for then the Court would exceed its competences in a *trias politica*, and act as lawmaker instead of a Court. Thirdly, Kortmann underlines that Article 94 Constitutional Act concerns all national legislation, regardless of the position in the hierarchy of legal rules in the Dutch legal order or the time of adoption. Finally, Kortmann argues that Article 94 Constitutional Act is not addressed to Courts specifically, and thus, as a consequence, to everyone in a public capacity who is competent in applying national standards.

Summarising the above, Kortmann argues that based on case law, the Dutch Constitution is of a monistic nature. Article 93 Constitutional Act limits this monism slightly by introducing the prerequisite of publication for domestic application of international standards that are binding on all persons. Article 94 Constitutional Act obliges the Courts (and other competent bodies) not to apply domestic law that is in contradiction with these international standards binding on all persons.

5.3.5 Vlemminx and Boekhorst[429]

Completely different from the above is the interpretation given by Vlemminx and Boekhorst to the Constitutional system embedded in Articles 93 and 94 Constitutional Act. Also Vlemminx and Boekhorst base their interpretation of both Articles on Constitutional history. The authors underline that during the Constitutional Reforms of 1953, the current Article 94 Constitutional Act was not included in the original Bill. The then-Government was of the opinion that a predecessor of the current Article 93 Constitutional Act needed to be included in the Constitution, to ensure legal security for citizens regarding the direct effect of international standards. The main purpose of including Article 93 in the Constitution was thus to ensure that citizens would be aware of directly applicable international standards before they would be applied to them. It is in this light that Vlemminx and Boekhorst stress that the original focus of Article 93 Constitutional Act was to ensure legal security for citizens concerning duties that are embedded in directly applicable international standards. According to the authors, the review of national standards against international treaty Provisions was a different issue. It was the original intention of the then-Government to leave this matter to the development in case law, instead of regulating it by a Constitutional Provision. Nevertheless, by Amendment Serrarens and co, the predecessor of the current Article 94 was adopted, albeit without the term 'binding on all persons'. Main motivation for introducing Article 94 Constitutional Act appeared to be the fear for uncertainty concerning the judicial review against international standards. According to Vlemminx and Boekhorst, the term 'binding on all persons' was later (during the Constitutional reform of 1956) added to Article 94, mainly to express that the Courts should show some restraint in reviewing national law against international standards. It is unclear however why the choice was made to use this particular phrasing.

Considering the above, Vlemminx and Broekhorst underline that historically, the functions of Articles 93 and 94 Constitutional Act differ fundamentally. In this light, the adding of the term 'binding on all persons' to Article 94 Constitutional Act, while the very same term is also used in Article 93, causes uncertainty about the meaning of this term, and the relation between Articles 93 and 94 Constitutional Act. According to the authors, the meaning of the concept 'binding on all persons' can be understood in two ways: either the term has an identical meaning in both Articles, or the term must be understood differently, depending on the Article in which it is embedded.

According to the authors, an identical understanding of the term, although generally accepted, leads to nothing good. In this understanding one starts from

---

[429] F.M.C. Vlemminx and M.G. Boekhorst, in: A.K. Koekoek (eds.), *De grondwet, een systematisch en artikelgewijs commentaar* (3e druk), Deventer: Tjeenk Willink, 2000.

the function of Article 94 Constitutional Act, that is to regulate the application of international law by the Judiciary, which concerns both duties and rights for citizens. Consequently, Article 93 Constitutional Act would then also concern rights and duties, and not merely duties, as originally intended. Such an identical interpretation leads to difficulties. In the first place, the authors argue that it was originally intended that due to its function Article 93 only concerned Provisions that were directly addressed to private persons.[430] On the other hand, the Judiciary, based on Article 94 CA, appears to be willing to grant direct effect to treaty Provisions that are not necessarily addressed to private persons. These different approaches to the term 'binding on al persons' are difficult to match. Furthermore, an identical interpretation leads to two methods of determining whether or not an international standard has direct effect in the domestic legal order. Firstly, based on Article 93 CA, the Government needs to decide beforehand and *in abstracto* whether or not to publish the standard, while secondly, it appears more convincing to decide *in concreto* whether a standard is binding on all persons based on Article 94 CA. Inevitably, tension between the two approaches exists. Although there seems to be agreement on the fact that it is up to the Courts to have a final say in determining whether a standard is directly applicable, legal practice proves otherwise. According to Vlemminx and Boekhorst, in the past, Governments did not hesitate to unbind a treaty that has been judged binding on all persons by the Judiciary in contrast to the opinion of the Government. In addition, theoretically, a Government could deny direct effect of an international Provision simply by not publishing the standard. This would be not much of a problem if the international standard would stipulate duties for citizens, but if Article 93 is understood as also concerning rights, it could lead to a situation in which a Government could (deliberately or not) deny these rights to citizens, and as a consequence, Article 93 may work against the citizens instead of ensuring legal security. The Government thus might have a strong influence in the discussion whether an international Provision is binding on all persons or not, which affects the divisions of power in *trias politica*, and might cause serious problems considering the increasing importance of international law.

As an alternative to this generally accepted identical interpretation, Vlemminx and Boekhorst suggest giving a different meaning to 'binding on all persons' per Article, based on the originally intended function in the first reading of the Constitutional reforms in 1953. In this understanding, Article 93 CA only concerns duties for citizens, and Article 94 both rights and duties. As a consequence, the Judiciary

---

[430] To underline this point of view, F.M.C. Vlemminx en M.G. Boekhorst refer to Parliamentary Reports, II 1951-1952, 2374, no. 3, p. 7. Although the following nuance does not necessarily detract from their main argument, it is unclear to us how it appears from the Explanatory Memorandum that the then-Government indeed referred to treaty Provisions that were literally addressed to private persons. The Government merely underlined that *'private persons are not bound to international agreements insofar they have not been published in any way.'* Original text in Dutch: *'private personen niet aan een internationale overeenkomst zijn gebonden, voor zover deze niet op een of andere wijze is bekend gemaakt.'*

and the Government do not obstruct one another. The Government will decide *in abstracto* when a treaty Provision that stipulates duties is specific enough that it is binding on all persons, doing justice to the principle of legality. The Courts may decide, without Governmental interference, whether an invoked treaty Provision has direct effect in a particular Court case. This would do most justice to the division of powers and leads to more coherence regarding the functioning of international law in the domestic legal order.

According to Vlemminx and Boekhorst, the identical interpretation led to confusion regarding the direct applicability of international standards. Courts are especially cautious at reviewing national standards against ECOSOC rights embedded in international treaties, for they are well aware of the Government's aversion against review against treaties that imply financial obligations for the Government. In addition, the authors observe that the case law appears to be inconsistent and confusing. It is therefore not surprising that in literature scholars struggle to capture the exact meaning of Article 94 CA, and its case law.

In addition, Vlemminx and Boekhorst argue that regarding international customary law, the Dutch Constitutional system is of a monistic nature, based on case law.

5.4 Questions

There appears to be a variety of opinions amongst de Dutch scholars on the exact meaning of the Constitutional system concerning the effect of international standards in the national legal order. There seems to be in particular uncertainty about the following issues: firstly, the origin of Dutch monism is unclear. Some authors claim that monism appears from case law, while others claim that monism was introduced by or reaffirmed in Article 93 CA. Secondly, the nature of monism is disputed: although most authors agree that Articles 93 and 94 CA somehow limit monism or introduce dualistic elements, they do not agree on the exact consequences of this. Thirdly, whereas most authors agree on the fact that the Courts have a final say in determining whether an international Provision is directly applicable, it is unclear what the exact relation is between Articles 93 and 94 CA, and thus the exact relation between the Legislature (Government) and the Judiciary. Fourthly, in this context, it is debated what legal sources the Courts then use in determining whether an international standard is binding on all persons. The railway-strike ruling appears to be an influential verdict here, especially when ECOSOC rights are involved, but is quoted in different contexts, basically to demonstrate that the intention of the contracting parties concerning direct effect of an international Provision hardly plays a role, and instead, the Courts consider the nature and content of the Provision. As demonstrated in Chapter 4, legal practice regarding the right to food shows a different approach.

As mentioned in the introduction to this Chapter, it is the purpose of this Chapter to fully understand how the Constitutional system of Articles 93 and 94 CA works. The authors of the most prominent literature on the matter appear to understand the Constitutional Provisions most differently, although there is some similarity in sources their conclusions are based on. Most authors quote the Parliamentary History of the three Constitutional reforms that finally resulted in the current Articles 93 and 94 CA: the reforms of 1953, 1956, and 1983. Therefore, in the next sections, an analysis of this Parliamentary History will be made, to try to reach a more unambiguous conclusion regarding the first three questions. The matter of the sources that Courts usually use to determine whether an international standard is directly applicable will be addressed further in this book, for it will appear that this is closely related not only to the exact relation between the Legislature and the Judiciary, but also intertwined with the considerations expressed in the Parliamentary History of the ratification Bills of international treaties. This will be explained more thoroughly in Chapters 7 and 8.

5.5 On parliamentary history/the legislative process

In the Netherlands, both the States General and the Government are involved in the legislation process. The States General has a bicameral system, consisting of a directly chosen House of Representatives and an indirectly chosen Senate, and their Parliamentarians are representing the people of the Netherlands. Although both the Government and the House of Representatives have the right to present a bill, in most cases this is done by the Government, who can use the expertise of its Ministries, and normally can count on a majority in both Chambers that will approve of the Bill. A Bill that ratifies an international treaty however, is naturally proposed by the Government, due to the fact that the treaty was negotiated for and signed in the international arena by (representatives of) the Government. Before the Government presents the Bill to Parliament, the supreme advisory body of the Crown (the Council of State) will give its recommendation on both the draft and the related Explanatory Memorandum, which may lead to some alterations in the proposed texts. The Government then will present both documents to the House of Representatives. Firstly, a committee consisting of a handful of Members of the House of Representatives (but a proportional representation of the political parties) will carry out the preparatory examinations of the Bill, and in a Preliminary Report formulate written questions to the Government, which will reply in writing as well. Sometimes two or three rounds of written communication between this committee and the Government are deemed necessary. Secondly, the Bill will be discussed in and by the House of Representatives. The regular procedure is that one (and sometimes two) representative(s) of each political party express(es) the opinion of the whole party, and ask(s) further questions to the Government. All Members of the House of Representatives and the Government have the right to amend the Bill. Finally, the present Members of the House of Representatives vote on the bill and the submitted amendments. After the Bill has passed the

House of Representatives, it will be presented to the Senatese. Now, a committee consisting of Members of the Senate will carry out the preparatory examinations. Thereafter, the Bill will be discussed in the Senate, and will be voted on. Members of the Senate have no right to make amendments however. All documents are publicly available, and published as Parliamentary Documents. The procedure for a Constitutional reform is similar to the above, although the Government is regularly not only advised by the Council of State, but also by a Commission of State, and occasionally by special advisory bodies that are temporarily established for that specific purpose. Also, a Constitutional reform must pass the States General twice. The first time with a normal majority, after which the House of Representatives will be dissolved and replaced through elections by a new one. The States General will have to adopt the reform a second time (without the possibility of the House of Representatives to make amendments) with a two-third majority.[431]

5.6 The Constitutional reform of 1953

In the explanatory documents written by the Dutch Government on the first reading of the Constitutional reforms of 1953, some contextual remarks were made concerning Dutch international relations.[432] According to the Government, there was a worldwide tendency to increase international cooperation. The Netherlands appeared to be more active in investing in international relations as well, and became a Member of, among others, the Western European Union, the NATO, the BENELUX, and the UN. The Government considered that the increasing international cooperation would eventually result in significant international rights and duties, which could possibly also affect the domestic legal order. Also, the Government did not preclude the possibility that in the near future international bodies with a Parliamentary nature would be installed, which was expected especially within the Council of Europe. These circumstances constituted the cause for the Dutch Government to propose a Constitutional reform regarding international relations.[433]

In the first draft, the Government only proposed to adopt a predecessor of the current Article 93 CA,[434] while the predecessor of the current Article 94 CA was adopted by amendment.[435] The debates in Parliament on the draft Bill mainly focussed on the question whether the Legislature or the Judiciary should decide whether an international Provision had direct effect.

---

[431] In this section, we thankfully made use of the terminology and explanation offered in: L. Prakke and C. Kortmann (eds.), *Constitutional law of 15 EU Member States,* Deventer: Kluwer, 2004, especially p. 621-625.

[432] Parliamentary Documents, II 1951-1952, 2374, no. 2.

[433] Parliamentary Documents, II 1951-1952, 2374, no. 3, in particular p. 2.

[434] Parliamentary Documents, II 1951-1952, 2374, no. 2, draft bill, Article 60d.

[435] Parliamentary Documents, II 1951-1952, 2374, no. 17, later replaced by no. 32.

5.6.1 The predecessor of Article 93 CA during the Constitutional reform of 1953

Originally, the Dutch Government proposed to adopt only the following Provision:

> Agreements are binding on citizens only insofar as they have been published.
> The law provides rules concerning the publication.[436]

De Preparatory Committee argued that it was difficult to exactly determine what individuals would fall under the scope of 'civilian', and therefore recommended to replace the term 'civilian' with 'all',[437] in line with an earlier report of the Constitutional Commission-Van Eysinga.[438] This suggestion was later accepted by the Government in a letter of amendment, replacing the original draft:

> The law provides rules concerning the publication of agreements. Agreements
> are binding on all persons insofar as they have been published.[439]

The Government however, offered no further explanation on the exact scope of the term 'civilian' or 'all.'

It seems that with this draft Bill the Government chose to introduce a monistic system in the Dutch Constitution, by underlining in its Explanatory Memorandum that: '*the Government thus rejects the view that international agreements can only bind citizens through a law.*'[440] On the other hand, the Government considered that it would not be desirable when a civilian or any private party could be bound to an international agreement while this agreement was not published first. In line with Vlemminx and Boekhorst,[441] it appears that the Government proposed the adoption of this Article to avoid a situation in which citizens and private persons would not be aware of the fact that they are bound by international treaties that impose duties on them. In the Preparatory Committee there were some doubts

---

[436] Parliamentary Documents, II 1951-1952, 2374, no. 2 Original text in Dutch: '*Overeenkomsten verbinden de burgers slechts voor zover zij zijn bekend gemaakt. De wet geeft regels omtrent de bekendmaking.*'

[437] Parliamentary Documents, II 1951-1952, 2374, no. 9, Preliminary Report, p. 26.

[438] The official Dutch name fort this committee was: '*Commissie nopens samenwerking tussen regering en Staten-Generaal inzake het buitenlandse beleid,*' freely translated: '*Committee concerning the cooperation between Government and Parliament on foreign policy.*' See the committee's final report: '*Eindrapport van de commissie nopens de samenwerking tussen Regering en Staten-Generaal inzake het buitenlandse beleid,*' 9 July 1951, NL-HaNA, BuZa/Code-Archief 45-54, 2.05.117, inv.no. 27297, p. 24. Available at: www.historici.nl.

[439] Parliamentary Documents, II 1951-1952, 2374, no. 26, letter of amendment, submitted on 18 March 1952. Original text in Dutch: '*De wet geeft regels omtrent de bekendmaking van overeenkomsten. De overeenkomsten verbinden een ieder, voorzover zij zijn bekend gemaakt.*'

[440] Parliamentary Documents, II 1951-1952, 2374, no. 3, Explanatory Memorandum, p. 7. Original text in Dutch: '*De regering verwerpt derhalve het standpunt, dat internationale overeenkomsten slechts via een wet de burgers kunnen binden.*'

[441] F.M.C. Vlemminx and M.G. Boekhorst, in: A.K. Koekoek (ed.), *De grondwet, een systematisch en artikelgewijs commentaar* (3e druk), Deventer: Tjeenk Willink, 2000.

on whether such a Provision would not stand in the way of the proper functioning of the international legal order. Especially when international Provisions would only have effect in the domestic legal order under certain conditions, in this case the prerequisite of publication. The Government responded that it was not the intention to limit the effect of international agreements between States, but merely to regulate the liability of the citizen towards the State. The Government argued that it would object to a situation in which *'nationals of the State could be bound to rules whose existence they could not know of'*[442], which would follow from the primacy of international law over domestic legislation without restrictions. Therefore, the Government responded that it would rather restrict the effect of international law to protect the legal security of the individual, than *'to deny the general principle of law that citizens cannot be considered to be bound to standards that are not known to them.'*

5.6.2 The predecessor of Article 94 CA during the Constitutional reform of 1953

5.6.2.1 The position of the Government

While there appeared to be hardly any disagreement on the primacy of international law over national law in general, the competence of the Judiciary to review anterior national legislation against international standards adopted later,[443] and the general prohibition of judicial review against the Constitution,[444] there was a fierce discussion on the question whether it should be the Legislature or the Judiciary who decides whether a subsequent domestic law is contrary to an earlier adopted international standard.

As stated above, in the first draft of the Constitutional reform proposed by the Government, no predecessor of the current Article 94 CA was included, which was in contradiction to the recommendations of both the Committee-Van Eysinga and the Constitutional Committee.[445] For instance, the Committee-Van Eysinga suggested adding a Provision that would stipulate that *'The thus proclaimed treaties are binding for everyone even if (domestic) regulations deviate.'*[446] Although the Government was of the opinion that the *'existing Constitutional rules give rise to*

---

[442] Original text in Dutch: *'onderdanen van de Staat gebonden zouden kunnen zijn aan regels van welker bestaan zij geen kennis hebben kunnen nemen.'*
[443] Parliamentary Documents, II 1951-1952, 2374, no. 9, p. 19; Parliamentary Documents, I 1951-1952, 2374, no. 113, p. 3.
[444] Parliamentary Documents, II 1951-1952, 2374, no. 7, letter of amendment: *'The Judiciary shall not enter into the review of the Constitutionality of agreements.'* Original text in Dutch: *'De rechter treedt niet in de beoordeling van de grondwettigheid van overeenkomsten.'* A view that is now expressed in Article 120 Constitutional Act.
[445] Both committees were installed to advise the Government on the draft of Constitutional reforms.
[446] Parliamentary Documents, II 1951-1952, 2374, no. 3, p. 3. Original text in Dutch: *'De aldus afgekondigde verdragen zijn voor een ieder verbindend ook indien wettelijke voorschriften ervan afwijken.'*

disagreement regarding the primacy of an international agreement over a subsequently adopted (national) law and that it would therefore in itself be considered desirable to to explicitly regulate this by Constitution'[447], it considered that it would be too early to take a fundamental decision on the matter. Especially due to the fact that within the Constitutional Committee there was certainly no agreement on the recommended Provision, which was expressed by a minority memorandum of some of its Members who did not agree with the proposed Article; a view that was apparently also widely supported outside the arena of the Committee. Also, until then the Judiciary seemed to be reluctant in taking a clear position on the matter.[448] The Parliamentary discussion therefore mainly focussed on the question whether it should be the Legislature or the Judiciary who decides whether a subsequent domestic law is contrary to an earlier adopted international standard, while the Government persisted to abstain from taking a position, considering the deep and fierce Parliamentary debates.[449]

5.6.2.2 *The review of anterior national law against posterior international standards*

The Parliamentary debate between those against and those in favour of judicial review can be subdivided into four main issues:

---

[447] Parliamentary Documents, II 1951-1952, 2374, no. 3, p. 3. Original text in Dutch: '...*dat bestaande Grondwetsvoorschriften aanleiding geven tot meningsverschil omtrent de voorrang van een internationale overeenkomst boven een later tot stand gekomen wet en dat het op zichzelf beschouwd derhalve wel gewenst zou zijn dit punt in de Grondwet uitdrukkelijk te regelen.*'

[448] The Preparatory Committee underlined that so far only in one verdict a Court ruled on the matter whether international law should have primacy over subsequent domestic law: the Special Council of Cassation held that '*in cases of doubt, the Court has the duty to determine whether a law is compatible with the international agreement, also in cases when the latter was adopted before the first.*' (Special Council of Cassation, 12 January 1949, NJ 1949, no. 87, 'Rauter' ruling; Original text in Dutch: '*de rechter in geval van twijfel heeft te onderzoeken, of de wet met de overeenkomst verenigbaar is, ook indien de laatste van een vroegere datum is dan de eerste*'. The Supreme Court however structurally refrained from ruling on the matter. See: Parliamentary Documents, II 1951-1952, 2374, no. 9, p. 19; Parliamentary Documents, II 1951-1952, 2374, no. 10, p. 28.

[449] For instance, Parliamentary Documents, II 1951-1952, 2374, no. 10, p. 28. However, it appeared that for a short moment Minister Beel broke with that attitude, by referring – albeit very cautiously – to a conclusion of Attorney-General Langemeyer. In this conclusion, it was argued that the basic idea of a prohibition on Constitutional review was that '*in case of serious doubt concerning the legality of a certain act of legislation, it is the Legislator who is best suited to settle the doubt, and there is in principle no room for judicial review. The Attorney-General finally concludes that the review of the law against international standards would only be appropriate in case of the absence of a serious review by the Legislator.*' In addition, the Minister argued that it would not be easy to interpret international standards due to the often vague formulation and the inclusion of a restriction clause that allows for deviation from international law. While instantly adding that due to the very different visions concerning the matter, the Government did not want to take a final decision, the Minister appeared to favour review of international standards by the Legislator over judicial review, or at least in cases in which the Legislator did not express an unambiguous view on direct effect of a Provision. See: Minister Beel, House of Representatives, 61st meeting, 14 March 1952. Mr Anema argued in the Senate that such an approach from a practical point of view would not be feasible. See: Mr Anema, Senate, 40th meeting, 6 May 1952.

Firstly, the question was raised how and to what extent judicial review and review by the Legislature would affect (national or international) legal security.

In Parliament, some supporters of a system of judicial review argued international legal security could not be guaranteed when Governments could intentionally, but more likely unintentionally in complex legal situations, undo the effect of treaty Provisions by adopting contradicting legislation.[450] In this light, the question was raised whether a Legislator can be aware of all the current international standards when adopting a domestic law.[451] The prohibition of judicial review of domestic law against international standards would then lead to case law in which the Court would continuously have to speculate about the intention of the Legislature. A wrong interpretation of the international standard then could lead to subordination of Dutch citizens to citizens of countries in which judicial review against treaty Provisions is allowed and such interpretation could be contested in Court.[452] Another consequence would be that the Dutch Judiciary applied the international standard differently – that is: as understood by the Dutch Legislature – compared to the understanding of this standard elsewhere. This would then lead to international legal insecurity, or inequality.[453] Nonetheless, according to some Parliamentarians, the argument that national judicial review of international standards in itself could also lead to legal insecurity by unequal application of the law not under dispute, but some Parliamentarians explicitly considered that international legal security was more important than national legal security.[454] In addition, it was considered that the assurance of justice was more important than the assurance of legal security: a Legislator would in its review be led by policy and State-interest arguments, while a Court would only be led by legal rules. In that light, it would be a better option to choose for judicial review instead of review by the Legislature.[455]

Supporters of review by the Legislature however argued that national legal security would best be guaranteed when the Legislature would review national law against international law, for it would result in an unambiguous nationally binding interpretation of the national law that was established before.[456] In case of judicial review, the Legislature, especially in case of any doubts concerning the compliance of an anterior domestic standard with a posterior international Provision, would only adopt that domestic law if it is most certainly in line with the

---

[450] For instance, Parliamentary Documents, II 1951-1952, 2374, no. 3, p. 3.

[451] See for instance: Parliamentary Documents, II 1951-1952, 2374, no. 3, p. 3; Parliamentary Documents, II 1951-1952, 2374, no. 9, p. 19.

[452] Parliamentary Documents, II 1951-1952, 2374, no. 3, p. 3.

[453] Parliamentary Documents, II 1951-1952, 2374, no. 9, pp. 19 and 20.

[454] See for instance: Serrarens, House of Representatives, 60th meeting, 13 March 1952; Mr Bruins-Slot, House of Representatives, 62nd meeting, 18 March 1952; Mr Wijers, Senate, 40th meeting, 6 May 1952.

[455] Parliamentary Documents, II 1951-1952, 2374, no. 9, p. 19; Parliamentary Documents, I 1951-1952, 2374, no. 113, p. 4, see also Mr Bruins-Slot, House of Representatives, 60th meeting, 13 March 1952.

[456] Parliamentary Documents, II 1951-1952, 2374, no. 9, p. 20.

international Provision in order to establish legal security. This could however lead to a situation in which the treaty Provisions are understood in a much broader way than was originally intended by the contracting parties.[457] Also, it was argued that national Legislators would easier come to a uniform understanding of international standards than national Courts, due to the fact that the national Governments are more capable to discuss the understanding of an international standard compared to (independently operating) national Courts.[458] Another argument was that the international legal system consisted of an obscure mixture of international bodies that produce different kinds of international standards, which makes it difficult to determine what standards exactly would fall under the scope of international law. judicial review then would lead to more national legal insecurity.[459]

Secondly, the question was raised whether or not the national ban on Constitutional review, currently embedded in Article 120 CA,[460] could be applied analogously to international law.

A group of Parliamentarians who were in favour of judicial review argued that analogous application of a national ban on Constitutional review in accordance with international standards would be incorrect, for there were too many differences between national and international law. For instance, it was underlined that the existing ban on Constitutional review was developed only within the national context, and had a historic background that was not comparable to the international situation.[461] In addition, it was argued that law development at an international stage could not be compared with national law development, and therefore, an analogous ban on judicial review of domestic law against international standards could not be justified by arguments based on the assumption that the ban on reviewing law against the Constitution encompasses a general principle.[462] In that light, Parliamentarian Weijers was of the opinion that a ban on Constitutional review not necessarily encompassed a basic idea not to review lower legal standards against higher legal standards, but rather that Constitutional review was not comparable with review against international standards. He considered that firstly a Constitution represents a relatively small set of rules which the Legislature is quite familiar with and therefore most capable of reviewing its own legislation against, while the Legislature is usually less familiar with international law. Secondly, a Constitution should be a rigid set of rules, not altered or differently understood, and therefore the review by the Legislature is the best option, while international law should be more flexible, in line with international developments. Thirdly and most importantly, the draft of a Constitution and other legislation completely

---

[457] Parliamentary Documents, I 1951-1952, 2374, no. 113a, p.8.
[458] Mr Oud, House of Representatives, 62nd meeting, 18 March 1952.
[459] Mr Burger and Mr Oud, House of Representatives, 60th meeting, 13 March 1952.
[460] During the Parliamentary discussion Article 124 CA.
[461] Parliamentary Documents, II 1951-1952, 2374, no. 9, p. 19.
[462] Parliamentary Documents, II 1951-1952, 2374, no. 9. p. 20.

falls under the authority of the Legislature, whereas the draft of international Provisions instead limits the sovereignty of the Legislature.[463] While Mr Weijers defended the position against a review ban against international law based on arguments that underlined the differences between national law and international law, Mr Romme defended the same position by analogously applying the national review competences of the Courts to the international setting. He argued that according to Dutch law, a Court was allowed to review a posterior law against an anterior law, and the lower legal standard against the higher legal standard. When applied to international law, this would thus imply that also here the Courts are competent to review the lower, posterior domestic law against the higher, anterior international Provision.[464]

Those supporting the idea of a review by the Legislature, as recommended by the Constitutional Commission,[465] argued *inter alia* that the existing ban on Constitutional judicial review could very well be applied analogously to the situation in which a posterior law was reviewed against an anterior international standard, due to the high degree of similarity between both situations.[466] It was argued that the ban on Constitutional review encompassed one of the pillars of Dutch Constitutional law and should therefore not be put aside too easily. Judicial review against international Provisions would be contrary to that principle.[467] Mr Oud explained that the principle embedded in the ban on Constitutional review implied that in a national setting, it is explicitly the competence of the Legislature to review its own legislation against the Constitution, and not of the Judiciary, effectively establishing immunity of the law. The increasing adoption of international Provisions that have effect in the domestic legal order does not change this principle, and also in this relation the law should be immune, and thus it should be the Legislature that reviews a national law against a treaty Provision.[468] Mr Oud disagreed with the analogous application of the review competences of the Judiciary as suggested by Mr Romme, arguing that in the Netherlands there is no such principle as the competence of the Judiciary to review a lower standard against a higher standard. He stated that there was rather the principle that it is normally the Court that belongs to the community that adopted the higher standard that is competent in reviewing the lower standard against the higher standard. In case of international law, such a Court that belongs to the international community did not yet exist. According to Mr Oud, it is preferable

---

[463] Mr Wijers, Senate, 40th meeting, 6 May 1952.

[464] Mr Romme, in the House of Representatives, 61st meeting, 14 March 1952.

[465] Parliamentary Documents, II 1951-1952, 2374, no. 8.

[466] Parliamentary Documents, II 1951-1952, 2374, no. 3, p. 3. See also: Mr Oud, House of Representatives, 62th meeting, 18 March 1952.

[467] Parliamentary Documents, II 1951-1952, 2374, no. 9. p. 20.

[468] Mr Oud, House of Representatives, 62th meeting, 18 March 1952.

to also apply the principle of law immunity to the relation between national law and international law, until such international Court exists.[469]

Thirdly, the question was raised whether the Judiciary or the Legislature would be most competent to review against international Provisions.

Those supporting judicial review argued that in case of a legal dispute the Legislature would not be very competent to perform an interpretative function regarding treaty Provisions.[470] Some Parliamentarians expressed great confidence in the interpretative capacity of the Judiciary.[471] Also, it was argued that when review competences would be entrusted to the Legislature, it would practically rule on its own case,[472] especially when the case concerned a duty of the Government towards a citizen. In that case, the citizen should be capable to enforce the proper fulfilment of that duty, in line with the international Provision.[473] In addition, it was argued that the Judiciary would be more competent to review against international law instead of a Legislator, who would be more driven by nationalist considerations.[474]

Those defending review by the Legislature argued that in matters regarding the review against international law not only concerned matters of law, but also matters of policy, which should normally be decided on by the Legislature and not by the Judiciary.[475]

Fourthly, the aforementioned discussions must be placed in the context of the perception of international law at the time. There were some huge expectations about the future development of international law: it was argued that in the long term eventually an international Court would be installed that would be authorised to rule on matters of international law. The question then was whether in the meantime the Legislature or the Courts would be responsible for the interpretation of international standards.[476]

Fifthly, among those who defended review by the Legislature, occasionally the fear was expressed that citizens might abuse the legal system when posterior

---

[469] Mr Oud, House of Representatives, 62th meeting, 18 March 1952.
[470] Parliamentary Documents, II 1951-1952, 2374, no. 9. p. 19.
[471] Mr Wijers, Senate, 40th meeting, 6 May 1952.
[472] Mr Bruins Slot, House of Representatives, 60th meeting, 13 March 1952.
[473] Mr Molenaar, Senate, 40th meeting, 6 May 1952.
[474] Mr Anema, Senate, 40th meeting, 6 May 1952.
[475] An argument for Minister Beel to defend the Government's position to advise against the adoption of the amendment proposed by Serrarens, which would accept judicial review against international Provisions. See: Minister Beel, House of Representatives, 61st meeting, 14 March 1952.
[476] For instance (favouring judicial review): Mr Romme, in the House of Representatives, 61st meeting, 14 March 1952; see also: Mr Wijers, Senate, 40th meeting, 6 May 1952. Expecting on the long term an international Judiciary but opposing judicial review by the national Legislator: Mr Oud, House of Representatives, 62nd meeting, 18 March 1952.

national standards could be reviewed against anterior international standards by the Judiciary, which could lead to an overload of the Courts' capacity.[477]

5.6.3 The Serrarens and co amendment

The discussion above was finally put to an end by the adoption of the Serrarens and co amendment, in which judicial review was recognised. Originally, the amendment was formulated as:

> In case of violation against legal rules in force within the Kingdom, the Provisions of agreements shall precede.[478]

In his (short) explanation during the Parliamentary debate in the House of Representatives, Serrarens argued under reference to the debate in which Mr Romme and Mr Oud already thoroughly defended the positions in favour or against judicial review, that he and his co-submitters were of the opinion that review against international standards should be entrusted to the Judiciary. Remarkably however, Serrarens and co decided to alter the text of the amendment on the day it was put to the vote in the House of Representatives. The earlier text was replaced by:

> Legal rules in force within the Kingdom do not apply if they are not compatible with agreements that are published, either before or after the enactment of the regulation, in accordance with Article 60f.[479]

The choice for the renewed text was deemed necessary '*to take away ambiguity concerning the interpretation of the original amendment (...).*'[480] Apparently, although Serrarens only briefly motivated the renewed version of the amendment, the ambiguity that is referred to was the question whether anterior international standards would have precedence over posterior national regulations as well.[481] Considering the preceding Parliamentary discussion it is remarkable that Serrarens observed ambiguity on that particular question: in Parliament, it was not significantly contested that international standards would have precedence over later adopted national regulations, but rather on whether it is the Legislature or the Judiciary that should review those national standards against the international

---

[477] Parliamentary Documents, II 1951-1952, 2374, no. 3, p. 3.
[478] Parliamentary Documents, II 1951-1952, 2374, no. 17. Original text in Dutch: '*In geval van strijd met binnen het Koninkrijk geldende wettelijke voorschriften hebben de bepalingen van overeenkomsten de voorrang.*'
[479] Parliamentary Documents, II 1951-1952, 2374, no. 32. Original text in Dutch: '*Binnen het Koninkrijk geldende wettelijke voorschriften vinden geen toepassing, wanneer deze niet verenigbaar zijn met overeenkomsten, die hetzij voor, hetzij na de totstandkoming der voorschriften zijn bekend gemaakt overeenkomstig artikel 60f.*'
[480] Parliamentary Documents, II 1951-1952, 2374, no. 32. Original text in Dutch: '*om twijfel aan de interpretatie van het oorspronkelijke amendement weg te nemen (...).*'
[481] Mr Serrarens, House of Representatives, 63rd meeting, 19 March 1952.

Provision. After the submission of the new version of the Amendment, a short and rather unstructured discussion was held in the House of Representatives in which the new text was discussed, from which does not follow that the adoption was well-considered. The amendment was adopted with a small majority only: 46 voted for, and 40 against.[482]

Until the amendment was put to the vote, the Government obviously was not supporting the adoption, and even discouraged the Parliament to introduce judicial review against international standards in the Constitution. It recommended considering the matter more thoroughly during the next Constitutional reform.[483] Only when the amendment was adopted, the Government was willing to express its trust in the Judiciary: 'In the meanwhile, the Government trusts (...) that the Dutch Judiciary will wield the exceptional powers given here in such manner that the risk of legal uncertainty will be reduced to the smallest possible proportions (...).'[484] Due to the fact that this predecessor of Article 94 CA was not designed by the Government, it cannot be distilled from the first reading of the Constitutional reform how the Government exactly understood the content of the Provision. However, during the second reading, the Government clarified its understanding of the Provision at some points. Until now, the discussion on the proposed Constitutional Articles focussed on the question whether the Judiciary or the Legislature should review domestic law against international standards. For the first time, the question was raised to what extent a civilian could extract rights from international standards through the Constitutional construction that was adopted in the first reading.[485] The Government responded that not all international standards would grant subjective rights to civilians, but a distinction should be made between international standards that implied instruction standards to the Legislature, and directly applicable standards, that would indeed imply subjective rights for civilians. In answer to the question who would have to determine whether a standard would be directly applicable or not, the Government underlined that it would not be desirable 'to regulate by law, (...), whether an agreement shall be deemed to include instructional standards, or have direct effect towards civilians. Ultimately, this concerns a question of interpretation, which the Court will have to decide on, who on the basis of Article 60e, and not the Legislature, will be able to bindingly interpret an agreement. Nevertheless, especially with regard to Article 60e, the Government will, in case of doubt, be able to make clear in its explanatory notes which intention it had during the adoption, for it

---

[482] House of Representatives, 63rd meeting, 19 March 1952.

[483] Minister Beel, House of Representatives, 63rd meeting, 19 March 1952.

[484] Parliamentary Documents, I 1951-1952, 2374, no. 113a, p. 8. Original text in Dutch: 'Inmiddels vertrouwt de regering (...) dat de Nederlandse rechter de hem hier gegeven uitzonderlijke bevoegdheid op zodanige wijze zal hanteren, dat het gevaar voor rechtsonzekerheid tot de kleinst denkbare proporties zal worden teruggebracht (...).'

[485] Parliamentary Documents, I 1952-1953, 2700, no. 63.

can be assumed that the Judiciary will assign great significance to such statements.'[486]
On the other hand, the Government recognised the possibility that the Judiciary would have an opposing view on the enforceability of an international standard to the intention of the Legislature. Also, the Government was of the opinion that it was the responsibility of the Judiciary not only to not apply domestic legislation that is contradicting internationally directly applicable standards, but also, following from Article 60f (the predecessor of Article 93 CA), to apply the international standard.[487]

5.7 The Constitutional reform of 1956

Only three years after the Constitutional reform of 1953, the then-Government decided to propose another Constitutional reform that also concerned the predecessors of Articles 93 and 94 CA. The Government emphasised that it was not the intention to alter the legal principles that were adopted in the previous Constitutional reform, but merely to amend the Constitution on some textual details.[488] The Government proposed to adopt the following Articles:[489]

V. Article 65 of the Constitution reads:

> Provisions of agreements which may be binding on all persons by virtue of their contents shall become binding after they have been published. The law provides rules on the publication of the agreements.[490]

---

[486] Parliamentary Documents, I 1952-1953, 2700, no. 63a, p. 3. Original text in Dutch: *'bij de wet, (…), vast te leggen, of de overeenkomst moet worden geacht instructienormen te behelzen, dan wel directe werking heeft ten opzichte van burgers. Uiteindelijk betreft het hier een interpretatievraag, die door de rechter zal moeten worden beslist, nu op de voet van artikel 60e deze en niet de wetgever overeenkomsten bindend zal kunnen interpreteren. Dit neemt niet weg, dat de regering, juist met het oog op artikel 60e, in twijfelgevallen in de toelichting op overeenkomsten zal kunnen duidelijk maken, welke bedoeling bij de totstandkoming heeft voorgezeten, aangezien mag worden aangenomen, dat de rechter aan zodanige uitlatingen grote betekenis zal toekennen.'*

[487] Parliamentary Documents, I 1952-1953, 2700, no. 63a, p. 3.

[488] Parlementaire Geschiedenis II, 1955-1956, 4133 (R 19), no. 3, memorie van toelichting, p. 3; Parlementaire Geschiedenis II, 1955-1956, 4133 (R 19), no. 7, memorie van antwoord, p. 1.

[489] Parlementaire Geschiedenis II, 1955-1956, 4133 (R 19), no. 2, ontwerp van rijkswet, p. 1.

[490] Original text in Dutch: *'V. Artikel 65 van de Grondwet wordt gelezen: Bepalingen van overeenkomsten, welke naar haar inhoud een ieder kunnen verbinden, hebben deze verbindende kracht nadat zij zijn bekend gemaakt. De wet geeft regels omtrent de bekendmaking van overeenkomsten.'*

VI. Article 66 of the Constitution reads:

> Statutory regulations in force within the Kingdom shall not be applicable if
> such application is in conflict with Provisions of agreements that are binding
> on all persons that are adopted either before or after the enactment of the
> regulations.[491]

The most important differences compared to the Provisions adopted during the
Constitutional reform of 1953 are threefold: firstly, the sequence of the Provisions
was reversed. The Government was of the opinion that the Constitution should
first stipulate to what extent a treaty is binding on citizens, before the relation
between a treaty and national law is clarified after that.[492]

Secondly, the words 'binding on all persons' were added to the predecessor of
Article 93 CA, to underline that *'the binding of contracts to everyone only relates to
agreements which by its nature are eligible for direct application.'*[493] The Government
explained that this was frequently pointed out during the Constitutional reforms
of 1953, and therefore the Provision should be understood this way. From this, it
would reasonably follow that a similar addition would be adopted in the predecessor
of Article 94 as well. The Government underlined that in case of not adding
'binding on all persons' to that Provision, it would attribute an improper task to the
Judiciary, when the primacy of international law over national law would also
be applied by the Judiciary in case of Provisions addressed to national executive
powers or the Legislature. In that case, the Judiciary would fulfil duties that were
originally imposed on a Legislator or executive powers. It appears here that the
Government wanted to confirm the more narrow understanding of the effect of
international Provisions for only those that are directly binding already discussed
in 1953, but also wanted to suggest a similar understanding of the phrase *'binding
on all persons'* in both Articles.[494] In this context, the question was raised who
would have a final say in determining whether a treaty Provision is binding on
all persons, considering this predecessor of Article 94 CA.[495] The Government
responded that *'the decision on whether Provisions of agreements are binding on
all persons by virtue of their contents ultimately shall be made by the Judiciary. The
focus of this Provision is that it instructs the judge to review national laws against*

---

[491] Original text in Dutch: *'VI. Artikel 66 van de Grondwet wordt gelezen: Binnen het Koninkrijk geldende
wettelijke voorschriften vinden geen toepassing, wanneer deze toepassing niet verenigbaar zou zijn met een
ieder verbindende bepalingen van overeenkomsten, die hetzij vóór, hetzij na de totstandkoming der voorschriften
zijn aangegaan.'*
[492] Parliamentary Documents II, 1955-1956, 4133 (R 19), no. 3, pp. 4-5.
[493] Parliamentary Documents II, 1955-1956, 4133 (R 19), no. 3, p. 5. Original text in Dutch: *'het verbindend
zijn van overeenkomsten voor een ieder slechts betrekking heeft op overeenkomsten die naar haar aard voor
rechtstreekse toepassing in aanmerking komen.'*
[494] Parliamentary Documents II, 1955-1956, 4133 (R 19), no. 3, p. 5.
[495] Parliamentary Documents II, 1955-1956, 4133 (R 19), no. 6. p. 6.

international agreements, which implies that the judgment of the Court, and not that of the Legislature, is decisive.'[496]

Thirdly, the phrase 'agreements', was replaced by 'Provisions of agreements'. The Government agreed with the reasoning of the Advisory Committee,[497] that there *'are agreements in which some Provisions have a self-executing nature while others have not.'*[498]

During the Parliamentary debates on the proposed Constitutional reform, the proposed amendments to the system that is currently embedded in Articles 93 and 94 CA were hardly discussed, and were adopted without further amendments.

Only once it was suggested that *'the amendments proposed by the Government grammatically considered might imply more than just a technical amendment, which is a restriction to the judicial review.'*[499] The Government merely answered that: *'the proposed amendment was not of a principal nature (...) for it is of the opinion of the undersigned that the current text already implied a restriction.'*[500] It must be concluded then that it was not the intention to principally alter the meaning of the Constitutional system by adding the phrase *'binding on all persons.'*

5.8 The Constitutional reform of 1983

For the third time, the Constitution was amended on the Provisions that regulate foreign relations. The Government remarked that it was the purpose of the Constitutional reform of 1953 to fulfil the need that existed after the Second World War to regulate the Dutch foreign relations in the Constitution. The reform of 1956 was deemed necessary, *'partly to meet some concerns, which had arisen meanwhile'.*[501] An interesting point of view, for this clearly does not match the intentions that were formulated in the Parliamentary documents of the Constitutional reform of 1956 to only alter the text on some technical points. In this light, the purpose of

---

[496] Parliamentary Documents II, 1955-1956, 4133 (R 19), no. 7, p.4. Original text in Dutch: *'de beslissing over de vraag of er sprake is van bepalingen van overeenkomsten, die naar hun inhoud een ieder kunnen binden, uiteindelijk bij de rechter ligt. Het zwaartepunt dezer bepaling ligt vooral daarin, dat zij de rechter opdraagt de wetten aan overeenkomsten te toetsen, hetgeen meebrengt, dat het oordeel van de rechter en niet dat van de wetgever beslissend is.'*

[497] Parliamentary Documents II, 1955-1956, 4133 (R 19), no. 3, p. 5.

[498] Parliamentary Documents II, 1955-1956, 4133 (R 19), no. 4. p. 14. Original text in Dutch: *'overeenkomsten zijn waarvan sommige bepalingen een self-executing karakter hebben en andere niet.'*

[499] Parliamentary Documents II, 1955-1956, 4133 (R 19), no. 6, p. 6. Original text in Dutch: *'de door de regering voorgestelde wijzigingen naar de letter wellicht meer betekent dan alleen een technische verbetering, namelijk een beperking van het toetsingsrecht van de rechter.'*

[500] Parliamentary Documents II, 1955-1956, 4133 (R 19), no. 7. p. 4. Original text in Dutch: *'de voorgestelde wijziging is niet van principiële aard (...) daar naar het oordeel der ondergetekenden ook de strekking van de huidige tekst reeds beperkt was.'*

[501] Parliamentary Documents II, 1977-1978 (R 1100), no. 3, p. 5. Original text in Dutch: *'mede om tegemoet te komen aan enige bedenkingen die inmiddels waren gerezen.'*

the Constitutional reform of 1983 might be read with some suspicion: '*Because the scope of those Articles is quite generally perceived as satisfactory, there is no reason to propose such changes that would change the principles of the present regime. However, an attempt is made to make a number of clarifications, especially from a terminology and editorial point of view, while an effort was made to grant the Legislature more leeway.*'[502] The Government explained that in its proposal the order of the Provision was reversed, and '*profoundly shortened by omitting unnecessary details.*'[503] The Government proposed the following amendments to the predecessor of Articles 93 and 94 CA:

Article 5.2.2.a.:

> Provisions of treaties and of resolutions by international institutions which may be binding on all persons by virtue of their contents shall become binding after they have been published.[504]

Article 5.2.2.b.:

> Statutory regulations in force within the Kingdom shall not be applicable if such application is in conflict with Provisions of treaties or of resolutions by international institutions that are binding on all persons.[505]

At first glance, there are four minor changes compared to the amendments of the Constitutional reform of 1956.

Firstly, in both Provisions, the word '*agreements*' was replaced by the word '*treaties*' for this word was regularly used in legal practice to refer to '*agreements that are binding to the State according to international law criteria, regardless of the form.*'[506] It was not the intention to alter the content or scope of the Provision by this alteration. The words '*agreement*' and '*treaty*' therefore have exactly the same meaning.[507]

---

[502] Parliamentary Documents II, 1977-1978 (R 1100), no. 3, p. 5. Original text in Dutch: '*Omdat de strekking van die artikelen vrij algemeen als bevredigend wordt ervaren, bestaat geen aanleiding zodanige wijzigingen voor te stellen dat een verandering zou worden aangebracht in de uitgangspunten van de huidige regeling. Wel is gepoogd een aantal verduidelijkingen aan te brengen, met name in terminologisch en redactioneel opzicht, terwijl ernaar is gestreefd aan de wetgever meer armslag te geven.*'

[503] Parliamentary Documents II, 1977-1978 (R 1100), no. 3, p. 5. Original text in Dutch: '*sterk bekort door weglating van overbodige details.*'

[504] Parliamentary Documents II, 1977-1978 (R 1100), no. 2, p. 2. Original text in Dutch: '*Artikel 5.2.2a. Bepalingen van verdragen en besluiten van volkenrechtelijke organisaties, die naar haar inhoud een ieder kunnen verbinden, hebben verbindende kracht nadat ze zijn bekendgemaakt.*'

[505] Parliamentary Documents II, 1977-1978 (R 1100), no. 2, p. 2. Original text in Dutch: '*Artikel 5.2.2.b. Binnen het Koninkrijk geldende wettelijke voorschriften vinden geen toepassing, indien deze toepassing niet verenigbaar is met een ieder verbindende bepalingen van verdragen en van besluiten van volkenrechtelijke organisaties.*'

[506] Parliamentary Documents II, 1977-1978 (R 1100), no. 3, p. 6.

[507] Parliamentary Documents II, 1977-1978 (R 1100), no. 7, p. 6.

Secondly, before the Constitutional reform of 1983, a separate Provision stipulated the analogous application of the former Article 65 (currently Article 93 CA), *mutatis mutandis* to resolutions of international institutions. In the Constitutional amendments of 1983 the separate Provision was deleted and embedded in the aforementioned amending Articles 5.5.2.a. and 5.5.2.b., resulting in a shorter wording but with the same content.[508]

Thirdly, a minor change in Dutch grammar was made, in which an unnecessary conjunction was omitted in the amending Article 5.5.2.a., that was already lost in the English translation.[509]

Fourthly, the phrase *'either before or after the enactment of the regulations'* was deleted, for *'this phrase was incorporated in 1953, in order to establish beyond doubt that a subsequent law cannot detract from a treaty. This is now sufficiently clear and also appears from the wording of the* (proposed [B.W.]) *Provision.'*[510]

Apparently thus the Government proposed only textual changes, without the ambition to alter the principles embedded in the existing Constitutional system. Nevertheless, the proposed amendments were cause for an extensive debate in Parliament regarding the exact meaning of the content. This was mainly due to the differences of opinion between the Commission of State-Cals/Donner, the Council of State (both advising the Government) and the Government on the exact wording of the amending Provisions. In Parliament, three major issues were debated.

First, the Commission of State had recommended to delete the phrase *'by virtue of their contents'*, in the predecessor of Article 93 CA, for it was unnecessary and incorrect, *'because the answer to the question whether a treaty Provision is generally binding – by virtue of the wordings, intentions of the parties, or scope – depends on the treaty itself and will be addressed during the interpretation of the treaty.'*[511] The Commission of State argued that by leaving the phrase *'by virtue of their contents'* in the text, unjustly a standard would be created that should be reviewed against when deciding whether a treaty Provision is binding on all persons or not. In the original draft of the Constitutional reform, the Government had adopted the

---

[508] Parliamentary Documents II, 1977-1978 (R 1100), no. 3, p. 10.

[509] Parliamentary Documents II, 1977-1978 (R 1100), no. 3, p. 10, the Dutch word *'deze'* was deleted in the phrase *'hebben deze verbindende kracht nadat zij zijn bekend gemaakt.'*

[510] Parliamentary Documents II, 1977-1978 (R 1100), no. 3, p. 14. Original text in Dutch: *'deze zinsnede werd in 1953 ten overvloede opgenomen, ten einde buiten twijfel te stellen, dat ook een latere wet geen afbreuk kan doen aan een verdrag. Dit staat thans wel voldoende vast en blijkt ook uit de redactie van de bepaling.'*

[511] Parliamentary Documents II, 1977-1978 (R 1100), no. 3, p. 11. Original text in Dutch: *'omdat het antwoord op de vraag of een verdragsbepaling algemeen verbindend is – krachtens tekst, partijbedoeling of strekking – afhangt van het verdrag zelf en bij de interpretatie van elk verdrag aan de orde komt.'* See also the original draft of the Explanatory Memorandum: Parliamentary Documents II, 1977-1978 (R 1100), no. 4, p. 32.

proposed Provisions by the Commission, and thus omitted the discussed phrase, and even combined the content of the current Articles 93 and 94 CA in one Provision:

> Statutory regulations do not apply, if such application is in conflict with Provisions of treaties or with resolutions by international institutions that are binding on all persons, after they are announced.[512]

On the advice of the Council of State however, the Government renounced the original draft. The Council was of the opinion that '*In particular, the omission of the phrase in the current Article 65 concerning Provisions of treaties that shall be binding by virtue of their contents, meets objections with the Council. The purpose of that Provision is not to establish a standard "that could be reviewed against, to determine whether a treaty is binding on all persons" as the Explanatory Memorandum* (old text, not the final [B.W.]) *states, but to establish that treaty Provisions, which can be (by virtue of their contents) substantially and generally binding, really possess that quality when certain conditions relating to their publication are fulfilled. The Council considers that this Provision is essential, because it is an important principle of Dutch Constitutional law that is not accepted in several other countries, for example in Great Britain.*'[513] The Government agreed with this reasoning of the Council of State, and almost literally copied the quoted passage in its final Explanatory Memorandum.[514] This reasoning is rather confusing, for it does not necessarily explain why '*by virtue of their contents*' is deemed to be a necessary phrase in the Constitution, but it rather (and somewhat cryptically) argues why the phrase '*binding on all persons by virtue of their contents shall become binding after they have been published*' – almost the entire proposed amending Article 5.2.2.a. – is indispensable in the Constitution. If I understand this rather cryptic passage correctly, the purpose of the phrase '*by virtue of their contents*' is to make sure that when a treaty has been published, the question whether it is also binding on all persons must only be determined by grounds based on its content. Regarding this assumption, no further evidence to the contrary is available.

---

[512] Parliamentary Documents II, 1977-1978 (R 1100), no. 4, p. 29, Article 5.1.10, mainly in line with the proposition of the Commission of State (Article 73 of that proposition, see: Parliamentary Documents II, 1977-1978 (R 1100), no. 4, p. 27). Original text in Dutch: '*Wettelijke voorschriften vinden geen toepassing, indien deze toepassing niet verenigbaar is met een ieder verbindende bepalingen van verdragen en van besluiten van volkenrechtelijke organisaties, nadat deze bekend zijn gemaakt.*'

[513] Parliamentary Documents II, 1977-1978 (R 1100), no. 4, p. 17-18. Original text in Dutch: '*Met name het weglaten van de in het bestaande artikel 65 voorkomende passage omtrent bepalingen van verdragen, die naar hun inhoud een ieder kunnen verbinden, ontmoet bij het college bezwaren. De strekking van die bepaling is niet zozeer een standard te geven, «waaraan getoetst zou kunnen worden, of een verdrag ieder bindt», zoals de memorie van toelichting stelt, doch om vast te stellen, dat verdragsbepalingen, die materieel (naar hun inhoud) algemeen verbindend kunnen zijn, die kwaliteit ook werkelijk bezitten, wanneer is voldaan aan een aantal voorwaarden met betrekking tot hun bekendmaking. De Raad acht deze bepaling onmisbaar, omdat daarin een belangrijk beginsel van Nederlands Constitutioneel recht wordt neergelegd, dat in verschillende andere landen – bijvoorbeeld in Groot-Brittannië – niet wordt aanvaard.*'

[514] Parliamentary Documents II, 1977-1978 (R 1100), no. 3, p. 11.

Some Members of the Preparatory Committee expressed their concern about the phrase '*binding on all persons by virtue of their contents*'. Firstly, the Supreme Court had given this phrase – that was added to the Constitution in 1956 – a very limited meaning in several cases, while in one other case the phrase was understood somewhat broader. Therefore, this wording had led to uncertainty in case law, and might therefore be ill-chosen.[515] The Government disagreed, argued that it could not see how the disputed phrase would lead to difficulties in case law, and referred to a case in which the Supreme Court clearly granted only a limited meaning to the words, an approach the Government also had in mind.[516] Secondly, Members of the Preparatory Committee argued that the case law of the European Court of Justice had shown a much broader understanding of the concept of directly applicable Provisions than was expected during the Constitutional reform of 1956: not only did the Court grant direct effect to treaty Provisions that were directly aimed at citizens, but also this effect was granted to regulations and decisions that were addressed to States. In that light, it was the opinion of the Members of the committee that it was unwise to maintain a standard in the Constitution that regulates the direct effect of an international legal standard, and it proposed to leave this matter to the Judiciary instead.[517] According to the Government, the disputed phrase would not stand in the way of such a broad interpretation, and the Dutch Judiciary is authorised, based on the proposed amending Articles, to consider the direct applicability of every international legal standard in every specific case.[518] Thirdly, with reference to the phrase quoted in the previous section, it was asked whether the phrase '*binding on all persons by virtue of their contents*' did not imply a standard that should be reviewed against by the Courts.[519] The Government answered that it was the intention to establish such a standard, but to make sure that an international Provision that is binding on all persons can only bind these persons '*after they have been published*'. The Provision thus has a guardian function, to prevent international standards to be binding on persons before they have been published, a reasoning that reminds us of the original function of Article 93 CA when it was adopted in 1953. The Government added that this did not imply that the Dutch Constitution would determine from what moment on a treaty Provision that is binding on all persons would have binding powers. Instead, this would be determined by international law and the treaty itself, albeit with consideration of the general legal principle of publication. The Government argued that with this principle of publication embedded in the Dutch Constitution only the meaning of treaty Provisions that are binding on all persons in the Dutch legal order is underlined.[520] It was thus not the purpose of the phrase

---

[515] Parliamentary Documents II, 1977-1978 (R 1100), no. 6, p. 13.

[516] Parliamentary Documents II, 1977-1978 (R 1100), no. 7, p.16.

[517] Parliamentary Documents II, 1977-1978 (R 1100), no. 6, p. 13.

[518] Parliamentary Documents II, 1977-1978 (R 1100), no. 7, pp.16-17.

[519] Parliamentary Documents II, 1977-1978 (R 1100), no. 6, pp. 13-14.

[520] Parliamentary Documents II, 1977-1978 (R 1100), no. 3, p. 11, and Parliamentary Documents II, 1977-1978 (R 1100), no. 7, p. 17.

'binding on all persons by virtue of their contents' to provide for a standard that the Courts would have to review against. However, the Government did underline that *'we consider it desirable to explicitly embed a point of reference for the Courts in the Constitution'.*[521] This was mainly considered in cases in which a treaty Provision explicitly obliges the Legislature to adopt measures within a certain time frame was under dispute.[522] The Government considered that it would be undesirable when Courts would rule in these matters, thereby disregarding the Legislature. On the other hand, the Government agreed that it *'was indeed not intended that only directly applicable Provisions would have binding power.'*[523] The fact that the phrase *'binding on all persons'* in relation to the role of the Courts was also (and even mostly) discussed in view of the current Article 93 CA at least suggests that the meaning of that phrase is similar in the current Article 94 CA, and thus based on the most resent Parliamentary History, there is no reason to understand the phrase in both Articles differently.[524]

Thus, also in the context of the current Article 94 CA, the phrase *'binding on all persons'* was discussed. The Government did not adopt the proposal of the Advisory Commission on International Legal Issues,[525] who advised to delete the words: *'By deleting the words…, new problems arise. When deleting the words "binding on all persons" also statutory regulations would be inapplicable if they are considered incompatible with Provisions of treaties and decisions of international organisations that are clearly intended to only bind the Government in its relation to other States.'*[526] This would appear *inter alia* problematic when such Provision requires further elaboration in domestic law. The Government therefore argued that *'deleting the phrase referred to may imply the risk that one assumes that there is no limitation at all regarding the nature of the Provisions of international law that are to be applied.'*[527] On the other hand, the Government nuanced this point of view by stating that in the end, the Judiciary has a final say in the matter: *'It may occur that the national Court, despite the fact that an implementation of the international legal Provisions*

---

[521] Parliamentary Documents II, 1977-1978 (R 1100), no. 7, p. 17. Original text in Dutch: *'Wij achten het gewenst uitdrukkelijk in de Grondwet in deze zin een aanknopingspunt voor de rechter te vermelden.'*

[522] Parliamentary Documents II, 1977-1978 (R 1100), no. 10, p. 10.

[523] Parliamentary Documents II, 1977-1978 (R 1100), no. 7, p. 17. Original text in Dutch: *'inderdaad niet bedoeld is dat alleen rechtstreeks werkende bepalingen van verdragen verbindende kracht hebben.'*

[524] As demonstrated above, F.M.C. Vlemminx and M.G. Boekhorst propose to understand the phrase differently, in line with the original purposes of the two Articles.

[525] Dutch name: de Commissie van Advies in Volkenrechtelijke Vraagstukken.

[526] Parliamentary Documents II, 1977-1978 (R 1100), no. 3, pp. 11-12. Original text in Dutch: *'Door het schrappen van deze woorden (...) zouden nieuwe problemen rijzen. Bij het vervallen van de woorden 'een ieder verbindende' zouden immers wettelijke voorschriften ook buiten toepassing blijven, indien zij onverenigbaar zouden moeten worden geacht met bepalingen van verdragen en van besluiten van volkenrechtelijke organisaties welke duidelijk bestemd zijn om alleen de overheid te binden in haar betrekking tot andere staten.'* See for a similar reasoning: Parliamentary Documents II, 1977-1978 (R 1100), no. 7, p. 20.

[527] Parliamentary Documents II, 1977-1978 (R 1100), no. 7, p. 20. Original text in Dutch: *'Het doen vervallen van de bedoelde uitdrukking zou het risico kunnen meebrengen, dat de gedachte postvat dat aan de aard van de toe te passen bepalingen van geschreven internationaal recht geen enkele beperking is gesteld.'*

seems to be suitable, still considers that the Provision has direct effect in the national legal order.[528] It appears that the Government struggled to find a proper balance in maintaining freedom for the Courts to rule on direct applicability of international standards, but simultaneously preventing the Courts to apply unlimitedly standards that should be further implemented by the Legislature.

Furthermore, in the context of the proposed amending Article 5.2.2.b., the Government declared that this phrase should not be explained grammatically, for it would encompass more than only duties to citizens: the phrase would cover *'also the granting of entitlements to all persons.'*[529]

One Member of Parliament (House of Representatives), Brinkhorst, submitted an amendment in which he proposed *inter alia* to delete the phrase *'binding on all persons'* in the predecessor of Article 94 CA.[530] He argued that firstly, legal practice showed a discrepancy between the broad interpretation of Provisions that are binding on all persons by the European Court of Justice, and the more narrow interpretation of other international Provisions by the national Courts. According to Brinkhorst, this discrepancy could be reduced by removing the phrase. Secondly, he underlined that also the Government had recognised the possibility that a Court could rule that a Provision has direct effect, even when it is not binding on all persons. The added value of such a standard in a Constitution then is unclear. And thirdly, he observed that the Netherlands had established a tradition of taking a very long time at ratifying a treaty. He considered that the Dutch Legislature is usually very keen on making sure that adequate implementation legislation is adopted to avoid a legal gasp when the treaty Provisions are ratified. Brinkhorst suggested a causal link between the long period before ratification and the phrase *'binding on all persons'*, and argued that the deletion of these words would contribute to a more relaxed an open attitude towards the implementation of treaties.[531] However, in the subsequent meeting of the House of Representatives, Brinkhorst withdrew his amendment, due to the fact that he had to conclude that he overlooked the fact that also in the predecessor of Article 93 the words *'binding on all persons'* were embedded, and should also be removed for in order to establish a coherent Constitutional system. He also observed that he would not convince a majority of the House of Representatives, and wanted to prevent his amendment to be

---

[528] Parliamentary Documents II, 1977-1978 (R 1100), no. 3, p. 12. Original text in Dutch: *'Het kan zijn, dat de nationale rechter ondanks het feit, dat een implementatie van de internationaalrechtelijke bepalingen aangewezen lijkt, niettemin een rechtstreekse 'binding' van die bepaling voor de nationale rechtsorde aanwezig acht.'*

[529] Parliamentary Documents II, 1977-1978 (R 1100), no. 3, p. 11. Original text in Dutch: *'ook het toekennen aan een ieder van aanspraken'*.

[530] Parliamentary Documents II, 1977-1978 (R 1100), no. 14.

[531] Mr Brinkhorst, House of Representatives, 62nd meeting, 18 March 1980.

counterproductive when it would be rejected and as a result, the phrase '*binding on all persons*' would be reaffirmed perhaps even more firmly than before.[532]

In the context of the effect of international standards in the domestic legal order, another extensive discussion was held on the question to what extent unwritten international standards were also binding in the Dutch legal order. The Government argued that unwritten international standards were binding in the Dutch legal order. However, it was added that the Constitution did not imply that national regulations that were contradictory to unwritten international standards would not be applied, in line with the recommendation of the Council of State.[533] The Government underlined that firstly, this would be unpractical, for unwritten international law would mostly be vague. Secondly, it would be undesirable if treaties that were not yet ratified were considered to be unwritten international law, and thus be applied accordingly. And thirdly, the Government argued that unwritten international law is often difficult to recognise for citizens.[534] The matter was extensively discussed during the subsequent Parliamentary discussions.[535]

A final conclusion was drawn by the Government, who summarised the Constitutional system of the current Articles 93 and 94 CA: the Legislature reviews the Constitution against international law, and (national) statutory regulations against unwritten international law. The Judiciary is competent to review inferior national legal rules against superior national legal rules, and (national) statutory regulations against international Provisions that are directly binding.[536]

5.9 Conclusion

As it appears from the above analysis of the Parliamentary History of the three Constitutional reforms that finally resulted in the current system embedded in Articles 93 and 94 CA, the Constitutional Legislature made certain fundamental choices regarding that system in 1953. Since then, it was not the ambition of the Legislature to alter the earlier adopted Constitutional principles, but merely to make some technical amendments. Despite this, it was deemed necessary to further enlighten the functioning of the Constitutional system and clarify some of its wordings in especially the Constitutional reform of 1983. The following conclusions on the meaning of Articles 93 and 94 CA as considered by the Legislature can be drawn:
- Monism
  In 1953, the Constitutional Legislator made an explicit choice for a monistic system regarding the working of international legal standards in the national

---

[532] Mr Brinkhorst, House of Representatives, 73rd meeting, 23 April 1980.
[533] Parliamentary Documents II, 1977-1978 (R 1100), no. 4. p. 19.
[534] Parliamentary Documents II, 1977-1978 (R 1100), no. 10, pp. 10-11.
[535] House of Representatives, 62nd meeting, 18 March 1980, 66th meeting, 25 March 1980, 73rd meeting, 23 April 1980.
[536] See especially: Parliamentary Documents II, 1977-1978 (R 1100), no. 19. pp. 1-3.

legal order. Basically, this monism was not intended to be '*qualified*' in any way, and with regard to the working of international standards in the domestic order not limited at all. The working of international standards in the domestic legal order even includes international customary law or other unwritten standards. However, when it comes to judicial review, domestic statutory regulations that are in contradiction with international standards do not apply only then when they have been published and are binding on all persons. According to the Legislature, this is not a limitation to monism as such, but is experienced as a limitation in literature, which explains the use of the term '*qualified monism*'.

- The function of Articles 93 and 94 CA

  The above analysis of the Parliamentary History of the Constitutional reforms shows that there can be no misunderstanding regarding the functions of Articles 93 and 94 CA. Article 93 determines what international standards are binding to citizens. Article 94 CA determines how these binding standards relate to domestic law. As a result, the Judiciary has the competence to review statutory regulations against these international standards that are binding on all persons, and has the obligation not to apply a conflicting national standard. In addition, as pointed out during the Constitutional reform of 1953, the Judiciary may – based on Article 93 CA – even apply the international standard instead. A point of view that has not been discussed ever since.

- 'Binding on all persons'

  The phrase '*binding on all persons*' was embedded in both Articles 93 and 94 CA in 1956. The amendment was perceived then to be a technical change only, and not as an amendment to the Constitutional principles adopted earlier in 1953. Already during the reforms of 1953, it was made clear that a national standard could only be not applied when it was in conflict with an international standard that is self-executing. Therefore, adding the phrase '*binding on all persons*' in both Articles did not change that principle. Although '*self-executing standards*' and '*standards that are binding on all persons*' might be understood slightly differently, it appears that throughout the years the Legislature did not intentionally differ between those terms. The fact that according to the Legislature from the introduction of the phrase in Article 93 CA it naturally followed that the same phrase was also added to Article 94 CA. Especially during the 1983 reform the phrase was explained as one concept in both Articles, suggesting that there is no significant difference in understanding those words in the context of each particular Article. The phrase '*binding on all persons*' was understood as a protection of the division of powers in the Dutch *trias politica,* and implies a basic principle – not a specific norm – that should be taken into consideration by the Courts when reaching a verdict, in order to prevent the Judiciary from implementing standards that are addressed to the Legislature. It is however, according to the Legislature, without doubt that the Judiciary has a final say in determining whether a standard is binding on all persons and consequently directly applicable.

- From duties to entitlements

    Originally, the purpose of the Constitutional system regarding the working of international law in the national legal order was to regulate the effect of international standards that would impose duties on citizens, and more or less expand the principle of legality to the territory of international law as well by implementing a publication demand. Gradually, this focus on duties for citizens shifted to a focus on both duties and entitlements for citizens in 1983. Legal practice however shows a strong focus on international standards that stipulate entitlements only.

- The Legislature, the Judiciary and *trias politica*

    During the Constitutional reform of 1953, the explicit choice was made to embed judicial review of national standards against international law in the Constitution, instead of review by the Legislature. The Government did not intend to regulate this at all in its original draft, but judicial review was adopted by a small majority by amendment, after an extensive debate in which both defenders of judicial review and defenders of review by the Legislature argued why the Constitution should not remain silent on the matter. The amendment was altered by the submitters on the same day as it was put to the vote, followed by a brief, hasty and unstructured Parliamentary discussion. Only a small majority voted for the amendment. During the Parliamentary discussion, two questions were most frequently addressed, and were also discussed during the subsequent Constitutional reforms, especially the 1983 reform. Firstly, the question was raised against what standards exactly the Judiciary was competent to review. It can be concluded that on national level, the Court reviews inferior statutory regulations against superior statutory regulations (including acts of Parliament), the Legislature reviews its own legislation against the Constitution. Concerning international law, the Judiciary reviews national statutory regulations (except for Constitutional Provisions) against international standards that are binding on all persons. The Legislature reviews the Constitution against international standards, and all national statutory regulations against unwritten international law. Implicitly, it can be concluded that the Legislature thus also reviews statutory regulations against international standards that are not binding on all persons. This leads to the second question, which is: who decides whether an international Provision is directly applicable or not? Already during the Constitutional reform of 1953 it was made quite clear that it is the Judiciary, and not the Legislature, who has a final say on determining whether a treaty Provision is directly applicable. However, there are two limitations to this embedded in the Constitution: firstly, the Provision must be published before it can have effect in the national legal order. This principle was embedded as a safeguard to prevent citizens from being bound by treaty Provisions unknowingly: a sort of principle of legality. Secondly, the Provision must be binding on all persons. This principle was not adopted to express a clear standard the Judiciary should review against before applying an international standard, but rather to express a point of reference to the Courts, and underline that international law cannot be directly applicable without any limits. The Constitutional system thus uncomfortably balances between recognising the final judgment of the Judiciary on the direct applicability of international Provisions, and not unlimitedly broadening these review powers resulting in Court verdicts that interfere with the competences of the Legislature.

6. Dutch Parliamentary History on the right to adequate food

6.1 Introduction

As demonstrated in Chapter 4, the Courts frequently refer to Parliamentary History in their considerations when deciding on individual appeals to the right to food, and seem to greatly value the opinion of the Legislature on the matter. The question whether the right to food is directly applicable or not is however not specifically discussed in Parliamentary History. Rather, in the Explanatory Memoranda and subsequent documentation, the obligation coming from ratifying the right to food was discussed per Provision, whereas the matter of direct applicability of ECOSOC rights in general is considered separately. Therefore it is the purpose of this Chapter to determine and analyse how Dutch Legislature interpreted the obligations that are implied when ratifying Provisions that stipulate the right to food, and in Chapter 7, the opinion of the Legislature on direct applicability of ECOSOC rights as expressed in Parliamentary History will be analysed.

As demonstrated in Chapter 3, the right to food is embedded in various ways in the international human rights system. In some treaty Provisions, the right is recognised as a more or less independent right. That means basically – considering the ratifications of the Netherlands – Article 11 of the International Covenant on Economic, Social and Cultural Rights (Section 6.2),[537] and in the treaties that aim at the protection of certain groups of individuals especially Article 27 (Section 6.5).[538] Furthermore, in the field of healthcare and pregnancy, the right to food is stipulated in Articles 12 (2) of The Convention on the Elimination of All forms of Discrimination against Women (Section 6.6) and 24 of the International Covenant for the Rights of the Child (Section 6.3).[539] Therefore, naturally the Parliamentary History of the ratification Bills will be analysed when it concerns these Provisions.

Besides, there is reason to also include the Parliamentary History on Article 26 in the analysis. The Netherlands made a reservation to Article 26, in which it stated that the Article '*shall not imply an independent entitlement of children to social security, including social insurance.*'[540] The reservation was a reflection of the general idea of the Dutch Government regarding the treaty: children cannot derive independent rights directly based on the treaty, for the Dutch legislation is thus formulated that children generally have only indirect entitlements, via the parents or legal

---

[537] A/RES/21/2200, 16 December 1966, *International Covenant on Economic, Social and Cultural Rights,* Article 11.
[538] A/RES/44/25, 20 November 1989, *Convention on the Rights of the Child,* Article 27.
[539] A/RES/34/180, 18 December 1979, *Convention on the Elimination of All Forms of Discrimination Against Women,* Article 12(2).
[540] Parliamentary Documents, II 1992-1993, 22855 (R1451), no. 2, Article 2.

guardians. This approach was more than once debated during the Parliamentary debates, also in the context of Article 27 ICRC. Also, the Government referred to the Dutch system of social security to substantiate the Dutch compliance with the obligations coming forth from Articles 26 and 27 ICRC in its Explanatory Memorandum. Therefore, the analysis of Parliamentary History on Article 26 ICRC cannot be excluded from this thesis (Section 6.4).

6.2 Article 11 ICESCR

Article 11 ICESCR stipulates:

1. The States Parties to the present Covenant recognise the right of everyone to an adequate standard of living for himself and his family, including adequate food, clothing and housing, and to the continuous improvement of living conditions. The States Parties will take appropriate steps to ensure the realisation of this right, recognising to this effect the essential importance of international cooperation based on free consent.
2. The States Parties to the present Covenant, recognising the fundamental right of everyone to be free from hunger, shall take, individually and through international co-operation, the measures, including specific programmes, which are needed:
    - To improve methods of production, conservation and distribution of food by making full use of technical and scientific knowledge, by disseminating knowledge of the principles of nutrition and by developing or reforming agrarian systems in such a way as to achieve the most efficient development and utilisation of natural resources.
    - Taking into account the problems of both food-importing and food-exporting countries, to ensure an equitable distribution of world food supplies in relation to need.

In its Explanatory Memorandum, the Government merely clarified why the Netherlands at the moment of ratification already fulfilled its duties that are implied in Article 11 ICESCR. Regarding the right to food, it was stated that no general regulation existed in the Netherlands that would regulate food supply, although there was a common EC policy on agriculture in place. In that light, Article 39 of the EEC treaty stipulated the goal to secure the food supplies within the EU. To this end, the organisation of the common market offered several instruments. Furthermore, a reference was made to emergency law on food supply.[541] Regarding the ensuring of food quality, a reference was made to the Quality Law

---

[541] In Dutch: Noodwet Voedselvoorzieningen (Stb. 1935, 793); Hamsterwet (Stb. 1962, 542).

on Agriculture,[542] the Commodities Act,[543] and the Product Board Regulation.[544] Also, a reference was made to agricultural education and research, offered and conducted by several institutions, some of them subsidised by the Government. In addition, the Government considered that the Netherlands contributed to the realisation of Article 11 ICESCR by supporting several programmes and institutions in the international arena, in the field of development cooperation.[545]

The Members of the Preperatory Committee of the Dutch political party Democratic Socialists '70 (DS '70) asked whether the general idea expressed in the Article of *'the continuous improvement of living conditions'* was a realistic goal that could be strived for endlessly, for *'the Treaty Provision on the continuous improvement of living conditions collides with the acquired insights on the limits to growth.'*[546] To this, the Government replied that *'This idea is indeed based on beliefs and expectations from the years in which the Treaties were drawn, while these beliefs and expectations in most recent years have changed, in which it has been understood at the global level that unlimited growth will not be unlimited.'*[547] The Government emphasised that especially for developing countries the continuous improvement of living conditions was an absolute necessity. Regarding developed countries however, the Government argued that *'if in developed countries limits must be imposed on growth, or if there should be a possible decline in the enjoyment of economic and social rights, the second Section of Article 2 and Article 4 of this Treaty then should be a guideline.'*[548] From this, it appears that the goal as stipulated in Article 11 ICESCR to continuous improvement of living conditions was understood by the Dutch Government as a goal for especially developing countries, while developed countries would have the obligation to justify any restrictions to the enjoyment of ECOSOC rights based on Articles 2 (2) and 4 ICESCR.

---

[542] In Dutch: Landbouwkwaliteitswet (Stb. 1971, 371).

[543] In Dutch: Warenwet (Stb. 1935, 793).

[544] In Dutch: Productschapsverordening.

[545] Parliamentary Documents, II 1975-1976 13932 (R 1037), no. 3, pp. 50-51.

[546] Parliamentary Documents, II 1975-1976 13932 (R 1037), no. 7, p. 16. Original text in Dutch: *'De verdragsbepaling inzake steeds betere levensomstandigheden botst met de verworven inzichten inzake de grenzen van de groei.'*

[547] Parliamentary Documents, II 1975-1976 13932 (R 1037), no. 8, p. 26. Original text in Dutch: *'Deze gedachte is inderdaad gebaseerd op opvattingen en verwachtingen uit de jaren waarin de Verdragen zijn opgesteld en welke opvattingen en verwachtingen in de meest recente jaren, waarin het besef is doorgedrongen dat er op wereldniveau geen sprake kan zijn van onbegrensde groei, tot een kentering zijn gekomen.'*

[548] Parliamentary Documents, II 1975-1976 13932 (R 1037), no. 8, p. 26. Original text in Dutch: *'Indien in de ontwikkelde landen grenzen moeten worden gesteld aan de groei, of indien sprake zou moeten zijn van een eventuele teruggang in het genot van economische en sociale rechten, zullen hierbij het tweede lid van artikel 2 en artikel 4 van het onderhavige Verdrag tot richtsnoer moeten dienen.'*

During the subsequent debates in Parliament, Article 11 ICESCR was hardly discussed.[549]

6.3 Article 24 ICRC

Article 24 ICRC stipulates that:

1. States Parties recognise the right of the child to the enjoyment of the highest attainable standard of health and to facilities for the treatment of illness and rehabilitation of health. States Parties shall strive to ensure that no child is deprived of his or her right of access to such healthcare services.
2. States Parties shall pursue full implementation of this right and, in particular, shall take appropriate measures:
   - to diminish infant and child mortality;
   - to ensure the Provision of necessary medical assistance and healthcare to all children with emphasis on the development of primary healthcare;
   - *to combat disease and malnutrition, including within the framework of primary healthcare, through,* inter alia, *the application of readily available technology and through the Provision of adequate nutritious foods and clean drinking-water, taking into consideration the dangers and risks of environmental pollution;*
   - to ensure appropriate pre-natal and post-natal healthcare for mothers;
   - to ensure that all segments of society, in particular parents and children, are informed, have access to education and are supported in the use of basic knowledge of child health and nutrition, the advantages of breastfeeding, hygiene and environmental sanitation and the prevention of accidents;
   - to develop preventive healthcare, guidance for parents and family planning education and services.
3. States Parties shall take all effective and appropriate measures with a view to abolishing traditional practices prejudicial to the health of children.
4. States Parties undertake to promote and encourage international co-operation with a view to achieving progressively the full realisation of the

---

[549] In the Senate, B. De Gaay Fortman (MP for the Dutch political party PPR, cf. Rainbow/Radical Party; now merged into a green party) acknowledged that ECOSOC rights do not encompass individual entitlements. However, he opposed the idea that the ICESCR would be promotional only. He argued for promotional duties, which oblige a Government to promote access for the individual to what he is entitled to, regardless of the economic situation. In this context De Gaay Fortman, particularly referred to the right to an adequate standard of living and the right to be free from hunger. He made the observation that: *'it seems as though we are now treading water with regard to the obligations the ICESCR bring us.'* See: B. de Gaay Fortman (PPR), Senate, 5th meeting, 21 November 1978. Original text in Dutch: *'Het lijkt erop alsof wij thans pas op de plaats willen maken in de verplichtingen die dat voor ons meebrengt.'*

right recognised in the present Article. In this regard, particular account shall be taken of the needs of developing countries.

In its Explanatory Memorandum, the Government only briefly discussed the contents of Article 24 ICRC, since it considered the contents to be equivalent to Article 12 ICESCR, which was already extensively discussed in the Explanatory Memorandum to the ICESCR.[550] The Government merely referred to the Collective Prevention of Public Health Act,[551] which entrusted the Communities with various responsibilities in the field of health prevention, especially regarding minors. Food-related issues were not discussed. The Government considered the contents of Article 24 ICRC to be equivalent to the contents of Article 12 ICESCR, 11 ESC and 22 (1) CA.[552]

6.4 Article 26 ICRC

Article 26 ICRC stipulates that:

1. States Parties shall recognise for every child the right to benefit from social security, including social insurance, and shall take the necessary measures to achieve the full realisation of this right in accordance with their national law.
2. The benefits should, where appropriate, be granted, taking into account the resources and the circumstances of the child and persons having responsibility for the maintenance of the child, as well as any other consideration relevant to an application for benefits made by or on behalf of the child.

The Government stated that Article 9 ICESCR, 12 and 13 ESC and 20 CA stipulated corresponding rights.[553] In its Explanatory Memorandum, the Government gave an overview of social security law, and underlined that children had mostly access to social benefits through their parents or legal guardians.[554] Therefore, a Reservation was proposed to Article 26 ICRC: '*The Kingdom of the Netherlands accepts the Provisions of Article 26 of the Convention with the reservation that these Provisions shall not imply an independent entitlement of children to social security,*

---

[550] In its Explanatory Memorandum on the ratification Bill of the ICESCR, the Government had indeed more extensively discussed the implementation of Article 12 ICESCR, mainly by referring to the various existing legislation in the field of prenatal care, health care, and a healthy living and working environment. Nutrition-related issues were not mentioned. See Parliamentary Documents, II 1975-1976 13932 (R 1037), no. 3, pp. 51-52.

[551] In Dutch: Wet Collectieve Preventie van Volksgezondheid.

[552] Parliamentary Documents, II 1992-1993, 22855 (R1451), no. 3, p. 34.

[553] Parliamentary Documents, II 1992-1993, 22855 (R1451), no. 3, third attachment, pp. 55-56.

[554] Parliamentary Documents, II 1992-1993, 22855 (R1451), no. 3. pp. 35-36.

including social insurance.'[555] In this context the Government had generally stated in its Explanatory Memorandum that the treaty does not imply any mechanisms through which the child could effectuate its rights individually, and that such a mechanism was also not necessary, for in principle, the parents or legal guardians would be responsible for the effectuation of the child's rights.[556] Despite this assumption, the Government explained that the reservation was deemed necessary for the wordings of Article 26 ICRC might imply that a child would have an independent right to social security, which is in contradiction with Dutch social security law: *'While in the Kingdom the child, in some cases, in the capacity of an employee or resident, may have independent claims to benefits under social security regulations, in practice usually the social security for the child is derived from the social security that belongs to the parents. The height of the latter is hereby determined so that the parental care and maintenance obligations towards children can be met. Independent social security benefits for the child only exist in the Netherlands to a limited extent, and there are no future plans to change this. Therefore, to preclude a different interpretation, it is proposed to make a Reservation for the Kingdom, meaning that the Government considers that Article 26 does not imply an independent right to social security for children, including social insurance.'*[557] In the Preparatory Committee however, both the Labour Party and Liberal Party raised the question whether such a Reservation would be necessary, for the wordings of Article 26 in itself would not imply an obligation to grant the child individual entitlements to social security. Furthermore, it was assumed by the Liberal Party that almost all other countries that signed the ICRC would have social security systems that were far less advanced compared to the Dutch system. Therefore, in those countries, the idea that a child would have individual entitlements would not even be considered, as long as adults would not yet have proper entitlements to social security.[558] The Government replied that it was unaware of the interpretation of other countries of the Article, and underlined that in some countries a dualistic system was embedded in their Constitution, and thus entitlements of citizens could only be

---

[555] Parliamentary Documents, II 1992-1993, 22855 (R1451), no. 2, Article 2. Original text in Dutch: *'Het Koninkrijk der Nederlanden aanvaardt het bepaalde in artikel 26 van het verdrag, onder het voorbehoud dat deze bepaling niet verplicht tot een zelfstandig recht van kinderen op sociale zekerheid, daarbij inbegrepen sociale verzekering.'*

[556] Parliamentary Documents, II 1992-1993, 22855 (R1451), no. 3, p. 12.

[557] Parliamentary Documents, II 1992-1993, 22855 (R1451), no. 3, pp. 35-36. Original text in Dutch: *'Hoewel in het Koninkrijk het kind in voorkomende gevallen, in de hoedanigheid van werknemer of als ingezetene, zelfstandige aanspraken kan hebben op prestaties krachtens de sociale zekerheid komt het er in de praktijk veelal op neer dat de sociale zekerheid voor het kind is afgeleid van de sociale zekerheid die toekomt aan de ouders. De hoogte van de aan laatstgenoemden verstrekte sociale zekerheidsuitkeringen is hierbij zodanig vastgesteld dat aan de ouderlijke zorg- en onderhoudsverplichtingen jegens de kinderen kan worden voldaan. Zelfstandige sociale zekerheidsaanspraken voor het kind bestaan in Nederland slechts in beperkte mate en er bestaan geen voornemens hierin voor de toekomst verandering te brengen. Derhalve wordt, teneinde een andere opvatting ter zake uit te sluiten, voorgesteld voor het Koninkrijk een voorbehoud te maken, inhoudend dat de regering van mening is dat artikel 26 niet verplicht tot een zelfstandig recht van kinderen op sociale zekerheid, daarbij inbegrepen sociale verzekering.'*

[558] Parliamentary Documents, II 1992-1993, 22855 (R1451), no. 4, p. 19.

derived from domestic legislation. Regarding the wordings, the Government stated that *'In our opinion, however, neither the wordings of Article 26, nor the "travaux préparatoire" provide sufficient certainty to conclude that from Article 26, no independent rights to social security for children, including social insurance, could result. It is also particularly uncertain whether the Dutch Court could come to such a conclusion.'*[559] The fact therefore that the Courts could decide on the applicability of the Article was a reason for the Government to maintain the Reservation. The Government explained, on a request of the Preparatory Committee for the Senate,[560] that the Reservation was not unnecessary, or an expression of timorousness, for *'also in view of experiences with other international instruments'* it was obvious *'to critically consider new obligations.'*[561] From the subsequent debates in Parliament, it can be deduced that especially the broad interpretation of the ECHR and the ICCPR amongst the Courts had not been expected by the then-Government, and was reason for caution regarding the ratification of the ICRC.[562] The Government stated that with this Reservation, the Netherlands could clearly express their interpretation of the Article, but also, it could be prevented that any undesired situation, especially in financial terms, would occur when the Courts would rule on the Provision, considering the system embedded in Articles 93 and 94 Constitutional Act, implying that the Courts have the sole authority to give a final interpretation on the meaning of international Provisions.[563] During the Parliamentary debates on the ratification Bill of the ICRC, the Reservation was a permanent cause for discussion, in which Articles 26 and 27 were sometimes linked to one another.

---

[559] Parliamentary Documents, II 1993-1994, 22855 (R1451), no. 6, p. 31. Original text in Dutch: *'Naar ons oordeel echter geven noch de tekst van artikel 26, noch de "travaux préparatoire" voldoende zekerheid om te kunnen concluderen dat er voor kinderen geen zelfstandig recht op sociale zekerheid, met inbegrip van sociale verzekeringen, uit artikel 26 zou kunnen voortvloeien. Het is met name ook onzeker of de Nederlandse rechter niet tot een dergelijke conclusie zou kunnen komen.'* See for a similar reasoning by the Government: House of Representatives, 84th meeting, 30 June 1994, second note.

[560] Parliamentary Documents, I 1994-1995, 22855 (R1451), no. 22, p. 1, question posed by the CDA (Christian Democratic Appeal) fraction in which the Government was asked to respond to a letter of Defence for Children (NGO), of 26 September 1994.

[561] Parliamentary Documents, I 1994-1995, 22855 (R1451), no. 22a, p. 1.

[562] See for instance the contribution of Mrs Soutendijk-van Appeldoorn (CDA, Christian Democratic Appeal), House of Representatives, 81st meeting, 23 June 1994 (regarding the ECHR), and Mr Koekoek (CDA, Christian Democratic Appeal), House of Representatives, 84th meeting, 30 June 1994, regarding the ICCPR. Koekoek stated with regard to Article 26 ICCPR that: *'The highest Court on social security has decided for many years that this Article is not binding on all persons, but at one point ruled that it had lasted long enough, and therefore the Article was now binding on all persons. This has cost the public treasury millions. We are thus warned. It is therefore wise to indeed make the reservation.'* Original text in Dutch: *'De hoogste sociale-zekerheidsrechter heeft hiervan vele jaren gezegd dat dit artikel niet een ieder verbindend was, maar op een gegeven moment heeft hij gezegd dat het nu lang genoeg geduurd had en dat het artikel nu wel een ieder verbindend was. Dat heeft de schatkist miljoenen gekost. Wij zijn dus gewaarschuwd. Het is dus verstandig om het voorbehoud wel te maken.'* See also the Note of the Government on direct effect: House of Representatives, 84th meeting, 30 June 1994, second note.

[563] Parliamentary Documents, I 1994-1995, 22855 (R1451), no. 22a, p. 1. See also: House of Representatives, 84th meeting, 30 June 1994, second note.

Two Members of Parliament proposed an amendment in order to withdraw the Reservation.[564] The Government advised not to support this Amendment.[565]

In the House of Representatives, some Members pleaded against maintaining the Reservation. Van der Burg (PvdA, Labour Party), argued that the reasoning of the Government was incorrect, and it was clear that Article 26 ICRC would not imply any direct entitlements to social security of the child. Firstly, she underlined that Article 27 ICRC clearly stipulates that the parents or legal guardians are primarily responsible for the child, and need to be enabled to take that responsibility, which is then a duty of the Member State. Only in occasional situations, a minor may apply for social benefits in the Netherlands, although it cannot be deduced from the Treaty that a child has an independent right to such social security. A strong interrelationship between Articles 26 and 27 is thus assumed. Secondly, Van der Burg could not understand the precaution of the Government regarding the reviewing competences of the Judiciary, and reminded the Government that they were of the opinion that *'the review competences of the judge in this case are not so obvious, because it is primarily the responsibility of a democratically elected Government to assess the possibility, taking into account the national circumstances and financial situation (...).'*[566] Apparently, she referred to the earlier replies of the Government to questions from the Preparatory Committee, in which the Government had stated that: *'Formulations in treaty Provisions that oblige States Parties to undertake appropriate measures to facilitate the implementation of the rights in accordance with national conditions and with the available resources, first and foremost intend to make clear that a Member State is not bound to the impossible. Such Provisions are without prejudice to the responsibility of the democratically elected Government, but rather appeal to its primary responsibility to do what is reasonably possible, considering the national (especially financial) situation. This is exactly why in these cases judicial review is unlikely.'*[567] Thirdly, Van der Burg was surprised that the Government anticipated to a possible direct effect of Article 26 regarding independent rights of minors, while the Netherlands had signed and ratified Treaties with a similar content, in some cases with even more extensive wordings compared to Article

---

[564] Parliamentary Documents, II 1993-1994, 22855 (R1451), no. 10, submitted by Van den Burg and Versnel-Schmitz on behalf of the PvdA (Labour Party) and D'66 (Democrats '66).

[565] House of Representatives, 84th meeting, 30 June 1994, second note.

[566] Mrs Van der Burg (PvdA, Labour Party), House of Representatives, 81st meeting, 23 June 1994. Original text in Dutch: *'de toetsingstaak van de rechter in dit geval niet voor de hand ligt, omdat het primair de taak is van een democratisch gekozen bestuur om te kunnen beoordelen of het mogelijk is, gelet op de nationale omstandigheden en de financiële situatie (...).'*

[567] Parliamentary Documents, II 1993-1994, 22855 (R1451), no. 6, p. 9. Original text in Dutch: *'Formuleringen in verdragsbepalingen die verdragsstaten verplichten passende maatregelen te nemen om te helpen rechten te verwezenlijken in overeenstemming met de nationale omstandigheden en met de middelen die hen ten dienste staan, beogen eerst en vooral duidelijk te maken, dat een verdragsstaat niet tot het onmogelijke is gehouden. Dergelijke bepalingen doen dus niet af aan de verantwoordelijkheid van het democratisch gekozen bestuur, doch appelleren veeleer aan diens primaire verantwoordelijkheid ter zake te doen wat redelijkerwijs mogelijk is, gelet op de nationale (vooral financiële) situatie. Juist daarom ligt in dezen een toetsingstaak van de rechter niet voor de hand.'* See for a further context of this phrase: *Section 7.6.2.*

26 ICRC, without a Reservation. She referred to the ICESCR and ESC, and stated that these treaties stipulated rights to everyone, including children.[568]

Also De Vries (VVD, Liberal Party), later supported by Van der Burg (PvdA, Labour Party),[569] emphasised the interrelation between Articles 26 and 27. She underlined that Article 26 (1) ICRC stipulated that States Parties shall recognise for every child the right to benefit from social security, including social insurance, and shall take the necessary measures to achieve the full realisation of this right in accordance with their national law, while in Aricle 26 (2) ICRC it was recognised that those benefits should be granted, taking into account the resources and the circumstances of the child and persons having responsibility for the maintenance of the child. Article 27 then explicitly stipulates the State's obligation to provide for material assistance to enable parents to realise an adequate standard of living for their child. In general, she remarked that the Treaty starts from the assumption that the parents are primarily responsible for their children. Considering all this, De Vries could not see how the Treaty would imply then independent rights for the child in the field of social security, and therefore, the Reservation could be withdrawn. In addition to that, she referred to the fact that countries with a similar level of social security did not make such a reservation to the treaty.[570] Mrs Versnel-Schmitz (D'66, Democrats '66) had a similar view on the interrelation between Articles 26 and 27 ICRC, and added that '*it is clearly intended in Articles 26 and 27 to work indirectly via the parents. In that sense, Article 26 does not oblige to establish an independent right.*'[571]

Other Members of Parliament however agreed with the Government, and especially the confessional parties underlined that the wordings of Article 26 did not sufficiently preclude that States were obliged to guarantee an independent right of the child to social benefits, and that such an independent right would be a violation of parental authority.[572] The fact that the ICRC specifically focussed on the rights of the child was considered to be a possible factor that could lead to an easier recognition of an independent right for children, contrary to the earlier ratified general human rights treaties: although in those treaties similar rights were recognised, the treaties had a much broader scope (each individual, instead of children), and consequently there would be more room for interpretation in the context of the national systems of social security, and thus also more room to

---

[568] Mrs Van der Burg (PvdA), House of Representatives, 81st meeting, 23 June 1994.

[569] Mrs Van der Burg (PvdA), House of Representatives, 84th meeting, 30 June 1994.

[570] Mrs De Vries, House of Representatives, 81st meeting, 23 June 1994.

[571] Mrs Versnel-Schmitz (D'66, Democrats '66), House of Representatives, 81st meeting, 23 June 1994. Original text in Dutch: '*Het is ook duidelijk de opzet van de artikelen 26 en 27 om via de omweg van de ouders te werken. Artikel 26 verplicht in die zin niet tot een zelfstandig recht.*' See also her contribution to the House of Representatives, 84th meeting, 30 June 1994.

[572] See for instance the contribution of Mrs Soutendijk-van Appeldoorn (CDA, Christian Democratic Appeal), House of Representatives, 81st meeting, 23 June 1994; Van den Berg (SGP, Political Reformed Party), House of Representatives, 84th meeting, 30 June 1994.

respect parental authority.[573] Also, the Government underlined that there would be a larger margin of appreciation for the Administration when a Provision was not aimed at a specific group.[574] Minister Kosto underlined as well that the ICRC was aimed at a specific group and *'because of that, the Courts could decide to recognise direct effect sooner than in case of a convention that is addressed to everyone.'*[575] The Government stated that despite the fact that they were of the opinion that Article 26 would not be suitable for direct effect, the Courts had a final say in that matter, and used their own criteria to determine the possibility of direct effect. To avoid a possible interpretation that would grant independent entitlements to social security to children, the Reservation was deemed necessary, and therefore not withdrawn.[576]

The Dutch Judiciary since then consistently ruled – line with the Reservation – that children had no direct entitlements to social benefits based on Article 26, and therefore most claims based on Article 26 were rejected. Consequently, the Courts did hardly rule on the possibility of direct applicability of the Provision.[577]

6.5 Article 27 ICRC

Article 27 ICRC stipulates that:

1. States Parties recognise the right of every child to a standard of living adequate for the child's physical, mental, spiritual, moral and social development.
2. The parent(s) or others responsible for the child have the primary responsibility to secure, within their abilities and financial capacities, the conditions of living necessary for the child's development.
3. States Parties, in accordance with national conditions and within their means, shall take appropriate measures to assist parents and others responsible for the child to implement this right and shall in case of need provide material assistance and support programmes, particularly with regard to nutrition, clothing and housing.
4. States Parties shall take all appropriate measures to secure the recovery of maintenance for the child from the parents or other persons having financial responsibility for the child, both within the State Party and from

---

[573] See for instance the contribution of Mrs Soutendijk-van Appeldoorn (CDA, Christian Democratic Appeal), House of Representatives, 81st meeting, 23 June 1994.

[574] House of Representatives, 84th meeting, 30 June 1994, second note.

[575] Minister Kosto, House of Representatives, 84th meeting, 30 June 1994. Original text in Dutch: *'waardoor eerder dan bij verdragen die zich tot een ieder richten, een rechter tot erkenning van rechtstreekse werking kan besluiten.'*

[576] House of Representatives, 84th meeting, 30 June 1994, second note. See also the contribution of Minister Kosto during the same meeting.

[577] See Section 4.5.

abroad. In particular, where the person having financial responsibility for the child lives in a State different from that of the child, States Parties shall promote the accession to international agreements or the conclusion of such agreements, as well as the making of other appropriate arrangements.

The Government stated in its Explanatory Memorandum that Article 27 ICRC stipulated equivalent rights compared to Article 11 ICESCR and Article 22 CA.[578] According to the Government, the added value of ratifying Article 27 ICRC on top of the already ratified Article 11 ICESCR was that the ICRC Provision recognises that children also, and in particular, have a right to an adequate standard of living, *'taking into account the specific needs of the child: the standard of living should indeed be sufficient in view of the physical, mental, spiritual, moral and social development of the child.'*[579] The Government emphasised that Article 27 (2) ICESCR stipulates that it is primarily the responsibility of the parents (or other legal guardians) to secure, within their ability and financial capacities, the conditions of living for the child. Only when the parents or legal guardians are not capable to take that responsibility, Article 27 (3) ICRC stipulates that it is the obligation of a Member State to take appropriate measures to assist parents and others responsible for the child to implement this right, in accordance with national conditions and within their means, by providing material assistance and support programmes, particularly with regard to nutrition, clothing and housing.[580] According to the Government, from the wording of Article 27 ICRC therefore, contrary to Article 26, it cannot be deduced that a child would have direct entitlement to social benefits.[581] Article 27 (4) ICRC then gives an international dimension to the rights enshrined in Article 27. In its Explanatory Memorandum, the Government basically refers to domestic legislation through which the State obligations enshrined in Article 27 were already fulfilled by the moment of ratification, such as the Social Assistance Act,[582] regulations regarding financial support for youth care (including the Youth Care Act[583]) and legislation to guarantee the maintenance of children abroad.[584]

As mentioned above, it is noteworthy here that a group of Parliamentarians who advised not to adopt a Reservation to Article 26 ICRC argued that from Article 27 ICRC, it already appeared that the parents or the legal guardians were primarily responsible for the adequate standard of livings of the child. Therefore, when reading Article 27 in conjunction with Article 26, it could be concluded that

---

[578] Parliamentary Documents, II 1992-1993, 22855 (R1451), no. 3, third attachment, pp. 55-56.

[579] Parliamentary Documents, II 1992-1993, 22855 (R1451), no. 3, p. 37. Original text in Dutch: *'rekening houdend met de specifieke behoeften van het kind: de levensstandaard moet immers toereikend zijn voor de lichamelijke, geestelijke, intellectuele, zedelijke en maatschappelijke ontwikkeling van het kind.'*

[580] Parliamentary Documents, II 1992-1993, 22855 (R1451), no. 3, p. 37.

[581] Parliamentary Documents, II 1993-1994, 22855 (R1451), no. 6, p. 31.

[582] In Dutch: Algemene Bijstandswet.

[583] In Dutch: Wet op de Jeugdzorg (before: Wet op de Jeugdhulpverlening).

[584] Parliamentary Documents, II 1992-1993, 22855 (R1451), no. 3, pp. 37-38.

the child has no direct entitlement to social security based on the Convention. Instead, it was the obligation of the Member State to only support those parents who could not take their primary responsibility. A child therefore has only indirect entitlement to social security, via its parents or legal guardians.

6.6 Article 12 CEDAW

Article 12 CEDAW stipulates:

1. States Parties shall take all appropriate measures to eliminate discrimination against women in the field of healthcare in order to ensure, on a basis of equality of men and women, access to healthcare services, including those related to family planning.
2. Notwithstanding the Provisions of Section 1 of this Article, States Parties shall ensure to women appropriate services in connection with pregnancy, confinement and the post-natal period, granting free services where necessary, as well as adequate nutrition during pregnancy and lactation.

With regard to the first Section, the Government emphasised in its Explanatory Memorandum that men and women had equal access to healthcare, although a majority of the medical staff in the Netherlands was male, which was a point for improvement. Furthermore, the Government concluded that women had adequate access to contraceptives, whose quality was sufficiently guaranteed through national legislation (e.g. the Drug Provision Act, and the Medical Device Act[585]).

With regard to the second Section, the Government stated that the services in connection with pregnancy and pre- and post-natal care were of a high quality in the Netherlands. The costs of these services are covered by insurances, or by the persons concerned, who could – if necessary – apply for financial assistance. With regard to adequate nutrition, the Government stated that: *'The attending physician or midwife and maternity assistant give due attention to adequate nutrition during pregnancy and the period in which the infant is fed by the mother. Also the clinics for baby and nursery care provide information on the subject.'*[586]

During the subsequent debates in Parliament, Article 12 CEDAW was not significantly discussed.

---

[585] In Dutch: Wet op de Geneesmiddelenvoorziening; Wet op de Medische Hulpmiddelen.
[586] Parliamentary Documents, II 1984-1985, 18950 (R 1281), no. 3. Original text in Dutch: *'Aan passende voeding gedurende de zwangerschap en de periode waarin de zuigeling door de moeder wordt gevoed, wordt door de behandelende arts c.q. verloskundige en kraamhulp de nodige aandacht besteed. Ook geven de consultatiebureaus voor zuigelingen- en kleuterzorg voorlichting ter zake.'*

6.7 Conclusion

From the above it can be concluded that in general, when ratifying the various international human rights treaties, the Government considered that the Netherlands already, and generously, fulfilled its obligations coming forth from Provisions that stipulate (or are related to) the right to adequate food. Especially in view of Article 11 ICESCR it was underlined that a country cannot be expected to continuously improve the standard of living of its population, and such duties were especially addressed to developing countries. Naturally, the Provisions in the specified treaties that relate to the right to food are discussed from the perspective of the target group and in the related context, although also here, the conclusion is drawn that the Netherlands fulfil their human rights obligations regarding the right to food. From this, it can be deduced, that the Netherlands, already at the moment of ratification, expressed hardly any ambition to improve its legislation or policies regarding the implementation of the right to food.

In view of Articles 26 (and also 27) ICRC, it was the clear intention of the Government that children would have no direct entitlement to social security. Although regarding both Articles the Government underlined that the Dutch social security system was of a high level and the obligations stipulated in the Articles were thus fulfilled, a Reservation to Article 26 ICRC was deemed necessary, for this Article did not – unlike Article 27 ICRC – stipulate that it was the primary responsibility of the parents or legal guardians to provide for an adequate standard of living, and therefore it might be understood by the Courts that children would have an independent right to social benefits. Some Parliamentarians considered the contents of Articles 26 and 27 complementary to one another. It is unclear whether this vision was shared by the Government.

7. Dutch Parliamentary History on the enforceability of human rights

7.1 Introduction

As demonstrated in Chapter 4, the Courts frequently refer to Parliamentary History in their considerations when deciding on individual appeals to the right to food, and seem to greatly value the opinion of the Legislature on the matter. Therefore, in this Chapter these viewpoints will be explored by analysing the relevant Parliamentary History. From the literature survey in Chapter 5, as well as from the analysis below, it appears that the Legislature (mostly represented by different Governments through the years) hardly expressed its view on the possibility of direct effect of the right to food in particular. Rather, human rights are subdivided in two categories: civil and political rights on the one hand, and economic, social and cultural rights on the other hand. The direct applicability is discussed for each category of rights altogether. Since the right to food is usually categorised as a right falling under the scope of economic, social and cultural rights, the analysis below will include all ratification Bills on international treaties that stipulate this category of rights, for in the Explanatory Memorandum of the Governments, and the subsequent Parliamentary debates, the matter of direct applicability of economic, social and cultural rights has been discussed several times into more detail. Also the ESC is included, since whereas in the treaty the right to food is not literally recognised, many rights that are stipulated are related. In addition, analysing the Parliamentary History on the ESC concerning the matter of direct effect contributes to a better understanding of the overall attitude of the Legislature towards direct applicability of ECOSOC rights. Both the ICESCR and the ESC have a counterpart that stipulates civil and political rights. It is especially the contrast between the viewpoints of the Legislature on the possibility of direct effect of civil and political rights on the one hand, and of economic, social and cultural rights on the other hand that leads to a better understanding. Therefore, ratification Bills to the treaties that stipulate both types of rights (the CEDAW and the ICRC) are especially important, for in that context, the Governments were forced to further clarify their viewpoints on the matter. Also, the ICCPR and the ICESCR were discussed based on the same ratification Bill, which had consequently a similar contrasting effect regarding the debate on direct applicability. To this end, the ratification Bill of the ECHR was included in the analysis as well, in order to be able to contrast this with the ratification Bills of the ESC (original and revised version). This analysis thus includes in chronological order – so that also the development in the debate over time can be observed – the ratification Bills of the ECHR, the ESC, the ICCPR and the ICESCR, the CEDAW, the ICRC and the revised version of the ESC.

In the analysis below, three recurring elements are permanently included per discussed treaty. Firstly, the motivation or perceived added value of the treaty in question will be discussed, and is included in the analysis. This gives an impression of the future expectations regarding the ratification of the treaties. In this light, to understand the context of the Bills better, I also felt the need to put those in their historical context as expressed in the Explanatory Memoranda. Secondly, of course the possibility of direct applicability of the treaty Provisions will be discussed. Furthermore, complaint procedures were frequently the center of fierce discussions. In that light, thirdly, the desirability of a complaint procedure, and – where relevant – individual complaint procedures, will be debated and is included in the analysis, for it tells us something about the viewpoints of the Legislature on the possibility to stimulate the effectuation of rights in the international human rights system. Finally, conclusions will be drawn.

7.2 The ECHR

The European Convention on Human Rights was signed by its first Member States on 4 November 1950[587] and the first Optional Protocol on 20 March 1952.[588] Both documents were discussed and ratified simultaneously by Dutch Parliament in the period 1952-1954.

7.2.1 Background and added value of the ECHR

In its Explanatory Memorandum, the Government explained some of the backgrounds of the drafting of the ECHR. While the treaty was inspired by the content of the Universal Declaration on Human Rights, it was the ambition to embed those rights in a convention with the characteristics of a multilateral agreement between States. It was the global intention to also create such agreements based on the UHDR, but during that period, these negotiations did not proceed very well,[589] and the need was felt to regionally adopt a legally binding instrument, including a monitoring procedure, at an earlier stage.[590] The fact that the ECHR was a convention instead of a declaration was portrayed by the Government as the most important added value for the Dutch legal order,[591] a vision that was generally shared by Parliament.[592]

---

[587] European Convention on Human Rights, 4 November 1950, Rome, in: Council of Europe Treaty Series, no. 5.

[588] Protocol to the Convention for the Protection of Human Rights and Fundamental Freedoms, 20 March 1952, Paris, in: Council of Europe Treaty Series, no. 9.

[589] Parliamentary Documents, I 1952-1953 3043, no. 162, p. 2; see: Parliamentary Documents, II 1975-1976 13932 (R 1037), no. 3, p. 9.

[590] Parliamentary Documents, II 1952-1953 3043, no. 3, p. 2.

[591] Parliamentary Documents, II 1952-1953 3043, no. 3, p. 2.

[592] See the discussions in general, held in: House of Representatives, 48th meeting, 3 March 1954; House of Representatives, 49th meeting, 4 March 1954; Senate, 50th meeting, 27 July 1954. See also: Parliamentary Documents, I 1952-1953 3043, no. 162, p. 2.

In the Preliminary Report, the observation was made that both the substantive and geographical scope of the ECHR were not as broad as the UDHR. In that context, especially the fact that the draft did not stipulate any ECOSOC rights, was underlined,[593] which was later also criticised in the House of Representatives by some Parliamentarians,[594] and discussed in the Preliminary Report of the Senate.[595] The Government explained that at that time, it was possible to reach an agreement on the wordings of the ECHR rights, but not yet on ECOSOC rights. However, the Government pointed out that in the context of the Council of Europe the possibility was explored to come to a European Social Charter, that *'in the field of social policy should complement of the present Convention.'*[596] Already here, a difference can be distinguished between civil and political rights, which are explicitly embedded in a legally binding Convention, and ESOCOC rights, which are possibly embedded in a Charter, and are seen as social policies, instead of rights.

The Government informed that during the negotiations preceding the final draft of the ECHR, there were two points of contention. The first issue was the question whether the rights enshrined in the treaty had to be defined into detail, or merely generally listed.[597] The Government had held the position that the rights should be clarified into detail, to ensure that Member States knew what the scope would be of the obligations they agreed upon.[598] During the negotiations, a compromise text was finally adopted, in which an attempt was made to find a balance between specifically defined rights and more general descriptions, and the acceptance of the jurisdiction of a Court of Human Rights was optional. On some rights it was harder to reach an agreement, and to avoid further delay they were not included in the Convention, but in an Optional Protocol. The rights involved were mainly the right to the protection of property, the right to education and the right to free elections.[599] It is in this light interesting to notice, especially with the later attitude of Dutch Governments towards ECOSOC rights in mind, that regarding the right to education, the Dutch representative – in line with the reasoning of the Government – had declared during the negotiations preceding the adoption of this Optional Protocol, that a right to parents to raise their children should not only imply the respect for this right, but a State should also *'ensure the possibility of exercising those rights by appropriate measures.'*[600] The second matter was the question whether a European Court for Human Rights should be installed, besides the already existing European Commission for Human Rights (the original treaty

---

[593] Parliamentary Documents, II 1952-1953 3043, no. 3, p. 2.
[594] See especially the contribution of: Mr Vink, Mr Welter and Mrs Lips-Odinot, House of Representatives, 48th meeting, 3 March 1954.
[595] Parliamentary Documents, I 1952-1953 3043, no. 162, p. 2.
[596] Parliamentary Documents, II 1952-1953 3043, no. 6. p. 3.
[597] Parliamentary Documents, II 1952-1953 3043, no. 3, p. 2.
[598] Parliamentary Documents, II 1952-1953 3043, no. 6. p. 2.
[599] Parliamentary Documents, II 1952-1953 3043, no. 3, p. 3.
[600] Parliamentary Documents, II 1952-1953 3043, no. 3, p. 3.

body).[601] Originally, the Government was of the opinion that one supervisory body would be sufficient, but later agreed on the installation of a European Court for Human Rights.[602]

7.2.2 Direct applicability of the ECHR

During the Parliamentary debate on the ratification of the ECHR, for the first time since the Constitutional reform of 1953, the matter of direct applicability of human rights was addressed. Members of the Preparatory Committee had raised the question how the national Courts would have to judge whether the national Legislature did not exceed its competence in limiting the rights enshrined in the ECHR as recognised in the limitation clauses. In particular, the question was raised how the phrase *'necessary in a democratic society'* – an expression used in those limitation clauses – relates to the review competences of the Judiciary, while it is the Legislature who usually decides by adopting national statutory regulations on what is deemed necessary in a democratic society.[603] The Government responded that: *'The undersigned acknowledge that the standards laid down in Articles 8 through 11 are vague, so with regard to the content different opinions are possible. They doubt whether it is possible, in a general regulation as the present one, to come to more specific formulations that would be acceptable to all Member States. They confirm in general that as suggested by the Committee Members the hypothesis that the national law, established in a democratic polity like ours, is the expression of what is "in a democratic society is needed." This will indeed generally come down to a situation in which the national Court will accept a considered judgment of the national Legislature on whether or not an exception may have been made. However, also in view of the new Constitutional Provisions, the national Court remains in principle authorised to pass a judgment dissenting of the national Legislature. Still, the undersigned think that the Court in such case may only verify whether the exception rule in general may be necessary in a democratic society, in the interests of safety, public order, etc., without being allowed to enter into the question whether the exception in the case is opportune. Indeed, in the opinion of the undersigned, the review duty of the judge, resulting from Article 65 of the Constitution Act, may in no way imply that the Courts rule in matters of opportuneness.'*[604]

---

[601] Parliamentary Documents, II 1952-1953 3043, no. 3, p. 2.
[602] Parliamentary Documents, II 1952-1953 3043, no. 6. p. 2.
[603] Parliamentary Documents, II 1952-1953 3043, no. 5, p. 2.
[604] Parliamentary Documents, II 1952-1953 3043, no. 6. p. 2. Original text in Dutch: *'De ondergetekenden geven toe dat de normen gesteld in artikelen 8 t/m 11 vaag zijn, zodat omtrent de inhoud daarvan verschil van mening mogelijk is. Zij betwijfelen echter of het mogelijk is, in een algemene regeling als de onderhavige, tot meer concrete formuleringen te komen, welke aanvaardbaar zouden zijn voor alle aangesloten landen. Zij beamen in het algemeen de door de aan het woord zijnde leden geopperde stelling, dat de nationale wet, tot stand gekomen in een democratisch staatsbestel als het onze, de uitdrukking is van datgene wat 'in een democratische samenleving nodig is.' Dit zal inderdaad in het algemeen hierop neerkomen, dat de nationale rechter zich zal neerleggen bij een weloverwogen oordeel van de nationale wetgever over de vraag of al dan niet een uitzonderingsregel mocht worden gemaakt. Mede gelet op de nieuwe grondwetsbepalingen blijft echter de*

The Government was of the opinion that the duties stipulated in the ECHR would not play a role of any significance, for it was assumed that the Netherlands already fulfilled these duties. In the Preperatory Committee, the question was raised whether the ratification would lead to any amendments in national legislation.[605] The Government responded that *'there is no reason for this, since they consider that the complex regulations of Dutch law, (...), offer sufficient guarantees that in our country these principles will find respect.'*[606] Minister Beyen even considered that: *'Despite the limited meaning that the convention, which we now discuss, obviously has, the Government considers the realisation of this Convention an important and gratifying fact.'*[607] A viewpoint that was criticised in Parliament more than once.

As already demonstrated above, one of the arguments for not recognising an individual complaints procedure, was the fact that according to the Government, individuals could adequately invoke the ECHR rights in the Courts. A second, parallel, complaint procedure would therefore be unnecessary, and would mostly lead to the abuse of such a procedure.[608] This approach seems to imply that it was assumed that the Provisions of the ECHR have direct effect in the domestic legal order.[609]

7.2.3 Individual complaints procedures

The Government informed that there was another matter that had led to extensive debates during the negotiations preceding the final text of the ECHR, namely the possibility to implement an individual complaints procedure. Member States could recognise the competence of the Commission of Human Rights to receive and

---

nationale rechter in beginsel bevoegd om een van de nationale wetgever afwijkend oordeel te geven. Toch menen de ondergetekenden dat de rechter in zulk een geval slechts mag nagaan of de uitzonderingsregel in het algemeen in een democratische samenleving in het belang van veiligheid, openbare orde e.d. noodzakelijk zou kunnen zijn zonder dat hij mag treden in de vraag of de betrokken uitzonderingsregel in het concrete geval doelmatig is. Immers, naar het oordeel van de ondergetekenden houdt de toetsingplicht van de rechter, voortvloeiende uit artikel 65 van de Grondwet, geenszins in dat de rechter zou moeten treden in doelmatigheidsvragen.'

[605] Parliamentary Documents, II 1952-1953 3043, no. 5, p. 2.

[606] Parliamentary Documents, II 1952-1953 3043, no. 6. p. 3. Original text in Dutch: *'dat hiertoe geen aanleiding bestaat, aangezien zij van oordeel zijn dat het complex regelingen van de Nederlandse wet,..., voldoende waarborgen biedt dat deze beginselen in ons land eerbiediging zullen vinden.'*

[607] Minister Beyer, House of Representatives, 48th meeting, 3 March 1954. Original text in Dutch: *'ondanks de beperkte betekenis, die de conventie, waarover wij thans beraadslagen, uiteraard heeft, acht de Regering het tot stand komen van deze conventie toch een belangrijk en verheugend feit.'*

[608] Parliamentary Documents, II 1952-1953 3043, no. 3, p. 4; Parliamentary Documents, II 1952-1953 3043, no. 6. p. 4; See also Minister Beyen, House of Representatives, 49th meeting, 4 March 1954.

[609] Mrs Vink even literally concluded this in Parliament (House of Representatives, 48th meeting, 3 March 1954), a view that was not contested: *'Ultimately, the individual is currently already offered a reasonable certainty that his rights and freedoms enshrined in the convention will be respected by the Court, due to the judicial review competences that are attributed to the Judiciary under Article 65 of the Constitution.'* Original text in Dutch: *'Uiteindelijk wordt het individu op dit moment reeds door het toetsingsrecht, dat de rechter krachtens artikel 65 van de Grondwet toekomt, een behoorlijke zekerheid geboden, dat zijn rechten en vrijheden in de conventie neergelegd, door de rechter zullen worden geëerbiedigd.'*

consider individual complaints by an official statement, as was stipulated in the former Article 25 of the ECHR.[610] This procedure existed besides a complaints procedure for States that was embedded in Article 24 ECHR, and not optional.[611] The Government had great expectations concerning the complaints procedure for States, but rejected, as will be demonstrated below, the possibility of an individual complaints procedure.[612]

By signing the ECHR the Member States had to make some choices. The complaints procedure for States and the competences of the European Commission on Human Rights were automatically recognised while signing the ECHR, but in addition, States could voluntarily recognise the individual complaints procedure and the competence of the European Court of Human Rights. According to the Government, the purpose of the Commission was to facilitate the friendly settlements of disputes, before a case was brought before the Court.[613] Another choice to be made was the recognition of the above mentioned Optional Protocol, in which the rights were stipulated that could not be agreed on during the negotiations. The Government's position was that the Netherlands should recognise the competences of the Court of Human Rights and ratify the Optional Protocol, but not recognise the individual complaints procedure.[614] Regarding the latter, the Government explained that firstly, in the domestic legal system, citizens could adequately invoke the rights embedded in the Convention, and therefore there was no reason to simultaneously open the possibility for individual complaints in an international forum. Secondly, the Government did not expect that such an individual complaints procedure would contribute to the realisation of the purpose of the treaty, while it could lead on the other hand to a cumbersome and expensive procedure. And thirdly, the Government called for cautiousness, for it should be taken into account that such a procedure might be abused because '*individuals or groups of persons could systematically attempt to provoke complaints.*'[615] Interesting detail in this light however, is that the Dutch Delegation that participated in the negotiations that preceded the adoption of the ECHR, had voted for the individual complaints procedure.[616] The rejection of the individual complaints procedure was heavily criticised in Parliament, but the Government maintained its position.

---

[610] Parliamentary Documents, II 1952-1953 3043, no. 3, p. 5.

[611] Parliamentary Documents, II 1952-1953 3043, no. 3, p. 5.

[612] See for instance Minister Beyen, in: House of Representatives, 49th meeting, 4 March 1954.

[613] Parliamentary Documents, II 1952-1953 3043, no. 3, p. 5.

[614] Parliamentary Documents, II 1952-1953 3043, no. 3, in particular p. 5.

[615] Parliamentary Documents, II 1952-1953 3043, no. 3, p. 4; Parliamentary Documents, II 1952-1953 3043, no. 6. p. 4; See also Minister Beyen, House of Representatives, 49th meeting, 4 March 1954. Original text in Dutch: '...*dat personen of groepen zich stelselmatig er op zouden toeleggen klachten uit te lokken.*'

[616] See for instance Mrs Tenderloo, House of Representatives, 48th meeting, 3 March 1954.

7.3 The ESC (earliest version)

The European Social Charter was adopted in Turin on 18 October 1961.[617] The Dutch ratification of the Charter was finally completed in the period 1966 (submission of the ratification Bill by the Government[618]) to 1978 (final adoption of the ratification by the Senate[619]). A part of the ratification process thus coincided with the ratification process of the ICESCR (see below in Section 7.4). In both the House of Representatives and the Senate, the ESC was discussed about one month earlier compared to the ICESC. This explains the large amount of references to the other treaty during both Parliamentary debates.

7.3.1 Background and added value of the ESC

The Explanatory Memorandum submitted by the Government can be characterised as short and concise. It was explained that the Charter embodies '*a codification of the in the Member States existing ideas in the social field and thus intends to provide for guidance for the national social policies of those Member States.*'[620] The added value for the Dutch legal order was not discussed in the Explanatory Memorandum, but it was obvious that the Government was very cautious to include all the ESC Provisions in the ratification Bill, most in particular regarding the fiercely discussed right to strike (Article 6 sub 4 ESC). In the original ratification Bill, the right was included,[621] later – as recommended by the Preparatory Committee due to insufficient clarity on the national legal status of the right to strike[622] – excluded,[623] and finally included with the limitation that it would only concern employees that are not in the service of the Government, in line with a national Bill that was expected to pass Parliament soon.[624] In the end however, Parliament did not agree on the recognition of the Article.

7.3.2 Direct applicability of the ESC

Regarding the possibility of direct effect of the ESC, the Government in its Explanatory Memorandum cannot be misunderstood: '*The Charter has no 'internal effect' in the Member States. Citizens of the Parties thus cannot invoke the Charter*

---

[617] European Social Charter, 18 October 1961, Turin, in: European Treaty Series 35.
[618] Parliamentary Documents, II 1965-1966, 8606 (R 533), no. 2.
[619] Senate, 3rd meeting, 31 October 1978.
[620] Parliamentary Documents, II 1965-1966, 8606 (R 533), no. 3, p. 2. Original text in Dutch: '(...)vormt een codificatie van de in de lid-staten levende gedachte op sociaal terrein en beoogt aldus een leidraad te zijn voor het nationaal sociaal beleid van elk dier Staten.'
[621] Parliamentary Documents, II 1965-1966, 8606 (R 533), no. 2, Article 2.
[622] The Preparatory Committee was the Commission for Social Affairs and Public Health, see: Parliamentary Documents, II 1965-1966, 8606 (R 533), no. 5, pp. 2-3.
[623] Parliamentary Documents, II 1965-1966, 8606 (R 533), no. 7.
[624] Parliamentary Documents, II 1965-1966, 8606 (R 533), no. 11.

before a national judicial institution.'[625] This point of view was critically received by the Preparatory Committee, which considered that indeed most Articles were directly addressed to State Parties (mostly by using the word 'undertake'), and not to citizens, although in some Provisions, the word 'recognise' was used (especially in Articles 6 sub 4 and 18 sub 4): a word that would suggest direct effect.[626] The Government responded that: *'The answer to the question whether an agreement or Provision of the ESC has direct effect depends in general on whether the parties to the agreement have intended the effect. This is not the case regarding the European Social Charter, as is illustrated by* inter alia *the Annex to Part III. Indeed, the Articles of the Charter referred to by the Committee stipulate on the one hand State duties with respect to each other to perform certain actions, and on the other hand the recognition of certain rights of individuals. This does not mean, however, that in the latter cases these persons may derive the rights directly from the Charter. Rather, by using the terminology, one tried to express that in one case the focus on the State Party's obligation is on the measures to be taken, while in the other case, apart from measures that are necessary, the focus of the Government policy will be on the right of these persons. The Provisions, however, are still addressed to the state.'*[627] In general, this view was not opposed in Parliament, except with regard to one Provision: Article 6 (4), stipulating the right to strike. The particularities of that discussion will be referred to into detail in Chapter 8, for it will appear that this Provision is the pivot of the debate on direct effect of ECOSOC rights, and therefore a separate Chapter is dedicated to the mater.

7.3.3 Complaints procedures

In the earliest version of the ESC, no complaints procedure existed, nor was the matter discussed during any Parliamentary debate.

---

[625] Parliamentary Documents, II 1965-1966, 8606 (R 533), no. 3, p. 2. Original text in Dutch: *'Het handvest heeft geen 'interne werking' in de deelnemende Staten. Onderdanen van Partijen kunnen derhalve geen beroep doen op het Handvest voor een nationaalrechtelijke instantie.'*

[626] Parlementaire Geschiedenis, II 1966-1967 8606 (R 533), no. 5, voorlopig verslag van de vaste Commissie van Sociale Zaken en Volksgezondheid.

[627] Parliamentary Documents, II 1965-1966, 8606 (R 533), no. 6. Original text in Dutch: *'De beantwoording van de vraag, of een overeenkomst of bepaling daaruit rechtstreekse werking heeft of hebben hangt in het algemeen hiervan af, of de partijen bij de overeenkomst deze werking hebben beoogd. Bij het Europese Sociaal Handvest is dat niet het geval, getuige onder meer de bijlage bij Deel III. Inderdaad stellen de door de commissie genoemde artikelen van het Handvest tegenover elkaar enerzijds verplichtingen voor de staten tot het verrichten van zekere handelingen, anderzijds de erkenning door hen van zekere rechten van personen. Dit wil echter niet zeggen, dat in de laatstbedoelde gevallen deze personen de genoemde rechten rechtstreeks aan het Handvest kunnen ontlenen. Veeleer is door de gebezigde terminologie getracht tot uitdrukking te brengen dat in het ene geval het accent van de aan de Verdragspartij opgelegde verplichting rust op de door haar te treffen maatregelen, terwijl in het andere geval, afgezien of maatregelen geboden zijn, het uitgangspunt van het te voeren regeringsbeleid het recht van de bedoelde personen zal zijn. De desbetreffende bepalingen blijven zich echter richten tot de staat.'*

7.4 The ICCPR and the ICESCR

In 1976, one ratification Bill[628] regarding the ICCPR (including its first Protocol), the ICESCR and certain regulations regarding the abolishment of the death penalty, were discussed in Parliament.

7.4.1 Background and added value of the treaties

The Government clarified some of the backgrounds of the discussed treaties in its Explanatory Memorandum. Both the ICCPR and the ICESCR constituted a step further in the development of a global human rights mechanism. As discussed in Chapter 3, the Universal Declaration on Human Rights (adopted on 10 December 1948)[629] was considered to be a first step in global realisation of respect for human rights. At that time, it was the intention to realise a second step soon, by de adoption of a treaty that would translate the UDHR into legally binding Provisions. However, this next step was only realised in 1966, by the adoption of two treaties instead of one. The Government explained that: *'It should be noted that it was a controversy whether all rights to the example of the Universal Declaration should be included in one treaty or the civil and political rights on the one hand and the economic, social and cultural rights on the other hand in two separate treaties. In 1951, the General Assembly decided on the last option, a decision which has been discussed frequently ever since, but never reconsidered.'*[630] The originally intended third step – a global monitoring mechanism – had been incorporated in the treaties: both the ICCPR and the ICESCR contain a reporting procedure and a States Complaints procedure. In an Optional Protocol, an individual complaints procedure was adopted to the ICCPR.[631]

In its Explanatory Memorandum, the Government frequently reminded of the difficult negotiations that preceded the adoption of both treaties, and underlined that therefore the final wordings of the Provisions were mostly a result of a compromise. As a result, *'not all the Provisions excel in precision and clarity.'*[632] Despite this rather negative perception of the content of the treaties, the Government

---

[628] Parliamentary Documents, II 1975-1976 13932 (R 1037), no. 2.

[629] 217 A (III), 10 December 1948, *The Universal Declaration on Human Rights.*

[630] Parliamentary Documents, II 1975-1976 13932 (R 1037), no. 3, pp. 7-8. Original text in Dutch: *'Hierbij zij aangetekend dat het een strijdpunt is geweest of alle rechten naar het voorbeeld van de Universele Verklaring moesten worden opgenomen dan wel de burgerlijke en politieke rechten enerzijds en de economische sociale en culturele rechten anderzijds in twee afzonderlijke verdragen. In 1951 besloot de Algemene Vergadering tot het laatste, een beslissing die daarna herhaaldelijk in discussie is geweest, maar waarop niet meer werd teruggekomen.'*

[631] A/RES/21/2200, 16 December 1966, *International Covenant on Economic, Social and Cultural Rights, International Covenant on Civil and Political Rights and Optional Protocol to the International Covenant on Civil and Political Rights.*

[632] Parliamentary Documents, II 1975-1976 13932 (R 1037), no. 3, p. 8. Original text in Dutch: *'niet alle verdragsartikelen uitmunten door precisie en helderheid.'*

regarded the adoption of the ICCPR and the ICESCR as important: '*Despite the defects and shortcomings that stick to the treaties in the Dutch opinion, the Dutch Government considers their realisation a positive development because thus the possibility was created to better found the promotion and protection of human rights on a global scale.*'[633] The Government came to this conclusion after exploring the negative and positive aspects of ratification of the treaties. Arguments against ratification were in the first place that both the ECHR and the ESC '*already function within a reasonable coherent legal community and offer sufficient additional security, in addition to the national legal protection, especially where the European monitoring system regarding the compliance with the treaty Provisions clearly goes beyond that of the global conventions.*'[634] Secondly, the Government noticed that a combination of regional and global treaties might lead to uncertainty and conflicts between the two systems. However, it was the arguments for ratification that finally convinced the Government to propose the ratification Bill. In the first place, the Netherlands had pledged itself to contribute to the international promotion and protection of human rights, which finds its origin in the UN Charter and the UDHR. The ICCPR and the ICESCR are therefore necessary further steps to this end. Secondly, the Government argued that occasionally, the protection of the UN treaties has a broader scope compared to the European equivalents. Thirdly, in case of a possible conflict between a regional and a global treaty, the Government underlined the principle stipulated in both UN treaties that one convention may not be invoked to detract from a Provision that is more favourable to a person.[635] In that light, a conflict between treaty Provisions seems unlikely.[636]

In its Explanatory Memorandum, the Government thoroughly explained the differences between the contents of the regional treaties and its international equivalents. In general, the added value of ratifying the UN treaties seems to be that both treaties offer an extra guarantee for the respect for and realisation of human rights: it was suggested that in case when hypothetically, due to whatever reason, the Netherlands would denounce one of the European treaties, the UN equivalent would still guarantee the substantive rights embedded in that treaty. In addition, the Government underlined that the UN treaties would provide for additional protection of the rights, '*because an alleged violation of a right may be brought before*

---

[633] Parliamentary Documents, II 1975-1976 13932 (R 1037), no. 3, p. 8. Original text in Dutch: '*Ondanks de gebreken en de tekortkomingen die naar Nederlands inzicht aan de verdragen kleven, acht de Nederlandse Regering de totstandkoming ervan een positieve ontwikkeling omdat aldus de mogelijkheid is geschapen de bevordering en bescherming van rechten van de mens op wereldschaal beter te funderen.*'
[634] Parliamentary Documents, II 1975-1976 13932 (R 1037), no. 3, pp. 10-11. Original text in Dutch: '*reeds binnen een redelijk coherente rechtsgemeenschap functioneren en voldoende additionele waarborg bieden naast de nationale rechtsbescherming, temeer waar het Europese stelsel betreffende het toezicht op de naleving van de verdragsbepalingen duidelijk verder gaat dan dat van de mondiale verdragen.*'
[635] The Government argued that this principle would also apply to a conflict between an international treaty and the Dutch Constitution. Parliamentary Documents, II 1975-1976 13932 (R 1037), no. 8, p. 5.
[636] Parliamentary Documents, II 1975-1976 13932 (R 1037), no. 3, pp. 10-11.

more forums'[637] Regarding the contents of the ICCPR and the ICSCR compared to the ECHR and the ESC, the Government argued that especially the ICCPR and the ECHR showed many similarities. This was due to the fact that the ECHR was adopted in a period in which the complex political relations within the UN led to difficulties in the process to come to global Human Rights treaties. Therefore, there was a growing need in Europe to adopt a human rights convention to *'guarantee in a regional, European context the democratic rights and freedoms.'*[638] The wordings of the ECHR were strongly inspired by the UDHR and the then existing draft texts of the ICCPR.[639] According to the Government, the European experiences with the ECHR also had a stimulating effect on the European contribution in the preparation of the ICCPR.[640] The Government explained that there were fewer similarities between the ICESCR and the ESC. Especially the monitoring mechanisms differed significantly, for the ESC monitoring mechanism *'contains elements derived from facilities that are applied by the International Labour Organisation.'*[641] In general, it must be observed that the Government took considerably greater effort in comparing the ECHR with the ICCPR, and was less comprehensive in comparing the ESC with the ICESCR.

7.4.2 The direct applicability of the ICCPR and the ICESCR

The Government explained in its Explanatory Memorandum the criteria used by the Courts to determine whether a treaty Provision would be binding on all persons: *'Criterion for answering the question whether treaty Provisions are 'binding on all persons by virtue of their contents' is according to the case law of the Supreme Court the nature and content of the Provision, while also the intention of the draftsmen of the Convention may be a guideline.'*[642] With regard to the ICESCR, the Government was of the opinion that the Provisions of the ICESCR would not be directly applicable, due to the nature and the content, as well as the wordings of the Articles. The Convention *'is focussed on the progressive and increasing realisation through legislation and other implementation measures.'*[643] The Government underlined that Article 2 sub 1 *'clearly expresses that the draftsmen of the Convention took into consideration*

---

[637] Parliamentary Documents, II 1975-1976 13932 (R 1037), no. 8, p. 6. Original text in Dutch: *'doordat een beweerde schending van een recht aan meer fora kan worden voorgelegd.'*
[638] Parliamentary Documents, II 1975-1976 13932 (R 1037), no. 3, p. 9. Original text in Dutch: *'in regionaal Europees verband de democratische rechten en vrijheden te verzekeren.'*
[639] Parliamentary Documents, II 1975-1976 13932 (R 1037), no. 3, p. 9; Parliamentary Documents, II 1975-1976 13932 (R 1037), no. 8, p. 6.
[640] Parliamentary Documents, II 1975-1976 13932 (R 1037), no. 8, p. 6.
[641] Parliamentary Documents, II 1975-1976 13932 (R 1037), no. 3, p. 9.
[642] Parliamentary Documents, II 1975-1976 13932 (R 1037), no. 3, p. 12. Original text in Dutch: *'Criterium voor de beantwoording van de vraag of verdragsbepalingen 'naar de inhoud eenieder kunnen verbinden' is volgens de jurisprudentie van de Hoge Raad de aard en de inhoud van de bepaling, terwijl tevens de bedoeling van de opstellers van het verdrag een richtsnoer kan zijn.'*
[643] Parliamentary Documents, II 1975-1976 13932 (R 1037), no. 3, pp. 12-13. Original text in Dutch: *'is afgestemd op geleidelijke en toenemende verwezenlijking door middel van wetgeving en andere uitvoeringsmaatregelen.'*

that the economic, social and cultural rights as regards to their nature and their contents in general do not provide a standard criterion for the pace and degree of the realisation of those rights.'[644] The pace and degree of the realisation thereby greatly depends on the economic and social conditions of the Member State. The Government underlined that Article 2 ICESCR had been adopted in a period in which it was assumed that economic stagnation or even recession was unlikely to happen. In the meantime however, this assumption had proven to be unrealistic, and the Government therefore was of the opinion that Article 2 ICESCR must be thus understood that its function was also to be a guideline in case of such economic stagnation or recession. Furthermore, any restriction on the enjoyment of the rights embedded in the ICESCR must be at least in compliance with the non-discrimination clause of Article 2 (2) but also with Article 4, that only allows such restrictions under certain conditions.[645]

How different was the perception of the Government regarding the possibility of direct applicability of the ICCPR Provisions! The Government referred in its Explanatory Memorandum to Article 2 ICCPR, which obliges the Member States to *'take the necessary steps, in accordance with their Constitutional processes and with the Provisions of the present Covenant, to adopt such laws or other measures as may be necessary to give effect to the rights recognised in the present Covenant.'*[646] According to the Government, the purpose of this Article is to concretise the responsibility of the Member States by indicating the States' duty to effectuate the contents of the treaty in the domestic legal order. It was the opinion of the Government that in the Netherlands, Articles 65 and 66 (currently, Articles 93 and 94) fulfil this obligation, for these Constitutional Provisions grant direct effect to the eligible international regulations. Whether or not the ICCPR Provisions are binding on all persons, and thus have direct effect through Articles 65 and 66 CA could not be directly deduced from a textual interpretation of the treaty Provisions.[647] The fact however that many analogous Provisions embedded in the ECHR were already considered to be directly applicable appeared to be a decisive factor for the Dutch Government to conclude that the corresponding ICCPR Provisions would also have direct effect. The Government concluded that: *'many Provisions concerning substantive rights, embedded in Part III of the International Covenant on Civil and Political Rights, have, due to the content and wording of these Provisions, and following the example of most Provisions stipulating substantive rights of the European*

---

[644] Parliamentary Documents, II 1975-1976 13932 (R 1037), no. 3, p. 45. Original text in Dutch: *'duidelijk doet uitkomen dat de opstellers van het verdrag in aanmerking hebben genomen dat de economische, sociale en culturele rechten wat betreft hun aard en inhoud in hun algemeenheid geen vaste maatstaf bieden voor het tempo en de graad van de verwezenlijking van die rechten.'*

[645] Parliamentary Documents, II 1975-1976 13932 (R 1037), no. 3, p. 45.

[646] Parliamentary Documents, II 1975-1976 13932 (R 1037), no. 3, p. 13. Original text in Dutch: *'langs de door het staatsrecht van de betreffende staat voorgeschreven weg en in overeenstemming met de bepalingen van het verdrag, alle maatregelen te nemen die nodig zijn om de in het verdrag erkende rechten tot gelding te brengen.'*

[647] Parliamentary Documents, II 1975-1976 13932 (R 1037), no. 3, p. 13.

Convention, direct effect and can be applied by the Courts without the need for further legislation.'[648] The Government added that *'partly due to this, civil and political rights on the one hand, and economic, social and cultural rights on the other, are embedded in two separate documents, because in general the first group of rights is suitable for direct application, while the realisation of the second group often requires implementing measures.'*[649] This reasoning seems to be in contrast with the explanation offered by the Government earlier in its Explanatory Memorandum, where instead the unity between the two treaties was emphasised. There, the Government underlined that both treaties were a further elaboration of the UDHR. The draftsmen of the treaties tried to underline the parallel character of the Covenants by *'formulating the preambles in almost identical wordings. They especially wanted to express in the preambles that, although civil and political rights on the one hand, and economic, social and cultural rights on the other hand are enshrined in two separate documents, the enjoyment of these rights altogether is essential for humans. Also part 1 of both Conventions, relating to the right of all peoples to self-determination, is defined in equal terms.'*[650] In general, the opinion of the Government regarding the direct applicability of the ICCPR and the ICESCR was hardly criticised in Parliament, although occasionally, the matter was discussed briefly, as will be demonstrated below in a short summary.

In the Preparatory Committee, no particular questions were raised concerning the direct applicability of both treaties. The Members of the Labour Party (PvdA) *'regarded the considerations on the direct applicability in itself as satisfactory. They stressed however, that fundamental rights isolated in a convention can never be abstractly expressed, independent of any Constitutional system. As long as there is no world legal order, there will always remain a certain tension between the granting of rights to individuals on the one hand and the claims of states on internal sovereignty on the*

---

[648] Parliamentary Documents, II 1975-1976 13932 (R 1037), no. 3, p. 13. Original text in Dutch: *'vele bepalingen, houdende materiële rechten, voorkomende in deel III van het Internationale Verdrag inzake Burgerlijke en Politieke Rechten, op grond van inhoud en formulering van deze bepalingen, en in navolging van de meeste bepalingen, houdende materiële rechten, van het Europese Verdrag, rechtstreeks werken en door de rechter kunnen worden toegepast zonder dat daarvoor nog enige wetgeving vereist is.'*

[649] Parliamentary Documents, II 1975-1976 13932 (R 1037), no. 3, p. 13. Original text in Dutch: *'de burgerlijke en politieke rechten enerzijds, en de economische, sociale en culturele rechten anderzijds, mede hierom in twee afzonderlijke akten zijn neergelegd, omdat in het algemeen de eerste groep rechten voor rechtstreekse toepassing vatbaar is, terwijl voor de verwezenlijking van de tweede groep veelal uitvoeringsmaatregelen zijn vereist.'*

[650] Parliamentary Documents, II 1975-1976 13932 (R 1037), no. 3, p. 8. Original text in Dutch: *'de preambules vrijwel geheel in identieke bewoordingen te formuleren. Zij hebben met name in de preambule doen uitkomen dat, hoewel de burgerrechten en politieke rechten enerzijds, en de economische, sociale en culturele rechten anderzijds in twee aparte documenten zijn opgenomen, het genot van het totaal van deze rechten voor de mens essentieel is. Ook deel 1 van beide verdragen, betrekking hebbend op het recht van alle volkeren op zelfbeschikking, is in gelijke termen vastgelegd.'*

other hand.'[651] In particular, this was demonstrated by the fact that not all the States involved were willing to recognise the complaints procedures to the ICCPR by signing the Optional Protocol. In this light, the Members of the Labour Party seemed to imply that the Netherlands contributed to the realisation of the treaties in a profound way.

In the House of Representatives a Member of the Liberal Party (VVD), Mrs Kapeyne van de Coppello, agreed that '*Although in the wordings of the treaty the economic, social and cultural rights are referred to as rights, they are essentially instructional standards to the Government and not rights citizens can invoke.'*[652] With regard to the ICCPR Provisions, she asked to be informed on what Articles exactly would have direct effect in the domestic legal order. In response, Minister Wiegel replied that: '*I can agree with her that the requested clarification would be a good thing, but I believe that the honourable representative also may realise...that the treaties were written for States around the entire globe, States with often divergent traditions and systems for the way in which treaty obligations are effectuated in their legal order.'*[653] He then referred to the Constitutional system of Articles 65 and 66 CA (the current Articles 93 and 94 CA), and argued that: '*It is thus our Constitutional system that entails that the question of the effect of the current Treaty Provisions within the Dutch legal system is determined by the interpretation of the nature and content of the treaty Provisions. I believe that one should not criticise the draftsmen of the treaty too much for insufficiently indicating which Provision is directly applicable to citizens and which is not. It is the bodies in our legal system which are responsible for the interpretation of the Treaty Provisions that should answer that question.'*[654]

---

[651] Parliamentary Documents, II 1975-1976 13932 (R 1037), no. 7, p. 7. Original text in Dutch: '*ervoer de beschouwingen over de directe werking op zich als bevredigend. Zij onderstreepten wel, dat in een conventie geïsoleerde grondrechten nooit abstract, derhalve los van enig Constitutioneel bestel, tot uitdrukking kunnen worden gebracht. Zolang een wereldrechtsorde ontbreekt, zal er altijd een zekere spanning blijven bestaan tussen de toekenning van rechten aan individuen enerzijds en de aanspraken van staten op interne soevereiniteit anderzijds.*'
[652] Mrs Kapeyne van de Coppello, House of Representatives, 3rd meeting, 21 September 1978. Original text in Dutch: '*Hoewel de economische, sociale en culturele rechten wel in de verdragtekst als rechten aangeduid worden, zijn zij in wezen instructienormen aan de overheid en niet rechten waarop de burger zich kan beroepen.*'
[653] Minister Wiegel, House of Representatives, 3rd meeting, 21 September 1978. Original text in Dutch: '*ik kan met haar instemmen dat die door haar gevraagde duidelijkheid een goede zaak zou zijn, maar ik meen dat de geachte afgevaardigde ook kan bedenken (...) dat de verdragen zijn geschreven voor staten over de gehele wereld, staten met vaak van elkaar afwijkende tradities en systemen voor de wijze, waarop verdragsverplichtingen in hun eigen rechtsorde tot uitwerking worden gebracht.*'
[654] Minister Wiegel, House of Representatives, 3rd meeting, 21 September 1978. Original text in Dutch: '*Het is dus ons grondwettelijk systeem dat meebrengt, dat die vraag naar de doorwerking van de voorliggende verdragsbepalingen binnen de Nederlandse rechtsorde wordt bepaald door de interpretatie van de aard en de inhoud van de verdragsbepalingen. Ik geloof dat men daarom de verdragsopstellers niet al te zeer mag verwijten, dat zij onvoldoende aangeven welke bepalingen rechtstreeks van toepassing zijn voor de burgers en welke niet. De organen die in onze rechtsorde met de interpretatie van de verdragsbepalingen zijn belast moeten die vraag beantwoorden.*'

Also the representative of the Rainbow/Radical Party (PPR), Waltmans, was of the opinion that '*the Provisions enshrined in the Covenant on Civil and Political Rights General will have an direct effect in the Dutch legal order. Equally essential are the 'promotional rights' recognised in the Covenant on Economic, Social and Cultural Rights. This means that it is the purpose of the treaty to achieve the progressive realisation of the rights embedded, through legislation and other implementing measures.*'[655]

However, representative Mommersteeg, Member of the Christian Democratic Appeal (CDA), emphasised the connection between the ICCPR and the ICESCR, and its shared origin in the UDHR. He argued that this '*it is not only about the enforcement of the spiritual, physical and political freedom of the individual and citizen, but also about the care of the social well-being of the human being as a matter of justice.*'[656] He specifically underlined the strong interrelation between the treaties to stress the link between human rights and development cooperation.

Representative Roethof, Member of the Labour Party (PvdA), advised to consider the UN treaties not too strictly in legal terms only. He underlined that in both treaties important moral standards are recognised that should be complied with, also when a Provision does not contain a legally enforceable rule. In that light, he considered that moral standards are superior to legal standards. Also, Roethof drew attention to the limited interest amongst lawyers in the working of international Provisions, and recommended legal educational institutions to pay more attention to the matter.[657]

In the Senate, Rainbow/Radical Party (PPR) Senator B. de Gaay Fortman argued that the ICESCR Provisions do not contain subjective rights, but was of the opinion that the term 'promotional rights' should be replaced by the term 'promotional duties' because '*that what should encompass international justice for the individual, is embodied in obligations for States. Required is that there is progress in the achievement of these obligations, thus the so-called "progressive implementations."*'[658] He thereby underlined that also in times of economic recession no deterioration with regard

---

[655] Waltmans, House of Representatives, 3rd meeting, 21 September 1978. Original text in Dutch: '*de in het Verdrag inzake Burgerlijke en Politieke Rechten opgenomen bepalingen generaal een rechtstreekse werking zullen hebben in de Nederlandse Rechtsorde. Evenzeer wezenlijk zijn de in het Verdrag inzake Economische, Sociale en Culturele Rechten opgenomen 'promotional rights'. Dat wil zeggen, dat het verdrag beoogt de daarin opgesomde rechten geleidelijk en in toenemende mate te laten verwezenlijken door middel van wetgeving en andere uitvoeringsmaatregelen.*'

[656] Mommersteeg, House of Representatives, 3rd meeting, 21 September 1978. Original text in Dutch: '*het niet alleen gaat om handhaving van de geestelijke, lichamelijke en politieke vrijheid van individu en staatsburger, maar ook om de zorg van het sociale welzijn van de mens als een zaak van gerechtigheid.*'

[657] Roethof, House of Representatives, 3rd meeting, 21 September 1978.

[658] De Gaay Fortman, Senate, 5th meeting, 21 November 1978. Original text in Dutch: '*datgene wat internationale rechtvaardigheid zou moeten inhouden voor het individu, wordt geconcretiseerd in verplichtingen van de staten. Vereist is dan, dat er vorderingen worden gemaakt in de verwezenlijking van die verplichtingen, die zogenaamde 'progressive implementations.'*'

to the ICESCR rights may be allowed. A principle that would apply as well to the level of development aid to the support of the right to an adequate standard of living and the right to be free from hunger elsewhere.[659]

Senator Wiebenga (Liberal Party, VVD), expressed his concern regarding the adoption of resolution 32/130 by the UN General Assembly, in which according to him the matter of social rights was unjustly separated from political rights.[660] He feared that this resolution recognised a hierarchical relation between the two types of rights, in which political rights would be inferior to social rights. In this light, Wiebenga referred to a speech of Professor Couwenbergh during a meeting of the Atlantic Commission in 1977, who made a distinction between three visions on human rights: a Western vision, focussing on classic human rights, assuming that the individual is an emancipated human being in a plural society, an Eastern-European vision, focussing on social human rights for the benefit of the common cause, in line with a socialist view on the State, and Developing Countries, with a focus on the distinction between human rights for people (such as the right to self-determination) and human rights for the individual.[661]

Minister Van der Klaauw responded to these concerns, by stating that according to him no such hierarchical relation between types of human rights existed nor would be desirable to exist. He underlined that '*regarding the resolution, the latest development in the debate that has been started in recent years within the United Nations concerns the approach towards the issue of human rights, especially concerning the relationship between the classical civil rights and the political rights on the one hand, and the social and economic rights on the other hand, and to the structural causes that stand in the way of the realisation of human rights.*'[662] He observed that a large number of countries emphasised the economic, social and cultural human rights, possibly for the purpose of hiding shortcomings concerning the realisation of civil and political rights, but also underlined that while in the Netherlands a strong focus existed on civil and political rights, in some countries the matter of economic, social and cultural rights was indeed urgent. In his view, the General

---

[659] De Gaay Fortman, Senate, 5th meeting, 21 November 1978.

[660] Most likely, this concern was caused by this phrase: '*The full realisation of civil and political rights without the enjoyment of economic, social and cultural right is impossible; the achievement of lasting progress in the implementation of human rights is dependent upon sound and effective national policies of economic and social development as recognised by the proclamation of Teheran of 1968.*' A/RES/32/130, 16 December 1977, *Alternative approaches and ways and means within the United Nations system for improving the effective enjoyment of human rights and fundamental freedoms, Section 1.b.*

[661] Wiebenga, Senate, 5th meeting, 21 November 1978.

[662] Minister Van der Klaauw, Senate, 5th meeting, 21 November 1978. Original text in Dutch: '*wat de resolutie betreft, gaat de meest recente ontwikkeling in de discussie die in de laatste jaren in de Verenigde Naties op gang is gekomen over de benadering van de problematiek van de mensenrechten, waarbij het vooral gaat om de relatie tussen de klassieke burgerlijke rechten, de politieke rechten aan de ene kant, en de sociale en economische mensenrechten aan de andere kant alsmede om de structurele oorzaken die de verwezenlijking van de mensenrechten in de weg kunnen staan.*'

Assmebly resolution 32/130 did not recognise a hierarchical relation between types of human rights, but rather emphasised that both types of rights should be given equal and urgent attention, for '*both categories are essential to the development of human beings as an individual and as a Member of society.*'[663]

7.4.3 Complaints procedures

According to the Government, the Netherlands contributed in a profound way to the realisation of both treaties. Especially the adoption of an individual complaints procedure in an Optional Protocol to the ICCPR was initiated by the Netherlands, when an earlier proposed monitoring mechanism was not accepted by the General Assembly.[664] It appears generally from the Parliamentary History of the ratification Bill of both treaties that the Dutch Government was not satisfied with the monitoring mechanisms as were finally adopted, especially regarding the individual complaints procedure of the ICCPR. It was the ambition of the Netherlands to strive for more profound and also more judicial monitoring mechanisms in the future.[665] However, the discussion on monitoring procedures and more specifically complaints procedures was basically held in the context of the ICCPR. A complaints procedure regarding the ICESCR was not adopted by the General Assembly, and therefore that possibility was hardly discussed during the Parliamentary debates on the adoption of the ratification Bill. In its Explanatory Memorandum the Government only pointed out that the ICESCR monitoring mechanism had a more modest extent compared to the ICCPR, due to '*the other kind of commitments, at least broadly, the treaties impose on the parties, that is an immediate obligation to respect and ensure on the one hand, and a commitment to progressively realise on the other.*'[666]

7.5 The CEDAW

The ratification Bill of the Convention on the Elimination of All Forms of Discrimination against Women[667] was discussed in Parliament in the period 1984-1985.

---

[663] Minister Van der Klaauw, Senate, 5th meeting, 21 November 1978. Original text in Dutch: '*Beide categorieën zijn essentieel voor de ontplooiing van de mens, als individu en als lid van de samenleving.*'

[664] Parliamentary Documents, II 1975-1976 13932 (R 1037), no. 3, pp. 39-40. In order to do so, the Netherlands closely cooperated with Nigeria, according to Mommersteeg, House of Representatives, 3rd meeting, 21 September 1978.

[665] See for instance: Minister van Agt, House of Representatives, 3rd meeting, 21 September 1978; and Minister Van der Klaauw, Senate, 5th meeting, 21 November 1978.

[666] Parliamentary Documents, II 1975-1976 13932 (R 1037), no. 3, p. 57. Original text in Dutch: '*aan de andersoortige verplichtingen die, althans globaal genomen, de verdragen aan de partijen opleggen, namelijk een verplichting tot onmiddellijke eerbiediging en verzekering enerzijds en een verplichting tot geleidelijke verwezenlijking anderzijds.*'

[667] A/RES/34/180, 18 December 1979, *Convention on the Elimination of All Forms of Discrimination Against Women.*

7.5.1 Background and added value of the CEDAW

In its Explanatory Memorandum, the Government explained that the Convention exclusively focuses on all discrimination against women, stipulating both civil and political rights on the one hand, and economic, social and cultural rights on the other hand. According to the Government, it was the duty of the Member States to *'implement a policy to eliminate discrimination with all their available resources.'*[668] The Committee on the status of women was founded in 1946, and drafted the declaration on the elimination of discrimination against women,[669] a document that would later be of great influence on the wordings of the CEDAW. It is remarkable here that in a footnote, the Government declared to prefer a translation of the name of this commission that deviates from the official Dutch translation in Article 21 of the CEDAW. In the Explanatory Memorandum, the Commission is referred to as 'Commissie inzake de Positie van de Vrouw' (freely translated: 'Committee on the Status of Women'), while the official translation of Article 21 CEDAW refers to 'Commissie inzake de Rechtspositie van de Vrouw' (freely translated: 'Committee on the Legal Position of Women'). However, no explanation for the preference of an alternate translation is given.[670]

In the Preparatory Committee, a Member of the Reformed Political Association (GPV) raised the question what the actual consequences would be for the Dutch legal order when ratifying the CEDAW, since it appeared that no direct results would follow from this ratification, especially due to the fact that non-compliance with the CEDAW would not be sanctioned, for no agreement on a complaint procedure had been reached, and many Member States had made reservations to Article 29 CEDAW, that recognises an arbitration procedure. In addition, the Government had admitted in its Explanatory Memorandum that there are many different views worldwide on how to realise emancipation, which implies that the effect of the treaty may differ per region. Also, the GPV Member underlined the increasing politicisation within the UN bodies, which could also be noticed during the negotiations preceding the adoption of the CEDAW. In that light, the question was raised whether ratifying the treaty would have any added value at all.[671] The Government replied that *'By ratifying the Convention, the Netherlands supports the establishment of global standards in this area and thus contributes to the development of international law (Article 90 CA). By recognising these standards for themselves the Netherlands can also expect other countries to comply with these*

---

[668] Parliamentary Documents, II 1984-1985, 18950 (R 1281), no. 3, p. 3. Original text in Dutch: *'zich met alle hun ter beschikking staande middelen een beleid te voeren dat gericht is op de uitbanning van discriminatie.'*
[669] A/RES/2263 (XXII), 9 November 1967, *Declaration on the elimination of discrimination against women.*
[670] Parliamentary Documents, II 1984-1985, 18950 (R 1281), no. 3, p. 4, 5th footnote.
[671] Parliamentary Documents, II 1984-1985, 18950 (R 1281), no. 4, p. 5.

standards. To this end, the reporting procedure enshrined in the Treaty may serve as an important tool.'[672]

In general, the Government highlighted certain aspects of the drafting history of the CEDAW. It underscored that considering the large cultural differences concerning the position and role of women, the agreement on the adoption of the CEDAW was an important achievement that could only be accomplished through the willingness to constructively negotiate. Especially concerning private law and family law issues there appeared to be large differences of opinion on the rights of women between Western-oriented and Islam-oriented countries, which had led to a number of compromises in the final wordings of the relevant CEDAW Provisions. The participation of new Member States – mainly developing countries – in the UN arena, and the resulting new balance of power within the UN bodies, led to the adoption of Provisions that stipulate aspects that were new compared to the existing human rights instruments, such as the adoption of Article 14, addressing in particular the position of rural women.[673]

It is interesting to note that in its Explanatory Memorandum, the Government summarised the most important contribution of the Netherlands in the realisation of the CEDAW, which included the drafting of resolution A/34/180[674] with the final version of the CEDAW as accepted by the General Assembly.[675]

7.5.2 The direct applicability of the CEDAW

The Government explained that in Articles 7-17 CEDAW the substantive regulations on the equal treatment of women are enshrined. According to the Government *'the nature and content of these substantive treaty obligations and the formulation of the duties to States lead to the conclusion that implementation is a task of the legislative and administrative authorities.'*[676] The Government underlined that the global eradication of all discrimination against women could not be realised immediately, and therefore the Convention allowed the progressive realisation of the rights in Article 2. This would mean then that a Member State, although having the obligation to immediately with all necessary measures focus its policies on the

---

[672] Parliamentary Documents, II 1984-1985, 18950 (R 1281), no. 6, p. 9. Original text in Dutch: *'Door het Verdrag te ratificeren levert Nederland een bijdrage aan de totstandkoming van wereldwijde normen op dit terrein en draagt daarmee bij tot de ontwikkeling van de internationale rechtsorde (artikel 90 van de Grondwet). Door deze normen voor zichzelf te aanvaarden, kan Nederland ook van andere landen verlangen deze normen na te leven. De in het Verdrag neer neergelegde rapportageprocedure vormt daarvoor een belangrijk hulpmiddel.'*
[673] Parliamentary Documents, II 1984-1985, 18950 (R 1281), no. 3, p. 5.
[674] A/RES/34/180, 18 December 1979, *Convention on the Elimination of All Forms of Discrimination Against Women.*
[675] Parliamentary Documents, II 1984-1985, 18950 (R 1281), no. 3, p. 6.
[676] Parliamentary Documents, II 1984-1985, 18950 (R 1281), no. 3, p. 7. Original text in Dutch: *'De aard en de inhoud van deze materiële verdragsverplichtingen en de formulering van de opdrachten aan staten leiden tot de conclusie dat de uitvoering ervan een taak is van de wetgevende en besturende overheid.'*

realisation of the treaty obligations, may achieve the realisation in itself gradually. Therefore, at the moment on which the Member State ratifies the Convention, not all implementation measures need to be adopted yet. However, according to the Government, the Netherlands already fulfilled most of its duties at the moment of ratification, and there were hardly any intentions to adjust existing or adopt new domestic Provisions or policy with the purpose to progressively realise the CEDAW rights. This was criticised by several Members representing various parties in the Preparatory Committee.[677] On the request of the Christian Democratic Appeal (CDA),[678] the Government eventually published an overview with intended legal reforms that would result from the adoption of the CEDAW.[679]

To establish whether a CEDAW Provision would be directly applicable, the Government referred to criteria used by the Supreme Court, which are *'the nature and content of the Provision as well as its wordings, while also the intention of the draftsmen of the Convention may be a guideline.'*[680] Concerning the latter criterion, the Government informed that the matter of direct applicability for those countries whose Constitution would allow this had not been explicitly discussed during the negotiations preceding the adoption of the CEDAW,[681] due to the fact that a majority of the UN countries is unfamiliar with a Constitutional construction that specifically regulates the direct effect of international Provisions.[682] The Member of the Reformed Political Association (GPV) raised the question in the Preparatory Committee whether the railway-strike ruling could be of influence on future case law concerning CEDAW Provisions, considering the fact that in that case, the Supreme Court held that concerning Article 6 (4) ESD, *'there is no evidence that the contracting parties have agreed that direct effect may not be granted'*[683], a consideration that was thus understood by this Member of Parliament that it would mean that in the absence of an agreement on direct applicability between contracting parties, the Supreme Court consequently would grant the Provision direct effect.[684] The Government responded that: *'it appears to us that the Provision of the European Social Charter on which the Supreme Court ruled (Article 6, Section 4) had such a clear content in view of the disputed question of law, that the Supreme Court in the application of that Provision could remain within the previously by him established criteria, which formed the basis of our observations in our reply on the*

---

[677] Parliamentary Documents, II 1984-1985, 18950 (R 1281), no. 4, especially p. 2 (Labour Party/PvdA), and 5 (Reformed Political Association/GPV).

[678] Parliamentary Documents, II 1984-1985, 18950 (R 1281), no. 8, p. 2.

[679] Parliamentary Documents, II 1984-1985, 18950 (R 1281), no. 9, pp. 1-3.

[680] Parliamentary Documents, II 1984-1985, 18950 (R 1281), no. 3, pp. 7-8. Original text in Dutch: *'de aard en inhoud van de bepaling alsmede de formulering daarvan, terwijl tevens de bedoeling van de opstellers van het verdrag een richtsnoer kan zijn.'*

[681] Parliamentary Documents, II 1984-1985, 18950 (R 1281), no. 3, p. 8.

[682] Parliamentary Documents, II 1984-1985, 18950 (R 1281), no. 6, p. 14.

[683] Parliamentary Documents, II 1984-1985, 18950 (R 1281), no. 8, pp. 4-5. Original text in Dutch: *'uit niets blijkt dat verdragsluitende partijen zijn overeengekomen dat geen rechtstreekse werking mag worden toegekend.'*

[684] Parliamentary Documents, II 1984-1985, 18950 (R 1281), no. 8, pp. 4-5.

issue of direct effect. ... However, we note that the ruling of the Supreme Court does not inevitably lead to the conclusion drawn by the Member of the Commission, that every treaty should be granted direct effect unless agreed otherwise. Rather, it appears from the first sentence of the above quoted phrase that in principle most significance can be attributed to the nature, content and wording of the Provision. In our view, the Supreme Court only expressed that there were no indications that the Contracting Parties intended to block the possible direct effect of Charter Provisions that can be granted in accordance with the domestic law of some States Parties.'[685]

On the one hand, as demonstrated above, the Government was of the opinion that the implementation of the Provisions was mainly a responsibility of the Legislature. On the other hand, the Government admitted that *'The decision and the conditions under which direct application is granted to a Provision would be, in our legal system mainly up to the Judiciary. Therefore, in this Explanatory Memorandum, a certain restraint would be appropriate.'*[686] Nevertheless, the Government underlined that in the CEDAW, Provisions were enshrined with equivalents in other human rights treaties that had been granted direct applicability before. It could therefore be reasonably expected that the CEDAW versions would also be granted direct effect. As an example, the Government referred to the equal treatment concerning the right to vote and to hold public office as stipulated in Article 7 CEDAW, but also in Article 2 (1) and (3) in conjunction with Article 25 and 26 ICCPR.[687] Several fractions of the Preparatory Committee however, asked for more clarity concerning the possibility of direct applicability of the treaty Provisions. The Members of the Christian Democratic Appeal (CDA) asked this clarity in particular in view of the role of the Explanatory Memorandum in the considerations of the Courts when they would be confronted with the question whether or not a treaty Provision would have direct effect.[688] Both the Communist Party of the Netherlands (CPN) and the Reformed Political Association (GPV) fractions asked whether the aforementioned statements of the Government must thus be understood that

---

[685] Parliamentary Documents, II 1984-1985, 18950 (R 1281), no. 9, pp. 8-9. Original text in Dutch: *'Het komt ons voor dat de bepaling van het Europees Sociaal Handvest waarover de Hoge Raad zich uitsprak (artikel 6, vierde lid) voor de in het geding zijnde rechtsvraag een zo duidelijke betekenis had, dat de Hoge Raad bij de toepassing van die bepaling kon blijven binnen de eerder door hem geformuleerde criteria, die de grondslag vormden voor onze beschouwingen in de memorie van antwoord over de problematiek van de rechtstreekse werking. (...) Wel merken wij op dat de uitspraak van de Hoge Raad niet dwingt tot de door het genoemde lid getrokken conclusie, dat aan iedere verdragstekst rechtstreekse werking moet worden toegekend tenzij de verdragssluiting anders zou zijn overeengekomen. Veeleer is uit de eerste volzin van het voornoemde citaat af te leiden dat in beginsel de meeste betekenis toekomt aan aard, inhoud en formulering van de bepaling. De Hoge Raad zou dan ook in onze opvatting slechts tot uitdrukking hebben gebracht dat er geen aanwijzingen waren dat de Verdragspartijen de eventueel door het nationale recht van sommige Verdragspartijen toe te kennen rechtstreekse werking van Handvestbepalingen hebben willen blokkeren.'*

[686] Parliamentary Documents, II 1984-1985, 18950 (R 1281), no. 3, p. 8. Original text in Dutch: *'De beslissing hierover en de voorwaarden waaronder die rechtstreekse toepassing aan een bepaling zou worden gegeven, zijn in ons rechtsbestel voornamelijk aan de rechter. In deze toelichting past daarom terughoudendheid.'*

[687] Parliamentary Documents, II 1984-1985, 18950 (R 1281), no. 3, p. 8.

[688] Parliamentary Documents, II 1984-1985, 18950 (R 1281), no. 4, p. 6.

only those CEDAW Articles would have direct effect when its equivalents in other treaties had been granted direct effect before.[689] In addition, Members of the CPN asked for more clarity concerning the Government's statement that the decision as to whether a treaty Provision would have direct effect would 'mainly' be up to the Judiciary, since 'mainly' would imply that there are exceptions to the rule.[690] The Government replied to the aforementioned questions that from the nature and content, as well as the wordings of the Provisions – due to the fact that these were addressed to States – it followed that it was the task of the Legislature and Administration to progressively realise the rights stipulated in the CEDAW. However, this does not stand in the way of the fact that the Constitutional system (Articles 93 and 94) authorised the Judiciary to not apply domestic legislation that is contrary to international Provisions, and it is the Judiciary alone that has a final say in the matter.[691] On the other hand, this does not preclude that also the Administration – to which the said Constitutional Provisions are also addressed – can be confronted with the same question whether a Provision is suitable for direct applicability.[692] It is for the first time (and certainly not for the last time) that in human rights-related Parliamentary History, the Government seemed to struggle with the exact meaning of Articles 93 and 94 in relation to the balance of power between the Legislature and Judiciary in deciding whether an international Provision has direct effect or not. A struggle that certainly did not lead to the most unambiguous explanation of this Constitutional system.

Eventually, the Government decided to explore what CEDAW Provisions, according to them, would be suitable for direct applicability, taken into consideration the criteria as developed by the Supreme Court: the nature and content of the Provisions, as well as the wordings thereof, while also the intentions of the contracting states might play a role.[693] Based on this, the Government discussed several CEDAW Provisions, and concluded that the introductory words of Article 2 (the non-discrimination Provision) are a principle that should be realised through 'all appropriate means', which implies a margin of discretion for Member States with regard to their policies. Also the period in which the principle should be realised was not specifically embedded in the Provision, which would suggest that the Member States should act within a reasonable period. Due to this lack of clarity, the Government was of the opinion that this part of the Provision would not be suitable for direct application by the Courts.[694] According to the Government, Article 2 (d), that recognises the prohibition for public authorities and institutions to discriminate against women, might be used in conjunction with Article 1 CEDAW, as a yardstick in civil proceedings based on tort law (in the Netherlands

---

[689] Parliamentary Documents, II 1984-1985, 18950 (R 1281), no. 4, p. 7.
[690] Parliamentary Documents, II 1984-1985, 18950 (R 1281), no. 4, p. 7.
[691] Parliamentary Documents, II 1984-1985, 18950 (R 1281), no. 6, p. 10.
[692] Parliamentary Documents, II 1984-1985, 18950 (R 1281), no. 6, p. 13.
[693] Parliamentary Documents, II 1984-1985, 18950 (R 1281), no. 6, p. 10.
[694] Parliamentary Documents, II 1984-1985, 18950 (R 1281), no. 6, pp. 10-11.

the current Article 6:162 Civil Code, in the Parliamentary Documents reference is made to the earlier Article 1401 Civil Code), but not likely as an independent standard in these proceedings.[695] With regard to Article 10 (Sections a, b, c, d, e, and g), the Government underlined that the Provision basically stipulated the obligation to take appropriate measures in order to eliminate discrimination against women in the field of education, implying a large margin of discretion, and therefore the Article would not be suitable for direct application.[696] The Government considered that Article 11, Section 1 sub d and e also implied a certain margin of discretion, and concluded (after additional comparison to existing case law in which some Members of the Preparatory Committee assumed equivalent communitary law was involved) that direct applicability was unlikely.[697] Regarding Article 11 Section 2, the Government argued that in most cases direct applicability would not be possible, due to the aforementioned margin of discretion, and the fact that the involved protective legislation should be reviewed periodically in the light of scientific and technological knowledge and shall be revised, repealed or extended as necessary, implies a permanent development in legislation that is not suitable for judicial review. However, the Government considered – without further explanation – that direct applicability of the Provision would be not inconceivable in case of the prohibition of dismissal on grounds of pregnancy or maternity leave and discrimination in dismissals on grounds of marital status.[698] Considering the nature, contents and wordings of Article 15, sub 3, the Government recognised that this Provision would be suitable for direct application by the Courts. However, this notion would only be of an academic relevance, since the essence of this Article was already embedded in the Dutch Civil Code.[699] Regarding Article 16 sub 2, the Government underlined that the Netherlands already amply fulfilled the obligations stipulated in that Provision, and thus its realisation is already guaranteed. The possibility of direct application is therefore not explored.[700] Earlier, as mentioned above, the Government had already recognised the possibility of direct application of Article 7 CEDAW.[701] Unfortunately, the Government did not discuss its views on the possibility of direct applicability of other CEDAW Provisions containing substantive treaty obligations, nor was this discussed significantly elsewhere in Parliament. The Government explained that it had discussed the matter concerning all treaty Provisions whose direct effect had been specifically discussed in Parliament.[702] This, and the fact that no further Parliamentary debate concerned the direct effect of treaty Provisions, implies that there was no specific

---

[695] Parliamentary Documents, II 1984-1985, 18950 (R 1281), no. 6, p. 11.
[696] Parliamentary Documents, II 1984-1985, 18950 (R 1281), no. 6, p. 11.
[697] Parliamentary Documents, II 1984-1985, 18950 (R 1281), no. 6, pp. 11-12.
[698] Parliamentary Documents, II 1984-1985, 18950 (R 1281), no. 6, p. 12.
[699] Parliamentary Documents, II 1984-1985, 18950 (R 1281), no. 6, p. 12.
[700] Parliamentary Documents, II 1984-1985, 18950 (R 1281), no. 6, p. 13.
[701] Parliamentary Documents, II 1984-1985, 18950 (R 1281), no. 3, p. 8.
[702] Parliamentary Documents, II 1984-1985, 18950 (R 1281), no. 9, p. 9.

interest in the matter of direct effect of other treaty Provisions than the Articles discussed.

Regarding the frequently mentioned margin of discretion in case of Provisions that are not suitable for direct application, the Government declared that this margin was not unrestricted, for Article 2 CEDAW stresses to adopt a policy '*with all appropriate means*' and '*without delay*'. Interestingly, the Government added that once these measures have been adopted, the position of women would be in compliance with the treaty, and seems to suggest that any further improvements or nuances would not fall under the scope and protection of the CEDAW.[703]

7.5.3 Complaints procedures

Since the Member States of the CEDAW did not agree on the establishment of complaints procedures, no such procedures were discussed during the Parliamentary debates concerning the adoption of the ratification Bill.

7.6 The ICRC

The ratification Bill of the Convention of the Rights of the Child[704] was discussed in Parliament in 1994.

7.6.1 Background and added value of the ICRC

In its Explanatory Memorandum, the Government underlined that it appeared from the *travaux préparatoires* that it was the intention of the State Parties to adopt a treaty that would go beyond a declaration on the rights of the child.[705] The Government considered that the already existing declarations on the rights of the Child of 1924[706] and 1959[707] were insufficiently far-reaching, because '*for the effective safeguarding of those rights it was not sufficient to embed these in a statement adopted as a resolution.*'[708] In addition, the Dutch Government considered that with regard to the rights of the child, the Universal Declaration on Human Rights was inadequate, for while these rights were recognised for each human being, they could, due to their declarative nature, not be invoked in Courts, and it concerned a general list of human rights, not focussed on the position of children.[709]

---

[703] Parliamentary Documents, II 1984-1985, 18950 (R 1281), no. 6, pp. 3-4.
[704] A/RES/44/25, 20 November 1989, *Convention on the Rights of the Child*.
[705] Parliamentary Documents, II 1992-1993, 22855 (R1451), no. 3, p. 1.
[706] *Geneva Declaration on the Rights of the Child*, adopted by the League of Nations, 26 September 1924.
[707] A/4354, 20 November 1959, General Assembly Resolution 1386 (XIV), *Declaration on the Rights of the Child*.
[708] Parliamentary Documents, II 1992-1993, 22855 (R1451), no. 3, p. 2. Original text in Dutch: '*het voor een effectieve waarborging van die rechten niet voldoende was deze neer te leggen in een bij een resolutie aangenomen Verklaring.*'
[709] Parliamentary Documents, II 1992-1993, 22855 (R1451), no. 3, p. 1.

Recognising the rights of the child in a treaty instead, would considerably strengthen these rights. However, during the negotiations preceding the adoption of the final draft of the ICRC, it appeared to be difficult to come to a common approach regarding the substantive rights of the treaty. The added value on top of the existing international legal instruments, in which human rights for each individual were already embedded, of which some rights specifically stipulate certain rights of the child, was frequently discussed. During the negotiations preceding the adoption of the ICRC, the Dutch Delegation had underlined that: 'a *lot of the principles in the present draft Convention are already embodied in the mandates of agencies like the ILO, WHO and UNESCO, or in more judicial wordings, in instruments like the Covenants on Civil and Political and on Economic, Social and Cultural Rights (...) In its view a draft convention on the rights of the child must consist of timely, up-to-date concrete principles, accompanied by practical guidelines for application, and supplementary to already existing instruments and activities, in order to avoid unnecessary duplication.*'[710] According to the Government, the contents of the ICRC could be limited to the mere recognition of the applicability of existing human rights to children, and of certain additional rights that would be applicable to children in particular. The Government noticed that this advice was not seriously considered in the international arena, and instead a comprehensive convention containing a wide variety of Provisions that may therefore overlap with existing human rights treaties was drafted.[711] Therefore, the Government published in an attachment to its Explanatory Memorandum an overview of overlapping, equivalent rights in several human rights treaties, as well as the Dutch Constitutional Act. For instance, the right to social security was considered to be recognised in Article 26 ICRC, but also in Article 20 CA, Article 9 ICESCR, and Articles 12 and 13 ESC. The right to an adequate standard of living was considered to be recognised in Article 27 ICRC, but also in Article 22 CA and Article 11 ICESCR.[712]

The Government explained the negotiations were based on the principle of consensus, in order to achieve the greatest possible degree of acceptance of its content. However, this led to the fact that the wordings of the Provisions were the result of a compromise between various viewpoints, while the Government had hoped that certain Provisions would have a more far-reaching scope.[713] This even led to an official declaration of vote to the Dutch vote for adoption of the final draft text, in which the Dutch Delegation declared that '*fully aware of the importance of the Convention on the Rights of the child, my Delegation did not wish to break the prevailing consensus on the draft resolution just adopted. As we all know, the birth of the Convention has been far from easy. The youngest among the United Nations*

---

[710] Parliamentary Documents, II 1992-1993, 22855 (R1451), no. 3, p. 6.
[711] Parliamentary Documents, II 1992-1993, 22855 (R1451), no. 3, p. 6.
[712] Parliamentary Documents, II 1992-1993, 22855 (R1451), no. 3, p. 55, third attachment.
[713] Parliamentary Documents, II 1992-1993, 22855 (R1451), no. 3, p. 54, second attachment. In this attachment, the Delegation furthermore focussed on the – in its view – limited meaning of Article 38 ICRC, and the protection of children with regard to armed conflicts.

treaty instruments came into existence after ten years of labour. Setting standards and particularly drafting binding obligations in the field of human rights is an exercise which should take place with a magnifying glass in hand. The Convention on the Rights of the Child bears the mark of extensive negotiations and thus of many a compromise. The Netherlands itself were actively involved in the work of the Working Group which prepared the Convention. We accepted in a spirit of compromise and with a view to progressively developing human rights law a number of Provisions which could or should have been – in the view of my Government – formulated differently and more consistent with human rights instruments.'[714]

Despite the overlap of the ICRC with other treaties, and the rather mild obligations recognised in the treaty, the Government considered that the treaty did have its added value, and approved of its adoption, for several reasons. In general, the Government underlined that the adoption of the ICRC was in line with the global tendency to adopt specific treaties on particular issues, alongside the already existing general human rights conventions.[715] The added value of the ICRC then could be found in firstly: the fact that in the Convention the rights already recognised in the UDHR, the ICCPR and the ICESCR, were not only reaffirmed, but also focussed on the particular situation of children. Secondly, certain rights recognised in the ICRC were only addressed as and applicable to children (for instance, the Provisions concerning adoption, education, and contact with the parents). Thirdly, in the ICRC also new rights were recognised, which were not embedded before in any human rights treaty, such as the regulations regarding the relation between the rights of the child and the authority of parents, and the (international cooperation in the field of) disabled children. Fourthly, the Government underlined that some States were not a Member of the more general human rights treaties, but might be interested to ratify the ICRC due to this new interpretation of human rights.[716] In this light, Minister Kosto suggested in the House of Representatives that the ICRC could be used as a basis to address other countries to maintain the rights stipulated in the Convention.[717] Interesting detail is that in their reply to the questions raised by the Preparatory Committee, the Government stated that the Netherlands were relatively late in ratifying the Convention (150 States already preceded the Netherlands), and this has led to questions in the international relations with regard to the Dutch attitude towards the treaty.[718]

---

[714] Parliamentary Documents, II 1992-1993, 22855 (R1451), no. 3, p. 54, second attachment.

[715] Parliamentary Documents, II 1992-1993, 22855 (R1451), no. 3, p. 3.

[716] Parliamentary Documents, II 1992-1993, 22855 (R1451), no. 3, p. 7. See also: Parliamentary Documents, II 1992-1993, 22855 (R1451), no. 6, p. 2.

[717] Minister Kosto, House of Representatives, 84th meeting, 30 June 1994.

[718] Parliamentary Documents, II 1992-1993, 22855 (R1451), no. 6, p. 2.

In Parliament, most Members expressed their approval regarding the added value of the ICRC.[719] However, there were also some critical notes, especially from the confessional parties, which expressed their concerns that ratification of the treaty would result in the interference in family relations, which should not be a duty of the Government.[720]

7.6.2 The direct applicability of the ICRC

The Government was pleased that in the ICRC, both civil and political rights on the one hand, and economic, social and cultural rights on the other hand were embedded, for this was in line with the idea coming forth from the UDHR, but also expressed in the ICCPR and the ICESCE, that all human rights are inextricably linked.[721]

This time, the Government was somewhat less confusing in its interpretation of the Constitutional system embedded in Articles 93 and 94 CA, and enlightened on the relation between the Legislature and the Judiciary regarding the decision whether a treaty Provision would be directly applicable or not. The Government underlined that it was stipulated in Article 94 CA[722] that *'ultimately, the Dutch Courts decide whether a Provision that is "binding on all persons" has direct effect or not.'*[723] The Legislature could offer a 'helping hand' to the Judiciary by altering domestic legislation that is – in the view of the Legislature – in contradiction with a directly applicable treaty Provision, but still the Judiciary has final say in the matter. However, the Government emphasised that to the Courts, the viewpoints of the Legislature regarding the direct applicability of a Provision would be *'of great importance, as demonstrated by the case law of the Supreme Court.'*[724] According to the Government, it appears from the case law of the Supreme Court that fixed criteria are used to determine whether a treaty Provision has direct effect or not: *'the nature, content and scope of the Provision, as well as the formulation (the wording) thereof, while also the intention of the national Legislature during the realisation of the ratification law of the relevant treaty could be a guideline, and – insofar as this is the case – the intention of the draftsmen of the treaty should be a guideline. Also of*

---

[719] See especially: Van den Burg (PvdA), and De Vries (VVD), House of Representatives, 81st meeting, 23 June 1994.

[720] See especially: Van den Berg (SGP), House of Representatives, 81st meeting, 23 June 1994, House of Representatives, 84th meeting, 30 June 1994; Schutte (GPV) and Rouvoet (RPF), House of Representatives, 81st meeting, 23 June 1994.

[721] Parliamentary Documents, II 1992-1993, 22855 (R1451), no. 3, p. 3.

[722] Parliamentary Documents, I 1994-1995, 22855 (R1451), no. 22a, p. 5.

[723] Parliamentary Documents, II 1992-1993, 22855 (R1451), no. 3, p. 8. Original text in Dutch: *'uiteindelijk bepaalt de Nederlandse rechter of een "een ieder verbindende" bepaling rechtstreekse werking heeft of niet.'*

[724] Parliamentary Documents, II 1992-1993, 22855 (R1451), no. 3, p. 8. Original text in Dutch: *'wel van grote betekenis is, getuige ook de rechtspraak van de Hoge Raad.'*

importance in this is the presence or absence of implementation legislation, and whether it is possible within the national legal system to directly apply the Provision.'[725]

These criteria can be subdivided into four categories, and will be discussed here into more detail:

1. The nature, content and scope of the Provision, as well as the formulation (the wording) thereof. The Government did not hesitate to express its opinion about this criterion concerning the treaty Provisions. The Government stated that '*Considering the nature, content and wording of most Provisions of the present Convention, the implementation of and compliance with a large number of substantive treaty Provisions implies a task for the Legislature and the governing authorities. This does not entirely exclude the possibility that a particular treaty Provision in a specific case is granted direct effect. The decision and the conditions under which direct effect would be given to a Provision, as stated in our legal system, is ultimately up to the Courts.'[726]* In fact, this statement is a general idea of the perception of the Legislature regarding the direct applicability of the treaty Provisions, based on the content of the Articles, but actually concerns the second criterion: the intention of the Legislature.

2. The intention of the national Legislature can most likely be distilled from the Parliamentary History on the ratification of the treaty. The Government specifically mentioned what Articles would, in its view, be suitable for direct application. Firstly, this would concern the Articles with equivalents in other treaties that had been granted direct effect before, basically civil and political rights: Articles 7 (1), 9 (1), (2), (3), (4), 10 (1, second sentence), 13, 14, 15, 16, 30, 37, and 40 (2). In addition, the Government did consider it likely that Article 12 (2) would have direct effect, and did not preclude the possibility of direct effect of some Articles that stipulate rights that were not embedded in other treaties yet, especially addressed to the child, and also with characteristics of

---

[725] Parliamentary Documents, II 1992-1993, 22855 (R1451), no. 3, p. 9. Original text in Dutch: '*de aard, inhoud en de strekking van de bepaling, alsmede de formulering (de bewoording) daarvan, terwijl tevens de bedoeling van de nationale wetgever bij de totstandkoming van de goedkeuringswetgeving van het desbetreffende verdrag een richtsnoer kan zijn en – zover hiervan sprake is – de bedoeling van de opstellers van het desbetreffende verdrag een richtsnoer dient te zijn. Eveneens is in deze van belang de aan- of afwezigheid van uitvoeringswetgeving en of het binnen het nationaal rechtelijk bestel mogelijk is de bepaling rechtstreeks toe te passen.'* The Government referred to the following case law of the Supreme Court: Supreme Court, 24 February 1960, NJ 1960, 483; 28 November 1961, NJ 1962, 90; 25 April 1967, NJ 1968, 63, 6 December 1983, NJ 1984, 557; 18 February 1986, NJ 1987, 62; 30 May 1986, NJ 1986, 688 (railway-strike ruling); 14 April 1989, NJ 1989, 469 ('Harmonisation Act' ruling), 20 April 1990, RvdW 1990,88. The Government also referred to these criteria in: Parliamentary Documents, I 1994-1995, 22855 (R1451), no. 22a, p. 5.
[726] Parliamentary Documents, II 1992-1993, 22855 (R1451), no. 3, p. 9. Original text in Dutch: '*gelet op de aard, inhoud, en formulering van de meeste bepalingen van het onderhavige verdrag houden de uitvoering en de naleving van een groot aantal materiële verdragsvoorschriften een taak van de wetgever en van de besturende overheden in. Dit behoeft overigens niet helemaal uit te sluiten dat aan een bepaald verdragsvoorschrift in een concrete casus rechtstreekse werking wordt toegekend. De beslissing hierover en de voorwaarden waaronder die rechtstreekse werking aan een bepaling zou worden gegeven, zijn, zoals gezegd, in ons rechtsbestel uiteindelijk aan de rechter.'*

civil and political rights: Articles 5, 8 (1) en 12 (1). In this light, hardly anything was mentioned concerning the rights with the characteristics of economic, social and cultural rights.[727] The Government argued that *'Formulations in treaty Provisions that oblige States Parties to undertake appropriate measures to facilitate the implementation of the rights in accordance with national conditions and with the available resources, first and foremost intend to make clear that a Member State is not bound to the impossible. Such Provisions are without prejudice to the responsibility of the democratically elected Government, but rather appeal to their primary responsibility to do what is reasonably possible, considering the national (especially financial) situation. This is exactly why in these cases judicial review is unlikely.'*[728] In its reply to questions of the Preparatory Committee, the Government even emphasised that only in extreme situations, judicial review against economic, social and cultural rights of the ICRC would be possible.[729]

3. The intention of the draftsmen of the treaty was also referred to as a possible criterion to determine whether a Provision is directly applicable. The Government stated that during the negotiations preceding the adoption of the ICRC, the matter of direct applicability had not been discussed by the draftsmen of the treaty. In the Preparatory Committee, the Democrats '66 (D'66) Members asked how this would relate to the intention of the draftsmen of the treaty.[730] The Government replied to this that therefore, no intention of the draftsmen of the treaty could be distilled from the *travaux préparatoires* at all.[731] The Reformed Political Association (GPV) fraction expressed a different understanding of the fact that the direct applicability of ICRC Provisions was not discussed during the negotiations, and asked whether this could thus be understood that, because the matter was not discussed, *'...at any rate, from the nature, content and scope of the Treaty Provisions no argument for direct effect can be distilled.'*[732] This considering, the GPV fraction asked whether that could mean that *'...the Government does not intend to grant direct effect insofar this does*

---

[727] Parliamentary Documents, II 1992-1993, 22855 (R1451), no. 3, p. 9.

[728] Parliamentary Documents, II 1992-1993, 22855 (R1451), no. 6, p. 9. Original text in Dutch: *'Formuleringen in verdragsbepalingen die verdragsstaten verplichten passende maatregelen te nemen om te helpen rechten te verwezenlijken in overeenstemming met de nationale omstandigheden en met de middelen die hen ten dienst staan, beogen eerst en vooral duidelijk te maken, dat een verdragsstaat niet tot het onmogelijke is gehouden. Dergelijke bepalingen doen dus niet af aan de verantwoordelijkheid van het democratisch gekozen bestuur, doch appelleren veeleer aan diens primaire verantwoordelijkheid ter zake te doen wat redelijkerwijs mogelijk is, gelet op de nationale (vooral financiële) situatie. Juist daarom ligt in dezen een toetsingstaak van de rechter niet voor de hand.'*

[729] Parliamentary Documents, II 1992-1993, 22855 (R1451), no. 6, p. 18.

[730] Parliamentary Documents, II 1992-1993, 22855 (R1451), no. 4, p. 6.

[731] Parliamentary Documents, II 1992-1993, 22855 (R1451), no. 6, p. 10.

[732] Parliamentary Documents, II 1992-1993, 22855 (R1451), no. 4, p. 6. Original text in Dutch: *'...in ieder geval aan aard, inhoud en strekking van de verdragsbepalingen geen argument ontleend kan worden voor een rechtstreekse werking.'*

not already follow from Provisions of other treaties.'[733] The Government disagreed, and stated that it had already expressed in its Explanatory Memorandum that it considered it possible that some of the Provisions of the ICRC that had no equivalents in other existing human rights treaties would be suitable for direct effect. Apart from that, the Government added that it follows from Articles 93 and 94 CA that it is up to the Judiciary to finally decide on the matter.[734] In general however, it is doubtful whether the matter of direct applicability will ever be discussed during the negotiations preceding the adoption of a treaty. As mentioned earlier, during the Parliamentary History of the ratification Bill of the CEDAW, the Government had emphasised that a majority of the UN countries is unfamiliar with a Constitutional construction that specifically regulates the direct effect of international Provisions.[735]

4. The presence or absence of implementation legislation, and whether it is possible within the national legal system to directly apply the Provision also may have influence on the decision whether a Provision is directly applicable or not. To demonstrate this, the Government referred to case law, including the 'Harmonisation Act' ruling.[736] It seems most likely that the following consideration of the Supreme Court inspired the Government to refer to this ruling: *'moreover, it concerns Provisions that imply performance duties of the Government towards citizens; such Provisions can hardly function in the legal order without further elaboration, so that direct effect is not likely.'*[737] It appears that the need for further clarification consequently leads to the conclusion that a Provision is not directly applicable. However, in this light it is interesting to note that it was the conclusion of the Government in the Explanatory Memorandum that when considering all Provisions of the ICRC, existing Dutch law for the major part already adequately implemented the rights embedded in the ICRC. It also seems that in case of doubt as to whether the Netherlands did fulfil the obligations coming forth from a certain treaty Provision, the Government had the tendency to make a reservation to that Article, instead of proposing adjustments in or additions to existing domestic law.[738] Also, on some points, the Government seemed to assess the consequences when a treaty Provision

---

[733] Parliamentary Documents, II 1992-1993, 22855 (R1451), no. 4, p. 6. Original text in Dutch: '*...de regering niet beoogt een rechtstreekse werking toe te kennen voor zover deze niet reeds voortvloeit uit bepalingen van andere verdragen.*'
[734] Parliamentary Documents, II 1992-1993, 22855 (R1451), no. 6, p. 10.
[735] Parliamentary Documents, II 1984-1985, 18950 (R1281), no. 6, p. 14.
[736] Parliamentary Documents, II 1992-1993, 22855 (R1451), no. 3, p. 9. Supreme Court 14 April 1989, AB 1989, 207/NJ 1989, 469 ('Harmonisation Act' ruling).
[737] Supreme Court 14 April 1989, AB 1989, 207/NJ 1989, 469 ('Harmonisation Act' ruling), consideration 5.3. Original text in Dutch: '*Bovendien gaat het hier om bepalingen die betrekking hebben op door de overheid jegens burgers te verrichten prestaties; dergelijke bepalingen kunnen in het algemeen bezwaarlijk zonder nadere uitwerking in de rechtsorde functioneren, zodat rechtstreekse werking niet voor de hand ligt.*' *In casu*, among others things, the direct applicability of Article 13 ICESCR was under dispute.
[738] For instance, the reservation to Article 37 ICRK. The Government could not guarantee that children would be detained separately from adults. Parliamentary Documents, II 1992-1993, 22855 (R1451), no. 3, pp. 45-46.

from which it did not consider it possible to be directly applicable, would nevertheless be directly applied by the Courts. Such an assessment was made regarding Article 26 ICRC, and also led to a reservation.[739]

In this light, the Minister of Justice Kosto responded in a written reply to several questions from Parliament, which included the matter of direct applicability of ICPR Provisions. Kosto also referred to the earlier mentioned criteria developed by the Supreme Court to determine whether an international Provision is directly applicable, but added that it appeared from case law that the Courts should be guided by the intention of the Legislature *'as much as possible'*.[740] This statement shows that the Government considers it of great importance that its view on direct applicability is taken into consideration by the Courts, but also refers to a certain practice in case law that the view of the Legislature is indeed seriously considered. It could be questioned whether the Minister here deliberately suggested a certain hierarchy between the criteria of the Supreme Court, in which the opinion of the national Legislature prevails to all other criteria. Whatever the intention of the Minister was, the phrase 'as much as possible' leads to questions and also seems to be in tension with other statements of the Minister concerning the criteria. For instance, Minister Kosto seems to suggest earlier in the same note that *'the effect of these Provisions is primarily determined by the nature, content, scope, as well as the wording of the Provision itself. Then also the intention of the draftsmen of the Convention and the national Legislature in the drafting of the implementing legislation could be a guideline.'*[741] The Minister appears to be struggling with the autonomous competence of the Judiciary to review national legislation against international legislation ex Article 94 CA: *'Of course, Judges will always base their interpretation on the current circumstances in which they live and on the circumstances arising in particular situations, so there is nothing else that can be done during the adoption of a Bill than indeed give an interpretation that contributes to the Parliamentary History, that could be taken into account by the Judge, when, if necessary, considering the intentions of the Legislature.'*[742] Indeed, while discussing the reservation to Article 26 ICRC, the Minister even argued that: *'In the interpretation of Article 26, the Parliamentary History of course may play a role. Judges may use this in their*

---

[739] Parliamentary Documents, II 1992-1993, 22855 (R1451), no. 3, p. 34.

[740] House of Representatives, 84th meeting, 30 June 1994, second note, p. 84-5783.

[741] House of Representatives, 84th meeting, 30 June 1994, second note. Original text in Dutch: *'De doorwerking van deze bepalingen wordt in de eerste plaats bepaald door de aard, de inhoud, de strekking, alsmede door de formulering van de bepaling zelf. Vervolgens kan tevens de bedoeling van de opstellers van het verdrag en van de nationale wetgever bij de totstandkoming van de uitvoeringswetgeving een richtsnoer vormen.'*

[742] Minister Kosto, House of Representatives, 84th meeting, 30 June 1994. Original text in Dutch: *'Rechters zullen natuurlijk altijd interpreteren naar de omstandigheden van de tijd waarin ze leven en naar de omstandigheden die zich in concrete situaties voordoen, zodat er nooit iets anders gedaan kan worden bij het behandelen van een wetsvoorstel dan inderdaad een interpretatie geven, die dan bijdraagt tot de geschiedenis van het wetsvoorstel, waar de rechter, eventueel de bedoelingen van de wetgever peilend, rekening mee kan houden.'*

considerations. However, the Parliamentary History is only one of the sources that can be used for the interpretation of Article 26. The text will remain of primary importance. This may lead to the interpretation that a child has an independent right to social security. Because it is uncertain how the Court will rule on this, the reservation was made.'[743] Concerning the relation between the Legislature and the Judiciary regarding the interpretation of international Provisions concerning their direct applicability, the Minister was very clear: *'that interpretation is not done in advance, as we know, by the Legislature, but only – I refer to Articles 93 and 94 of the Constitutional Act – at some unknown time in the future by the independent Judiciary.'*[744] It seems that the Minister on the one hand is very well aware of the fact that Parliamentary History is a source for the Judiciary that is used to determine whether a treaty Provision has direct effect or not, and tries to emphasise the importance of this source. On the other hand, the Minister is also bound by the competence of the Judiciary to have a final say in this matter, as embedded in Articles 93 and 94 CA.

It is this competence of the Judiciary, and the consequential uncertainty as to what Provisions would have direct effect in the future and the following consequences to that in the domestic legal order, that caused some political fractions to be cautious towards adopting the ratification Bill. For instance, Members of the Political Reformed Party (SGP) and the Reformed Political Association (GPV), both in the Preparatory Committee and in Parliament, expressed their concern that ratification of the ICRC could result in unpredictable case law. The SGP fraction referred to the legal development after ratification of the ECHR that had led to verdicts that were contrary to the original intentions of the Legislature.[745] In this light, Schutte (GPV) even referred to this as *'the problem of the direct effect.'*[746] Soutendijk-van Appeldoorn (Christian Democratic Appeal, CDA) underlined that the Judiciary increasingly became less hesitant in granting direct effect (and also horizontal effect) to international Provisions. She agreed that during the adoption of the ratification Bill of the ECHR, indeed a wrong assessment had been made regarding the consequences of ratification of the treaty for the Dutch legal order, although she emphasised that these consequences were not necessarily a bad thing. According to her, this demonstrates that an Explanatory Memorandum

---

[743] Minister Kosto, House of Representatives, 84th meeting, 30 June 1994. Original text in Dutch: *'Bij de interpretatie van artikel 26 kan de wetsgeschiedenis uiteraard een rol spelen. Rechters kunnen dit in hun beoordeling betrekken. De wetsgeschiedenis is echter slechts een van de interpretatiemiddelen van artikel 26. De tekst blijft primair van belang. Zij kan aanleiding geven tot de interpretatie dat een kind zelfstandig recht op sociale zekerheid heeft. Omdat het niet zeker is hoe de rechter hierover zal oordelen, is dit voorbehoud gegeven.'*
[744] Minister Kosto, House of Representatives, 84th meeting, 30 June 1994. Original text in Dutch: *'die uitleg geschiedt niet op voorhand, zoals wij weten, door de wetgever maar uitsluitend – ik verwijs naar de artikelen 93 en 94 van de Grondwet – op enig nog onbekend moment in de toekomst door de onafhankelijke rechter.'*
[745] Parliamentary Documents, II 1992-1993, 22855 (R1451), no. 4, pp. 4 and 6. See also: Schutte (GPV), House of Representatives, 81st meeting, 23 June 1994.
[746] Schutte (GPV), House of Representatives, 84th meeting, 30 June 1994. Original text in Dutch: *'het probleem van de rechtstreekse werking.'*

can hardly meet the requirements of reality, for law is a dynamic and continuing process.[747] Rouvoet (Reformed Political Federation, RPF), argued that until then, the system embedded in Articles 93 and 94 CA had not led to any significant problems yet, although *'especially in the field of social security and in the sphere of justice it has brought us eventually some more or less unpleasant surprises.'*[748] In this light, he asked whether it is *'to be feared that, given the power of the Courts to review laws against the present Convention, the Netherlands may be confronted with surprises and consequences that they do not really want? If so, are there any specific areas where problems are (can be) expected?'*[749] In response, the Government argued that the Netherlands already complied with the obligations enshrined in the treaty, but also that it could never be precluded that in the future, there would be a broader interpretation of the Articles than originally foreseen, for instance due to suggestions of the Committee on the Rights of the Child, or on a national level, due to the reviewing powers of the Judiciary as embedded in Articles 93 and 94 CA.[750] The Government underlined elsewhere that *'Given the experience of recent years with (broad) interpretations of Provisions of human rights treaties by the Courts, we have, based on the current and, as far as now foreseen, future legal positions, on page 9 of the Explanatory Memorandum, generally indicated what Provisions of the Convention, in our estimation, are without doubt or with some probability, suitable for direct effect; this to avoid surprises in the future. On the basis of that assessment, the present reservations are proposed.'*[751]

The CDA fraction in the Preparatory Committee asked to what extent the direct effect of treaty Provisions could interfere with policy in the field of for instance education, healthcare and social security: a broad area in which the treaty recognised certain rights. The fraction underlined that certainly regarding parental authority in relation to financial consequences, choices had been made in the political arena, in which the Judiciary normally does not play a role.[752] The Government responded that while it cannot be predicted with absolute certainty, it can be

---

[747] Soutendijk-van Appeldoorn (CDA), House of Representatives, 81st meeting, 23 June 1994.

[748] Rouvoet (RPF), House of Representatives, 81st meeting, 23 June 1994. Original text in Dutch: *'met name op het terrein van de sociale zekerheid en in de justitiële sfeer plaatste dat ons uiteindelijk voor een aantal min of meer onaangename verrassingen.'*

[749] Rouvoet (RPF), House of Representatives, 81st meeting, 23 June 1994. Original text in Dutch: *'Bestaat de vrees dat, gezien de macht van de rechter om wetten ook aan dit verdrag te toetsen, Nederland wel eens voor verrassingen en gevolgen geplaatst kan worden die het eigenlijk niet wil? Zo ja, zijn er specifieke sectoren waar problemen verwacht (kunnen) worden?'*

[750] Parliamentary Documents, II 1992-1993, 22855 (R1451), no. 6, p. 7.

[751] Parliamentary Documents, I 1994-1995, 22855 (R1451), no. 22a, p. 5. Original text in Dutch: *'Gelet op de ervaringen van de laatste jaren met rechterlijke (ruime) interpretaties van bepalingen uit mensenrechtenverdragen hebben wij op grond van de huidige en de, voor zover nu te voorzien, toekomstige rechtsopvattingen op blz. 9 van de memorie van toelichting globaal aangegeven aan welke bepalingen van het verdrag naar onze inschatting rechtstreekse werking zonder meer of met enige waarschijnlijkheid zou kunnen worden toegekend; een en ander om verrassingen in de toekomst te voorkomen. Op grond van genoemde inschatting zijn de voorliggende voorbehouden voorgesteld.'*

[752] Parliamentary Documents, II 1992-1993, 22855 (R1451), no. 4, pp. 5-6.

assessed with some accuracy whether a treaty Provision will be granted direct effect or not. As already discussed before, this would for the major part concern ICRC Provisions with equivalent Provisions in other treaties that had been granted direct effect before. But in addition, the Government underlined that it had also made an assessment of the consequences when a treaty Provision unexpectedly would be granted direct effect, and based on that assessment, where necessary, proposed a reservation (such as the reservation to Article 26 concerning social security). The Government once again emphasised that it did not expect treaty Provisions that stipulated duties to undertake appropriate measures to facilitate the implementation of the right in accordance with national conditions and with the available resources to be directly applicable.[753]

As mentioned earlier, especially some Christian parties were hesitant in approving the ratification Bill, due to the fact that the Treaty might cause Governmental interference in family relations.[754] The Senate's fraction of the Political Reformed Party (SGP) eventually decided not to support the adoption, due to the possibility that some treaty Provisions would be directly applicable: '*my fraction does not wish to bear the co-responsibility for possible consequences within our domestic law on which they cannot exert influence anymore. Based on this consideration, they will abstain from supporting the Bill.*'[755]

7.6.3 Individual complaints procedures

In the ICRC, a monitoring procedure is embedded in Article 44 that obliges Member States to submit periodic reports to the Commission on the Rights of the Child on the implementation of the rights enshrined in the Convention. No complaints procedures were adopted. In this light, according to the Government, it was a general practice to include – besides a reporting mechanism – an individual complaints procedure in treaties that stipulate civil and political rights. However, regarding treaties in which economic, social and cultural rights were embedded, usually only a reporting mechanism was adopted. Since the ICRC enshrined both types of human rights, the draftsmen had included, on top of the reporting mechanism, as some sort of compromise between the two approaches, a special role for the Committee on the Rights of the child to generally oversee that States comply with the treaty, and the possibility for specialised agencies of the UN (especially

---

[753] Parliamentary Documents, II 1992-1993, 22855 (R1451), no. 6, p. 9.

[754] See for the viewpoints of the Political Reformed Party (SGP) and the Reformed Political Association (GPV): Parliamentary Documents, II 1992-1993, 22855 (R1451), no. 4, p. 4-5. See for the viewpoints of the Christian Democratic Appeal (CDA): Parliamentary Documents, II 1992-1993, 22855 (R1451), no. 4, p. 8. See also: Schutte (GPV) and Rouvoet (Reformed Political Federation (RPF)), House of Representatives, 81st meeting, 23 June 1994.

[755] Holdijk (SGP), Senate, 6th meeting, 22 November 1994. Original text in Dutch: '*Mijn fractie durft de medeverantwoordelijkheid voor mogelijke consequenties binnen onze nationale rechtsorde waarop zij geen invloed meer kan uitoefenen, niet aan. Op grond van deze overweging zal zij haar steun aan het wetsvoorstel moeten onthouden.*'

UNICEF) and NGOs to support and advise the Commission on the Rights of the Child (Article 45 ICRC)[756]. According to the Government, at that time, this was a revolutionary approach within the UN, and: *'Therefore, the absence of an individual complaints procedure is not considered a deficiency.'*[757] In addition, the Government argued that an individual could nevertheless effectuate ICRC rights with equivalents in the ICCPR indirectly, through the individual complaints procedure adopted in the first Optional Protocol to the ICCPR. In both the Preparatory Committee and in Parliament, the viewpoint of the Government was criticised by several fractions, especially PvdA and D66. Basically, it was criticised that the ICRC also stipulated rights that were not embedded in existing treaties yet, and this could not be effectuated by individual complaints procedures via equivalent Provisions elsewhere.[758] The Government replied that it would undesirable when also economic, social and cultural rights could be effectuated through an individual complaints procedure, due to the margin of discretion of the Administration and Legislature, and it would in case of the ICRC lead to unwanted juridification of relations within the family. The Government underlined that the effectuation by individual complaints procedures of those rights embedded in the ICRC from which it would be desirable that they could be effectuated as such, was already adequately guaranteed in other treaties.[759] The question was raised whether an Optional Protocol could be expected to the ICRC, in which an individual complaints procedure would be embedded.[760] The Government replied that this could not be expected soon.[761] The discussion on individual complaints procedures was – especially in Parliament – intertwined (and sometimes incorrectly mingled) with another discussion about whether the treaty would imply that children would be

---

[756] In one of their responses to questions from the Preperatory Committee (see: Parliamentary Documents, II 1992-1993, 22855 (R1451), no. 6, p. 3), the Government further explained that from Article 45 ICRC and the *travaux préparatoires*, it *'can be concluded that other competent bodies, which the Committee may consider appropriate, including NGOs: (a) may provide expert advice on the implementation of the Convention in areas falling within the scope of their respective mandates; (b) may receive any reports from States Parties that contain a request, or indicate a need, for technical advice or assistance; (c) may be involved in undertaking studies on specific issues relating to the rights of the child; (d) through their information provided to the Committee, also provide the basis for the draft of recommendations and suggestions by the Committee to States and the General Assembly.'* Original text in Dutch: *'...kan worden afgeleid dat andere bevoegde instellingen, die het comité passend acht, dus ook NGOs: (a) het toezichthoudend Comité voor de rechten van het kind van deskundig advies kunnen dienen op die gebieden, waarop zij werkzaam zijn; (b) elk rapport van een verdragsstaat zullen ontvangen waarin een staat te kennen geeft op (technisch) advies of ondersteuning prijs te stellen; (c) ingeschakeld kunnen worden bij het verrichten van onderzoek op gebieden die verband houden met de rechten van het kind; (d) door middel van door hen aan het comité verschafte informatie, mede de basis kunnen leggen voor door dit comité op te stellen aanbevelingen en suggesties aan staten en de Algemene Vergadering.'*

[757] Parliamentary Documents, II 1992-1993, 22855 (R1451), no. 3, p. 4. Original text in Dutch: *'Het ontbreken van een individueel klachtenrecht derhalve niet als een gemis ervaren.'*

[758] Parliamentary Documents, II 1992-1993, 22855 (R1451), no. 4, p. 2-3. Van der Burg (PvdA), House of Representatives, 81st meeting, 23 June 1994.

[759] Parliamentary Documents, II 1992-1993, 22855 (R1451), no. 6, p. 4.

[760] Parliamentary Documents, II 1992-1993, 22855 (R1451), no. 4, p. 3.

[761] Parliamentary Documents, II 1992-1993, 22855 (R1451), no. 6, p. 6.

capable to exercise their own juridical remedies, instead of their parents or legal guardians. The Government together with some (mostly confessional[762]) parties was of the opinion that this was not the case, in contrast with some other fractions. The Government explained in its Explanatory Memorandum that the treaty itself did not imply juridical remedies for children, which was also not necessary, for it is normally the parent or the legal guardian that will act therefore on behalf of the child. In some cases however, the Government emphasised that the child has the right to express its own opinion in the proceedings, or to start proceedings when it concerns the interest of the child.[763] In this context it is however unclear what exactly is referred to by legal remedies for children. It is our understanding that the political parties that expressed their concerns about the possibility of a legal remedy for children did so in a context of uncertainty about future rulings of the Courts regarding possible direct effect of rights that would affect the relations within a family: it was their view that in the first place, a Court should not rule on family relations (mostly embedded in economic, social and cultural rights), but if a Court would do so nevertheless (due to the uncertainty that comes from Articles 93 and 94 CA), it should certainly not be possible that a child could start its own proceedings based on those rights. The fractions of the Labour party (PvdA), GreenLeft (GL) and Democrats '66 (D'66), and the liberal party (VVD) on the contrary did not consider the discussion on possible legal remedies for children in view of family relations, but rather considered that in society, there was an increasing demand for juridical remedies for children.[764]

7.7 The European Social Charter (revised version)

The ratification Bill of the Additional Protocol to the European Social Charter Providing for a System of Collective Complaints and the ESC (revised version) was discussed in Parliament in the period 2004-2005.

7.7.1 Background and added value of the ESC (revised version)

In its Explanatory Memorandum, the Government explained that the original ESC, originally intended to be the counterpart of the ECHR, was never as influential as the ECHR, due to interpretation problems and an ineffective monitoring mechanism.[765] Therefore, a substantive revision of the ESC and the adoption of

---

[762] See for the viewpoints of the Political Reformed Party (SGP) and the Reformed Political Association (GPV): Parliamentary Documents, II 1992-1993, 22855 (R1451), no. 4, p. 10. See for the viewpoints of the Christian Democratic Appeal (CDA): Parliamentary Documents, II 1992-1993, 22855 (R1451), no. 4, p. 8. See also: Schutte (GPV), House of Representatives, 81st meeting, 23 June 1994.
[763] Parliamentary Documents, II 1992-1993, 22855 (R1451), no. 3, p. 12.
[764] Parliamentary Documents, II 1992-1993, 22855 (R1451), no. 4, pp. 8-10. See also: Van den Burg (PvdA) and Versnel-Schmitz (D'66). From the speech of De Vries (VVD), House of Representatives, 81st meeting, 23 June 1994.
[765] Parliamentary Documents, II 2004-2005, 29941, no. 3, pp. 1-2.

an improved monitoring mechanism were deemed necessary. To this end, the European Committee on Social Rights had earlier drafted a Protocol amending the European Social Charter,[766] which was already ratified by the Netherlands in 1993, to improve the reporting procedure. Since then, the European Committee on Social Rights had drafted an Additional Protocol to the European Social Charter Providing for a System of Collective Complaints,[767] and also revised the substantive Provisions of the original version of the ESC.[768] The last two documents were discussed together in Parliament based on one ratification Bill. According to the Government, the most important criteria that were used to formulate a substantively renewed version of the ESH were '*the need to take account of the major changes in the development of social rights, the content of other international instruments and the will to deal with issues that are not covered by these other international instruments.*'[769]

7.7.2 Direct applicability of the ESC (revised version)

In its Explanatory Memorandum, the Government was quite clear regarding the possibility of direct applicability of the renewed ESC Provisions: '*The revised Charter is a codification of the current ideas in the social field that live in the Member States of the Council of Europe and thus aims to provide guidance for national social policies of each of these States and a common basis for the development of Europe in social matters.*'[770] Indeed, as mentioned earlier, the Government considered the ESC to be complementary to the ECHR: '*Where the latter Convention in the first place aims to protect the citizen against impermissible infringements of the Government on private life, the Charter demands of that same Government to deploy measures that are necessary to realise that the standard of living and the social-cultural well-being of the population meet a (European) basic level, and are further improved.*'[771] Regarding these performance duties of the Member States recognised in the ESC, the Government argued that '*A number of Provisions of the Revised Charter are so*

---

[766] Amending Protocol of 1991 reforming the supervisory mechanism, 21 October 1991, Turin, in: European Treaty Series 142.

[767] Additional Protocol to the European Social Charter Providing for a System of Collective Complaints, 9 November 1995, Strasbourg, in: European Treaty Series 158.

[768] European Social Charter (revised), 3 May 1996, Strasbourg, in: European Treaty Series 163.

[769] Parliamentary Documents, II 2004-2005, 29941, no. 3, p. 2. Original text in Dutch: '*de noodzaak tot het in aanmerking nemen van de belangrijkste veranderingen in de ontwikkeling van sociale rechten, de inhoud van andere internationale instrumenten en de wens thema's te behandelen die niet door deze andere internationale instrumenten worden bestreken.*'

[770] Parliamentary Documents, II 2004-2005, 29941, no. 3, p. 2. Original text in Dutch: '*Het herziene handvest vormt een codificatie van de in de lidstaten van de Raad van Europa levende gedachte op sociaal terrein en beoogt aldus een leidraad te zijn voor het nationaal sociaal beleid van elk van deze lidstaten en een gemeenschappelijke basis voor de ontwikkeling van Europa op sociaal gebied.*'

[771] Parliamentary Documents, II 2004-2005, 29941, no. 3, p. 2. Original text in Dutch: '*Waar laatstgenoemde verdrag in de eerste plaats ziet op de bescherming van de burger tegen ontoelaatbaar te achten inbreuken van de overheid op de individuele levenssfeer, vraagt het Handvest van diezelfde overheid activiteiten te ontplooien welke noodzakelijk zijn om de levensstandaard en het sociaal-cultureel welzijn van de bevolking op een (Europese) basisniveau te brengen en verder te verbeteren.*'

generally formulated, that they are mainly of a programmatic nature, allowing the contracting States a very large margin of discretion regarding the implementation of those principles. However, the adoption of these principles can be appreciated, since they form, inter alia *in the exchange of views between the European states in view of the monitoring procedure, a basis to come to a shared vision on the realisation of these principles.'*[772] It was the opinion of the Government that the Appendix to the revised ESC can be understood in such way that in the ESC only duties of an international nature were embedded, and the compliance with those duties was subject to the ESC monitoring mechanism only. Considering the Appendix, read in conjunction with the wording and the contents of the Charter Provisions, the Government concluded that *'the undersigned assume that the Treaty Provisions are generally not suitable for direct effect in the Dutch legal system. If that what is stipulated in those Provisions is not proven to be realised in the Netherlands, the Provisions should be further implemented through acts of legislation and administration. The monitoring process...is also focussed on this.'*[773]

However, the Council of State expressed in its recommendations to the Government regarding the proposed Explanatory Memorandum a different view on the matter of direct applicability of the revised ESC Provisions. The Council argued that the Government seemed to suggest that direct effect of the Articles is not possible, based on *inter alia* the Appendix to the ESC. According to the Council of State, it is stipulated in the appendix that the compliance with the Charter Provisions is supervised by the corresponding monitoring mechanism to the ESC only, to exclude other international monitoring agencies, and not to exclude direct effect of the substantive Provisions. The Council then stated that for instance through Article E (Part V), that stipulates the principle of non-discrimination, an direct effect might be granted to other substantive Provisions that would in itself not be suitable for direct effect. In addition, the Council reminded that, due to Articles 93 and 94 CA, it is the Judiciary that eventually decides on the matter of direct effect, and argued that therefore, excluding direct effect beforehand by the Legislature, was in contradiction with these Constitutional Provisions. That is why the Council of State recommended deleting the text that is referred to.[774]

---

[772] Parliamentary Documents, II 2004-2005, 29941, no. 3, p. 2. Original text in Dutch: *'Een aantal bepalingen van het herziene Handvest zijn dusdanig algemeen geformuleerd, dat deze in de hoofdzaak een programmatisch karakter hebben, waarbij zij aan de verdragssluitende staten een zeer grote beleidsvrijheid laten met betrekking tot de uitwerking van deze beginselen. Nochtans is ook de vaststelling van deze beginselen positief te waarderen, aangezien deze, onder meer in de gedachtewisseling tussen de Europese staten in het kader van de toezichtprocedure een basis vormen om tot een gemeenschappelijke visie te komen over concretisering van deze beginselen.'*

[773] Parliamentary Documents, II 2004-2005, 29941, no. 3, p. 4. Original text in Dutch: *'gaan de ondergetekenden er van uit dat de verdragsbepalingen zich in het algemeen niet lenen voor rechtstreekse werking in de Nederlandse rechtsorde. Indien niet vaststaat dat hetgeen in die bepalingen is neergelegd in Nederland wordt gerealiseerd, zullen zij hun vertaling moeten krijgen in daden van wetgeving en bestuur. De toezichtprocedure (...) richt zich daar ook op.'*

[774] Parliamentary Documents, II 2004-2005, 29941, no. 4, p. 7.

The Government responded to this that it is indeed the Judiciary that has a final say in the matter of direct applicability, but that the Government is free to express its own views thereon. The Government then referred to the criteria as developed by the Supreme Court to determine whether a treaty Provision is suitable for direct effect which are *'the intention of the contracting Parties, as well as the wordings, the nature and the content of the present Provisions, while also the intention of the Legislature may be of importance.'*[775] According to the Government, it was the clear intention of the contracting parties not to introduce any monitoring mechanism besides the mechanism to the ESC. Whether or not the Government then considered the national Judiciary as a monitoring mechanism besides the ESC mechanism, and therefore rejects the possibility of direct effect, remains rather unclear, and has not been discussed further. The Government emphasised that many of the substantial ESC Provisions, considering their wordings, stipulated duties basically for the Legislature, to adopt measures to further implement the present treaty Provisions. The wordings 'appropriate', 'reasonable', 'as necessary', 'as possible', etc. seem to suggest that there is a large margin of discretion for the Member States. *'In the past, only in respect of Article 6, fourth Section, of the Charter... read in conjunction with Article 31 (original ESH) the Supreme Court has accepted direct effect.'*[776] The Government added that of course, within the margins of the law, the Dutch national States institutions would have to consider the ESC in their decision making processes.[777] The matter of direct effect was not significantly discussed in Parliament.

7.7.3 Complaints procedures

One of the improvements concerning the monitoring mechanism to the ESC was the adoption of a collective complaints procedure, supervised by the Committee on Social Rights. In the Additional Protocol to the European Social Charter Providing for a System of Collective Complaints, the right was recognised for international and national organisations of employers and trade unions, and other international non-governmental organisations of relevance, to submit complaints to the Secretary General of the Council of Europe.[778] In its Explanatory Memorandum, the Government expressed its approval of the procedure.[779] In Parliament, the protocol has not been discussed significantly. The possibility of an individual complaints procedure was not discussed at all.

---

[775] Parliamentary Documents, II 2004-2005, 29941, no. 4, p. 8. Original text in Dutch: *'de bedoelingen van verdragsluitende partijen, alsmede de formulering, de aard en de inhoud van de betrokken bepalingen, terwijl ook de bedoeling van de wetgever van belang kan zijn.'*
[776] Parliamentary Documents, II 2004-2005, 29941, no. 4, p. 9. Original text in Dutch: *'Slechts ten aanzien van artikel 6, vierde lid, van het Handvest (...) gelezen in samenhang met artikel 31 (ESH oud) heef de Hoge Raad in het verleden rechtstreekse werking aangenomen.'*
[777] Parliamentary Documents, II 2004-2005, 29941, no. 4, p. 9.
[778] Additional Protocol to the European Social Charter Providing for a System of Collective Complaints, 9 November 1995, Strassbourg, in: European Treaty Series 158, especially Article 1.
[779] Parliamentary Documents, II 2004-2005, 29941, no. 3, pp. 38-42.

7.8 Conclusion

Based on the analysis above, conclusions can be drawn on three elements, which each contribute to a better understanding of the viewpoints of the Legislature concerning the direct effect of ECOSOC rights, and thus the right to food. Firstly, what was the motivation to ratify the present treaty; secondly, are the Provisions suitable for direct effect according to the Legislature; and thirdly, what are the viewpoints on complaints procedures.

At the moment of ratification, the Dutch Government is, without exceptions, convinced that the Dutch legal order is already in confirmation with the analysed treaties. In case of any doubt, the choice is usually made to make a reservation, to guarantee compliance with the treaty, instead of proposing implementation legislation. From this it appears that throughout the years, the Dutch Legislature did not have any ambition to improve the domestic legal order in the field of human rights, based on the treaties. Remarkably, realising human rights in the Netherlands seems hardly to play a role in the motivation to ratify the treaties. Rather, arguments to ratify mostly focus on foreign relations. For instance, the fact that a treaty would be legally binding, or would have an (improved) monitoring system, is many times expressed as a motivation to ratify, but mostly discussed in view of other countries. The Netherlands could then use the treaty as a basis to appeal to other countries to respect human rights, an argument that is expressed in for example the Explanatory Memorandums to the ratification Bills of the CEDAW and the ICRC. On the other hand, the Legislature consequently expressed its disappointment in the final result of most treaties, for they are usually sketched as compromises after long negotiations, and quite often the Governments regretted that the treaties or monitoring mechanisms were not further-reaching, or even declared that in the international arena. Both approaches appear to be at odds.

Regarding the matter of direct applicability, a similar approach, albeit technically differently and sometimes rather clumsily formulated, can be observed throughout the years. It appears that the different Governments, each time when a ratification Bill is discussed in Parliament, struggle with the system laid down in Articles 93 and 94 CA. On the one hand, they want to underline the importance of the opinion of the Legislature regarding the possibility of direct applicability of treaty Provisions, which is stressed in various degrees of urgency, and seem to be well aware of the fact that the Parliamentary Documents are used in the considerations of the Courts. On more than one occasion, the possibility of legal developments beyond the reach of the Legislature is considered with suspicion. On the other hand, the Government feels bound by especially Article 94 CA, and each time, sometimes after being reminded by Parliament, almost reluctantly recognises the fact that the Judiciary has a final say in deciding whether an Article is suitable for direct application. In the Parliamentary Documents, the Government often refer to case law of the Supreme Court, in which three, and sometimes four, criteria are described that the Courts usually refer to when deciding on direct applicability. These criteria are described in most detail in the Explanatory Memorandum to the ratification Bill of the ICRC: '*the nature, content and scope of the Provision, as well as the formulation (the wording) thereof, while also the intention of the national Legislature during the realisation of the ratification law of the relevant treaty could be a guideline, and – insofar as this is the case – the intention of the draftsmen of the treaty should be a guideline. Also of importance in this is the presence or absence of implementation legislation, and whether it is possible within the national legal system to directly apply the Provision.*'[780] The first three criteria are referred to more often in the various Parliamentary Documents throughout the years, while the last criterion (the presence or absence of implementation legislation, and whether it is possible within the national legal system to directly apply the Provision) was only referred to in the Explanatory Memorandum to the ratification Bill of the ICRC. With respect to a possible hierarchy, the Legislature expresses confusing viewpoints – especially in the Parliamentary Documents that concern the ratification Bill of the ICRC – that are more than once contrary to one another, and flow between the notion that the intention of the draftsmen of the treaty (sometimes confused or with the intention of the national Legislature) is the primary source to establish whether a Provision is suitable for direct applicability on the one hand, and that the nature, content and scope of the Provision, as well as the formulation (the wording) thereof, are the most important source of interpretation on the other hand.

In general, it can be concluded that civil and political rights are usually suitable for direct effect, whereas economic, social and cultural rights are not, due to the fact that the Provisions in which those rights are recognised are mostly addressed to the Judiciary and Administration, concern a duty to progressively realise and therefore imply a large margin of appreciation for them. The discussion on direct applicability of ECOSOC rights was more detailed in view of the CEDAW and ICRC, for in those Treaties also civil and political rights are stipulated. The Governments then were challenged more intensely to explain their viewpoints on the matter, and it can be observed that the Governments did not reject the possibility of direct effect of the ECOSOC rights in the specialised treaties, the CEDAW and the ICRC, as strongly compared to the ICESCR and ESH Provisions. One notable exception to the rule that ECOSOC rights are not suitable for direct application appears to be Article 6 (4) ESC.[781] In Chapter 5, it was discussed that the case law on this is frequently referred to in the literature on the functioning of Articled 93 and 94 CA in relation to the division of powers between the Legislature and the Judiciary concerning the decision on direct applicability of international standards. It is used as an example to underline that also in legal practice, the Judiciary, based on Articles 93 and 94 CA has a final say in deciding whether a Treaty Provision is directly applicable, because the Supreme Court, in its case

---

[780] Parliamentary Documents, II 1992-1993, 22855 (R1451), no. 3, p. 9.
[781] Under certain conditions, in literature, also Article 7 (a) (i) ICESCR is considered to have direct effect.

law, would be opposing the view of the Legislature. Since regarding the right to adequate food the Judiciary seems to follow the views of the Legislature (see Chapter 4), it may be of interest in this research to further explore to what extent indeed, and if so why, the Judiciary decided to oppose the views of the Legislature in the case of Article 6 (4) ESC. This will be done in Chapter 8.

Regarding the view on monitoring procedures, the different perception of civil and political rights on the one hand, and economic, social and cultural rights on the other hand, are also visible. The position of the Dutch Legislature throughout the years can be summarised as follows: regarding the ECHR and the ICCPR, both stipulating civil and political rights, the Legislature appears to have strongly stimulated the adoption of individual complaints procedures in the negotiations preceding the adoption of the treaties. Remarkable detail regarding the ECHR is that the Government in Parliament advised against the ratification of an individual complaints procedure, while the Dutch Delegation voted in favour of this procedure during the negotiations. Regarding the ICESCR and ESC, the Governments did not always discuss the possibility of complaints procedures, for they were usually not adopted in the treaties. Regarding the ESC, the Government appeared not to oppose a collective complaints procedure. However, in the Parliamentary debate on the ratification Bill of the ICRC, the Government did discuss the possibility of an individual complaints procedure, and argued that the possibility of an individual complaint regarding ECOSOC rights would be undesirable due to the margin of discretion of the Legislature and Administration, and the fact that individual complaints procedures might possibly lead to juridification of family relations.

In general however, it must also be concluded that throughout the years, with some noticeable exceptions, the Parliamentarians have a poor knowledge on the functioning of Articles 93 and 94 CA, and more than once seem to discuss manners in a way that could be characterised as '*ad hoc*', and not so well informed.

8. Intermezzo: the right to strike

8.1 Introduction

From the previous Chapters, especially Chapter 4, it appeared that the Courts are generally not willing to grant direct effect to ECOSOC rights, and more in particular the right to food. The most notable exception to that rule appears to be the case law concerning Article 6 (4), stipulating the right to strike: in their verdicts, the Courts consequently consider that the Provision has direct effect in the Dutch legal order. As outlined in Chapter 5, in the leading literature on the matter, this case law is often considered as proof that it is solely the Judiciary that has a final say in determining whether an international Provision has direct effect or not. Some authors suggested that it appears from this case law, that the nature, scope as well as its wordings are the only or primary source for the Courts used in their decisions, for the Supreme Court, in the railway-strike ruling,[782] would have decided, contrary to the view of the Legislature as expressed in Parliamentary History, that Article 6 (4) ESC is directly applicable. If that is indeed true, it would not match with the conclusions drawn from Chapter 4, after analysing the case law on the right to food, in which I observed that generally speaking the Courts follow the lead of the Legislature, in particular considering the Parliamentary History of the ratification Bill, and mostly not even consider the nature, scope and wordings of the Provision itself. In this light, it appeared from especially Chapter 7, that the Dutch Legislature, in their Parliamentary Documents, generally considers the direct applicability of ECOSOC rights that are stipulated in international Human Rights Treaties, as one general issue, and does not discuss the matter separately per treaty Provision. In most Explanatory Memorandums thus direct applicability of ECOSOC rights is generally rejected. Apparently, there is something special about the right to strike, as stipulated in Article 6 (4) ESC. Therefore, in this 'intermezzo', I will explore why the case law on Article 6 (4) ESC appears to be so different compared to case law on ECOSOC rights in general. To this end, I will firstly explore to what extent Article 6 (4) ESC is the only exception to the rule that ECOSOC rights are generally not directly applicable (Section 8.2). Then, to get more insight in the particularities of the case, I will discuss the railway-strike ruling into more detail (Section 8.3). Furthermore, I will explore the Parliamentary History on the ESC more thoroughly regarding the direct applicability of Article 6 (4). Since the Article was excluded from the first draft Bill, as discussed in Chapter 7, the Government did not discuss the direct applicability of the Article in its Explanatory Memorandum, but mainly during the discussions in Parliament (Section 8.4). Remarkably, Article 8 ICESCR, in which also the right to strike is enshrined, has no direct effect – a fact that has been

---

[782] Supreme Court, 30 May 1986, NJ 1986, 668 (railway-strike ruling).

noted by other authors as well.[783] Therefore, the Parliamentary History on Article 8 ICESCR will also be explored (Section 8.5), for the differences between Article 6 (4) ESC and 8 ICESCR may give more insight in what elements are decisive for the granting of direct effect. Of course, in Section 8.6, conclusions will be drawn.

8.2 The direct applicability of Economic, Social and Cultural Rights in general

As demonstrated in Chapter 5, in literature Article 6 (4) ESH and Article 7 ICESCR are often referred to as (the) two examples of ECOSOC rights in international treaties that have direct effect on the Netherlands.[784] However, it must be observed that Article 7 ICESCR does not have a similar effect compared to Article 6 ESH. In the case law in which Article 7 ICESCR plays a role, usually in cases concerning discrimination in remuneration, the Article in itself has no direct effect, but is rather read in conjunction with Article 26 ICCPR (a non-discrimination Provision), recognising a principle of non-discrimination in remuneration. The Courts then consider that: '*...all this means that this principle – which, given the fact that it is also supported by treaty Provisions such as Article 26 ICCPR and Article 7 ICESCR, may be granted a heavy weight – is not conclusive but should be involved in the judgment, alongside other circumstances of the case.*'[785] Article 7 ICESCR therefore may have a certain (indirect) effect in Court cases, but certainly not an direct effect as compared to Article 6 (4) ESC.

Regarding the ICRC, there appears to be a slightly different regime concerning the effect of ECOSOC rights, compared to the rights stipulated in the ICESCR. As demonstrated in Sections 4.4-4.6, focussing on Articles 24, 26 and 27 ICRC, occasionally – although the Courts generally deny their direct applicability –

---

[783] A.K. Koekoek (ed.), *De Grondwet, een systematisch en artikelgewijs commentaar* (3ᵉ druk), *Deventer:* Tjeenk Willink, 2000, p. 467. Reference was made to: Supreme Court, 6 December 1983, NJ 1984, 557 and Supreme Court, 12 February 1984, NJB 1984, 45. See also: C.A.J.M. Kortmann, *Constitutioneel Recht* (6ᵉ druk), Deventer: Kluwer, 2008, p.467. Also here, reference was made to: Supreme Court, 6 December 1983, NJ 1984, 557.

[784] See for the literature review Chapter 5.

[785] Original text in Dutch: '*een en ander betekent dat dit beginsel – waaraan, gelet op het feit dat het ook steun vindt in verdragsbepalingen als artikel 26 IVBPR en artikel 7 IVESCR, een zwaar gewicht kan worden toegekend – niet doorslaggevend is maar dat het naast andere omstandigheden van het geval moet worden betrokken in de afweging.*' See for instance: Central Court of Appeal, 16 February 1989, AB 1989, 164; Central Court of Appeal, 3 July 1986, TAR 1986, 215; Supreme Court, 30 January 2004, LJN AM2312; Court of First Instance of the Netherlands Antilles, 28 January 2009, EJ 2008/497; Court of Appeal, 's-Gravenhage, 28 April 2009, LJN BI3564; Court of First Instance of the Netherlands Antilles, 30 March 2010, EJ 2009/579. However, direct effect of Article 7 ICESCR is also explicitly denied in some cases. For instance: Supreme Court, 20 April 1990, NJ 1992/636; Supreme Court, 07 May 1993, NJ 1995, 259; The Joint Court of Justice of Aruba, Curaçao, Sint Maarten and of Bonaire, Saint Eustatius and Saba, 27 October 2009, EJ 497/08. In one older case, however, the Article was incidentally granted direct effect by the Public Service Tribunal that even concluded a violation with Article 7 ICESCR: Public Service Tribunal, Amsterdam, NJCM-Bulletin 1984, pp. 245-252.

the Courts are willing to use the analysed ICRC Provisions as an interpretative standard in their verdicts, more or less comparable to the case law on Article 7 ICESCR. This seems to be a development that also includes many other ICRC Provisions, as observed in a very extensive analysis by J.H. de Graaf, M.M.C. Limbeek, N.N. Bahadur, and N. van der Mey.[786] However, no evidence could be found that the Courts consider any ICRC Provisions stipulating ECOSOC rights to be directly applicable.

As already concluded in Chapter 4, in the Netherlands hardly any case law on the direct effect of CEDAW Provisions exists.

8.3 The railway-strike ruling

The Supreme Court granted direct effect to Article 6 (4) ESC for the first time in their landmark railway-strike ruling. *In casu*, the national railway company (NS) had claimed that the district Court of Utrecht would force the trade unions involved to call an end to the collective actions of the NS personnel, and in the future, not call for such actions again. The District Court rejected the claim, and also the Court of Appeal ruled likewise. Therefore, the NS appealed in cassation to the Supreme Court. According to NS, the Court of Appeal wrongly ruled, *inter alia*, that Article 6 (4) ESC would have direct effect, and thus the personnel could invoke this Article to justify their collective actions.[787] The Supreme Court however considered that '*to judge this legal remedy...in the first place, it should be determined whether the Supreme Court may review the ruling of the Court of Appeal against Article 6 (4) of the (...) European Social Charter.*'[788] To this end, the Court held that it would be competent to review '*when this Treaty Provision would fall under the scope of the Provisions that are binding on all persons, referred to in art. 93 CA.*'[789] To determine this, the Supreme Court argued in its famous passage: '*Whether or not the contracting States intended to grant direct effect to Article 6 ESC is not relevant, whereas neither from the text, nor the history of the formation of the treaty follows that they have agreed that Article 6 ESC is directly applicable. In that situation, according to Dutch law, only the content of the Provision is decisive: does this oblige the Dutch Legislature to implement a national measure with a certain content or scope, or is this of such nature that the Provision can be applied directly to national legal order as positive law? Article 6 Section 4 is – contrary to most other Provisions*

---

[786] J.H. de Graaf, M.M.C. Limbeek, N.N. Bahadur, and N. van der Mey, *De toepassing van het Internationaal Verdrag inzake de Rechten van het Kind in de Nederlandse Rechtspraak,* Nijmegen: Ars Aequi Libri, 2012.

[787] Supreme Court, 30 May 1986, NJ 1986, 668 (railway-strike ruling), legal appeal.

[788] Supreme Court, 30 May 1986, NJ 1986, 668 (railway-strike ruling), consideration 3.2. Original text in Dutch: '*ter beoordeling van dit middel (...) dient in de eerste plaats te worden vastgesteld of de Hoge Raad 's Hofs oordeel kan toetsen aan het bepaalde in art. 6 lid 4 van het (...) Europees Sociaal Handvest.*'

[789] Supreme Court, 30 May 1986, NJ 1986, 668 (railway-strike ruling), consideration 3.2. Original text in Dutch: '*indien deze verdragsbepaling behoort tot die in art. 93 Gr.w. bedoelde bepalingen die naar haar inhoud een ieder kunnen verbinden.*'

of the ESC and in particular also to the other Sections of this Article – thus worded that the States Parties concerned are not obliged to adopt measures, but instead that the said employees and employers may certainly invoke the law as recognised, within certain limits, by the concerned States in the domestic legal order.'[790] Furthermore, the Supreme Court considered that: *'Also from the 'Gloss' to Article 6 (4), attached to the 'Appendix to the ESC', it appears that the States are not required to regulate "the right to strike"...by law: Member States that leave the application and other limitation of Article 6 Section 4 and Article 31 to case law, therefore do not violate the Convention, provided that the Judiciary respects the limitations stipulated in the latter Article. This is also assumed in the authoritative 'Conclusions' of the Committee, referred to in Article 25 ESC (Conclusions 1, page 38 under (e) and VIII, page 97). All this justifies regarding Article 6 Section 4 ESC as a Provision which, by its content, can be binding on all persons, ex Article 93 CA.'*[791]

8.4 The Parliamentary History concerning Article 6 (4) ESH

As discussed in Chapter 7, the Government considered that the ESC Provisions were not suitable for direct applicability. This general view was slightly contested in the Preparatory Committee,[792] but not significantly opposed further in Parliament, except with respect to one Provision: Article 6 (4) ESC. The actual inclusion of Article 6 (4) ESC was not undisputed. In the first draft of the ratification Bill, the Government had included the Article,[793] and explained that the current situation regarding the right to strike was unsatisfactory. Although there was a slightly milder approach towards striking in the private sector by the Supreme Court, still striking was considered to be in essence a breach of contract by the employees. Therefore, the Government had proposed a draft Bill that would more adequately

---

[790] Supreme Court, 30 May 1986, NJ 1986, 668 (railway-strike ruling), consideration 3.2. Original text in Dutch: *'Of de verdragsluitende Staten al dan niet hebben beoogd aan artikel 6 lid 4 ESH directe werking toe te kennen, is niet van belang nu noch uit de tekst, noch uit de geschiedenis van de totstandkoming van het Verdrag valt af te leiden dat zij zijn overeengekomen dat aan artikel 6 lid 4 die werking mag worden toegekend. Bij deze stand van zaken is naar Nederlands recht enkel de inhoud van de bepaling zelf beslissend: verplicht deze de Nederlandse Wetgever tot het treffen van een nationale regeling met een bepaalde inhoud of strekking, of is deze van dien aard dat de bepaling in de nationale rechtsorde zonder meer als objectief recht kan functioneren? Artikel 6 lid 4 is – in tegenstelling tot de meeste andere bepalingen van het ESH en met name ook tot de overige leden van dit artikel – zo geredigeerd dat de betrokken verdragsstaten niet een verplichting tot regelgeving wordt opgelegd, maar integendeel aldus dat werknemers en werkgevers zich op het door de betrokken verdragsstaten binnen zekere grenzen erkende recht in de nationale rechtsorde zonder meer moeten kunnen beroepen.'*

[791] Supreme Court, 30 May 1986, NJ 1986, 668 (railway-strike ruling), consideration 3.2. Original text in Dutch: *'Ook uit de in de 'Bijlage bij het ESH' opgenomen 'glosse' bij art. 6 lid 4 blijkt dat de staten niet verplicht zijn 'het recht op staking' (...) bij de wet te regelen: verdragsstaten die de toepassing en andere begrenzingen van artikel 6 lid 4 en artikel 31 aan de jurisprudentie overlaten, schenden dusdoende – mits de rechtspraak de door laatstgenoemde artikel getrokken grenzen eerbiedigt – het verdrag niet. Van dit laatste gaan ook de gezaghebbende 'Conclusions' van de in artikel 25 ESH bedoelde Comité van deskundigen uit (Conclusions 1, blz. 38 onder (e) en VIII, blz. 97). Een en ander wettigt de bepaling van artikel 6 lid 4 ESH aan te merken als een bepaling die naar haar inhoud een ieder kan verbinden in de zin van artikel 93 Gr.w.'*

[792] Parliamentary Documents, II 1966-1967, 8606 (R 533), no. 5, p.1.

[793] Parliamentary Documents, II 1965-1966, 8606 (R 533), no. 2, Article 2.

regulate the right to strike by national law.[794] At the insistence of the Preparatory Committee however, Article 6 (4) was removed from the draft Bill.[795] According to Members of the Committee, the domestic status concerning the right to strike was unclear, and in particular, the status of the right to strike of civil servants. Until then, it appeared to most Members of the Committee to be wise to exclude Article 6 (4) from the ratification Bill, until there was more clarity about the Dutch status concerning the right to strike, which was expected to be obtained through the future adoption of a national regulation on the matter.[796] The Government agreed to these viewpoints, decided to await the advice of the national Social and Economic Council (SER) on the draft Bill on the right to strike that was submitted earlier, and excluded the Provision from the ratification Bill.[797] However, the draft of a domestic Bill on the right to strike appeared to be not that easy. The Parliamentary debate on the draft Bill 10111, which was intended to regulate the right to strike, was delayed several times and was cause for intense discussions. During the Parliamentary Debate on the ratification Bill of the ESC, draft Bill 10111 was still not adopted by Parliament. Especially the exact demarcation concerning professions that should be excluded from the scope of the right to strike, due to the severe consequences it would bring to society when its practitioners would exercise the right (basically civil servants), proved to be a problem. When it appeared to be impossible to reach agreement on a domestic measure, draft Bill 10111 was withdrawn.[798] Instead, in a new draft Bill on the ratification of the ESC which included Article 6 (4) ESC, it was proposed with a reservation *'with regard to the employees not being civil servants.'*[799] Still, the exact content and scope of the right to strike as embedded in Article 6 (4) was not clear, and as appeared from the subsequent Parliamentary Debate, highly controversial.[800]

---

[794] Parliamentary Documents, II 1965-1966, 8606 (R 533), no. 3, p. 5.

[795] Parliamentary Documents, II 1966-1967, 8606 (R 533), no. 6, p. 3; Parliamentary Documents, II 1965-1966, 8606 (R 533), no. 7.

[796] Parliamentary Documents, II 1966-1967, 8606 (R 533), no. 5, p. 2-3.

[797] Parliamentary Documents, II 1966-1967, 8606 (R 533), no. 6, p. 3.

[798] Parliamentary Documents, II 1979-1980, 10111 no. 5.

[799] Parliamentary Documents, II 1970-1971, 8606 (R 533), no. 11, Article 2; later also in Parliamentary Documents, II 1976-1977, 8606 (R 533), no. 14.

[800] For instance, regarding the ratification of Article 6 (4) ESC, two Amendments and one Motion were submitted, with various (and opposing) viewpoints regarding the ratification. See: Parliamentary Documents, II 1977-1978, 8606 (R 533), no. 15, Amendment of Parliamentarian Verburg, in which the proposition was made to exclude Article 6 (4) completely from the ratification Bill. Contrary to that Amendment, Roethof and co proposed to include Article 6 (4) without any reservation: Parliamentary Documents, II 1977-1978, 8606 (R 533), no. 18. In a motion, Parliamentarians De Voogd and Van Dam invited the Government to *'submit, no later than 1 January 1981, further proposals to ensure the proper functioning of the public service on essential parts, also in the situation in which Article 6 (4) ESC will be fully ratified, so that the present reservation can be withdrawn.'* Parliamentary Documents, II 1977-1978, 8606 (R 533), no. 21. Original text in Dutch: *'om uiterlijk op 1 januari 1981 nadere voorstellen aan het parlement te doen ter voorziening in het ongestoord functioneren van de openbare dienst op essentiële onderdelen ook in de situatie, waarin artikel 6 (4) van het Handvest onverkort zal zijn geratificeerd en het aldus mogelijk te maken,*

The Government as well as the Preparatory Committee had thus focussed their debate on Article 6 (4) on the matter whether the Provision should be included in the ratification Bill or not. The possibility of direct applicability of the Provision was not discussed at all in the Explanatory Memorandum and subsequent Parliamentary documents. Only in the final report of the Preparatory Committee the matter was introduced. The fraction of the Liberal Party (VVD) assumed that Article 6 (4) would be considered by the Courts to be self-executing. Therefore, the fraction was of the opinion that the Provision could not be ratified without a reservation, for they considered that an unlimited right to strike for civil servants would be undesirable.[801] The fraction of the Rainbow/Radical Party (PPR) instead argued that Article 31 ESC already adequately allowed restrictions to ESC Provisions by law and necessary in a democratic society, and therefore, an unlimited ratification of Article 6 (4) would be appropriate.[802] The Government did not literally discuss the assumption of the VVD fraction that Article 6 (4) would be directly applicable, but appeared to have reasoned a similar way: the Government argued that in case law, the matter of the right to strike regarding non-civil servants was already normalised, but not regarding civil servants. To avoid a legal vacuum that could lead to unacceptable consequences, the Government considered that the restrictions allowed by Article 31 ESC should first be adequately regulated in the Dutch legal order, before the Provision could be ratified without a reservation. Interestingly, the Government made a comparison with the ICESCR, and stated: '*also, the ESH does not encompass, unlike the International Covenant on Economic, Social and Cultural Rights, a Provision regarding the possibility of progressive realisation of the rights stipulated in the Charter.*'[803] The Government did not unambiguously take a position regarding direct effect of Article 6 (4) ESC, but seemed to take into account the possibility that the right could be effectuated by individuals in Court. The fact that the ESC did not encompass a general Provision allowing progressive realisation seems to influence that possibility.

---

dat het thans op dit stuk gemaakte voorbehoud zal kunnen worden ingetrokken.' See for the Parliamentarian debate: House of Representatives, 37th meeting, 4 April 1978; House of Representatives, 38th meeting, 5 April 1979; Senate, 3rd meeting, 31 October 1978.

[801] Parliamentary Documents, II 1976-1977, 8606 (R 533), no. 12, pp. 2-3.

[802] Parliamentary Documents, II 1976-1977, 8606 (R 533), no. 12, p. 3.

[803] Parliamentary Documents, II 1976-1977, 8606 (R 533), no. 13. Original text in Dutch: '*voorts houdt het ESH, anders dan het Internationaal Verdrag inzake Economische, Sociale en Culturele Rechten, geen bepaling in omtrent de mogelijkheid van geleidelijke verwezenlijking van het in het Handvest bepaalde.*' Later, during the Parliamentary debate, a similar reasoning was expressed by Minister De Ruiter. The Minister argued that, when hypothetically the Courts would consider that the right as embedded in Article 6 (4) ESC would have direct effect, there would certainly be a legal vacuum concerning the right to strike of civil servants. But even when the right would not be directly applicable, ratification without reservation might be cause for the Courts to be more eager to involve this fact in their ruling, for the ESC, unlike the ICESCR, has no general Article that allows for a progressive realisation. The absence of such a general Provision thus was reason for the Minister to be very cautious in ratifying the Provision without the reservation to exclude civil servants from the right. See: Minister De Ruiter, House of Representatives, 38th meeting, 5 April 1979.

However, earlier, during the interpellation-Bakker (Communist Party of the Netherlands, CPN) on the draft Bill 10111, the Government was a little more specific about the possibility of direct effect of Article 6 (4) ESC. Minister of Justice, Van Agt, stated that by ratifying the ESC, the legitimacy of striking would be implemented in the Dutch legal order, and thus no further national legislation would be required. In response, Parliamentarian Roethof (Labour Party, PvdA) argued that *'Unfortunately, to my regret, judges often experience some difficulty with the international conventions. Therefore, I consider it appropriate to embed the issue directly in national legislation.'*[804] In response, Minister Van Agt made a statement that would lead to complex discussions later, during the debates on the ratification Bill of the ESC: *'I think that it would be sufficient if we as Legislators should make clear that we consider the Provisions of Article 6 (4) as self-executing.'*[805] This statement of Minister Van Agt brought his successor, Minister De Ruiter, in a difficult position during the debates in Parliament on the ratification Bill of the ESC. In Parliament, Minister de Ruiter was asked several times whether he considered Article 6 (4) ESC suitable for direct application. He stated that: *'In the Explanatory Memorandum and Response, the position is taken that the Articles of the Charter have no direct effect. Thereby, reference was made to the intention that was leading in the drafting of the Charter, as expressed in the Annex to the Charter relating Part III. This Annex stipulates: "It is understood that the Charter contains legal obligations of an international character, the application of which is submitted solely to the supervision provided for in Part IV thereof".'*[806] In his view, the Articles were addressed to the State, including Article 6 (4). The fact that in the wordings of the Provision the phrase 'recognise' was used, did not automatically lead to the conclusion that the Provision would be suitable for direct application. He admitted however, that there appeared to be different views on the matter. When he was reminded of the statements of his predecessor, Van Agt (who was now Prime Minister), he argued that there was no reason to come to a different conclusion than the one that was expressed in the written documents before. He added that: *'It remains however a matter of interpretation of a treaty Provision in the context of the Convention in which that Provision is embedded. A decisive judgment – I really have to point this out – is not well-given from a Government's position. It is ultimately a matter in respect of which the decision is up to the Court. But even if one starts from the view that Article 6, fourth Section, of the Charter, has no direct effect, it does of course not mean that*

---

[804] Roethof (Labour Party, PvdA), House of Representatives, 51st meeting, 10 February 1977 (Interpellation-Bakker).

[805] Minister Van Agt, House of Representatives, 51st meeting, 10 February 1977 (Interpellation-Bakker). Original text in Dutch: *'ik meen dat wij er voldoende aan zouden doen als wij als wetgever duidelijk maken dat we de bepalingen van artikel 6 (4) als self-executing beschouwen.'*

[806] Minister De Ruiter, House of Representatives, 38th meeting, 5 April 1979. Original text in Dutch: *'In de memorie van toelichting en die van antwoord is het standpunt ingenomen dat aan de artikelen van het handvest geen rechtstreekse werking toekomt. Daarbij is verwezen naar de bedoeling die bij het opstellen van het handvest heeft voorgezeten, welke bedoeling tot uitdrukking is gebracht in de bijlage bij het handvest omtrent deel III. Die bijlage luidt: Het handvest houdt juridische verplichtingen van internationale aard, welker toepassing uitsluitend aan het in deel IV onderworpen toezicht is onderworpen.'*

approval and ratification of that Provision would be meaningless, also not to the Dutch legal order.'[807] The Minister argued that in case law, usually the legality of a strike will be considered in the context of tort law (Article 1401 Civil Code[808]). In their judgment, the Courts usually will consider whether the strike is *'appropriate in society'*[809] and, even if direct applicability of Article 6 (4) is not recognised, *'(…) will take into account the fact that Article 6 (4) ESC is adopted. Recognition of the right to strike, resulting from the adoption of Article 6 (4), may indirectly play a role in the application of Article 1401 of the Civil Code. When one looks at the matter in this way, one may wonder whether there is indeed an absolute contrast between whether or not direct effect is possible.'*[810] The Minister was interrupted by Parliamentarian Bakker, who argued that *'This view can be accepted. To make it abundantly clear, I would like to point out again that the Minister, also here, links the ratification of the Charter to the withdrawing of Bill no. 10111. This link is so closely established by the Government that it is absolutely impossible to deny the direct effect. The Minister has directly connected this with a piece of domestic legislation and therefore, there is no doubt.'*[811] The Minister replied: *'I have no problem with that viewpoint. However, I still would like to point out that the direct effect of such an open and general stated principle as Article 6 (4) is very different compared to the direct effect of a more or less detailed regulation. For me, the effect here – disregarding the words "directly" or "indirectly" – is that there is no doubt that the right to strike is recognised in the Netherlands.*[812]

Parliamentarian Roethof (Labour Party, PvdA) was very active in the debate on direct effect, and argued that in the first place, he agreed with the Minister that

---

[807] Minister De Ruiter, House of Representatives, 38th meeting, 5 April 1979. Original text in Dutch: *'Het blijft echter een kwestie van uitleg van een verdragsbepaling in de samenhang van het verdrag waarin die bepaling is opgenomen. Een beslissend oordeel – ik moet daarop wel even de aandacht vestigen – kan niet goed van achter de regeringstafel worden gegeven. Het is uiteindelijk een kwestie ten aanzien waarvan de beslissing aan de rechter toevalt. Ook al gaat men uit van de zienswijze dat artikel 6, vierde lid, van het Handvest, geen rechtstreekse werking heeft, dan wil dat uiteraard nog niet zeggen dat goedkeuring en ratificatie van die bepaling geen enkele betekenis hebben, ook voor de Nederlandse rechtsorde.'*

[808] Currently Article 6:162 CC.

[809] Original text in Dutch: *'in het maatschappelijk verkeer betaamd.'*

[810] Minister De Ruiter, House of Representatives, 38th meeting, 5 April 1979. Original text in Dutch: *'(…) acht zal slaan op het gegeven dat artikel 6 (4) ESH is goedgekeurd. Erkenning van het stakingsrecht, welke uit de goedkeuring van artikel 6 (4), voortvloeit, zal dan langs indirecte weg een rol kunnen spelen bij de toepassing van artikel 1401 BW. Wanneer men de zaak op deze manier beziet, kan men zich afvragen of er wel sprake is van een absolute tegenstelling tussen wel en geen rechtstreekse werking.'*

[811] Bakker (Communist Party of the Netherlands, CPN), House of Representatives, 38th meeting, 5 April 1979. Original text in Dutch: *'Deze zienswijze is wel te accepteren. Ten overvloede wijs ik er nog eens op dat de Minister ook nu weer de invoering van het handvest koppelt aan de intrekking van wetsontwerp no. 10111. Die koppeling is door de Regering zo nauw gelegd, dat het volstrekt onmogelijk is, de rechtstreekse werking te ontkennen. De Minister heeft het direct verbonden met een stuk binnenlandse wetgeving en daarmee ligt de zaak vast.'*

[812] Minister De Ruiter, House of Representatives, 38th meeting, 5 April 1979. Original text in Dutch: *'Daarmee heb ik ook geen moeite. Ik moet er nog wel even op wijzen, dat de rechtstreekse werking van zo'n open en algemene beginselverklaring als artikel 6 (4) heel iets anders is dan de rechtstreekse werking van een min of meer gedetailleerde regeling. Voor mij betekent hier de werking – ik laat de woorden 'rechtstreeks' of 'indirect' even buiten beschouwing – dat er geen twijfel over bestaat, dat het stakingsrecht in Nederland is erkend.'*

it is indeed up to the Courts to decide whether a Provision is suitable for direct effect. Secondly, he also stated that the opinion of the Legislature might play a role in the decision of the Courts as a guideline. However, he observed that this opinion of the Legislature was subject to change: *'the opinion of the Legislature...* *was that there certainly was direct effect. This is now slightly reversed by our Minister of Justice.'*[813] He then observed that in the past, the Government had recognised the direct effect of many Provisions in the field of Civil and Political rights; a reasonable approach, for with regard to those rights, in general, a Government is not demanded to act, but rather to *'tolerate'*[814] something. This was, logically, different in the case of economic, social and cultural rights, for with regard to those rights, generally, the Government is asked to undertake, instead of tolerate, something. However, with regard to the right to strike, as embedded in Article 6 (4), the wordings indicate – unlike most other treaty Provisions – that the Government were to 'recognise' the right, meaning that the Government should 'tolerate' instead of 'undertake' something. Roethof emphasised that he valued the fact that the Legislature here expressed a position on the direct applicability of the Provision, considering the role of Parliamentary History for Courts in their verdicts, and understood the Minister thus that he clearly agreed that by ratifying the ECS, the right to strike will be recognised in the Netherlands.[815] To this, the Minister responded that *'The essence of my argument was actually that the answer to the question of direct effect depends on the intention of the contracting parties. I explained that this intention clearly goes in the direction of non-direct effect.'*[816] Roethof then observed that he and the Minister had different views, reminded that it was eventually up to the Courts to decide on the matter, but underlined that it was interesting to establish the opinion of the Legislature in this. The Minister referred to the intention of Roethof to contribute to the argumentation on the matter. The Minister then decided to focus on one argument in particular, that is the fact that in the wordings of Article 6 (4) the word 'recognise' was used, in contrast to most other ESC Provisions, and that in itself was remarkable, and indeed would at least more easily tend to direct effect compared to other Articles. However, De Ruiter also underlined that preferably, when possible, direct application of Articles should be deduced from the treaty itself, and stated that the intention to adopt the Annex to the treaty was expressed by the involved Ministers of the signatory countries: *'They decided to add that Annex with the motivation that the Provisions of the Charter cannot be directly invoked by individuals before national*

---

[813] Roethof (Labour Party, PvdA), House of Representatives, 38th meeting, 5 April 1979. Original text in Dutch: *'de mening van de wetgever (...) was dat er wel degelijk rechtstreekse werking was. Dit is nu door onze Minister van Justitie weer enigszins teruggedraaid.'*
[814] Roethof used the Dutch word 'dulden'.
[815] Roethof (Labour Party, PvdA), House of Representatives, 38th meeting, 5 April 1979.
[816] Minister De Ruiter, House of Representatives, 38th meeting, 5 April 1979. Original text in Dutch: *'De kern van mijn betoog was eigenlijk dat het antwoord op de vraag naar de rechtstreekse werking afhankelijk is van de bedoeling van de verdragsluitende partijen. Ik heb uiteengezet dat die bedoeling overduidelijk gaat in de richting van de niet rechtstreekse werking.'*

Courts.'[817] Earlier, he had declared that in the Annex, direct effect was clearly excluded from the treaty.

The above demonstrates that a rather unstructured and confusing discussion preceded the adoption of Article 6 (4). It is hard to distract a clear viewpoint on the matter that can be attributed to the Legislature. Especially Minister De Ruiter, representing the Government, seems to weave between almost recognising direct effect to almost denying the possibility of direct effect of Article 6 (4) ESC. This also led to a variety of different interpretations in Parliament on the matter. Nevertheless, in Parliament the explanation of the Minister led to no further discussions, and in the Senate, the matter of direct applicability was hardly addressed.

8.5 The Parliamentary History concerning Article 8 ICESCR

The right to strike is also stipulated in Article 8 ICESCR. As demonstrated in Chapter 7, the Government underlined in its Explanatory Memorandum, that the rights enshrined in the ICESCR in general would not be suitable for direct applicability.[818] The direct applicability of Article 8 ICESCR therefore has not been discussed in particular in Parliamentary History.

However, some other related particularities were discussed and are worth mentioning here. With regard to the backgrounds of the Provision, the Government noted that the wordings were heavily inspired by ILO convention 87[819] that had already been ratified by the Netherlands, and observed that there was a certain overlap with Article 22 ICPR. However, as it appears, both legal documents mostly concerned the freedom of association, and not the right to strike. Concerning this right, the Government underlined that it was preparing the ratification of the ESC, and had submitted the draft Bills 10110 (Provisions relating to committees of inquiry on strike[820]) and 10111 (statutory Provision relating to strike[821]) to the States-General. The Government argued that restrictions to the right to strike when it concerned certain categories of civil servants could be justified, based on the wordings of Article 8 ICESCR.[822]

---

[817] Minister De Ruiter, House of Representative, 38th meeting, 5 April 1979. Original text in Dutch: '*Zij hebben besloten die Annexe te doen toevoegen met de motivering, dat de regelingen van het handvest niet direct kunnen worden ingeroepen door individuen voor de nationale rechtbanken.*'
[818] See especially: Parliamentary Documents, II 1975-1976 13932 (R 1037), no. 3, pp. 12-13.
[819] See: ILO Convention 87, 17 June 1948, *Freedom of Association and Protection of the Right to Organise Convention*. Although it must be noted that the right to strike is not stipulated in this ILO convention.
[820] Original name in Dutch: '*regelen met betrekking tot commissies van onderzoek inzake de werkstaking.*'
[821] Original name in Dutch: '*wettelijke bepaling met betrekking tot werkstaking.*'
[822] Parliamentary Documents, II 1975-1976 13932 (R 1037), no. 3, pp. 48-49.

Also here this opportunity was taken, in this case by the fraction of the Labour Party (PvdA) in the Preperatory Committee, to criticise the slow progress that was made on the draft Bills. Also, the fraction referred to the intended reservation to the ESC, Article 6 (4) concerning civil servants, and argued that such discrimination was unacceptable.[823] The Government in turn, referred to the interpellation-Bakker, during which it explained why the Draft-Bills were withdrawn. Basically, reference was made to the fact that the right to strike of employees in the private sector was already normalised in case law, and the right to strike of civil servants was about to be regulated in a separate measure. Therefore, ratifying Article 8 ICESCR would not lead to any difficulties. Remarkably, not a single reference was made to the fact that in that same interpellation-Bakker, it became clear that Article 6 (4) was considered to be a substitute for especially Draft-Bill 10111, except concerning civil servants.

It is interesting to note within the context of the reservation made to Article 6 (4) ESC, why the Government explained why Article 8 ICESCR could be fully ratified, without reservation: '*In determining our position that with regard to the Netherlands, Article 8 can be ratified without reservation, we assumed that indeed the legislation on the right to strike of civil servants shall be realised. From various elements of the Convention it can be deduced that the right to strike, mentioned in Article 8, may be gradually achieved. This possibility of progressive realisation, for which the ESC contains no Provision, means that the legislation of the right to strike of civil servants will not have to be realised immediately at the moment when the treaty enters into force. We therefore consider it unnecessary to make a reservation to the right to strike of civil servants.*'[824]

With regard to the scope of Article 8 ICESCR, the Government considered that the basic difference with Article 6 ESC was that Article 8 ICESCR focusses more on the right to form trade unions, while Article 6 ESC concentrates more on the right to collective bargaining and the right to strike.[825]

---

[823] Parliamentary Documents, II 1975-1976 13932 (R 1037), no. 7, pp. 15-16.

[824] Parliamentary Documents, II 1975-1976 13932 (R 1037), no. 8, p. 25. Original text in Dutch: '*Bij de bepaling van ons standpunt, dat artikel 8 voor Nederland zonder voorbehoud geratificeerd kan worden, zijn wij ervan uitgegaan dat een wettelijke regeling van het stakingsrecht van ambtenaren inderdaad tot stand gebracht zal kunnen worden. Uit diverse elementen van het verdrag valt op te maken dat het stakingsrecht, vermeld in artikel 8, geleidelijk verwezenlijkt mag worden. Deze mogelijkheid van geleidelijke verwezenlijking, ten aanzien waarvan het ESH geen bepaling bevat, betekent dat de wettelijke regeling van het stakingsrecht van ambtenaren niet terstond gereed hoeft te zijn als het verdrag in werking treedt. Het maken van een voorbehoud over het stakingsrecht van ambtenaren achten wij daarom niet noodzakelijk.*'

[825] Parliamentary Documents, II 1975-1976 13932 (R 1037), no. 8, p. 25.

8.6 Conclusion

In the first place, it can be concluded that the Supreme Court in the railway-strike Ruling did not necessarily oppose the viewpoints of the Legislature concerning the direct applicability of Article 6 (4) ESC. When one would only consider the Explanatory Memorandum, it might be concluded that the Government argued that the ESC Provisions in general are not suitable for direct applicability. However, the debate on Article 6 did not focus on direct applicability, but on the question whether or not the Provision should be included in the ratification Bill. Only later during the Parliamentary History, especially during the debates in the House of Representatives, the possibility of direct effect of Article 6 (4) was discussed in detail. Indeed, as Roethof already observed, the Government appeared to have altered its position regarding the possibility of direct effect, starting from literally recognising its direct effect (Minister Van Agt) for the Provision was considered to be a substitute for the withdrawn Draft-Bill 10111, to the statements of Minister De Ruiter, who visibly struggled with the matter, and balanced somewhere between almost agreeing on direct effect and strongly rejecting the idea. It appears thus that the opinion of the Legislature as a criterion for determining whether Article 6 (4) ESC is suitable for direct applicability, is an unclear criterion and therefore hardly practicable for the Judiciary. Therefore, the case law regarding Article 6 (4) is not necessarily in contrast with the conclusion drawn in Chapter 4, which is that the Courts, when determining whether Articles in which the right to food is stipulated, have direct effect, follow the lead of the Legislature.

In the second place, it can be questioned to what extent the intention of the contracting parties to a Convention seriously were considered by the Supreme Court in its railway-strike ruling, for it appears from the Parliamentary History that during the drafting of the ESC – or at least in the perception of the Minister, representing the Government – the intention was expressed amongst the signatories that the Treaty Provisions could not be invoked in the national Courts by individuals. In the railway-strike ruling, the Supreme Court considered that *'whether or not the contracting States intended to grant direct effect to Article 6 ESC is not relevant, whereas neither from the text, nor the history of the formation of the treaty follows that they have agreed that Articles 6 ESC is directly applicable.'*[826] Indeed it can be considered that in general, contracting parties will not express their intentions about direct effect of treaty Provisions in the international arena, for due to the different Constitutional systems of each country this would be pointless. However, according to Minister De Ruiter, the matter had been explicitly discussed.

---

[826] Supreme Court, 30 May 1986, NJ 1986, 668 (railway-strike ruling), consideration 3.2. Original text in Dutch: *'Of de verdragsluitende Staten al dan niet hebben beoogd aan artikel 6 lid 4 ESH directe werking toe te kennen, is niet van belang nu noch uit de tekst, noch uit de geschiedenis van de totstandkoming van het Verdrag valt af te leiden dat zij zijn overeengekomen dat aan artikel 6 lid 4 die werking mag worden toegekend.'*

In the third place, it can be concluded that the fact that the invoked Provision did not oblige Member States to adopt further legislation, unlike most other ESC Articles, seems to be a decisive element. In that respect, as demonstrated above, it was already frequently referred to in the Parliamentary History that Article 6 (4) obliged States to 'recognise', which implied generally to abstain from action, instead of adopting measures. The Supreme Court seems to agree as well, for *'in that situation, according to Dutch law, only the content of the Provision is decisive: does this oblige the Dutch Legislature to implement a national measure with a certain content or scope, or is this of such a nature that the Provision can be applied directly in the national legal order as positive law? Article 6 Section 4 – contrary to most other Provisions of the ESC and in particular also to the other Sections of this Article – is thus worded that the States Parties concerned are not obliged to adopt measures, but instead that the said employees and employers may certainly invoke the law as recognised, within certain limits, by the concerned States in the domestic legal order.'*[827]

Fourthly, it appears that it can be explained why Article 6 (4) ESC is considered directly applicable, and Article 8 ICESCR, stipulating an equivalent human right, is not, for two main reasons. One reason concerns the opinion of the Legislature, the second concerns the wordings of the treaty Provisions. Firstly, it can be observed that in the context of the ICESCR Provision, the Legislature has always coherently denied the possibility of direct effect, while this is certainly not the case regarding Article 6 (4) ESC. In the latter, it remains rather unclear what the vision of the Legislature was. On the one hand, ratifying Article 6 (4) ESC was considered to be an adequate substitute for national legislation by one Government, and on the other hand, the next Government did seem to express considerably more cautiousness in recognising the direct effect, but did not come to an unambiguous conclusion. Secondly, there is a difference in the wordings between both Treaties. In the ICESCR, a general Provision is embedded that recognises the possibility to progressively realise a right, implying that the right should not be implemented immediately. This is not the case concerning the ESC. The ESC seems to imply that at the moment of ratification the rights must already be realised, while this is not the case concerning the ICESC. In the ESC Provision the word 'recognise' is used, while in the ICESCR Provision, in line with the principle of progressive realisation, the word 'ensure' is used. Generally speaking, it can be observed that indeed Article 6 (4) ESC seems to stipulate a duty of the Government to abstain

---

[827] Supreme Court, 30 May 1986, NJ 1986, 668 (railway-strike ruling), consideration 3.2. Original text in Dutch: *'Bij deze stand van zaken is naar Nederlands recht enkel de inhoud van de bepaling zelf beslissend: verplicht deze de Nederlandse Wetgever tot het treffen van een nationale regeling met een bepaalde inhoud of strekking, of is deze van dien aard dat de bepaling in de nationale rechtsorde zonder meer als objectief recht kan functioneren? Artikel 6 lid 4 is – in tegenstelling tot de meeste andere bepalingen van het ESH en met name ook tot de overige leden van dit artikel – zo geredigeerd dat de betrokken verdragsstaten niet een verplichting tot regelgeving wordt opgelegd, maar integendeel aldus dat werknemers en werkgevers zich op het door de betrokken verdragsstaten binnen zekere grenzen erkende recht in de nationale rechtsorde zonder meer moeten kunnen beroepen.'*

from acting, while Article 8 ICESCR seems to focus more on Governmental action. (However, this did not prevent the Government to proudly report to the Committee on Economic, Social and Cultural Rights with regard to the implementation of Article 8 ICESCR that Article 6 (4) ESC is directly applicable in the Netherlands, thereby implying that both Provisions have the same scope.)[828]

---

[828] E/1990/6/Add.11, 5 August 1996, Sections 106-107, see also the further elaboration of the Government: E/C.12/NLD/q/3/Add.1, 23 August 2006, Section 18.

9. The Dutch periodic country reports

9.1 Introduction

As stated in Section 2.6.3, this chapter will focus on the question: *What does the Government of the Netherlands communicate in its reports in the United Nations arena regarding the enforceability of the right to food in its domestic legal order?* To this end, the Dutch reports submitted on the implementation of the ICESCR (Section 9.2), the ICRC (Section 9.3) and the CEDAW (Section 9.4) will be analysed. Furthermore, when relevant, the two reporting cycles in view of the Universal Periodic Review is discussed (Section 9.5). The Dutch reporting behaviour is partly determined by several reporting guidelines or formats provided for by the relevant treaty bodies, which will when relevant be discussed below. From the analysis it appears that in the reports the matter of direct applicability is mostly discussed separately from the substantive explanation of the implementation of the specific treaty rights. The same applies for the matter of aliens policies. Both issues are therefore addressed in this Chapter as a separate topic. In the section on the ICRC reports, no sub-section is dedicated to the direct effect of its Provisions, for this was not significantly debated during the various reporting cycles.

9.1.2 General (geographical) remarks on the Dutch reports

Since the Kingdom of the Netherlands signs and ratifies international treaties, all countries within the Kingdom have the obligation to implement the treaty. As already pointed out in Section 1.3, the Kingdom of the Netherlands does not only consist of the European Country, baring almost the same name (The Netherlands), but also includes overseas territories. Therefore, the Kingdom of the Netherlands, as signatory of international treaties, has the duty to include the overseas territories in its reports as well. This is done in different ways, without any coherence. Especially in the earlier reports, the choice was made to separately hand in reports on the implementation of a treaty on the Netherlands (European part of the Kingdom), Aruba and the Netherlands Antilles. Later, occasionally, one report with sub-parts covering the separate countries was submitted. In the subsequent procedures, the reports were discussed sometimes separately and sometimes altogether by the relevant treaty bodies. Since 2010 however, the Netherlands Antilles was dissolved and the Kingdom of the Netherlands reformed: the former Antilles islands of Curaçao and St. Maarten became sovereign States within the Kingdom in 2010 (as Aruba already was since 1986), while the islands Bonaire, St. Eustatius and Saba became special municipalities within the Netherlands, together referred to as 'the Caribbean Netherlands'. This reform occurred in some cases during a reporting cycle, making matters even more complex.[829] In general, it can be concluded that the quality, coherence and profundity of the reports varies considerably. Also the access to reports through official databases is not coherently organised, neither by the Member State nor the relevant UN institutions.

In this Chapter, the reports of the overseas countries and municipalities are included in the analysis for two reasons. Firstly, although this research focuses on the European part of the Kingdom (see Chapter 1, in particular Section 1.3) it is technically the Kingdom of the Netherlands that has a reporting duty, on behalf of all the countries within the Kingdom. Since there is no coherent approach in reporting and in some occasions the Kingdom of the Netherlands is generally discussed by the different Committees, the choice was made to – where relevant – include the reporting and subsequent procedures of the overseas territories as well. Secondly, since the matter of direct applicability is eventually determined by the Courts, and the Supreme Court is the Court of Last Instance in the entire Kingdom, the matter of direct applicability naturally concerns all countries within the Kingdom in a similar way. No evidence could be found that in the overseas territories a different legal doctrine exists regarding the enforceability of the right to food. Therefore, the reporting on direct applicability of treaty Provisions is not necessarily bounded by State borders within the Kingdom, but will be more or less imply one concept shared throughout the Kingdom.

9.2 The ICESCR reports

9.2.1 Introduction

Before 2013, the Netherlands had submitted 5 periodic reports to the Committee on Economic, Social and Cultural Rights: an initial report was submitted in 1984,[830] a second in 1996,[831] a third in 2005,[832] and a combined fourth and fifth report in 2009.[833] The Netherlands reported mostly separately on the implementation of the

---

[829] For instance, during the reporting cycle of the combined fourth and fifth report on the implementation of the ICESCR the Kingdom of the Netherlands was reformed, leading to three reports on the implementation of the ICESCR regarding the Netherlands, Aruba and the Netherlands Antilles, but later, the Kingdom of the Netherlands was represented by different delegations from the newly formed autonomous countries within the Kingdom, in one series of sessions with the Committee on Economic, Social and Cultural Rights. See: Mr Abath, E/C.12/2010/SR.43, 18 November 2010, Section 2.

[830] In three stages: E/1980/6/Add.33, 8 November 1983; E/1982/3/Add.35, 17 January 1986; E/1984/6/Add.14, 20 February 1986; E/1982/3/Add.44, 19 January 1988; E/1984/6/Add. 20, 19 August 1988. The reports are not available anymore in official databases of both the UN and the Kingdom of the Netherlands.

[831] E/1990/6/Add.11, 5 August 1996.

[832] E/1994/104/Add.30, 23 August 2005.

[833] E/C.12/NLD/4-5, 17 July 2009.

ICESCR in Aruba and the Netherlands Antilles.[834] However, the combined fourth and fifth report of the Netherlands, and the fourth reports of the Netherlands Antilles and Aruba were mostly discussed altogether in the List of Issues and written replies, and during the sessions with the Committee, with a combined Delegation, due to the dissolving of the Netherlands Antilles.[835]

It is obvious that the Netherlands rarely met the deadlines for submitting the reports, for the States are requested to submit an initial report within two years after ratification, and from then on a periodic report every five years.[836] Because of the delay, the Netherlands submitted in 2009 a combined fourth and fifth report, instead of two separate ones. Often, no explanation was given for such (huge) delays, but in the first Universal Periodic Report, the Dutch Government stated that: *'The Netherlands consider it very important that reports are thorough, accurate and submitted on time.'* On the other hand, it added that *'a great deal of time and energy is devoted to ensuring reports are drafted carefully and coherently and are very informative, e.g. with regard to the implementation of recommendations made by treaty bodies on previous occasions. Since relevant data on a given reporting period, including statistical data, often only becomes available after a deadline, processing the data can cause delays. Sometimes reports are merged.'*[837]

9.2.2 The quality of the ICESCR reports

In general, it can be observed that in its reports the Government usually has the tendency to conclude – in line with the Explanatory Memorandum to the ratification Bill on the ICESCR[838] – that the Netherlands already fulfil their obligations that come forth from the ICESCR. The Government then usually provides with a list of facts, policies, legislative measures and other initiatives that have been undertaken, to substantiate this. There is hardly any hint that there are points for improvement, and no intention at all is ever expressed to come to a further or better realisation of the right in the future. However, this approach has been criticised by the Committee on Economic, Social and Cultural rights on several occasions. During the sessions with the Committee on Economic, Social and Cultural Rights on the second periodic report, one Member reminded the Dutch Delegation that *'when considering the initial report of the Netherlands, the Committee had requested*

---

[834] Second report: E/1990/6/Add.12, 28 August 1996 (Netherlands Antilles), E/1990/6/Add.13, 28 August 1996 (Aruba); Third Report: E/C.12/ANT/3, 13 February 2006 (Netherlands Antilles); Fourth report: E/C.12/NLD/4/Add.1, 13 July 2009 (Netherlands Antilles); E/C.12/NLD/4/Add.2, 19 October 2010 (Aruba).

[835] During the reporting process of the fourth periodic report of the Netherlands Antilles, Curaçao and St. Maarten became sovereign States within the Kingdom in 2010 (as Aruba was since 1986), while the Islands Bonaire, St. Eustatius and Saba became special municipalities within the Netherlands.

[836] Economic and Social Council Resolution 1988/4, 24 May 1988, Section 6.

[837] A/HRC/WG/6/1/NLD/1, 7 March 2008, *National Report submitted in accordance with Section 15(a) of the annex to Human Rights Council Resolution 5/1,* Section 83.

[838] See especially Chapter 6.

that subsequent reports should specify areas in which the Government was encountering difficulties. As no action seemed to have been taken on that recommendation, he asked the Delegation to rectify the omission.'[839] A Dutch Delegate replied that 'he fully understood his concern, but it went somewhat against the grain for a civil servant to criticise the policy of his employer, the State.'[840] In a shadow-report from two NGOs on the third periodic report, this approach criticised again: 'when reading the report, one is left with the impression that the situation in the Netherlands concerning the implementation of economic, social and cultural rights is rather rosy. However the report does not refer sufficiently to the obstacles on the realisation of economic, social and cultural rights.'[841] The Committee seems to be inspired by this shadow report for one Committee Member stated that: 'the Delegation had pointed a rosy picture of the Netherlands' performance in implementing the Covenant. He could not believe, however, that there was any country that did not face challenges, crises or problems in that process and he asks the Delegation to provide some information on any difficulties encountered.'[842] In response, a Dutch Delegate emphasised that it was not the intention to 'to paint an overly positive picture in his country's report', and elaborated, on the spot, on two difficulties the Netherlands faced, which were the relatively high unemployment levels of ethnic minorities, and the integration of foreigners.[843] Apparently, this was not sufficient to the Committee, for in their Concluding Observations, in the Section on 'factors and difficulties impending the implementation of the Covenant', it was made quite clear that 'the Committee notes the absence of any factors or difficulties preventing the effective implementation of the Covenant in the State Party.'[844] However, the Dutch reporting behaviour did not alter significantly,[845] and also in the combined fourth and fifth report, there was hardly any reference to factors or difficulties in the implementation process. It is remarkable that in their List of Issues, and during the sessions, the Committee did not address the issue. Instead however, there was considerably more criticism on several aspects of the implementation in the Kingdom of the Netherlands, that became apparent during the reporting procedure, and resulted in a list of principal subjects of concern and recommendations of 37 (out of 45) Sections, which included issues related to discrimination against foreigners, disabled persons, women and prisoners, the direct applicability of ICESCR Provisions, a national

---

[839] Mr Riedel, E/C.12/1998/SR.14, 15 September 1998, Section 8.

[840] Mr Potman, E/C.12/1998/SR.14, 15 September 1998, Section 9.

[841] Contribution of the Dutch Section of the International Commission of Jurists (NJCM) and the Johannes Wier Foundation (JWS) to the Committee on Economic, Social and Cultural Rights, Provisional reaction to the Third periodic report submitted by the Netherlands under Articles 16 and 17 of the Covenant (E/1994/104/Add.30), Section B.1.

[842] Mr Sadi, E.C.12/2006/SR.33, 22 November 2006, Section 27.

[843] Mr De Klerk, E.C.12/2006/SR.33, 22 November 2006, Section 45-46.

[844] E/C.12/NLD/CO/3, 24 November 2006, C, Section 10.

[845] As will appear below, the reporting on the right to food as embedded in Article 11 ICESCR is even for the major part a copy of the previous report.

plan and institution on human rights, awareness raising on human rights, and some issues related to the overseas territories.[846]

9.2.3 The right to food in the ICESCR reports

9.2.3.1 The second periodic report

As stated above, the Dutch reports do not excel in self-reflection or critically reviewing the implementation status concerning the rights embedded in the ICESCR. This certainly also applies to the reporting on the right to food. In their second periodic report, the Netherlands basically reported that the food production in the Netherlands greatly exceeds the population's requirements, the product quality is high, the prices of foodstuffs are relatively low, and thus also affordable for low-income groups. Furthermore, the Netherlands reported that their food and nutrition policy was generally focussed on food quality and the promotion of healthy eating habits. It was pointed out that on a regular basis, research is conducted on the eating habits of the population, showing that in general, there was an adequate supply of micro- and macronutrients, although overconsumption of fat appeared to be an issue. Therefore, the Government aimed a campaign against over-consumption, resulting in a positive downward trend of fat consumption. Furthermore, it was reported that nutritional education was the most important policy instrument, in which the Nutrition Information Office played a vital role. Finally, it was reported that environmental considerations were of growing importance in food production and consumer choice.[847] During the sessions with the Committee, the right to adequate food was not addressed in particular, although one Member of the Committee '*noted that, according to the Netherlands Section of the International Commission of Jurists, 240,000 households, or almost 1 million persons, were living under an income below the social minimum and some 250,000 children belonging to poor families participated very rarely in recreational and cultural activities. He was surprised that a country as wealthy as the Netherlands was unable to solve the problems.*'[848] The response of the Delegation was not very constructive: '*There seemed to be no basis for the allegation that 250,000 children were unable to exercise their rights provided for in the Covenant. The Netherlands Delegation would subsequently describe the steps taken by the Government to alleviate poverty and*

---

[846] E/C.12/NDL/CO/4-5, 19 November 2010, 5-43.
[847] E/1990/6/Add.11, 5 August 1996, Sections 235-242.
[848] Mr Ahmed, E/C.12/1998/SR.14, 15 September 1998, Section 23.

improve the situation of children from low-income families.'[849] In the Committee's Concluding Observations,[850] the right to food was not further addressed.

In the second periodic report of the Netherlands Antilles[851] the right to food was not discussed at all, and in the report of Aruba, it was only concluded that there was adequate food supply, but also an overweight problem.[852]

9.2.3.2 The third periodic report

In their third periodic report, the Netherlands reported that it has a *'comprehensive system of social benefits guaranteeing its citizens an adequate minimum income.'*[853] Regarding the right to food, it was again underlined that the amount of food available in the Netherlands *'greatly exceeds domestic demand.'*[854] Therefore, Governmental policies mainly aimed at food safety, quality, the consequences of food production for the environment, and consumption patterns. It was emphasised that Dutch food policies were increasingly determined by European Union policies and legislation, such as the General Food Law Regulation. In that light, the supervisory role of the Food and Consumer Safety Authority[855] was mentioned. Some caution was expressed toward s the production of Genetically Modified Organisms (GMOs).[856] Also here, the importance of nutrition education was stressed, in view of unhealthy eating habits, and reference was made to the Food and Nutrition Information Office,[857] an organisation that launched campaigns on healthy eating habits.[858] Furthermore, reference was made to the Dutch role in the international arena concerning the equitable distribution of the world's food supplies:

> The Netherlands' efforts in this respect are focussed on:
> * strengthening the capacity of developing countries to analyse, monitor and address the food insecurity of vulnerable groups;
> * financing food aid targeted at vulnerable groups.[859]

---

[849] Mr Potman, E/C.12/1998/SR.14, 15 September 1998, Section 31. This conversation was also noted by B. van der Meulen and F.M.C. Vlemminx, *Chapter 2, An adequate right to food? The Netherlands abundant in food, wanting in law*, in: O. Hospes and B. van der Meulen (eds.), *Fed up with the right to food*, Wageningen: Wageningen Academic Publishers, 2009.

[850] E/C.12/1998/26, 27 April; 15 May 1998; 16 November; 4 December 1998; CESCR report on the eighteenth and nineteenth sessions.

[851] E/1990/6/Add.12, 28 August 1996. In Sections 109-119, some other aspects falling under the scope of Article 11 ICESCR were discussed, such as income and housing.

[852] E/1990/6/Add.13, 28 August 1996, Sections 110-115.

[853] E/1994/104/Add.30, 23 August 2005, Section 329.

[854] E/1994/104/Add.30, 23 August 2005, Section 331.

[855] In Dutch: Voedsel en Waren Authoriteit (www.vwa.nl).

[856] E/1994/104/Add.30, 23 August 2005, Sections 332-338.

[857] In Dutch: Voedingscentrum (www.voedingscentrum.nl).

[858] E/1994/104/Add.30, 23 August 2005, Section 339.

[859] E/1994/104/Add.30, 23 August 2005, Section 340.

In this light, the Netherlands referred to their support to – mostly in financial terms – the FAO, WFP, and the international Red Cross. It was reported that the Netherlands were convinced that *'food insecurity is a problem of food access rather than food availability.'*[860]

In its List of Issues, the Committee on Economic, Social and Cultural Rights with regard to the implementation of Article 11 ICESCR, asked: *'since the main objective of the overall development cooperation policy of the Netherlands is to combat extreme poverty, please indicate whether this policy has attained its objective, especially with regard to women, ethnic minorities, persons with disabilities, and the disadvantaged and marginalised groups.'*[861] In their reply, the Netherlands underlined that the Millennium Development Goals formed the core of their development cooperation policy, which was continuously and systematically monitored by the Ministry of Foreign Affairs. According to the Netherlands it could be concluded that the Dutch development cooperation activities *'generally contribute to the fight against poverty. However, the effectiveness of ODA[862] remains the subject of an ongoing debate within the Netherlands (...) and throughout the international community (...)'*[863]

During the sessions with the Committee on Economic, Social and Cultural Rights, for the first time in-depth questions were raised about the Dutch asylum policy. The Dutch Delegation informed the Committee that in the Netherlands, a distinction was made between foreigners legally residing in the Netherlands, who had equal entitlements to ECOSOC rights as Dutch citizens, and illegally residing foreigners, including asylum-seekers, who had only entitlements to basic needs. The latter includes basic healthcare, education and legal assistance.[864] The Committee then asked whether foreigners without a residence permit enjoyed the same ECOSOC rights as the rest of the population.[865] The Dutch Delegation responded that *'the vast majority of people who reside illegally in the Netherlands are asylum-seekers whose application has been rejected. These people do not have access to economic, social and cultural rights, and the State has only a limited responsibility towards them, which relates primarily to emergency medical assistance, education of minors and legal assistance.'*[866] This statement is indeed generally in confirmation with the

---

[860] E/1994/104/Add.30, 23 August 2005, Sections 340-343.

[861] E/C.12/NLD/Q/3, 2 March 2006, Section 26.

[862] Official Development Assistance.

[863] E/C.12/NLD/Q/3/Add.1, 23 August 2006, Section 25.

[864] De Klerk, E/C.12/2006/SR.34, 10 January 2007, Section 4.

[865] Mrs Barahona Riera, E/C.12/2006/SR.34, 10 January 2007, Section 18.

[866] De Klerk, E/C.12/2006/SR.34, 10 January 2007, Section 28. Original text in French: *'La grande majorité des personnes qui se trouvent illégalement aux Pays-Bass sont des demandeurs d'asile dont la demande a été rejetée. Ces personnes n'ont pas accès aux droits économiques, sociaux er cultureless et l'Etat n'a envers elles qu'un nombre limité d'obligations, qui touchent principalement à l'aide médicale d'urgence, à l'éducation des mineurs et à l'aide juridictionelle.'*

analysed case law in Chapter 4. The matter was not referred to in the Committee's Concluding Observations.[867]

In the third periodic report of the Netherlands Antilles, the right to food was not significantly discussed.[868]

9.2.3.3 *The combined fourth and fifth report of the Netherlands/the fourth report of the Netherlands Antilles and Aruba*

Remarkably, the Sections dedicated to the right to food in the combined fourth and fifth report are for the major part an exact copy of the text in the third report.[869] Perhaps due to the fact that the Netherlands did not change its approach in reporting, as requested several times by the Committee, or perhaps due to the extensive involvement of NGOs, addressing many shortcomings of the Netherlands in their implementation of the ICESCR rights,[870] the Committee appears to be considerably more critical. In three Sections of the List of Issues, the right to food was addressed.[871]

> 25. Given that there are reportedly some 140 food banks in the State Party, please indicate what concrete measures have been taken by the State Party in relation to food security, and their outcome.

The Netherlands replied that the Netherlands have an extensive system of social security, guaranteeing a minimum income, '*that enables people to meet their basic needs, including food, clothing and shelter.*' Reference was made to a recent study, that demonstrated that the income, although being '*lagged behind the general increase in prosperity*' was '*sufficient for single people, rather high for the elderly and single-parent families, and sometimes too low for households with several children.*' However, it was reported that '*some people get into debt and the repayment arrangements they make may force them to live below the minimum level. There are also people who fail to exercise their right to social provision. The authorities inform people, e.g. those on social assistance benefit, if they are entitled to crisis payments. These can be used for essential investments, such as the purchase of a fridge.*' Regarding the food banks, the Netherlands reported that the Government had agreed with the Dutch

---

[867] E/C.12/NLD/CO/3, 24 November 2006.

[868] E/C.12/ANT/3, 13 February 2006. See for the reporting on Article 11 ICESCR: Sections 90-124.

[869] E/C.12/NLD/4-5, 17 July 2009, Sections 219-232. Besides some textual rephrasing and some minor additions, the only significant addition was that: '*The Dutch Government brought the importance of the right to food to the attention of a wide audience at a 2006 seminar on the subject at Wageningen University and Research Centre, and during the national activities to mark the World Food Day last year. Recently, the Government supported a workshop that explored the similarities and differences between the concept of food sovereignty and the right to food.*'

[870] Five NGO shadow-reports had been submitted, and a special session with the Committee on Economic, Social and Cultural Rights was held with NGOs (see: E/C.12/2010/SR.30, 8 November 2010).

[871] E/C.12/NLD/Q/4-5, 22 December 2009, Sections 25-27.

municipalities to improve the cooperation between the food banks and the local authorities.[872]

> 26. Please provide information on the extent of homelessness and malnutrition among asylum-seekers and immigrants without legal residence. Please also indicate whether such persons are entitled to benefits under the Work and Social Assistance Act (Wet Werk en Bijstand, Stb. 2003, 375).

The Netherlands replied that the provisions to which asylum-seekers are entitled to certain benefits in kind, such as *'money to cover the costs for food, or provision of food in kind, a clothing allowance and pocket money"* are regulated by the *'Benefits in kind for asylum-seekers and other categories of aliens scheme 2005 (RVA)'*.[873] It was added that *'the amount of money to asylum-seekers available for food is determined in accordance with the standards of the National Institute for Family Finance Information (Nibud).'*[874] Besides that, it was reported that under this Scheme, asylum-seekers are *'insured against the costs of medical care, and therefore able to have access to guidance on healthy eating.'* Furthermore, asylum-seekers were offered accommodation. *'They are not entitled to social assistance until they have obtained a residence permit, however.'* The Netherlands informed that the general Act on social benefits, the Work and Social Assistance Act,[875] would provide a guideline *'for the nature and scale of Provisions available under the RVA 2005 scheme.'* Asylum-seekers that reside illegally in the Netherlands *'are entitled to accommodation and other facilities pending repatriation in order to allow them to prepare for their departure. The deadline for return is 28 days, which can be extended by three months in order for them to effectuate their departure.'*[876]

> 28. Please provide detailed and updated information on the nutritional status, in particular in the Netherlands Antilles, of immigrants, homeless people, single-parent families, children, unemployed people, low-income earners, older persons, persons with disabilities, persons living in rural areas, refugees and asylum-seekers, and their ability to access adequate, affordable and appropriate food and water.

The Dutch reply consisted mainly of a summary of the average food consumption in the Netherlands. There was hardly any information on the requested categories of persons. With regard to the data concerning the Netherlands Antilles, the Netherlands reported that the Delegation would answer the question during the upcoming sessions with the Committee on Economic, Social and Cultural Rights.[877]

---

[872] E/C.12/NLD/Q/4-5/Add.1, 14 October 2010, Sections 148-150.
[873] In Dutch: Regeling verstrekkingen asielzoekers en andere categorieën vreemdelingen 2005.
[874] In Dutch: Nationaal Instituut voor Budgetvoorlichting.
[875] In Dutch: Wet Werk en Bijstand.
[876] E/C.12/NLD/Q/4-5/Add.1, 14 October 2010, Sections 151-153.
[877] E/C.12/NLD/Q/4-5/Add.1, 14 October 2010, Sections 156-161.

It appears thus, that the Netherlands do not get away with only reporting positive facts anymore. However, it can be questioned whether the Dutch Government is seriously willing to answer the questions of the Committee. Regarding question 26, it is remarkable that the Netherlands do not fully answer the question, since no information is given at all concerning homelessness and malnutrition among asylum-seekers and immigrants without legal residence. It must also be noted that the response to question 28 is hardly an answer to the question, for no specific data on the categories of persons that is requested is provided, and with regard to the Netherlands Antilles, no information is provided at all. Interestingly, the representatives of the former Netherlands Antilles did not provide the promised data during the sessions with the Committee, and matters regarding question 28 were not even discussed. Regarding the status of asylum-seekers however, the Committee asked several questions, which will be discussed in the next section.

9.2.3.4 The Dutch asylum policies in the ICESCR reports

In general, the Dutch procedures and policies concerning the residence of foreigners are discussed in view of Article 2 ICESCR, and furthermore more specifically in view of the substantive ICESCR rights. In the second periodic report, the Netherlands reported that everyone in the Netherlands enjoyed the same rights: '*This does not only apply to those who are born in the Netherlands and brought up in a family, but also to the increasing number of under-age unaccompanied asylum-seekers and to some 13,000 (...) who grow up in institutional care,* i.e. *a home or residential facility (...).*'[878] Furthermore, the Dutch procedures on housing legally residing asylum-seekers is explained,[879] some statistical data on asylum-seekers is provided,[880] and it is underlined that foreigners who were granted the status of asylum-seeker have access to compulsory healthcare.[881] During the sessions with the Committee however, the Dutch Delegation further explained some details on the Linkage Act, that linked the entitlements to social security to the residence status of a foreigner, and explained that '*the Linkage Act did not aim to withdraw minimum subsistence rights from all persons without valid residence status. The Act was based on three principles: firstly, that regular social security schemes should be open only to foreign nationals admitted to the Netherlands unconditionally; secondly, that special arrangements and reception facilities should be established to provide social security for foreign nationals admitted to the Netherlands provisionally, including those whose applications were still being processed; and, lastly, that social security arrangements for foreign nationals who faced expulsion, and were able to leave the country, should be limited to elementary provisions such as medical care, free legal aid and childre''s education.*'[882] In response to this explanation, one Committee Member asked

---

[878] E/1990/6/Add.11, 5 August 1996, Section 134.
[879] E/1990/6/Add.11, 5 August 1996, Sections 191-195.
[880] E/1990/6/Add.11, 5 August 1996, Section 200.
[881] E/1990/6/Add.11, 5 August 1996, Section 265.
[882] Mr Potman, E/C.12/1998/SR.13, 7 May 1998, Section 85.

whether there were special Provisions for cases with an urgent need for foreigners without a residence permit.[883] Similar questions were raised, in the context of Article 11 ICESCR, concerning the living conditions of asylum-seekers awaiting a residence permit,[884] or rejected asylum-seekers.[885] The Dutch Delegation only answered the latter question, by stating that *'the Government was required to house rejected asylum-seekers until their departure from the country.'*[886] The Committee however, in their Concluding Observations, expressed their concern regarding *'the living conditions of asylum-seekers in some reception centres in the country.'*[887]

In their third periodic report, the Netherlands explained once again the working of the Linkage Act in the context of the entitlements of foreigners residing in the Netherlands to social benefits: *'the basic rule is that foreigners are only eligible (under certain conditions) for social services if they are in the possession of a valid residence permit. Anyone who is awaiting a decision on his or her application for a permit has no entitlement, with the exception of asylum-seekers and a few other categories of foreigners. Without a residence permit, foreigners can only claim education, medically necessary care and legal aid.'*[888] Furthermore, the specific procedures regarding unaccompanied under-age asylum-seekers, in which the best interest of the child should prevail, was explained.[889] Also, the Netherlands reported on the procedures regarding the housing of asylum-seekers, in which they are divided into two categories: first, those awaiting a decision on their application for a residence permit, who were housed in reception centres and falling under the authority of the Government, and second, those who were granted a residence permit, who were housed in mainstream housings, falling under the authority of the communities.[890] In their List of Issues, the Committee, referring to the explanation in the report on the Linkage Act, asked to explain why *'asylum-seekers and a few other categories of foreigners are granted preferential treatment over other foreigners'*[891] The Netherlands replied that indeed illegally residing foreigners were excluded from any entitlements to social benefits, except for the fact that *'all foreign nationals, with or without legal residence status, have the right to services such as education, necessary medical assistance and legal aid.'*[892] Furthermore, the Netherlands somewhat cryptically explained that the Linkage Act however does not apply to some categories of foreigners awaiting the outcome of their asylum application, but instead have

---

[883] Mrs Jimenez Butragueño E/C.12/1998/SR.13, 7 May 1998, Section 87. See for a similar question, in the context of healthcare, Mr Ahmed, E/C.12/1998/SR.14, 15 September 1998, Section 70.
[884] Mr Pillay, E/C.12/1998/SR.14, 15 September 1998, Section 42.
[885] Mr Riedel, E/C.12/1998/SR.14, 15 September 1998, Section 43.
[886] Mr Potman, E/C.12/1998/SR.14, 15 September 1998, Section 46.
[887] E/C.12/1998/26, 27 April; 15 May 1998; 16 November; 4 December 1998; CESCR report on the eighteenth and nineteenth sessions, Section 184.
[888] E/1994/104/Add.30, 23 August 2005, Section 110.
[889] E/1994/104/Add.30, 23 August 2005, Sections 324-327.
[890] E/1994/104/Add.30, 23 August 2005, Sections 373-377.
[891] E/C.12/NLD/Q/3, 2 March 2006, Section 8.
[892] E/C.12/NLD/Q/3/Add.1, 23 August 2006, p. 5.

a legal residence status for the duration of the procedures. Other foreigners are also allowed to remain in the Netherlands awaiting their procedures, for instance for the purpose of family reunification, under specific circumstances. Therefore, *'These rules illustrate that this is not a matter of some categories of foreign nationals receiving preferential treatment with regard to benefits, but rather a question of who is allowed to remain in the Netherlands.'*[893] However, while it is clear that some categories of aliens do not fall under the scope of the Linkage Act and therefore may have certain entitlements to social benefits, it cannot be deduced from the Dutch answer whether those entitlements are equal to foreigners with a valid residence permit.[894] During the Sessions with the Committee, especially the status of illegally residing aliens was questioned, *inter alia* in view of entitlements to social benefits.[895] The Dutch Delegation further elaborated on the Dutch procedures and policies in that field, but this time, the Delegate was somewhat clearer about the entitlements to social benefits of illegally residing aliens and those who were legally residing while awaiting the outcome of a procedure: *'The Netherlands make a distinction between persons who are legally resident in the country and as such enjoy all the rights enshrined in the Covenant, and illegal immigrants, including asylum-seekers and those whose application has been rejected, who have more rudimental rights (access to basic healthcare, education and legal aid). It goes without saying that Dutch with a foreign origin and migrant workers who have a residence permit may claim and exercise the same rights as a native Dutch national.'*[896] It was asked by one Member of the Committee whether the access to basic healthcare of foreigners without a legal residence permit would include secondary and tertiary healthcare, including treatment for HIV/AIDS.[897] The Dutch Delegation replied that *'the law on health insurance only covers immigrants in an emergency and in special cases, for example when they are affected by HIV/AIDS.'*[898] Oddly enough, regardless the critical questions posed during the Sessions, no further recommendations on the matter were made in the Committee's Concluding Observations.[899]

---

[893] E/C.12/NLD/Q/3/Add.1, 23 August 2006, p. 5.

[894] Not to be confused with a legal residence status, for the temporary purpose of allowing a foreigner to legally reside in the Netherlands while awaiting a procedure.

[895] See especially the questions asked by Mr Riedel, Mrs Barahona Riera, E.C.12/2006/SR.34, 10 January 2007, Sections 11 and 18.

[896] Mr De Klerk, E.C.12/2006/SR.34, 10 January 2007, Section 4. Originally published in French: *'Les Pays Bas font une distinction entre les personnes qui résident légalement dans le pays et jouissent à ce titre de tous les droits consacrés par le Pacte et les personnes en situation irrégulière, y compris les demandeurs d'asile et ceux déboutés de leur demande, qui ont des droits plus rudimentaires (accès aux soins de santé de base, à l'éducation et à l'aide juridictionnelle). Il va sans dire que les Néerlandais d'origine étrangère et les travailleurs migrants au bénéfice d'un permis de séjour peuvent revendiquer et exercer les mêmes droits que les Néerlandais de souche.'* See for a similar reply: Mr. De Klerk, Section 28.

[897] Mr Riedel, E.C.12/2006/SR.34, 10 January 2007, Section 11.

[898] Mrs Nicolai, E.C.12/2006/SR.34, 10 January 2007, Section 31. Originally published in French: *la loi sur l'assurance maladie ne couvre les immigrés qu'en cas d'urgence et dans des cas particuliers, par exemple lorsqu'ils sont atteints par le VIH/sida.'*

[899] E/C.12/NLD/CO/3, 24 November 2006.

In their combined fourth and fifth periodic report, the Netherlands stated in almost identical wordings used in the fifth periodic report on the implementation of the CEDAW[900] that it had no intention to sign and ratify the Convention on the Protection of the Rights of All Migrant Workers and Members of their Families, as recommended by the Committee,[901] due to the (earlier mentioned) Linkage Act, in which entitlements to social security were connected to residence status, and therefore a distinction was made between foreigners with and without a legal residence status.[902] In the report, no further particularities on procedures and policies regarding foreigners in view of entitlements to social benefits were discussed. During the sessions with the Committee, the asylum procedures and policies were discussed extensively. The Linkage Act in relation to the entitlements of illegally residing aliens to social benefits was debated once more.[903] The tone of the Committee had clearly changed, and the Delegation was forced to explain and justify into more detail the Dutch rules: a Committee Member underlined that *'Undocumented migrants were not provided with the minimum essential level of housing and were deprived of shelter, clothing and food, despite being under the jurisdiction of the State Party. The State Party thus appeared to have forgotten its core obligations under the Covenant, and the Committee's general comments seemed to have been disregarded by the Courts. He asked whether undocumented migrants were provided with free essential medical care.'*[904]Another Member of the Committee *'wished to know whether it was true that thousands of undocumented immigrants were held in detention centres alongside asylum-seekers, including victims of human trafficking and torture, and whether unaccompanied minors were also held in detention centres. In addition, she requested statistics on the number of women, children and elderly people detained in such centres.*[905] The Dutch Delegation replied that *'the Central Agency for the Reception of Asylum-seekers provided housing, healthcare and work for asylum-seekers and that they received financial assistance until they were granted a residence permit, which gave them the same rights as all other lawful residents. Since 1998, the provision of social security benefits had been linked to residence status. However, undocumented children were enrolled in schools and migrants had access to legal assistance and essential medical treatment.*[906] With regard to asylum-seekers whose application was rejected, the Delegate replied that they would have to leave the Netherlands within 28 days. *'If that period was insufficient, they could then be housed in facilities, in which their freedom of movement was restricted, for another 12*

---

[900] CEDAW/C/NLD/5, 24 November 2008, pp. 31-32.

[901] E/C.12/NLD/CO/3, 24 November 2006, Section 18.

[902] E/C.12/NLD/4-5, 17 July 2009, Sections 8-9.

[903] On behalf of the Dutch delegation: Mr Beets, E/C.12/2010/SR.44, 25 March 2011, Section 9; On behalf of the Committee: Mr Kedzia, E/C.12/2010/SR.44, 25 March 2011, Sections 52 and 54.

[904] Mr Pillay, E/C.12/2010/SR.43, 18 November 2010, Section 27.

[905] Mrs Bonoan-Dandan, E/C.12/2010/SR.43, 18 November 2010, Section 30. See also Mr Texier, E/C.12/2010/SR.43, 18 November 2010, Section 25.

[906] Mr Beets, E/C.12/2010/SR.44, 25 March 2011, Section 9.

weeks'. However, as a result of case law,[907] *'the 12-week limit did not apply to families with children whose asylum applications had been denied so long as they cooperated in preparations for their return to their country of origin (...) Children placed in such facilities received schooling, as did unaccompanied minors due to leave the country who were temporarily housed in special facilities for juveniles.'* He added that *'Only people legally residing in the Netherlands were entitled to housing. Asylum-seekers received shelter during their application process and housing if their application for residence was approved.*[908] With regard to the detention of asylum-seekers, the Delegate responded that detention was only a last resort, and could last for a maximum of six months. He added that *'the Government endeavoured to find alternatives to detention, especially in the case of families with children and people who cooperated in the preparations for their departure. Unaccompanied minors could be detained only in special juvenile detention centres.'*[909] The Committee however seems to have been rather sceptical about the Delegates' replies, for in its Concluding Observations, the Committee expressed that it is *'deeply concerned that asylum-seekers and unaccompanied minors in the Netherlands have been held in detention for long periods of time. The Committee also regrets that undocumented migrants, including families with children, are not entitled to a basic right to shelter and are rendered homeless after their eviction from reception centres. The Committee is also concerned that, although undocumented migrants are entitled to healthcare and education, in practice they cannot always have access to either.'* Therefore, *'the Committee urges the State Party to: (a) ensure that the legislation in the Netherlands guarantees that asylum-seekers are detained only when it is absolutely necessary and that the time which rejected asylum-seekers and irregular migrants spend in detention is limited to a strict minimum; and (b) meet its core obligations under the Covenant and ensure that the minimum essential level relating to the right to housing, health and education is respected, protected and fulfilled in relation to undocumented migrants.'*[910]

9.2.4 Direct applicability of the ICESCR rights in the ICESCR reports

9.2.4.1 The second periodic report

The matter of direct applicability was frequently discussed during the reporting procedures. During the sessions on the second periodic report, the Dutch Delegation underlined that *'the Netherlands Government was deeply committed to the object and purpose of the Covenant and to the Universal Declaration of Human Rights, and shared the belief that economic, social and cultural rights should have the same status as civil*

---

[907] Most likely, Mr Beets referred to the Court of Appeal of 's-Gravenhage, 11 January 2011, JV 2011, 91, which was later in cassation confirmed by the Supreme Court, 21 September 2012, NJ 2013, 22. The case law is also further discussed in Section 4.6, and involves the use of Article 27 ICRC as an interpretative norm.

[908] Mr Beets, E/C.12/2010/SR.45, 24 November 2010, Sections 24-25.

[909] Mr Beets, E/C.12/2010/SR.43, 18 November 2010, Section 37.

[910] E/C.12/NDL/CO/4-5, 19 November 2010, Section 25.

and political rights. Although the obligations of a Government with regard to civil and political rights were clear, however, they were less so with regard to economic, social and cultural rights. ICJ had moreover argued that the Covenant was not directly applied under the Dutch legal order. Indeed it was not. Article 93 of the Dutch Constitution established the possibility for international treaties to be directly applied. In case of the Covenant, the Government had expressly chosen not to invoke that Provision, on the grounds that many of the Covenant's clauses represented obligations and commitments calling for Government action that went beyond mere Government guarantees. In the view of the Netherlands, Government action in a democratic State should be based on choices that arose from the political will of all. (...) The way in which those rights were implemented was political rather than judicial, and therefore dynamic rather than static. That meant that although rights were fully recognised, the way in which they were implemented could change over time. Emphatically, it did not mean that the rights enshrined in the Covenant carried less importance than did those which could be invoked before Courts; they were simply implemented differently, and their implementation called for an active Government.'[911] This reasoning was critically received by the Committee. The question was raised whether there were *'any Court decisions which invoked the Covenant.'*[912] Also, the question was raised whether there was *'any reason why the Covenant should not be treated in the same way as other international legal instruments? A number of Dutch jurists have argued that the Covenant established different sorts of obligations, and to simply state that they were non-self-executing did not suffice.'*[913] According to the Committee, referring to General Comment 3,[914] at least seven rights enshrined in the ICESCR were directly applicable.[915] It was asked whether it was true, as suggested by the Dutch Section of the International Commission of Jurists, *'that the Covenant was only marginally, if at all, taken into consideration in the legislative and policy-making process at the national level.'* How then, it was asked, could the Netherlands Government *'be said to be complying with its obligations under the Covenant.'*[916] Also, the Committee asked *'for elucidation of the statement just made by the head of the Delegation to the effect that the precise obligations for Governments in relation to economic, social and cultural rights were "less clear" than those relating to civil and political rights.'*[917] Furthermore, it was asked how the different attitude towards ECOSOC rights *'could be regarded as compatible with the United Nations concept of the indivisibility of all human rights.'*[918] The Dutch Delegate replied to the above that *'his Government was fully committed to the concept of indivisibility of all human rights and accorded them the same status*

---

[911] Mr Potman, E/C.12/1998/SR.13, 7 May 1998, Sections 10-11.

[912] Mr Riedel, E/C.12/1998/SR.13, 7 May 1998, Section 17.

[913] Mr Riedel, E/C.12/1998/SR.13, 7 May 1998, Section 17.

[914] E/1991/23, annex III at 86 (1991), 14 December 1990, CESCR, *General Comment 3, The Nature of States Parties' Obligations.*

[915] Mr. Riedel, E/C.12/1998/SR.13, 7 May 1998, Section 18.

[916] Mr. Pillay, E/C.12/1998/SR.13, 7 May 1998, Section 21.

[917] Mr. Pillay, E/C.12/1998/SR.13, 7 May 1998, Section 21.

[918] Mr. Antanovich, E/C.12/1998/SR.13, 7 May 1998, Section 23.

in terms of their implementation. The fact that, for essentially technical reasons, it did not think that economic, social and cultural rights had direct applicability certainly did not mean that it attached less importance to those rights.'[919] And in addition that: *'(...) unlike civil and political rights, which were essentially rights-oriented, economic, social and cultural rights were principally oriented towards the obligations of the State. The Maastricht guidelines (...) allowed a certain margin of discretion with regard to the implementation of the latter category of rights. In his country's system, the question of the effectiveness of steps taken to discharge an obligation was considered a matter for Parliament rather than for the Judiciary. Of course, if the State failed to discharge its obligation altogether, an individual who believed that his or her human rights had been violated as a result could bring the matter before the Courts. That did not mean, however, that all the Provisions of the Covenant could be regarded as self-executing.'*[920] In response, a Member of Committee asked whether in the Netherlands it was possible, similar to other Western European countries, for the Courts to *'pass judgment in principle on the Government's compliance with an instrument to which it was party, but the question of precisely how that obligation was discharged was left to the relevant Government department.'*[921] Apparently, the Member of Committee referred to a type of reviewing that is known in especially Dutch Administrative Law as a 'marginal review'[922], in which an Administrative Court judges whether an administrative body could reasonably had reached a certain decision, taken into account all justified interests, without judging the content of that decision. The Dutch Delegate responded that *'(...) his Government recognised the direct applicability of many economic, social and cultural rights, which, in addition to being set forth in the Constitution, were also protected by various international instruments to which the Netherlands were a party. The difficulty arose where a caring obligation was imposed upon the Government. The right to health was a case in point. An individual could hardly complain about a violation of human rights simply because he or she was not enjoying good health. A violation certainly existed if the Government provided no healthcare at all, but where the system was not working very efficiently the matter became political rather than judicial.'*[923] Eventually, the Committee seems to have disagreed with the Dutch Delegation, and stated in its Concluding Observations that *'...in line with its General Comment 3, the Committee considers that, at a minimum, certain Provisions of the Covenant are potentially able to be directly applied both in law and in policy. It therefore cannot accept the assertion by the representative of the State Party that "for essentially technical reasons" the Covenant...is not directly applicable.'*[924] It is hard not to notice a slight sarcastic undertone in the last sentence, where the Dutch Delegate is quoted. The Committee is quite clear when it considers that certain Provisions of the ICESCR are potentially directly applicable, and is not

---

[919] Mr. Potman, E/C.12/1998/SR.13, 7 May 1998, Section 24.
[920] Mr. Van Rijssen, E/C.12/1998/SR.13, 7 May 1998, Section 25.
[921] Mr. Riedel, E/C.12/1998/SR.13, 7 May 1998, Section 27.
[922] In Dutch: 'marginale toetsing.'
[923] Mr Van Rijssen, E/C.12/1998/SR.13, 7 May 1998, Section 28.
[924] E/C.12/1/Add.25, 16 June 1998, Section 11.

willing to accept the arguments of the Dutch Delegation, especially when direct applicability is portrayed to be impossible mainly due to technical, instead of substantive reasons, a reasoning that almost seems to marginalise the fact that ICESCR Provisions are not directly applicable in the Netherlands.

9.2.4.2 The third periodic report

In their third periodic report, the Netherlands responded to the criticism of the Committee, expressed in its Concluding Observations on the second periodic report, on the fact that ICESCR Provisions were not directly applicable. Firstly, a translation was given of Articles 93 and 94.[925] According to the Government, '*this conveys the fact that some international legal standards have direct application and that a directly applicable international standard has priority over national law.*'[926] It is our understanding that the first halve of this sentence refers to the function of Article 93 CA, and the latter half to the function of Article 94 CA. Furthermore, it was reported that in Case Law, criteria had been established to determine whether a treaty Provision is directly applicable: '*in judging whether a treaty Provision can be directly applicable in the national legal order, consideration is given to the nature and content of the Provision, as well as to its wording. If a particular Provision is directed at citizens and gives rise to a claim without any further implementation laws being required, it can be directly applied in the Dutch legal order. If the Provision in question is directed at the State, and includes phrases such as 'bear responsibility for' 'take steps' or 'guarantee', citizens cannot then, generally speaking, directly base any claim on such a Provision and hence the said Provision cannot be directly applicable to the national legal order. A Provision directed at the State can only have direct application if its nature and content clearly allow citizens to base a claim thereon. One example is a State-directed ban on discrimination, on which a citizen can base a claim for equal treatment.*'[927] The Government stated that in this way, the direct effect of '*eligible treaty Provisions is sufficiently guaranteed.*'[928] However, with regard to the ICESCR, it was reported that '*The nature and content of the Covenant, as well as the wording of the Articles, indicate that it is aimed at the gradual and increasing achievement of objectives by means of legislation and further implementation measures. As a result, most Provisions cannot be applied directly. All the more because, where further implementation laws are required, this implies a certain freedom of choice for the national Legislature regarding the way in which the rights to be guaranteed are given substance. Simply accepting direct application of Provisions that need to be worked out in greater detail would mean that it would be left to the national Courts to put the*

---

[925] Article 93 CA: '*Provisions of treaties and of resolutions by international institutions which may be binding on all persons by virtue of their contents shall become binding after they have been published.*' Article 94 CA: '*Statutory regulations in force within the Kingdom shall not be applicable if such application is in conflict with Provisions of treaties or that are binding on all persons or of resolutions by international institutions.*'
[926] E/1994/104/Add.30, 23 August 2005, Section 5.
[927] E/1994/104/Add.30, 23 August 2005, Section 6.
[928] E/1994/104/Add.30, 23 August 2005, Section 7.

objectives set down in the Covenant into practice. As already indicated in the second report, the Dutch Government is set on implementing the obligations imposed on the Government under the Covenant within a democratic process. This will enable a better response to social developments.'[929] Not surprisingly, this explanation was not sufficient to the Committee, which asked in its List of Issues: *'please, explain the domestic legal status of the Covenant and whether its Provisions are considered by the State Party to be 'specific and precise enough' to be directly applicable.'*[930] The written reply was that *'The International Covenant on Economic, Social and Cultural Rights contains international obligations with which national legislation should not conflict. In the Government's opinion, all of the obligations under the Covenant are adequately incorporated in national legislation. Nevertheless, citizens who believe their rights under the Covenant are being infringed by Dutch law may invoke the relevant Provisions of the Covenant in Court insofar as the substance of the Provisions lends itself to direct application (Article 94 of the Constitution). It is then up to the Court to determine whether the latter condition is fulfilled and, if so, whether the Provision has been violated.'*[931] The Dutch Delegation, during the sessions with the Committee on Economic, Social and Cultural Rights, even held that it was *'the responsibility of the Courts to decide whether a Provision of the Covenant should be directly applicable in a given case, with case law providing guidance in determining the issue of direct effect.'*[932] Regarding this statement, a Committee Member asked what then the procedure was of the Courts to determine whether a Provision would have direct effect. Also, the question was raised *'whether economic, social and cultural rights were given as much importance as civil and political rights.'*[933] Another Committee Member asked for information regarding case law in which the Court considered an ICESCR Provision not suitable for direct application.[934] To this, the Dutch Delegation responded that *'under the Constitution, any directly applicable rule of international law took precedence over domestic law, even the Constitution. However, in the case of economic, social and cultural rights, it was for the individual Courts to decide whether a particular Provision was specific and precise enough for an individual to invoke. The Supreme Court had, in a number of decisions, referred to the statement by the Government, on ratification of the Covenant, that various Provisions would not be considered directly applicable by individuals. Nonetheless, in some of those cases the Supreme Court or the highest Administrative Court had de facto taken account of the substance of the Provisions, such as Articles 9, 11, and 15 when drafting its decisions. However, the domestic Courts had ruled that, generally speaking, an individual could not invoke Articles 2 (2), 6, 7, 8, 12, and 13 in order to claim specific rights. It was difficult to draw general conclusions on the status of case law, as only a few dozen*

---

[929] E/1994/104/Add.30, 23 August 2005, Section 8.
[930] E/C.12/NLD/Q/3, 2 March 2006, Section 1.
[931] E/C.12/NLD/Q/3/Add.1, 23 August 2006, Section 1.
[932] Mr De Klerk, E/C.12/2006/SR.33, 22 November 2006, Section 4
[933] Mr Malinverni, E/C.12/2006/SR.33, 22 November 2006, Section 13.
[934] Mr Atangana, E/C.12/2006/SR.33, 22 November 2006, Section 15.

cases had been brought before the highest Courts between 1979 and 2000.'[935] It is remarkable here that the Delegation made reference to *inter alia* Article 11 ICESCR. In Chapter 4, it is demonstrated that the Provision has no effect at all on legal proceedings, and our conclusions in that Chapter do not match with the alleged 'indirect' effect as suggested by the Delegate, in which a Court takes the content of the Provision into account. In this light, a Committee Member, asked what exactly '*the State Party's current position might be on direct applicability and thus on the proposed Optional Protocol, and he would welcome a clear statement on that position.*'[936] The Committee Member obviously could not match the fact that the Netherlands was actively involved in the negotiation process preceding the adoption of an Optional Protocol to the ICESCR containing an individual complaint procedure, whereas the country was simultaneously rejecting direct effect of the ICESCR Provisions. The Dutch Delegate considered that '*The Government's position with regard to the proposed Optional Protocol on an individual complaints mechanism had been rather cautious, although it fully supported the expansion of the mandate of the open-ended working group on that instrument. The Government considered the implementation of rights under the Covenant to be essentially a political question and, indeed, several of those rights were currently under discussion in the political arena prior to the upcoming elections. Although the Government wished to be involved in the negotiations on the proposed mechanism, the position it had adopted when ratifying the Covenant, namely that it considered the rights contained in the Covenant not to be of direct applicability, had not changed.*'[937] However, the Delegate had to acknowledge that '*there was a contradiction in acknowledging the indivisibility of human rights while at the same time having different implementation mechanisms for certain rights.*'[938] Again, the Committee had to conclude that the Dutch replies were insufficient, and stated in its Concluding Observations, this time in somewhat firmer wordings, that '*The Committee is concerned that the Courts in the State Party applies the Provisions of the Covenant only to the extent that they consider that these are directly applicable and that most Provisions of the Covenant cannot be applied directly.*'[939] In addition, the Committee recommended '*that the State Party reassess the extent to which the Provisions of the Covenant might be considered to be directly applicable. It urges the State Party to ensure that the Provisions of the Covenant are given direct effect by its domestic Courts, as defined in the Committee's General Comment 3, and that it promotes the use of the Covenant as a domestic source of law. It invites the State Party to include, in its fourth periodic report, information on case law concerning the rights recognised in the Covenant.*'[940]

---

[935] Mr Kuijer, E/C.12/2006/SR.33, 22 November 2006, Section 51.
[936] Mr Riedel, E/C.12/2006/SR.33, 22 November 2006, Section 26.
[937] Mr De Klerk, E/C.12/2006/SR.33, 22 November 2006, Section 47.
[938] Mr De Klerk, E/C.12/2006/SR.33, 22 November 2006, Section 47.
[939] E/C.12/NLD/CO/3, 24 November 2006, Section 11.
[940] E/C.12/NLD/CO/3, 24 November 2006, Section 19.

9.2.4.3 The combined fourth and fifth report

In its combined fourth and fifth report, the Government indeed refers to the matter of direct applicability with the following short phrase: *'The Netherlands would refer to previous reports for information on the Netherlands' position on the direct applicability of the Covenant. Information on case law concerning the rights recognised in the Covenant is attached to this report.'*[941]

It is obvious that the Dutch Government did not reassess its viewpoints on the issue of direct applicability of ICESCR Provisions, and was not attempting to further clarify this. The Committee on Economic, Social and Cultural Rights naturally considered the short explanation in the report inadequate, and asked in its List of Issues: *'Please indicate the ways in which the domestic legal system provides an effective remedy to persons whose rights under the Covenant have been violated.'*[942] And: *'While noting the State Party's reply in its report (E/C.12/NLD/4-5, Section 11), please indicate whether, in accordance with the Committee's recommendations issued in 2006, there have been any recent developments to ensure that the Provisions of the Covenant are given effect by domestic Courts.'*[943] In their written replies, the Government answered that *'In principle, any claimant is entitled to invoke a Provision of the Covenant in legal proceedings (e.g. an action for tort brought against the State before a civil Court under Article 6:162 of the Criminal Code).'*[944] According to the Government, this was possible due to Constitutional system stipulated in Articles 93 and 94 CA. Again, the function of both Articles was clarified: Article 93 CA stipulates that treaty Provisions that are binding on all persons by virtue of their content are *'binding under national law.'* In addition, Article 94 stipulates that national Provisions that are in conflict with treaty Provisions that are binding

---

[941] E/C.12/NLD/4-5, 17 July 2009, Section 11. In the first Appendix, reference is made to 31 cases in the period 1979-2007 in which ICESCR Articles were invoked. Indeed, in 24 of these cases, the direct applicability of the invoked Provisions is denied by the various Courts. In 5 cases, all rulings before 2000, some kind of non-direct effect of the Provision was considered by the Courts, which includes a method that will later, by the Dutch delegation, be referred to as the 'incorporation doctrine', in which the ICESCR treaty, although not directly applied, plays an interpretative role, sometimes in conjunction with another Provision (for instance Article 26 ICCPR, or Article 2 ICESCR). It is remarkable however, that in all cases the invoked Provision was considered not to be violated (including Central Appeals Court for Public Service and Social Security Matters, 31 March 1995, JB 1995, 161; 22 April 1997, JB 1997, 158). In only one of these five cases, an (indirect) violation of the ICESCR Article (in this case Article 7) can be found. In two incidental cases, the Court had considered an Article directly applicable, and ruled that the invoked Article was violated. One case concerned Article 7 ICESCR, and dates from 1984. The second concerned Article 11 ICESCR, and was ruled in 1979. However, from the same appendix it can be concluded that the Judiciary generally altered its approach to more permanent case law, in which direct effect of both Articles is denied (although Article 7 ICESCR is later used in conjunction with Article 26 ICCPR in several rulings).
[942] E/C.12/NLD/Q/4-5, 22 December 2009, Section 2.
[943] E/C.12/NLD/Q/4-5, 22 December 2009, Section 3.
[944] E/C.12/NLD/Q/4-5/Add.1, 14 October 2010, Section 9. Note that the Government is most likely mistaken, for Article 6:162 is a Provision of the Civil Code, instead of the Criminal Code. In this Article, a wrongful Act, or tort law, is embedded.

on all persons shall not be applied.[945] However, the Government admitted that *'it is indeed the case that Provisions in the Covenant are not generally regarded by the national Courts as 'binding on all persons'.*[946] On the other hand, the Government stressed that *'the question of whether the State is bound by the Covenant should be distinguished from the question of whether Provisions of the Covenant have direct effect within the State. The Netherlands are bound by international law to realise the rights set forth in the Covenant for persons within its jurisdiction. However, the question of whether a Provision has direct effect is ultimately determined in individual cases by the Dutch Courts.'*[947] Again, it appears that the Government attributes the responsibility for the direct applicability of ICESCR Provisions to the Courts. The Government argued furthermore *'that the influence of the Covenant's Provisions is not limited to those cases where the national Courts have declared a Provision "binding on all persons".'* The Government clarified that Courts might give an interpretation to national Provisions so that it is in compliance with an international Provision. The Government stated that *'It is important to note that the Dutch Courts adhere to the "incorporation doctrine", whereby the interpretation of a Provision of the Covenant given by the supervisory committee is "read into" the Provision.'*[948] Reference was made to a Case in which the Supreme Court ruled that although Article 7 ICESCR had no direct effect, *'the guarantee of equal pay for equal work in the Covenant is a goal that should be worked towards. To assume too readily that a reasonable and objective justification exists for a pay difference that is at odds with the principle of equal pay for equal work would not be consistent with this view.'*[949] Again, during the session with the Committee on Economic, Social and Cultural Rights, the fact that ICESCR rights were not directly applicable was heavily and extensively criticised by various Committee Members.[950] Also in the NGO shadow-reports, the matter was frequently stressed.[951] The Dutch Delegation once more referred to the fact that the Courts were responsible for the direct application, and that the Government could not interfere in that process.[952] Nevertheless, the Delegation appears to have attempted to present the facts somewhat more positive on the matter, by informing that *'a Constitutional review commission was about to publish a report on a possible amendment to enhance the direct applicability of international*

---

[945] E/C.12/NLD/Q/4-5/Add.1, 14 October 2010, Section 9.

[946] E/C.12/NLD/Q/4-5/Add.1, 14 October 2010, Section 10.

[947] E/C.12/NLD/Q/4-5/Add.1, 14 October 2010, Section 11.

[948] E/C.12/NLD/Q/4-5/Add.1, 14 October 2010, Section 12.

[949] E/C.12/NLD/Q/4-5/Add.1, 14 October 2010, Section 12. Reference was made to the Supreme Court, 7 May 1993, NJ 1995, 259.

[950] Mr Abashidze, Section 19; Mr Kedzia, Section 20; Mr Atangana, Section 23; Mr Pillay, Section 27; Mr Riedel, Section 28, in: E/C.12/2010/SR.43, 18 November 2010.

[951] Joint Parallel Report to the Combined Fourth and Fifth Periodic Report of the Netherlands on the International Covenant on Economic, Social and Cultural Rights, 28 October 2009, Section 2.1; Addendum to the Joint Parallel Report to the Combined Fourth and Fifth Periodic Report of the Netherlands on the International Covenant on Economic, Social and Cultural Rights (ICESCR) (28 October 2009), 16 September 2010, p. 24; Shadow report on the right to food in the Netherlands, 17 November 2009, European Institute for Food Law.

[952] Mr Beets and Mr Versluis, E/C.12/2010/SR.43, 18 November 2010, Sections 33-34.

human rights instruments. Those instruments and their applicability were included in the training of future judges. Pursuant to Article 2 of the Covenant, the country actively supported the work of the United Nations to codify the right to drinking water and sanitation and promoted other economic, social and cultural rights such as the right to food.'[953] While the latter sentence has nothing to do with the matter of direct applicability, it seems to be intended to place the Dutch attitude towards the ICESCR in a better daylight. Even more interesting was another remark of a Dutch Delegate Member: *'under the Constitution, universally binding Provisions were directly applicable. He then pointed out the criteria for application of international instruments established by case law. One such criterion was the intention of the authors; with the Covenant dating back to 1966, judges tended to interpret some of their Provisions according to modern standards. Moreover, because judges were often required to apply the Provisions of international human rights instruments, they were well versed in them and were equally familiar with general comments, which they also applied.'*[954] As observed in Chapter 4, the latter is simply not true.[955] How credible the information given by the Delegation may or may not be, a Committee Member *'deemed that answer unsatisfactory because, for the Committee, all the Provisions of the Covenant were directly applicable in their most stringent interpretation.'*[956] Therefore, in their Concluding Observations, in firmer wordings than before, the Committee considered that *'Given the fact that the State Party follows a monist system whereby international treaties are directly applicable, the Committee reiterates its concern that some Provisions of the Covenant are not self-executing and enforceable in the State Party and that they have not been admitted by Courts to substantiate legal claims relating to economic, social and cultural rights.'*[957] Again, *'The Committee reiterates its recommendation that the State Party has the obligation to give effect to the rights contained in the Covenant in each territory so that individuals can seek enforcement of their rights before national Courts and tribunals. Moreover, in view of the numerous decisions of the State Party's Courts to the effect that the Provisions of the Covenant are not self-executing and thus not binding in accordance with Articles 93 and 94 of the Constitution of the State Party, the Committee urges the State Party to consider all remedial measures, legislative or otherwise, to ensure that the Covenant rights are applicable and justiciable in all its constituent countries. In this regard, the Committee refers the State Party to its General Comment No. 9 on the domestic application of the Covenant (1998). The Committee also requests the State Party to continue to provide in its periodic reports detailed information on case-law from all its territories regarding the implementation of the Covenant.'*[958]

---

[953] Mr Versluis, E/C.12/2010/SR.44, 25 March 2011, Section 4.

[954] Mr Versluis, E/C.12/2010/SR.44, 25 March 2011, Section 11.

[955] This was also observed in the Joint Parallel Report to the Combined Fourth and Fifth Periodic Report of the Netherlands on the International Covenant on Economic, Social and Cultural Rights, 28 October 2009, Section 2.1.

[956] Mr Marchan Romero, E/C.12/2010/SR.44, 25 March 2011, Section 12.

[957] E/C.12/NDL/CO/4-5, 19 November 2010, Section 6.

[958] E/C.12/NDL/CO/4-5, 19 November 2010. Section 6.

9.2.4.4 The reports of the Netherlands Antilles

The matter of direct applicability was not only discussed in the context of the Netherlands reports, but also, albeit less detailed, in the context of the reports of the Netherlands Antilles. In its Concluding Observations to the second periodic report of the Netherlands Antilles (and a similar Observation was made regarding Aruba[959]), the Committee regretted that no ICESCR Provision has direct effect in the Member State, which is in violation of General Comment 3.[960] In their third report, the Netherlands Antilles reported in response that *'the Government will take the issue of direct applicability into consideration.'*[961] The Committee, during the sessions with the Netherlands Antilles Delegation, therefore stressed *'that the question of direct applicability of the Covenant's Provisions has been raised by the Committee in 1996 and that a decade later the Government was saying only that it would take the issue into consideration.'*[962] The Delegation replied that the Netherlands Antilles Government would *'consider the issue of direct applicability of the Provisions of the Covenant and would include in its fourth periodic report information on case law concerning rights recognised by the Covenant.'*[963] Again, in its Concluding Observations, the Committee stated that it regretted *'that little progress has been made with regard to the direct applicability of the rights set out in the Covenant. It takes note of the information that these rights may only be applied directly by the national Courts to the extent that the Courts deem such application possible. The Committee is concerned that the State Party thus considers that most economic, social and cultural rights are not directly applicable.'*[964] However, in the next report, the promise of the Delegation was not fulfilled, for it was reported that: *'as far as could be ascertained, there have been no new developments on the issue.'*[965]

9.3 The ICRC reports

9.3.1 The right to food and the ICRC reports

Before 2013, the Netherlands had submitted three,[966] the Netherlands Antilles two,[967] and Aruba also two[968] reports on the implementation of the International Convention on the Rights of the Child. During the reporting procedures, the

---

[959] E/C.12/1998/26, 27 April; 15 May 1998; 16 November; 4 December 1998; *CESCR report on the eighteenth and nineteenth sessions*, Section 200.
[960] E/C.12/1/Add.25, 16 June 1998, Section 49.
[961] E/C.12/ANT/3, 13 February 2006, Section 2.
[962] Mrs Atangana, E/C.12/2007/SR.9, 10 May 2007, Section 15. See also in this respect the related questions of Mr Ridel (Section 10) and Mr Pillay (Section 11).
[963] Mrs Ray, E/C.12/2007/SR.9, 10 May 2007, Section 19.
[964] E/C.12/NLD/CO/3/Add.1, 31 January 2008, Section 9. See in this light also Section 23.
[965] E/C.12/NLD/4/Add.1, 13 July 2009, Section 4.
[966] CRC/C/51/Add.1, 24 July 1997; CRC/C/117/Add.1, 5 June 2003; CRC/C/NLD/3, 23 July 2008.
[967] CRC/C/61/Add.4, 4 October 2001; CRC/C/NLD/3, 23 July 2008 (Part 3).
[968] CRC/C/117/Add.2, 17 June 2003; CRC/C/NLD/3, 23 July 2008 (Part 2).

right to food was not only discussed in the context of Article 27 ICRC, but also in the context of Article 24 (health and nutrition). As discussed in Chapter 6, the Netherlands assume a very close interrelationship between Articles 27 and 26 ICRC. Therefore, when relevant, debates on Article 26 ICRC will also be included in the analysis below, especially when it concerns the reservation to the Provision. Due to the format adopted by the Committee on the Rights of the Child, the Dutch asylum procedure, as earlier demonstrated closely related to the right to food, is discussed as a separate issue not necessarily linked to a particular Convention Provision. Thus, below, the debates on the Dutch asylum policy will also be dealt with separately.

9.3.1.1 Article 24 ICRC

In view of Article 24 ICRC, the reports frequently address the issue of health and nutrition. Both the Netherlands[969] and Aruba[970] addressed the issue of excessive fat consumption and other unhealthy eating habits amongst children, and the kind of measures taken by the authorities to improve the situation, in their initial report. Another prominent topic was the provision of data on breastfeeding,[971] which was occasionally reason for debate, for especially regarding the Netherlands the Committee criticised the fact that the percentage of women that choose to give breastfeeding to their infants was lower compared to other European countries.[972] The Netherlands furthermore reported on several initiatives related to health and nutrition in the context of cooperation with other countries,[973] including the promotion of the right to drinking water and sanitation.[974]

9.3.1.2 Article 26 ICRC

Article 26 is, based on the format as adopted by the Committee on the Rights of the Child,[975] mainly discussed in view of *'disability, basic health and welfare'*, and more in particular in the context of *'social security and childcare services and facilities'*, altogether with Article 18 Section 3. Despite the frequent insistence of the Committee on the Rights of the Child to withdraw the Dutch reservation

---

[969] CRC/C/51/Add.1, 24 July 1997, Sections 215-216.
[970] CRC/C/117/Add.2, 17 June 2003, Sections 190-191; CRC/C/NLD/3, 23 July 2008 (Part 2), Sections 196-198.
[971] CRC/C/51/Add.1, 24 July 1997, Sections 217; CRC/C/NLD/3, 23 July 2008, Sections 191-193 (the Netherlands); CRC/C/61/Add.4, 4 October 2001, Sections 183-184; CRC/C/NLD/3, 23 July 2008 (Part 3), Section 286-287, 319-322 (the Netherlands Antilles); CRC/C/117/Add.2, 17 June 2003, Section 188 (Aruba).
[972] CRC/C/SR.580, 18 November 1999, Section 43; CRC/C/SR.1377, 21 January 2009, Section 16; CRC/C/NLD/CO/3, 27 March 2009, Section 59-60.
[973] CRC/C/117/Add.1, 5 June 2003, Sections 131-142; CRC/C/NLD/3, 23 July 2008, Sections 224-227.
[974] CRC/C/117/Add.1, 5 June 2003, Section 142.
[975] CRC/C/58/Rev.2, 23 November 2010, in particular Section III B and Annex.

to Article 26,[976] the Dutch Government consequently persisted in maintaining the reservation, for '*in the opinion of the Netherlands, the Article implies that States Parties should grant social security rights to a child itself since the Article talks of recognition of the rights of a child to benefit and of applications for them made by or on behalf of the child.*'[977] In their reports the Netherlands explained that, apart from a few minor exceptions, all entitlements to social security of the child are guaranteed via the parents, meaning that the child has no direct entitlements to social benefits.[978] In their initial report, it was stated that '*the amount of social security to the parents is fixed in such a way that the obligation of the parents towards the child in terms of care and maintenance can be paid from the benefits. Independent rights of the child to social security exist in the Netherlands only to a limited extent, and there are no proposals of changing this in the future.*'[979] The major part of the reports however, naturally dealt with the Dutch social security system related to the position of children.[980]

9.3.1.3 Article 27 ICRC

Article 27 is, based on the format as adopted by the Committee on the Rights of the Child,[981] discussed in two different contexts. Firstly, in view of the '*family*

---

[976] CRC/C/90, 7 December 1999, Section 38; CRC/C/NLD/CO/3, 27 March 2009, Sections 10-11 (the latter also concerns the Netherlands Antilles and Aruba). See also: CRC/C/118, 3 September 2002, Section 534 (the Netherlands Antilles), and: CRC/C/15/Add.226, 26 February 2004, Sections 10-11 (Aruba). See furthermore the Committee's lists of issues: CRC/C/Q/NET.1, 30 June 1999, Section 1; CRC/C/Q/NET-ANT/1, 8 February 2002, Section B1 (the Netherlands Antilles); CRC/C/Q/NLD/2, 27 October 2003, Section B2 (the Netherlands and Aruba). See also during the sessions with the Committee and the Dutch delegation: Mrs Sardenberg, Mrs Karp, Mrs Tigerstedt-Tähtelä and Mrs Mboi, CRC/C/SR.578, 16 November 1999, Sections 17-18, 25 and 47-48; Mrs Sardenberg, CRC/C/SR.580, 18 November 1999, Section 61; Mr Kotrane and Mr Citarella, CRC/C/SR.928, 23 January 2004, Sections 9 and 16; Mr Citarella and Mrs Smith, CRC/C/SR.1376, 20 January 2010, Section 22 and 37; Mrs Herczog, CRC/C/SR.1377, 23 January 2009, Section 93.

[977] CRC/C/51/Add.1, 24 July 1997, Section 223.

[978] CRC/C/51/Add.1, 24 July 1997, Section 223; CRC/C/117/Add.1, 5 June 2003, Section 143; CRC/C/RESP/48, 16 December 2003, Section B2; CRC/15/Add.227, 30 January 2004, Section 10; CRC/C/NLD/3, 23 July 2008, Section 243; Mr Halff, CRC/C/SR.579, 8 October 1999, Section 7; Mrs Blom, CRC/C/SR.928, 23 January 2004, Section 29.

Mr Rouvoet, CRC/C/SR.1376, 20 January 2010, Section 41. See also: CRC/C/61/Add.4, 4 October 2001, Sections 194-195; CRC/C/NLD/3, 23 July 2008 (Part 3), Sections 57-58 (the Netherlands Antilles). The reservation to Article 26 ICRC was also discussed during the first cycle of the Universal Periodic Review. Russia recommended withdrawing the reservation (A/HRC/8/31, 13 May 2008, Section 30). The Netherlands replied that '*The Kingdom of the Netherlands will not withdraw its reservation with regard to Articles 26, 37 sub-section c, and 40 of the Convention on the Rights of the Child. The reasons for these reservations remain relevant and the Kingdom can therefore not support this recommendation*' (A/HRC/8/31/Add.1, 25 August 2008, Section 14).

[979] CRC/C/51/Add.1, 24 July 1997, Section 223.

[980] See: CRC/C/51/Add.1, 24 July 1997, Sections 224-227; CRC/C/117/Add.1, 5 June 2003, Sections 144-147. CRC/C/NLD/3, 23 July 2008, Sections 234-242. See also: CRC/C/61/Add.4, 4 October 2001, Sections 196-197 (Netherlands Antilles). CRC/C/117/Add.2, 17 June 2003, Section 105-108 (Aruba).

[981] CRC/C/58/Rev.2, 23 November 2010, in particular Section III B and Annex.

environment and alternative care', more in particular in the context of *'the recovery of maintenance for the child'* (Article 27 (4)). Secondly, in view of *'disability, basic health and welfare',* more in particular in the context of the *'standard of living and measures, including material assistance and support programmes with regard to nutrition, clothing and housing, to ensure the child's physical, mental, spiritual, moral and social development and reduce poverty and inequality'* (Article 27, Sections 1-3). In the initial report of the Netherlands, the primary responsibility of the parents towards their children is emphasised: *'Article 245, Section 2, of Book 1 of the Netherlands Civil Code provides that parents have a general obligation to look after and raise their children. In accordance with Article 404 of Book 1 of the Civil Code, parents are also obliged to bear the costs of caring for and raising their minor children (whether they are legitimate or illegitimate children). The parents should fulfil their financial obligations to the best of their ability given their financial means. This obligation therefore consists not only of the costs of such things as food and clothing but also of the costs of upbringing in general. If the parents lack the financial means to pay for some or all of the costs of subsistence, they may claim benefit under the new National Assistance Act.'*[982] In the reports, especially the legislation on the duty of care towards children in complex family relations, such as divorces, is explained,[983] and the Dutch system of social security, when relevant in relation to children, is explained.[984] In their third report, the Netherlands stated that: *'The Netherlands have a comprehensive system of social insurance and other Provisions that guarantee residents a minimum income. A minimum income does not equal poverty; it is enough to cover a person's living costs, provided they use the income support available and exercise financial discipline.'*[985]

9.3.1.4 The rights of foreign children in the ICRC reports

On more than one occasion, the Committee asked critical questions about the Dutch policy towards foreigners. It is not without reason, that the status of refugee children is a separate issue in the reporting format as adopted by the Committee on the Rights of the Child, in which the Member States are requested to report on the status of *'children outside their country of origin seeking refugee protection (art. 22), unaccompanied asylum-seeking children, internally displaced children, migrant children and children affected by migration.'*[986] As demonstrated in Chapter 4, it is especially in view of the Dutch asylum policy that ICRC Provisions, in particular

---

[982] 27 ICRC: CRC/C/51/Add.1, 24 July 1997, Section 246. See in the context of Article 27 also: CRC/C/117/Add.1, 5 June 2003, Section 75. See furthermore: CRC/C/61/Add.4, 4 October 2001, Section 209 (Netherlands Antilles). See also: CRC/C/117/Add.2, 17 June 2003, Section 215 (Aruba).

[983] CRC/C/51/Add.1, 24 July 1997, Sections 138-139; CRC/C/117/Add.1, 5 June 2003, Sections 76-79.

[984] CRC/C/51/Add.1, 24 July 1997, Sections 247-257; CRC/C/117/Add.1, 5 June 2003, Sections 175-197; CRC/C/NLD/3, 23 July 2008, Sections 294-311. See also: CRC/C/61/Add.4, 4 October 2001, Sections 129-133 (Netherlands Antilles). See furthermore: CRC/C/117/Add.2, 17 June 2003, Sections 215-219 (Aruba).

[985] CRC/C/NLD/3, 23 July 2008, Section 294.

[986] CRC/C/58/Rev.2, 23 November 2010 Section B and Appendix.

Article 27 ICRC, were invoked in the Courts, on behalf of a minor. Therefore, this Section focusses on the various debates on the Dutch asylum policies in view of the right of the child to an adequate standard of living.

In their initial report, the Netherlands described, *inter alia*, the procedures regarding the entitlements of foreign children to social benefits. In short, a distinction can be made between accompanied and unaccompanied asylum-seekers. In the first case, the parent(s) will ask asylum on behalf of the child, and the request of both the parent(s) and the child will be assessed altogether. In the second case, the child will request asylum on its own behalf. In both cases, the minor will be housed in a reception centre, and has entitlements to basic needs such as education and medical treatment. In its List of Issues[987] however, and more in particular during the sessions, the Committee expressed its concern regarding discrimination against asylum-seekers in the Netherlands, and the well-being of unaccompanied minor asylum-seekers.[988] The Dutch Delegation basically further clarified the asylum procedures, but could not take the concerns of the Committee away.[989] Although it noted the efforts of the State to deal with unaccompanied minor asylum-seekers, *'the Committee is concerned that they may need to receive increased attention. The Committee recommends that the State Party strengthen measures so as to provide immediate counselling and prompt and full access to education and other services for refugee and asylum-seeking children. Furthermore, the Committee recommends that the State Party take effective measures for the integration of these children into its society.'*[990]

Therefore, in their second periodic report, the Netherlands stated that *'unaccompanied minor asylum-seekers receive special care and attention.'*[991] In general however, it was underlined that the Dutch asylum policy concerning unaccompanied minor asylum-seekers focussed on the return of those minors to their country of origin, due to the increase of unaccompanied minors requesting for asylum, and the abuse of facilities when the minor reached adulthood: *'In this new policy, the possibility of return will be investigated far more vigorously than before, while naturally taking account of the minor's position and the situation in the country of origin.'*[992] In its List of Issues, the Committee requested additional data concerning *'the number of unaccompanied minors and asylum-seeking and refugee children, as well as the number of children awaiting expulsion'*, and asked for further information regarding *'discrimination in particular with respect to immigrants, undocumented migrants, refugees and asylum-seeking children in the Netherlands and Aruba'* and *'refugee and asylum-seeking children in the Netherlands, and their access to services*

---

[987] CRC/C/Q/NET.1, 30 June 1999, Sections 11 and 27.

[988] Mr Fulci, CRC/C/SR.578, 16 November 1999, Section 40; Mrs Tigerstedt-Tähtelä, CRC/C/SR.580 18 November 1999, Section 24.

[989] Mr Jansen, CRC/C/SR.580, 18 November 1999, Sections 33-35 and 38.

[990] CRC/C/90, 7 December 1999, Section 254.

[991] CRC/C/117/Add.1, 5 June 2003, Section 236.

[992] CRC/C/117/Add.1, 5 June 2003, Section 237.

including education and housing'[993] Besides providing the requested statistics, the Netherlands responded to the latter request that '*In November 2002, a pilot project was launched at a centre in Vught with the aim of promoting the return of minors to their countries of origin and preparing them for that event. The pilot involves the creation of a "campus" in which specific rules and standards of behaviour are enforced. The centre runs an intensive full-time programme of organised activities designed to prepare the children for return to the country of origin by developing transcultural competence. If there is no one to care for them in their countries of origin, the children are prepared instead for integration into Dutch society. In any case, the children attend school and are prepared for life in the Netherlands.*'[994] During the sessions with the Committee, the Committee expressed its concern regarding the new Dutch policy, and it was noted with regret '*that the definition given by Dutch law for "asylum-seeker" is not consistent with Article 22 of the Convention and other relevant international standards, and with regard to the reception of unaccompanied minors seeking asylum, the State Party has adopted a stricter policy favoring the return of children to their country of origin.*'[995] The Dutch Delegation disagreed, and argued that the definition used in the Aliens Act was in fact in accordance with the relevant international Provisions. Furthermore, the Delegation claimed that the Netherlands successfully decreased the number of unaccompanied minor asylum-seekers crossing the Dutch border, and was able to significantly speed up the asylum procedures.[996] The Delegation explained that '*Unaccompanied minors whose application has been rejected may be granted, due to their status as a minor, the temporary status of refugee. They are then placed in an institution in which, in accordance with the laws on compulsory education, they receive education equivalent in content and quality to that of mainstream schools but taught in English, with the purpose to establish a period in which concerned parties may return them to their country of origin. (...) In concern for the well-being of these children, they are offered a variety of activities, sports and others. The dialogue with refugee organisations is constant and they are in good contact with Parliamentarians to express their views.*'[997] This explanation was not satisfactory, for in its Concluding Observations, the

---

[993] CRC/C/Q/NLD/2, 27 October 2003, Sections A11 (c) and BIV.

[994] CRC/C/RESP/48, 16 December 2003, Section BVIII (a) (1).

[995] Mr Kotrane, CRC/C/SR.929, 23 January 2004, Section 55, original text published in French: '*que la définition que donne le droit néerlandais d'un «demandeur d'asile» n'est pas conforme à l'Article 22 de la Convention ni aux autres normes internationales pertinentes, et qu'en matière d'accueil des mineurs non accompagnés demandeurs d'asile l'État partie a adopté une politique plus sévère privilégiant le retour de ces enfants dans leur pays d'origine.*'

[996] Mr Ter Kuile, CRC/C/SR.929, 23 January 2004, Section 69.

[997] Mr Ter Kuile, CRC/C/SR.929, 23 January 2004, Section 70. Original text published in French: '*Les mineurs non accompagnés dont la demande a été rejetée peuvent se voir accorder, en raison de leur qualité de mineur, le statut de réfugié à titre provisoire. Ils sont alors placés en institution où, conformément à la loi sur l'enseignement obligatoire, ils suivent un enseignement équivalent en contenu et en qualité à celui des établissements ordinaires mais dispensé en anglais, l'objectif recherché étant de permettre à terme aux intéressés de retourner dans leur pays d'origine. ... Dans le souci du bien-être de ces mineurs, on leur propose diverses activités, sportives et autres. Le dialogue avec les organisations de réfugiés est constant et elles se sont ainsi entretenues avec les parlementaires pour exposer leurs vues.*'

Committee again expressed its concern '*that in the Netherlands the definition of an unaccompanied minor seeking asylum does not conform to international standards and it may make access to basic services more difficult for the child while in the country...*'[998] Therefore, among other things, the Committee urged the Netherlands to '*review the Aliens Act of 2001 and its application to ensure full conformity with international standards applicable to refugees and to the Convention.*'[999]

In their third report, the Netherlands reported that indeed some changes were made in the wordings of the Aliens Act. For instance, the definition of an 'unaccompanied minor' was revised, and now defined more broadly, so that any minor asylum-seeker '*who is not accompanied by a parent or legal guardian is eligible for special leave to remain if it is found that he or she is not eligible for refugee status.*'[1000] Also, some other asylum procedure details were adapted for the benefit of unaccompanied minor foreigners, especially concerning the duration of the asylum procedures.[1001] Elsewhere, reference was made to a Court ruling that led to changes in legislation regarding the entitlements of a legally residing foreign child in the Netherlands to social assistance: '*On 24 January 2006 the Central Appeals Tribunal of Public Service and Social Security Matters*[1002] *made a decision to the effect that children residing legally in the Netherlands with no residence permit are entitled to social assistance from the Government. With effect from 1 January 2007 these children are eligible for an allowance equal to the standard amount of social assistance for single young people below the age of 21. The Benefits for Specific Categories of Aliens Order has been amended to this effect. This is a specific scheme for certain categories of aliens, not general social assistance.*'[1003] Furthermore, in the third report, reference was made to several initiatives to improve the situation of minor asylum-seekers, including the establishment of reception centres in developing countries to house children who unsuccessfully requested asylum in the Netherlands, in which the minors received food, education and healthcare until they reach adulthood.[1004] It was on the other hand reported that in the asylum procedure, as a last resort, aliens could be held in detention.[1005] While unaccompanied minor asylum-seekers below twelve would not be placed in detention, and it was strived for that accompanied minors would not be held in detention, possibly together with their parent(s), the possibility could not be entirely ruled out.[1006] The Committee appreciated the

---

[998] CRC/C/15/Add.227, 30 januari 2004, Section 53.

[999] CRC/C/15/Add.227, 30 januari 2004, Section 54.

[1000] CRC/C/NLD/3, 23 July 2008, Section 389.

[1001] CRC/C/NLD/3, 23 July 2008, Sections 390-393.

[1002] Apparently, the reference concerns the ruling of the Central Court of Appeal, 24 January 2006, LJN AV0197, discussed in Section 4.6.

[1003] CRC/C/NLD/3, 23 July 2008, Section 311.

[1004] CRC/C/NLD/3, 23 July 2008, Sections 394-398.

[1005] CRC/C/NLD/3, 23 July 2008, Section 423.

[1006] CRC/C/NLD/3, 23 July 2008, Sections 423-428; see also: Mr Rouvoet, CRC/C/SR.1377, 23 January 2009, Section 87.

changes in legislation concerning unaccompanied minor asylum-seekers,[1007] but expressed its ongoing concerns about the discrimination against foreigners and asylum-seekers,[1008] and the occasional detention of minor asylum-seekers.[1009]

9.4 The CEDAW reports and the right to food

Before 2013, the Netherlands had submitted five,[1010] Aruba five[1011] and the Netherlands Antilles four[1012] reports on the implementation of the CEDAW, in accordance with Article 18 CEDAW. Although Article 12 CEDAW is closely related to the right to food (Chapter 3), the right is not very often discussed in particular in the periodic reports and subsequent procedures (Section 9.4.1). However, one closely related issue, that is the Dutch policies towards asylum-seekers and foreign residents, was discussed in detail, also in the context of ECOSOC rights as embedded in the CEDAW (Section 9.4.2).

9.4.1 The right to food in the CEDAW reports

In view of Article 12 (2) CEDAW, the focus of the reports was not on issues that are directly related to the right to food. The reporting guidelines as adopted by the Committee on the Elimination of all forms of Discrimination Against Women do not necessarily request Member States to discuss food related issues.[1013] Also, the wordings of the Provision imply a focus on healthcare. Consequentially, the majority of the reporting concerned issues such as the quality of healthcare, financial support in the field of healthcare, equal treatment of women in healthcare, the access of ethnic minorities to healthcare, and the promotion and use of contraceptives for birth control purposes and/or in the fight against AIDS. Only occasionally, food-related issues were discussed. For instance, policies and data on (the promotion

---

[1007] CRC/C/NLD/CO/3, 27 March 2009, Section 12.

[1008] CRC/C/NLD/CO/3, 27 March 2009, Sections 26-27.

[1009] Ms. Khattab, CRC/C/SR.1377, 23 January 2009, Section 8; CRC/C/NLD/CO/3, 27 March 2009, Sections 67-68.

[1010] CEDAW/C/NET/1, 7 April 1993 (with an updating supplement CEDAW/C/NET/1/Add.3, 18 October 1993); CEDAW/C/NET/2, 15 March 1999; CEDAW/C/NET/3, 22 November 2000; CEDAW/C/NLD/4, 10 February 2005; CEDAW/C/NLD/5, 24 November 2008. It should be noted that the Committee on the Elimination of all Discrimination Against Women discussed and considered the second and third periodic report of the Netherlands, the Netherlands Antilles and Aruba in one session.

[1011] CEDAW/C/NET/1/Add.2, 30 June 1993; CEDAW/C/NET/2/Add.2, 15 March 2000; CEDAW/C/NET/3/Add.1, 27 October 2000; CEDAW/C/NLD/4/Add.1, 7 June 2005; CEDAW/C/NLD/5/Add.1, 3 September 2009.

[1012] CEDAW/C/NET/1/Add.1, 17 September 1993; CEDAW/C/NET/2/Add.1, 15 March 1999; CEDAW/C/NET/3/Add.2, 22 November 2000; CEDAW/C/NLD/4/Add.2, 19 May 2009; CEDAW/C/NLD/5/Add.2, 19 May 2009.

[1013] HRI/GEN/2/Rev.1/Add.2, 5 May 2003.

of) breastfeeding,[1014] nutrition during (and after) pregnancy,[1015] unhealthy eating habits, and education on/the promotion of a healthy diet[1016] were recurring topics that were discussed in the various reports of the Netherlands, Aruba and the Netherlands Antilles.

9.4.2 The rights of foreign women in the CEDAW reports

9.4.2.1 The initial report

Also in the context of the CEDAW, the status of foreign women is frequently discussed. In the initial report, the Netherlands stated that *'as regards granting Dutch nationality to foreigners (naturalisation) or granting residence rights to stay in the country, Dutch law does not make a distinction between men and women.'*[1017] Nevertheless, it was reported that increasing attention was paid to the situation of women in application procedures for residence permit, especially regarding family reunion of migrant workers, the abuse of asylum procedures for human trafficking, and unawareness amongst foreign women of the possibility to apply for a residence permit in their own right.[1018] Furthermore, the report focussed on the policies regarding women refugees who were victim of sexual violence. *Inter alia* the appointment of female liaison officials for interviewing the refugee women was reported, and the policy on the evaluation of motives for fleeing the country of origin included that sexual violence *'could constitute grounds for admission as a refugee if the individual has a well-founded fear of persecution by the authorities of the country of origin, or of the authorities failing to be able or willing to offer protection against serious acts of violence or persecution by the individual's fellow nationals.'*[1019] In addition, women who were the victim of sexual violence but could not be granted a refugee status could be *'given exceptional leave to remain in the country if there are pressing humanitarian reasons which militate against an individual being sent back to her country of origin.'*[1020] In addition, it was reported

---

[1014] CEDAW/C/NET/1, 7 April 1993, Sections 535, 543-550 (updated: CEDAW/C/NET/1/Add.3, 18 October 1993, p. 7-8); CEDAW/C/NLD/5, 24 November 2008, p. 30; the Netherlands Antilles: CEDAW/C/NET/2, 15 March 1999, p. 68; CEDAW/C/NET/2/Add.1, 15 March 1999, p 103; CEDAW/C/NET/3/Add.2, 22 November 2000, pp. 28-29; CEDAW/C/NLD/5/Add.2, 19 May 2009, p. 47; Aruba: CEDAW/C/NET/1/add.2, 30 June 1993, especially p. 63; CEDAW/C/NET/3/Add.1, 27 October 2000, p. 24; CEDAW/C/NLD/4/Add.1, 7 June 2005, p.12; CEDAW/C/NLD/5/Add.1, 3 September 2009, p. 10.

[1015] CEDAW/C/NET/1, 7 April 1993, Section 575; the Netherlands Antilles: CEDAW/C/NET/1/Add.1, 17 September 1993, p. 56.

[1016] CEDAW/C/NET/2/Add.1, 15 March 1999, p.42; CEDAW/C/NLD/4, 10 February 2005, p. 75; CEDAW/C/NLD/5, 24 November 2008, p. 90, 95; the Netherlands Antilles: CEDAW/C/NET/2/Add.1, 15 March 1999, p. 42; CEDAW/C/NLD/5/Add.2, 19 May 2009, p. 39; Aruba: CEDAW/C/NET/2/Add.2, 15 March 2000, p.45; CEDAW/C/NLD/4/Add.1, 7 June 2005, p.13.

[1017] CEDAW/C/NET/1, 7 April 1993, Section 453.

[1018] CEDAW/C/NET/1, 7 April 1993, Sections 453-548.

[1019] CEDAW/C/NET/1, 7 April 1993, Section 277, see also Section 456 concerning female asylum-seekers in general.

[1020] CEDAW/C/NET/1, 7 April 1993, Section 277.

that The Refugee Healthcare Centre of the Ministry of Welfare, Health and Cultural Affairs was responsible to ensure adequate healthcare facilities, with a focus on the field of sexual violence.[1021] Also, measures to prevent circumcision, aimed at refugees and asylum-seekers were discussed in the supplement to the initial report.[1022] During the sessions with the Committee[1023] as well as in its General Observations, the status of asylum-seekers/foreign women was not significantly discussed.[1024]

9.4.2.2 The second periodic report

In their second periodic report, the Netherlands reported on a new assimilation policy that includes the obligation of newcomers to sign an assimilation contract, with the purpose *'to promote the assimilation of newcomers and in particular to improve their position in the labour market.'* The policy is especially aimed at women who enter the Netherlands for the purpose of founding or reuniting a family, for they were often facing more difficulties in assimilating compared to men.[1025] Furthermore, the report introduces the new Linkage Act that links the immigration policy with social services policies. One of the results from this new legislation is that *'The Linkage Act bars illegal immigrants from claiming welfare benefits.'*[1026] The Government then summed up the different consequences in the field of social benefits for different types of aliens residing in the Netherlands:

> The bill to amend the Aliens Act in order to take account of the Linkage Act defines three categories of aliens:
> 1. legal aliens: in principle entitled to all social services;
> 2. illegal aliens: entitled under the Linkage Act only to education for their minor children and medical assistance in emergencies;
> 3. lawfully resident aliens: people whose applications for residence are still being processed. Under the new Linkage Act people applying for residence for the first time will not be entitled to welfare benefits while waiting for their applications to be processed. People who have previously been lawfully resident in the Netherlands will be entitled to benefits during the procedure.[1027]

[1021] CEDAW/C/NET/1, 7 April 1993, Section 278.
[1022] CEDAW/C/NET/1/Add.3, 18 October 1993, pp. 9-10.
[1023] CEDAW/C/SR.234, 25 January 1994; CEDAW/C/SR.239, 4 February 1994.
[1024] A/49/38, 12 April 1994.
[1025] CEDAW/C/NET/2, 15 March 1999, p. 47.
[1026] CEDAW/C/NET/2, 15 March 1999, pp. 47-48.
[1027] CEDAW/C/NET/2, 15 March 1999, p. 48.

However, with regard to these three basic rules, some exceptions exist regarding certain types of benefits (mainly education, healthcare and legal assistance) for certain individuals, mostly children:

- Children of compulsory school age cannot be refused access to schooling on the ground that they are not lawfully resident.
- Healthcare safeguards: children of illegal immigrants are fully entitled to be vaccinated under national vaccination programmes, and both children and adults are entitled to medical assistance against infectuous diseases. Under the human rights conventions, the Dutch authorities are obliged to provide the best possible healthcare and to do so without charge where the recipients of the care cannot pay for it themselves.
- Legal assistance under the Legal Aid Act: the question of whether or not a person is legally resident has no bearing on the granting of free legal assistance under the Legal Aid Act. In other words, an immigrant is entitled to free legal assistance not only in connection with an asylum application or appeal against expulsion but also for other civil, criminal and administrative proceedings.[1028]

Furthermore, the Netherlands reported on the particularities of the status and asylum procedure regarding women with a foreign partner or husband; women with a dependent right of residence; women asylum-seekers and women without a right of residence.[1029] Regarding the latter category, it was indeed reported, in line with the aforementioned amendments to the Aliens Act, that these women have the same legal position compared to men in that situation, meaning thus that they have no entitlements to any social benefits. It was reported that women in that position are more likely to be victim of abuse or ill-treatment, such as human trafficking. Therefore, it was underlined that a separate Chapter of the Aliens Act focusses on the issue of human trafficking, with the purpose to facilitate the investigation and prosecution of cases of trafficking women and to provide assistance to the victims.[1030] The above was not discussed during the sessions with the Committee[1031] or in its Concluding Observations.[1032]

[1028] CEDAW/C/NET/2, 15 March 1999, p. 48.

[1029] CEDAW/C/NET/2, 15 March 1999, pp. 48-53.

[1030] CEDAW/C/NET/2, 15 March 1999, p. 53. See also the related discussion in: CEDAW/C/SR.513, 7 September 2001, Sections 3 and 10, and the Concluding Observations of the Committee: A/56/38, 31 October 2001, Sections 211-212.

[1031] CEDAW/C/SR.512, 5 September 2001; CEDAW/C/SR.513, 7 September 2001.

[1032] A/56/38, 31 October 2001.

9.4.2.3 The third periodic report

In the third periodic report, no significant additional information was provided on the status of asylum-seekers in the Netherlands, nor was there any subsequent debate concerning the matter that is worth mentioning.[1033]

9.4.2.4 The fourth periodic report

Also in the fourth periodic report, the Netherlands reported on their asylum policies, and discussed into more detail the particularities concerning the procedures for asylum applications of women, especially in the context of a resident who is dependant on the residence permit of the partner.[1034] In the List of Issues and subsequent replies, the Dutch asylum procedures were not discussed.[1035] During the sessions with the Committee however, the Dutch asylum procedures were discussed into detail. It was underlined by the Dutch Delegation that in the Netherlands, an illegally residing alien had no entitlements to social benefits at all, except in cases of a medical emergency, or a minor (who had the right to education, regardless of his/her residence status). Also, it was underlined that aliens who were the victim of honour-related violence could apply for residence, and could be granted access to shelter.[1036] Members of the Committee were especially worried about the number of illegally residing women in the Netherlands that were engaged in prostitution. In that context, the question was raised '*how illegal migrants who reported on trafficking to the authorities could enjoy the protection and assistance recommended in human rights standards.*'[1037] The Dutch Delegation replied that victims of human trafficking could, based on that, apply for a residence permit, although their cooperation in Court proceedings against the perpetrators was required. Also, the Delegate discussed several measures to prevent the relatively vulnerable underaged asylum-seekers from being exploited.[1038] In this context, there appeared to be doubts however, regarding the fact that sexual abuse was truly a recognised ground for granting the status of refugee.[1039] The Dutch Delegation replied to this by stating that '*while the Aliens Act 2000 did not recognise fear of sexual violence or of honour killings as specific grounds for asylum, several Sections of the Aliens Act dealt with gender-related persecution and provided for a gender-inclusive approach to asylum applications that was fully in compliance with United Nations gender guidelines.*'[1040] In its Concluding Observations, the Committee took

---

[1033] CEDAW/C/NET/3, 22 November 2000.

[1034] CEDAW/C/NLD/4, 10 February 2005, pp. 49-50.

[1035] CEDAW/C/NLD/Q/4, 4 August 2006 (see for the written replies: CEDAW/C/NLD/Q/4/Add.1, 27 October 2006).

[1036] Mrs Bleeker, CEDAW/C/SR.767 (B), 5 March 2007, Section 27.

[1037] Mrs Chutikul, CEDAW/C/SR.767 (B), 5 March 2007, Section 51.

[1038] Mrs Bleeker, CEDAW/C/SR.767 (B), 5 March 2007, Sections 62-64. See also the question posted by Mrs Begum, Section 55.

[1039] Mrs Tavares da Silva, CEDAW/C/SR.768 (B), 2 February 2007, Section 22.

[1040] Mrs Van der Zaal-Van Bommel, CEDAW/C/SR.768 (B), 2 February 2007, Section 30.

a mild approach towards the Dutch asylum policies. Besides several requests for additional data on the status of asylum-seekers in the Netherlands, the Committee expressed its concern about the procedure regarding the application for a residence permit in cases of human trafficking, and *'calls upon the State Party to provide for the extension of temporary protection visas, reintegration and support services to all victims of trafficking, including those who are unable or unwilling to cooperate in the investigation and prosecution of traffickers.'*[1041]

9.4.2.5 The fifth periodic report

In their fifth periodic report, the recommendation of the Committee was recalled, and the Netherlands explained that any foreign individual who claims to be a victim of human trafficking, and who does not derive a permit to reside in the Netherlands based on his/her nationality (mostly Dutch or European Community nationalities: they are protected and have entitlements to social benefits under different legislation), could make use of the so called B9-regulation. Based on this regulation, the victim has a consideration period of three months, during which he or she will receive aid and support, in which he/she must decide whether to *'cooperate with the investigation and prosecution of the human traffickers.'*[1042] When the victim does not choose to cooperate, only *'in extremely distressing cases there is also the option of the State Secretary of Justice granting a victim of human trafficking residence on purely humanitarian grounds, in other words, in cases in which it is clear that the person is a victim, but the victim cannot cooperate or is scared to do so.'*[1043] In addition, it was reported that *'The option of invoking this so-called discretionary power existed before, but was rarely used. In 2007 organisational changes were implemented at the IND (Immigration and Naturalisation Service) and agreements were made with aid providers to guarantee that these humanitarian cases are actively and more broadly submitted to the State Secretary so that he can give his opinion on each case.'*[1044] In case the victim would decide to cooperate, a residence permit was granted for the duration of the period. Then, when the prosecution procedure would lead to the conviction of the human trafficker, *'the victim can be offered continued residence in the Netherlands.'*[1045] Apparently, this is not a standard norm, for later it was stated that *'the rule applies that after expiration of the B9 permit it will be evaluated whether a return to the country of origin is an option.'* In case of a prosecution procedure that would take longer than three years, the victim could also apply for a continued residence, even when the case was not closed yet.[1046] In this context, the Netherlands reported that increasingly underaged asylum-seekers entered the Netherlands who were suspected to be a victim of human

---

[1041] CEDAW/C/NLD/CO/4, 2 February 2007, Sections 23-24.

[1042] CEDAW/C/NLD/5, 24 November 2008, p. 63.

[1043] CEDAW/C/NLD/5, 24 November 2008, p. 63.

[1044] CEDAW/C/NLD/5, 24 November 2008, pp. 63-64.

[1045] CEDAW/C/NLD/5, 24 November 2008, p. 64.

[1046] CEDAW/C/NLD/5, 24 November 2008, p. 64.

trafficking. A number of them however then disappeared from the reception centres, most likely returning back in the hand of their traffickers. To this end, '*a policy has gradually been developed for receiving minor foreign nationals who are at risk of human trafficking in a closed reception facility, which reduces the risk of them disappearing. A 'closed reception' pilot has now being started up for all minor foreign nationals of whom it is suspected that they were brought to the Netherlands for the purpose of being traded.*'[1047] Also, improvements were reported on the application procedures for women who were victims of (threatened) honour-related violence, and victims of domestic violence without a residence permit: based on an amended regulation, '*they may be eligible for financial support and health insurance, on the condition that they submit an application for a residence permit (which gives them lawful residence) and reside in a women's shelter. The intention of opening up the Regulation to this group is to eliminate a possible obstruction, in the financial sense, with regard to access to women's support.*'[1048] Furthermore, in the report several foreign policies in which the Netherlands strived to promote equality between men and women abroad were discussed,[1049] and in that context it was proudly reported that '*for many years the Netherlands have played a leading role when it comes to the promotion of women's rights*'.[1050] In addition, regarding illegally residing foreigners, the Netherlands explained that the fact that the (earlier mentioned) Linkage Act (Dutch: *Koppelingswet*) in which entitlements to social security were connected to residence status, was one of the reasons not to sign and ratify the Convention on the Protection of the Rights of All Migrant Workers and Members of their Families, as recommended by the Committee,[1051] for the Linkage Act does make a distinction between legally and illegally residing foreigners.[1052] It is obvious that the Committee was not convinced that alien women that were victim of sexual abuse or human trafficking were treated in line with the CEDAW and other international human rights documents. The Committee expressed its concern that in view of the revised, accelerated, asylum procedures '*women are often unable to relate experiences of rape or other traumatic incidents.*'[1053] Also, the Committee asked to '*please indicate the measures taken to ensure that women victims of trafficking are properly identified and are no longer held in alien detention without assistance and access to the protective services to which they are entitled. Please also indicate whether minors who are victims of trafficking and sexual exploitation have access to specialised institutions.*'[1054] To this, the Netherlands responded by underlining that '*there are various ways in which victims of trafficking, domestic violence and honour-*

[1047] CEDAW/C/NLD/5, 24 November 2008, p. 64.

[1048] CEDAW/C/NLD/5, 24 November 2008, p. 24.

[1049] CEDAW/C/NLD/5, 24 November 2008, pp. 27-32.

[1050] CEDAW/C/NLD/5, 24 November 2008, p. 29.

[1051] CEDAW/C/NLD/CO/4, 2 February 2007, Section 43.

[1052] CEDAW/C/NLD/5, 24 November 2008, pp. 31-32. As stated above, almost identical wordings as used in the combined fourth and fifth report on the implementation of the ICESCR: E/C.12/NLD/4-5, 17 July 2009, Sections 8-9.

[1053] CEDAW/C/NLD/Q/5, 13 March 2009, Section 22.

[1054] CEDAW/C/NLD/Q/5, 13 March 2009, p. 28.

related violence can obtain residence permits to protect them from further violence, irrespective of their immigration status and ethnic origin.' The Netherlands explained that in general, an applicant for asylum who was (or fears to become) a victim of such violence, would have to demonstrate that his/her own Government is incapable of providing the required protection, but that *'the Aliens Act Implementation Guidelines specifically mention domestic violence as a ground of asylum for immigrants from certain countries where there is a link between domestic violence and honour-related violence, discrimination against women or the absence of protection by the local authorities.'*[1055] A further specification is then provided with the description of the possible procedures for the application of regular (non-asylum) residence permits, depending on the sort of violence and residence status in order to receive a residence permit.[1056] Furthermore, the Netherlands responded to the concern expressed by the Committee that, especially in the newer, accelerated procedures, women who were victim of (sexual) violence could not adequately refer to their situation. The Netherlands in response stated that when the first asylum application, that lasted 48 hours, was rejected, it was always possible to submit a second application. Besides that, it was expected that the 48-hour procedure would be replaced by a procedure of 8 days. Regarding the psychological or medical status of the women involved in the procedure, it was reported that it was standard procedure to offer all asylum-seekers a medical check *'so that their condition can be taken into account during the asylum procedure.'* In addition, the Netherlands underlined that the Dutch asylum procedures were gender-sensitive, in the sense that violence against women was a standard ground for granting asylum or a normal residence permit.[1057] During the sessions with the Committee, only some formal aspects of the procedures were briefly discussed.[1058] In its Concluding Observations, the Committee, now in some firmer wordings, *'considers that even if extended to eight days, as envisaged by the Netherlands, the short length of the*

[1055] CEDAW/C/NLD/Q/5/Add.1, 19 October 2009, p. 15.

[1056] This list is provided in the replies, although most particularities of the procedures have already been discussed above: *'(a) victims of domestic violence or honour-related violence who have a dependent right of residence may be granted a residence permit in their own right within three years on contacting the police and producing confirmation of the violence (for example from a women's shelter, a doctor or social services); after three years of residence a dependent residence permit may be replaced by a permit for continued residence, in which case no proof of violence is necessary; (b) victims of domestic violence who are resident illegally in the Netherlands may apply for a residence permit only on humanitarian grounds based on the specific and individual circumstances of the case; (c) victims of honour-related violence who are resident illegally in the Netherlands may be granted a residence permit if information provided by a specialised police unit establishes that they are under threat of violence in the Netherlands and cannot safely return to their country of origin; (d) victims of trafficking who are resident illegally in the Netherlands may be granted a residence permit (a 'B9 permit') if they cooperate with the judicial authorities for the duration of the proceedings against the perpetrator.'* See: CEDAW/C/NLD/Q/5/Add.1, 19 October 2009, p. 16.

[1057] CEDAW/C/NLD/Q/5/Add.1, 19 October 2009, p. 25.

[1058] See the questions of Mrs Chutikul, Mrs Patten, CEDAW/C/SR.916, 3 March 2010, Sections 49-50, 53, and the reply of Mrs Bleeker, CEDAW/C/SR.916, 3 March 2010, Section 61. See furthermore, the questions of Mrs Awori, CEDAW/C/SR.917, 18 March 2010, Sections 57-58, and the reply of Mrs Dijksma, CEDAW/C/SR.917, 18 March 2010, Section 63.

accelerated asylum procedure remains unsuitable for vulnerable groups, including women victims of violence and unaccompanied children, and therefore urges the State Party to introduce in the procedure the possibility for women victims of violence and unaccompanied minors to fully explain their claims and to present evidence on their situation at a later stage. The Committee also urges the State Party to provide asylum-seekers with suitable accommodation during the entire review of their case, including during the appeal phase.'[1059] Also, it was urged to include (sexual) violence against women as an official ground for asylum, since currently they are still subdued to a margin of appreciation of the authorities involved. Furthermore, the Committee expressed its concerns regarding the access to healthcare of undocumented female asylum-seekers, due to a lack of information concerning the procedures.[1060]

9.4.3 Direct applicability of ECOSOC rights in the CEDAW reports

The matter of direct applicability was hardly discussed during the reporting procedures of the Netherlands Antilles[1061] and Aruba, but was discussed in detail concerning several periodic reports and the subsequent procedures of the Netherlands.

9.4.3.1 The second and third periodic report

The direct applicability of the CEDAW was discussed for the first time during the sessions with the Committee, concerning the second and third periodic report of the Netherlands.[1062] One Committee Member asked the question whether an individual could bring a case that was transmitted to the Committee on Equal Treatment *'before a Court in order to obtain a binding decision, if they were dissatisfied*

[1059] CEDAW/C/NLD/CO/5, 5 February 2010, Section 41.

[1060] CEDAW/C/NLD/CO/5, 5 February 2010, Section 47.

[1061] In their fourth periodic report, the Netherlands Antilles, in the context of Article 11 CEDAW, made reference to one Court case of the Supreme Court, involving Antilleans, in which the direct effect of an equivalent Provision – Article 7 ICERSCR – was discussed: *'In a case involving Antilleans, the Supreme Court (Hoge Raad; HR) held that discriminating between married and unmarried persons with respect to payment was incompatible with Article 7 of the ICESCR. It emerges from case law (HR 7 May 1993, no. 259) that Article 7 is not directly applicable, but that guaranteeing equal payment for equal work is an objective that the Government must work to achieve. It should not therefore be too readily assumed that there is a reasonable and objective justification for differences in payment.'* A seemingly contradicting Section, in which it is not entirely clear to what Court case is referred in which the Supreme Court ruled that a certain discrimination was incompatible with Article 7 ICESCR. As demonstrated in Chapter 8, there are numerous cases in which a not direct effect was accepted of Article 7 ICESCR, and not only ruled by the Supreme Court, but also by the Court of First Instance of the Netherlands Antilles. It is remarkable that in the report furthermore a reference was made to Supreme Court, 7 May 1993, NJ 1995, 259, in which the direct effect was explicitly denied.

[1062] While the second and third periodic reports were submitted separately, the Committee discussed both reports during one series of sessions with the Dutch delegation.

with the non-binding decision of the Commission.'[1063] The Dutch Delegation simply replied 'that individuals had the right to do so.'[1064]

9.4.3.2 The fourth periodic report

During the sessions with the Committee on the fourth periodic report, one Committee Member was 'interested in knowing the status of the Convention in the Courts, since some ambiguity seemed to prevail with regard to the application of the Convention in domestic law.'[1065] The Dutch Delegation replied, with similar arguments as used in the discussions of direct applicability of the ICESCR that 'the Netherlands did not challenge the concept that the Convention was legally binding. However, the question of its direct application to individuals was a separate issue. According to the Constitution of the Netherlands, the nature of a Provision was the decisive factor in determining whether it was binding on all persons. Depending on their content, international Provisions might be directly applied if they were binding on all persons. The Constitution further provided that international standards that were directly applicable to individuals had priority over national law. Certain criteria had been established under case law to determine whether treaty Provisions had direct application. National Courts had the final say on issues of direct effect and were responsible for decisions on the direct applicability of international law. In such cases, the Courts could decide whether a national regulation could be deemed inapplicable if it conflicted with international treaty Provisions. The Government was of the opinion that, by virtue of their content, only some of the Provisions of the Convention could be considered binding on all persons. The Convention as a whole was, of course, binding on the Netherlands as a State Party.'[1066] A Committee Member understood this statement as that the Netherlands considered the Convention binding in nature, but the direct applicability of the specific Provisions then was depending on the nature of the Provisions, that is that according to the Dutch Constitution the Provision must be binding on all persons to be directly applicable. Therefore, she asked to 'clarify which Provisions of the substantive Articles of the Convention were directly applicable. She suggested that the Provisions that were not directly applicable should be incorporated into the domestic legislation.'[1067] Interestingly, the Dutch Delegation seemed to make a stance that suggests that the CEDAW Provisions will likely be not directly applicable at all, by stating that 'the Government believed that the Articles of the Convention implied duties for the State and were aimed at the executive and legislative branches of States Parties. National policy and legislation were needed for the full implementation of the Convention. Because the Provisions of the Convention were not aimed at persons or individuals, they had no direct effect. It was

---

[1063] Mr Melander, CEDAW/C/SR.513, 7 September 2001, Section 3.

[1064] Mrs Verstand-Bogaert, CEDAW/C/SR.513, 7 September 2001, Section 9.

[1065] Mrs Patten, CEDAW/C/SR.767 (B), 5 March 2007, Section 17. See also Mrs Šimonović with a similar question in Section 20.

[1066] Mrs Dopheide, CEDAW/C/SR.767 (B), 5 March 2007, Section 28.

[1067] Mrs Šimonović, CEDAW/C/SR.767 (B), 5 March 2007, Section 43.

therefore up to national Courts to decide on the application of those Provisions.'[1068] It was added that *'it was the obligation of States Parties to implement the Convention through national legislation and to follow up on that obligation in full.'*[1069] In its Concluding Observations, the Committee expressed to remain deeply concerned about the Dutch viewpoints on the possibility of direct applicability of the CEDAW, and therefore *'calls upon the State Party to reconsider its position that not all the substantive Provisions of the Convention are directly applicable within the domestic legal order and to ensure that all of its Provisions are fully applicable in the domestic legal order. The Committee points out that by ratifying the Convention and its Optional Protocol, States Parties assume the obligation to provide for domestic remedies for alleged violations of any rights guaranteed to individuals by the Convention.'*[1070]

9.4.3.3 The fifth periodic report

Indeed, in their fifth periodic report the Netherlands recalled the recommendation of the Committee, and once again explained their view on the direct effect of the CEDAW, without adding any significantly different arguments or new information.[1071] However, it was reported that *'As a result of the criticism from the CEDAW Committee and the General Meeting with the House of Representatives, the Minister for Emancipation sent a letter to the House of Representatives on the subject of the legal application of the Convention on 5 November 2007. The letter addresses the question whether stipulations in the UN Women's Convention have direct effect in the Netherlands.'*[1072]A reference was made to Aricle 93 CA that *'stipulates that stipulations in Conventions have binding force if they can bind all individuals through their content. The more generally stipulations are formulated and the more active action on the part of the state they require, the less the question of direct effect. The question whether a stipulation binds everyone and therefore has direct effect is, in the final instance, determined by the Dutch Courts in individual cases.'*[1073] Interesting to note was the reference made to the SGP case that was in progress at the moment the report was submitted, and concerned Article 7 CEDAW.[1074] Clearly, the explanation was unacceptable to the Committee that noted in its List of Issues *'that the State Party continues to consider that it is the responsibility of the Judiciary to determine whether a particular Provision of the Convention is directly applicable in the legal order, although, by signing the Optional Protocol to the Convention, the State Party has recognised the individual right of complaint in relation to all the rights set forth in the Convention.*

[1068] Mrs Dopheide, CEDAW/C/SR.767 (B), 5 March 2007, Section 47.

[1069] Mr Licher, CEDAW/C/SR.767 (B), 5 March 2007, Section 48.

[1070] CEDAW/C/NLD/CO/4, 2 February 2007, Sections 11-12.

[1071] CEDAW/C/NLD/5, 24 November 2008, pp. 13-14.

[1072] CEDAW/C/NLD/5, 24 November 2008, p. 14.

[1073] CEDAW/C/NLD/5, 24 November 2008, p. 14.

[1074] CEDAW/C/NLD/5, 24 November 2008, p. 14. See also Chapter 8, and the following Court rulings: Court of Appeal, 's-Gravenhage, 7 September 2005, NJ 2005, 473; Court of Appeal, 's-Gravenhage, 20 December 2007, NJ 2008, 133; Council of State, 5 December 2007, AB 2008, 35; Supreme Court, 9 April 2010, LJN BK4549.

In this context, please indicate how the State Party envisages overcoming this legal inconsistency. Please also provide the Committee with the English translation of the letter sent on 10 December 2008 to Parliament on the applicability of the Convention.'[1075] The Dutch response however, implied the same type of arguments once more, although this time in some richer wordings. It was responded that *'Contrary to the assumption underlying the question, there is not necessarily a legal inconsistency between considering it to be the responsibility of the Judiciary to determine whether a particular Provision of the Convention is directly applicable in the legal order and signing the Optional Protocol to the Convention, and thereby recognising the individual right of complaint in relation to all the rights set forth in the Convention. (...) the question of whether the State is bound by the United Nations Women's Convention, should be distinguished from the question of whether Provisions of the Convention have direct effect within the State. The Netherlands are bound by international law to realise the rights set forth in the Convention for persons within its jurisdiction. (...) Although the question of whether a Provision has direct effect is ultimately determined in the Netherlands by the Dutch Courts in individual cases, this does not interfere with the individual right of complaint introduced by the Optional Protocol. If a Court decides in national proceedings that a particular Provision of the Convention is not directly applicable in the legal order, this would not prevent the individual concerned from filing a complaint with the Committee.'*[1076] The requested translation of the letter to Parliament was indeed added, but basically contained the same arguments that would justify not unambiguously recognising the possibility of direct effect of the CEDAW Provisions.[1077] Again, during the sessions with the Committee, several Committee Members asked critical questions about the Dutch attitude regarding the direct applicability of the CEDAW.[1078] Especially the explanation in the periodic report on the working of Article 93 was considered to be unclear. It must however generally be noted that the questions asked by the Committee Members were mostly repetitions of earlier asked (and answered) questions, and did not necessarily contribute to a more thorough or different response of the Dutch Delegation.[1079] The Delegation however tried to convince the Committee that the CADEW had some effect in the Dutch domestic legal order, by referring to the SGP case that was still in progress and involved CEDAW rights.[1080] It was underlined that *'the State Party attached great importance to the Convention, some Articles of which were more directly applicable than others.'*[1081] In addition, one Member made a remarkable statement, by arguing *'that the Charter for the Kingdom of the Netherlands required the Netherlands Antilles to harmonise its laws with those of the Netherlands, including international instruments ratified by the Netherlands,*

[1075] CEDAW/C/NLD/Q/5, 13 March 2009, Section 4.

[1076] CEDAW/C/NLD/Q/5/Add.1, 19 October 2009, p. 4.

[1077] CEDAW/C/NLD/Q/5/Add.1, 19 October 2009, Annex 3.

[1078] Mrs Jansing, Mrs Ameline, Mrs Šimonović, CEDAW/C/SR.916, 3 March 2010, Sections 20, 21, and 68.

[1079] See for instance: Mrs Dreesen, CEDAW/C/SR.916, 3 March 2010, Section 37.

[1080] Mrs Dijksma, CEDAW/C/SR.916, 3 March 2010, Section 75.

[1081] Mrs Dijksma, CEDAW/C/SR.916, 3 March 2010, Section 75.

while ensuring respect for the country's autonomy and culture. Most of the Provisions of the Convention were directly applicable in the Netherlands Antilles, as evidenced by a Court decision in the 1990s establishing equal salaries for men and women. However, the Courts had a tendency to invoke European standards, and she suggested that efforts to promote awareness of the Convention might change that situation.[1082] This statement is rather obscure: it is unclear what is suggested by the fact that the Netherlands Antilles needs to harmonise their law with that of the Netherlands, including the ratified international instruments in relation to the direct effect of such instrument. In addition, it cannot possibly be deduced from one Court ruling, in which one Article was directly applied, that most CEDAW Articles are directly applicable. It is remarkable that during the sessions no further questions were asked about the direct effect of the CEDAW, especially considering the standard replies, and in addition the latter confusing statement. On the other hand, not surprisingly, in its Concluding Observations, the Committee *'reiterates its concern that as a consequence of the position of the State Party, the Judiciary is left with the responsibility of determining whether a particular Provision is directly applicable and that consequently, insufficient measures have been taken to address discrimination against women and to incorporate all the Convention's substantive Provisions into domestic laws'*[1083], and therefore *'reiterates its call on the State Party to reconsider its position and to ensure that substantive Provisions of the Convention are fully applicable in the domestic legal order, in compliance with the obligation of the State Party to adopt measures against discrimination (including within the political party SGP) and to provide for domestic remedies for alleged violations of any rights guaranteed to individuals by the Convention.'*[1084]

9.5 The Universal Periodic Review

Although the right to adequate food has not been significantly discussed in the two reporting cycles of the UPR, the Dutch asylum policy – which can be considered a related issue considering the Dutch case law on the right to food – was discussed extensively, touching similar aspects compared to the treaty reports and subsequent procedures.[1085]

However, in the context of the Universal Periodic Review, the matter of direct applicability was discussed occasionally. In their first national report, the Netherlands emphasised that the country traditionally attached great value to human rights, within their borders as well as abroad.[1086] Therefore, many human

---

[1082] Mrs Leeflang, CEDAW/C/SR.916, 3 March 2010, Section 80.

[1083] CEDAW/C/NLD/CO/5, 5 February 2010, Section 12.

[1084] CEDAW/C/NLD/CO/5, 5 February 2010, Section 13.

[1085] See especially their reports: A/HRC/WG/6/1/NLD/1, 7 March 2008, pp. 13-15; A/HRC/WG.6/13/NLD/1, 8 March 2012, pp. 17-20. Based on those reports, frequent questions were posed regarding the asylum procedures and treatment of foreigners in general during both cycles.

[1086] A/HRC/WG/6/1/NLD/1, 7 March 2008, Section 1.

rights are embedded in the Constitutional Act.[1087] A second source for human rights is EU legislation and case law, for due to its supranational character, these standards are part of the Dutch legal order. Furthermore, '*A third major source is the UN human rights conventions to which the Netherlands is party, along with, for example, the Convention for the Protection of Human Rights and Fundamental Freedoms of the Council of Europe.*'[1088] Regarding the direct applicability of international human rights, the Netherlands reported that their Government had been criticised on more than one occasion in the past, for the fact that certain human rights would have no direct effect in the domestic legal order. This criticism was expressed by different actors, such as CSOs to UN treaty bodies, and basically concerned CEDAW and ICESCR rights. In their report, the Netherlands stated that:

> A distinction should be made between the direct effect of Provisions of international law and the binding nature of Provisions of international treaties. The latter is beyond dispute. However the Netherlands have a qualified monistic legal system. In a monistic system the Courts must, in principle, apply not only national rules but also the Provisions of treaties and resolutions of international institutions, with the latter two categories of law prevailing if the domestic legislation is incompatible with them. The Dutch system is characterised as a 'qualified' monistic system because the Provisions of treaties and resolutions of international institutions can only be applied if they (a) are binding on all persons; and (b) have been published.'[1089]

Also here, the responsibility of the Courts was underlined: '*Because Dutch Courts can apply the Provisions of treaties and the resolutions of international institutions without the need for implementing legislation they also have the authority to determine whether such Provisions or resolutions are formulated in general terms and the more action they require on the part of the state, the less likely the Courts are to rule that they have direct effect. Provisions of this kind give the state more latitude to make a variety of policy choices.*'[1090] Despite these statements of the Dutch Government, and the fact that NGOs also expressed their concerns about the issue in the UPR procedure,[1091] the matter was not further discussed during the first cycle. In the second cycle, the issue of direct applicability was not addressed at all.[1092] It must be noted here that the formats of the country reports and further procedure, as adopted by the Human Rights Council do not necessarily encompass the issue of direct applicability.[1093]

[1087] A/HRC/WG/6/1/NLD/1, 7 March 2008, Sections 2 and 11.

[1088] A/HRC/WG/6/1/NLD/1, 7 March 2008, Section 11.

[1089] A/HRC/WG/6/1/NLD/1, 7 March 2008, Section 18.

[1090] A/HRC/WG/6/1/NLD/1, 7 March 2008, Section 19.

[1091] A/HRC/WG.6/1/NLD/2, 13 March 2008, Sections 3-4.

[1092] See for the second national report: A/HRC/WG.6/13/NLD/1, 8 March 2012.

[1093] See: A/HRC/DEC/6/102, 27 September 2007 (first cycle) and A/HRC/DEC/17/119, 19 July 2011 (second cycle).

9.6 Concluding remarks

Based on the above analysis, the following conclusions can be drawn.

Firstly, in general, when it concerns the reports on the implementation of the human rights treaties of the European part of the Kingdom, the major tendency is to report, in line with the intention expressed in the Parliamentary History on the ratification Bills of the treaties, the Netherlands, from the start, generously fulfil their implementing duties that come from ratifying the treaties. It must also be noted that from the reports and subsequent procedures, hardly any ambition was expressed to improve the implementation of the human rights concerned.

Secondly, most criticism expressed by the various treaty bodies on the Dutch reports, is not related to the implementation of ECOSOC rights in itself. Let alone some exceptions, the sound policies and regulations in the field of ECOSOC rights, or the high standard of living in the Netherlands, or – more in particular – the system of social security is not necessarily a reason for concern. Most criticism however is aimed at the fact that one of the most vulnerable groups in society does not have (full) access to this system of social security: foreigners that are currently awaiting their application for a residence permit, or illegally residing foreigners.

Thirdly, it is unambiguously clear that the Netherlands do not recognise the direct applicability of ECOSOC rights. Although the Dutch Delegations once in a while tried to convince the Committee that there is a difference between an obligation to implement a treaty and recognising its direct effect, or explained that it is – due to the Dutch Constitutional system – not the Government's responsibility to recognise direct applicability, but a responsibility of the Courts instead, the Committees steadily urged the Member States to ensure the direct applicability of the treaty Provisions. Since the matter of direct effect was not discussed in view of the ICRC, it is a pity that the development of the Courts to use *inter alia* Article 27 ICRC as an interpretative standard was not discussed in the UN arena. It would be interesting to hear the opinion of the Committee in the Rights of the Child on this case law.

Fourthly, it can be questioned how serious the Netherlands is taking their reporting responsibilities. Besides the fact that the Dutch have the tendency to submit reports generously overdue, the recommendations of the various Committees are taken rather lightly, leading to an increasingly stiffer atmosphere during the reporting cycles, in which the Committees (especially the CESCR) use increasingly firmer wordings. In general, some questions of the Committees are answered only partially, or are even ignored.

10. Evaluation and comparison

10.1. Introduction

In this Chapter, the findings of Chapters 4-9 will be summarised. Furthermore, the legal practice in the Netherlands will be compared with the reporting behaviour of the Dutch Government on the enforceability of the right to adequate food (Comparison II).

10.2 An obscure Constitutional system

Since 1953, the Netherlands have quite an obscure Constitutional system regulating the position and effect of Provisions under international law. This so-called 'qualified monistic system' embedded in Articles 93 and 94 of the Dutch Constitution, enables the Courts to apply international Provisions directly, but only when they are binding on all persons and have been published. To be more precise, the function of Article 93 Constitutional Act is to determine what international legal standards are binding on individuals, while the function of Article 94 Constitutional Act is to regulate the relation between these internationally directly applicable standards and national legislation.

Especially the prerequisite that a Provision must be binding on all persons has led to much confusion, in particular with regard to the question whether the Government or the Courts have a final say in the matter. Officially, both the Government and the Courts agree that it is up to the Judiciary to finally decide whether or not a Provision is binding on all persons. It was however the intention of the Constitutional Legislator to establish a legal practice in which the Legislature would voice his opinion on direct applicability of international legal standards in its Parliamentary Documents. This is indeed done ever since by the Legislature when adopting a Bill on the ratification of human rights treaties.

10.3 The right to food invoked in the Dutch Courts

The international right to food is not a directly applicable human right in the Netherlands. Article 11 ICESCR seems to be completely non-enforceable. However, a case law seems to emerge in which ICRC standards – including Article 27 – are used as an interpretative norm that is used in the treaty conform interpretation of national legislation. As a result, the ICRC provisions had a significant effect on the verdicts. The claimants are usually minor asylum-seekers who lawfully reside in the Netherlands without a residence permit, and are in distressing humanitarian circumstances. Furthermore, Article 27 ICRC was used in a landmark ruling of the Supreme Court as an interpretative standard, in which the Court decided that the COA could not remove illegally residing children from its facilities, and thus provide for the child's basic needs. However, these cases are still exceptions to the rule that

Article 27 is not applied, because it is considered not to be directly applicable. More than once, in rejecting the direct applicability of ECOSOC standards, reference was made to the Parliamentary Documents in which the Government, in its function of Legislator, expressed that in its view, direct applicability of the ICESCR treaty will not be possible. It seems thus that the Courts follow the lead of politics in order to determinate whether these rights could be directly invoked or not.

10.4 The Dutch position with regard to the right to food

At the time of ratification of the right to food in several treaty Provisions, the Dutch Government underlined that the Netherlands already fulfilled the obligations that follow from the ratification of this right. A practice that seems to be standard procedure in case of the ratification of any economic, social and cultural rights: a treaty Provision will only be ratified if the Dutch Legislature is of the opinion that the Netherlands already meet the obligations resulting from the human rights in question. If there is any doubt, the Provision will not be ratified, or ratified with reservations or interpretative declarations.

Also, during the process of the various reporting procedures that constitute part of the obligations coming forth from the ratification of international human rights treaties, the Netherlands seem to have no ambition to improve the recognition of human rights within their borders, which seems to be contradicting with the obligation of progressive realisation that can be found in various human rights treaties. Especially in the context of the ICESCR, this attitude led to increasing differences of opinion between the Dutch Government and the Committee on Economic, Social and Cultural Rights. In addition, the Dutch Government does not seem to bother to seriously answer the questions of the treaty-based bodies that consider these reports, while the quality of the reports is more than once considered to be doubtful. In this light, it is noteworthy to emphasise the fact that the Section on the right to food in the combined 4th and 5th ICESCR report is almost entirely copied from the 3rd report.

The Dutch Government reported that within the Netherlands, there is far enough food of good quality and affordable for everyone, basically due to a sound system of social security. However, most cases in which the right to food is invoked concern asylum-seekers, who have no automatic access to this system of social security. This fact is therefore often discussed with and criticised by the Committees during the report cycles.

In general, it can be concluded that the Netherlands, in their reports, make a very clear distinction between on the one hand, civil and political rights, and on the other hand, economic, social and cultural rights, especially when it concerns the possibility of direct applicability: it is unambiguously clear that the Netherlands do not recognise the direct applicability of ECOSOC rights. An approach that led to fierce debates during the reporting cycles.

It is in this light remarkable that the Dutch human rights ambition seems to be focussed externally, which is *inter alia* expressed in the Parliamentary Documents, where more than once the added value of the ratification of a human rights treaty was perceived to be that after ratification, the Netherlands could with more conviction address human rights issues in other countries.

10.5 The direct applicability of economic, social and cultural rights in the Netherlands

The Dutch Judiciary seems to follow the lead of politics when a decision must be made on the direct applicability of ECOSOC rights. According to the Legislature, civil and political rights in general are suitable for direct applicability, and ECOSOC rights are not, due to the fact that for the realisation of ECOSOC rights policy decisions must be made, which is the responsibility of politics instead of Courts. This point of view is quite persistently expressed by the Government in the various Explanatory Memorandums on the Bills on the ratification of the international human rights treaties, but also during the various reporting procedures. The Government expresses similar opinions in case of international treaties for specific groups that address at the same time civil and political as well as economic, social and cultural rights, albeit not always as specific as for instance the Explanatory Memorandum to the ratification Bill of the ICESCR. It appears that when the Government did express itself clearly on the direct applicability of human rights Provisions, normally, the Judiciary doesn't hesitate to apply this view in its rulings. When the Government however is not so specific, the Courts appear to take the liberty in reaching their own conclusions with regard to the matter, based on arguments that are of a more practical nature and closer related to the case. This is clearly demonstrated with regard to the right to strike (Article 6 ESC), which is often referred to in literature when scholars address the issue of direct applicability of ECOSOC rights, or human rights in general.

Four main criteria can be distinguished in order to determine whether an international Provision containing human rights is directly applicable: (1) the intentions of the State Parties of the international document expressed during the negotiations preceding its final draft; (2) the nature and content of the Provision, as well as its wording; (3) the intentions of the national Legislature expressed in its Explanatory Memorandum; and (4) the existence of legislation that has the purpose of fulfilling the obligations coming forth from the Provision in question. The fourth criterion was only mentioned in the Parliamentary Documents of the Act of approval of the ICRC. It was hardly explained and has not been mentioned elsewhere since then. Furthermore, due to the various constitutional systems worldwide it seems unlikely that state parties will express their vision on direct applicability during the negotiations preceding and international human rights document. This leaves criteria (2) and (3) as the criteria that truly matter. There seems to be a difference between on the one hand the hierarchy as explained by both the Dutch Legislature and the Judiciary on the one hand, and the hierarchy that is actually used by the Courts with regard to direct applicability of ECOSOC rights. Officially, the Courts should look at the nature and content of the Provisions, as well as their wording in the first place, and voluntarily use the expressed intention of the Legislature as a source of inspiration only. This research demonstrates that in situations in which the Legislature clearly expressed its view on the possibility of direct applicability of international ECOSOC standards, the Courts duly follow this interpretation without giving due consideration to the nature, content or wording of the Provision. Only when the Legislature is unclear in its vision, the Courts feel free to base their verdict on the nature, content and wording of the Article. This is most clearly demonstrated in the 'railway-strike' ruling, concerning the right to strike ex Article 6 ESC. This led to a very clear and consequent approach by the Judiciary, basically due to the fact that the right to strike as formulated in this particular Provision has characteristics of a civil or political right (due to the phrase 'recognise'), and was clearly used as a substitute for failed attempts of the Legislature to agree on a Dutch Act that would regulate this right.

It is in this light noteworthy to mention that with regard to individual complaint procedures for international treaties in which ECOSOC rights are embedded, the Netherlands take a very cautious position, but in their pursuit for being known as a human rights defender in the international arena do not necessarily reject the idea, and officially support most initiatives. For instance, the Netherlands signed the OP-ICESCR. This leads to the contradicting situation in which the Netherlands officially support individual complaint procedures internationally, but nationally persist in rejecting the idea of direct applicability of international standards containing ECOSOC rights.

10.6 Comparing the legal practice with the reporting behaviour

In general, it can be concluded that the Netherlands, in their reports, are very clear about the distinction between on the one hand, civil and political rights, and on the other hand, economic, social and cultural rights, when it concerns the possibility of direct applicability. An approach that led to fierce debates during the reporting cycles. However, this being said, the Dutch reports are full with window-dressing when they try to paint a positive image of the Dutch implementation attempts of the ICESCR rights, including the substantive right to food. Simultaneously, the Dutch do not seem to be eager to answer critical questions, or to reflect on points for improvement. The Dutch attitude led to some remarkable statements on their domestic case law which are hardly in line with legal reality. It is for instance unclear, how the Dutch Government could come to the conclusion that the Dutch judges are familiar with the General Comments of the CESCR, and also apply them, or that Article 11 ICESCR has any effect in Dutch case law. While in general the Committees will probably have a rather accurate impression of the status of Dutch case law, the Dutch reporting behaviour cannot be said to be in full conformity with legal reality.

Part 3
Belgium

11. The Belgian Constitutional System and the applicability of international legal standards

11.1 Introduction

Before analysing the Belgian Court law regarding the direct applicability of the right to adequate food, it is first necessary to explore the particularities of the Belgian Judiciary. As it will appear below, the organisation of the Judiciary is heavily influenced by the complexity of the Belgian *trias politica*, with three main Legislators sharing equal and exclusive powers. Due to this Constitutional organisation, one Court was established to rule in matters of competence, and evolved into a Constitutional Court, dealing with *inter alia* matters of human rights law. The right to food – and ECOSOC rights in general – is considered in the case law of especially this Court, which shows many characteristics of a Court of Last Instance. This Chapter will be dedicated to thoroughly explore the Constitutional context in which the Belgian Judiciary operates in relation to the Legislature. Therefore, firstly, the Belgian Legislature (Section 11.2) and the legislative process (Section 11.3) will be analysed, to get a better understanding in what context the Belgian Judiciary operates. Furthermore, of course, the Belgian Judiciary will be explored (Section 11.4), in which both the subjective litigation (Section 11.4.1), and the objective litigation (Section 11.4.2) will be analysed, with special attention to the functioning of the Constitutional Court (Section 11.4.2). Also, the differences between the subjective and objective litigation will be discussed (Section 11.4.3). The functioning of international standards in the domestic legal order will be examined in Section 11.5, and finally, some conclusions will be drawn in Section 11.6.

11.2 The Belgian Legislature

Belgium has four language areas (Article 4 CA): the Dutch, French, German and the Brussels-Capital Language Area (the latter is both Dutch and French). Within the boundaries of the Dutch, French and German language area, the Government is obliged to communicate with its citizens in this language. In the Brussels-Capital Language Area however, all general communication is done in both Dutch and French, and communication with an individual will be in the language of his/her choice.

Since the Constitutional reforms of 1970, the Federation is subdivided in Communities on the one hand and Regions on the other. The Communities were established to maintain the cultural identity of the different language groups in Belgium, and therefore geographically overlap with the language areas, except for the Brussels-Capital Language Area, which falls under the authority of both the Dutch and the French Community.

Due to rather huge economic differences throughout Belgium, three Regions were established in order to facilitate a suitable economic approach in each different Region: the Flemish, Walloon and the Brussels-Capital Region. Geographically, the Flemish Region and the Dutch Language Area are the same, as well as the Brussels-Capital Region and the Brussels-Capital Language Area. The Walloon Region includes both the French and the German Language Area.

The Federation, the Communities and the Regions have legislative powers that are based on the principles of equality and exclusiveness. The Federal Legislator enacts Statutes, the Community and Region Legislators enact decrees, except for the Brussels-Capital Region Legislator, who enacts ordinances. These Acts require a majority in the relevant representation body (in most cases the Parliament involved, see Chapter IV, CA), except for the Brussels-Capital Region Legislator, who may only enact a law with a majority in both language groups. All Legislators may under circumstances make laws that require a $2/3^{rd}$ majority, which are naturally of a higher order compared to Statutes, decrees and ordinances that require a normal majority (or a majority in both language groups). There is no hierarchy amongst Statutes, decrees and ordinances.

The federated areas are divided into provinces (except for the Brussels-Capital Region) and municipalities, that make provincial and community regulations concerning provincial and community matters that do not fall under the authority of the Federation, the Communities or the Regions.

11.3 The Belgian legislative process

A general principle is applied in Belgian law, which is that lower legislation may not be in violation with higher legislation. Therefore, it is important to understand the hierarchy of Belgian legislation, which is as follows: (1) international law; (2) Statutes, decrees and ordinances adopted with a $2/3^{rd}$ majority; (3) normal Statutes, decrees and ordinances; (4) executive decisions from the Federation (Royal Decrees), the Communities and the Regions; (5) provincial regulations; and (6) municipal regulations. The Judiciary is therefore obliged not to apply lower legislations that conflict with higher legislation. However, the Constitutional Court is exclusively competent in reviewing Statutes, decrees and ordinances against Articles 10, 11, and 24 CA and other Constitutional Provisions defined by law (Article 142 CA, Sections 2 and 3).

With regard to international relations, the Federation, Communities and Regions are entitled to establish their own foreign policy and enter into international obligations as long as it falls within the scope of their substantive competences.

Belgium is however considered to be a single entity in international law, and will therefore be represented by the federation in international forums.

On the Federal level, the Legislator consists of the King and the Parliament. The King himself has very limited powers, but the cabinet carries the responsibilities for the King in the legislative process. Therefore, the Council of Ministers will speak *'in the name of the King.'* The Parliament consists of the House of Representatives (150 representatives that are elected directly by the electorate) and the Senate (40 Senators that are directly elected, 21 community Senators, 10 co-opted Senators and a varying number of Senators appointed by law, which are normally the children of the King, see Article 67 Constitution). In the procedures of installment of both Houses, there is a strong emphasis on a proportional representation of all language groups.

The legislative process consists of four stages. In the first stage, the King (the cabinet), or one of the Members of Parliament use their right to initiative (Article 75 Constitution). When the King (one or more Ministers) takes the initiative, and the preliminary draft is approved by the Council of Ministers, the Bill has to be submitted to the Council of State, which will make recommendations. When a Member of Parliament takes the initiative, a consultation of the Council of State is only under certain circumstances mandatory, for instance when one third of the relevant Chamber votes for such consultation. In such a case, the Chairman of the Chamber will ask the Council of State to submit its advice.

In the second stage, the Bill is discussed in Parliament (in both the House of Representatives and the Senate). This will be done in one of the specialised commissions first, consisting of a number of Parliament Members. In both Houses such commissions are established. The Members of these commissions may propose amendments to the bill. Hereafter, the Bill will be discussed in a plenary session of the Chamber in question, and all Parliamentarians may propose amendments. Finally, the Chambers will vote on the bill and the proposed amendments. To pass a Chamber in case of a normal Bill, a majority of the Chamber must vote, and the majority of the voters must vote in favor of the bill. Members that vote have three options: 'yes', 'abstain' and 'no.' To pass a Chamber in case of a bill requiring a $2/3^{rd}$ majority (also referred to as 'special majority' or 'communitarian majority'), a majority of every language group in the Chamber must approve the Bill. From each language group, a majority must participate in the voting, and a majority of the voters must approve with the bill. If this is the case, then $2/3^{rd}$ of all voters must approve of the Bill in order to let it pass the Chamber. In general, Bills that affect the relationship between the Federation, the Communities and the Regions will need a $2/3^{rd}$ majority to pass a Chamber.[1094]

[1094] See *inter alia* Articles 4 and 5 CA.

The distribution of competences between the House of Representatives and the Senate during this stage in the legislative process differs per legal issue. There are three ways in which the legislative competences are divided. Firstly, the House of Representatives and the Senate have equal legislative powers (Article 77 Constitution). A Bill will have to pass both Chambers. If one of the Chambers does not approve with the Bill, it will amend the Bill and send it back to the other Chamber, and vice versa, until both Chambers approve. For instance, ratification Bills of international treaties fall under this category. Secondly, the House of Representatives has sole competences, while the Senate has no powers (Article 74 Constitution). For instance, the Federal Finance Act and Federal social security legislation fall under the competence of the House of Representatives only. Thirdly, the Senate serves as a reflection Chamber, but has no legislative powers (Article 78 CA, see also: Articles 79-82 CA for procedural matters). In case the Senate decides to propose amendments to a bill, the House of Representatives may decide whether to accept these or not.

In the third stage, the Bill has been approved by Parliament, and will be presented to the King, who will ratify the Bill, together with a Minister of the cabinet (since the King himself has no or limited legislative responsibilities).

In the fourth stage, the King – this time in his capacity as Member of the executive power – will proclaim the Bill, ordering that the Bill shall be implemented in the Belgian legal order (Article 109 Constitution).

Finally, the proclaimed Bill will be published in the Belgian State Gazette (*Belgisch Staatsblad*). Unless arranged for otherwise, the Bill is a legally binding Act from the tenth day after its publication.

11.4 The Belgian Judiciary

In Belgium, the Judiciary consists of the Courts recognised in the Constitutional Act (Article 40), and bodies that are established by the Legislator (Article 146 CA) which are the Administrative Courts, the Council of State (Article 160 CA) and the Constitutional Court (Article 142 CA).

The Belgian Judiciary operates in two areas: a subjective litigation, in which the verdicts apply *inter pares*, and an objective litigation, in which the verdicts apply *ergo omnes*.

11.4.1 The Subjective Litigation

The Courts within the subjective litigation rule on disputes concerning civil and political rights. According to the Constitutional Act (Articles 144 and 145), the common Courts have exclusive jurisdiction over disputes concerning civil rights.

Depending on how one defines political rights, the Courts have also jurisdiction over disputes concerning political rights when the Legislature did not explicitly authorise another judicial body. The Legislature established several Administrative Courts and tribunals for specific disputes concerning political rights (Article 145 Constitution). The difference between civil and political rights is not easy to make. In both academic literature and within the Judiciary, the characteristics of civil and political rights are discussed, but no clear conclusion can be drawn yet. Basically, two approaches are defended in order to determine what civil and political rights are. The first is a substantial approach, in which rights are labeled 'civil' or 'political' based on their nature. In this approach, it can be considered that civil rights *'include all subjective rights in relation to the status and capacity of the person, his freedom and ability and also the most fundamental rights and freedoms'*[1095], which includes matters of private law and criminal law.[1096] Traditionally, political rights can be considered to include *inter alia* rights in which civilians contribute to or participate in the execution of public powers, and rights where civilians enjoy services and other (social) benefits from the Government.[1097] These lists are not exhaustive, and it is hardly possible to clearly and exhaustively define the characteristics of these two groups of rights. The second approach is the so-called classification based on an 'organic criterion.' In this context, civil rights can be understood as all subjective rights that are not political rights, and political rights can be understood as all rights whose disputes are entrusted to an administrative body by the Legislature (Article 145 Constitution). This circular reasoning seems to be more or less accepted,[1098] for it appeared impossible to use any substantive criteria. According to the Constitution, the Court of Cassation is authorised to judge in attribution conflicts (Article 158 Constitution), but the Constitutional Court considered that it is competent in judging whether the choice of the Legislature to entrust certain issues to an administrative body is not subdued to arbitrariness, and therefore in violation with Articles 10 and 11 of the Constitution.[1099] In his ruling, the Constitutional Court appears to return to a 'substantive criterion' in order to determine whether a right is of a civil or political nature, but in doing so giving a rather broad interpretation to 'political rights' that resembles the legal

---

[1095] K. Rimanque, *De grondwet toegelicht, gewikt en gewogen*, Antwerp: Intersentia, 2005, pp. 324-326. Original text in Dutch: *'omvat alle subjectieve rechten in verband met de staat en de bekwaamheid van de persoon, zijn vrijheden en vermogen en ook de meeste fundamentele rechten en vrijheden.'*

[1096] A. Alen and K. Muylle, *Compendium van het Belgisch staatsrecht, Syllabusuitgave* (2ᵉ uitgave), Mechelen: Kluwer 2008, p. 320.

[1097] J. Vande Lanotte, S. Bracke and G. Goedertier, *België voor beginners*, Brugge: Die Keure, 2010, p. 94. In Belgium, the Latin terms *ius suffragii* (active and passive suffrage), *ius honorum* (the right to be eligible for appointment to public service), *ius militiae* (the right and the obligation to be called to military service) and *ius tribute* (the right and the obligation to pay tax) are traditionally used to further subdivide political rights. See: A. Alen and K. Muylle, *Compendium van het Belgisch staatsrecht, Syllabusuitgave* (2ᵉ uitgave), Mechelen: Kluwer 2008, p. 320; K. Rimanque, *De grondwet toegelicht, gewikt en gewogen*, Antwerp: Intersentia, 2005, p. 327.

[1098] K. Rimanque, *De grondwet toegelicht, gewikt en gewogen*, Antwerp: Intersentia, 2005, pp. 327-328.

[1099] Constitutional Court, 14/97, 18 March 1997.

practice. It is to be expected therefore, that this Court will be very cautious in ruling that a right is not of a political nature when the Legislature established an Administrative Court for issues related to that right.[1100]

11.4.1.1 Civil rights

In Belgium, the Courts of First Instance are spread throughout the country and deal with the more common cases: the Peace Courts have jurisdiction over simple civil cases and trade disputes, while the Police Courts have jurisdiction over violations and traffic offences, including claims for compensation between the parties involved coming forth from traffic offences.

For the less common cases, one needs to turn to a specialised Court that then serves as a Court of First Instance, while on the other hand these Courts may also serve as a Court of second appeal for those cases dealt with by the Peace and Police Courts. These Courts are in the first place the so-called Courts of First Appeal, with three departments: the Civil Court (for civil disputes), the Correctional Court (for cases regarding misdemeanors[1101]) and the Juvenile Court (for all disputes regarding minors). In the second place, there are Commercial Courts that have jurisdiction over trade issues and disputes on companies. Thirdly, there are Labour Courts that deal with labour disputes and disputes regarding social affairs. It is noteworthy here that most cases with regard to economic, social and cultural human rights are dealt with by the Labour Courts in first appeal. Fourthly, there are District Courts that deal with issues related to the jurisdiction of Courts. In the fifth place, there are Sentencing Courts that deal with issues relating to sentencing modalities.

Third in hierarchy are the Courts of Appeal and the Court of Assizes. The Courts of Appeal consist of three departments: the Civil Court, the Correctional Court and the Juvenile Court. Furthermore there is a special Court of Appeal for the cases dealt with by the Labour Courts. The Court of Assizes has jurisdiction over criminal matters, political crimes and press offences, and is a Court of First Instance.

The Court of Cassation is the highest Belgian Court in the subjective litigation that deals with civil rights (and some political rights, depending on one's reasoning). This Court considers whether the lower Courts acted in accordance with the relevant legal standards and formal requirements, but does not reconsider the case in itself (Article 147 CA).

[1100] A. Alen and K. Muylle, *Compendium van het Belgisch staatsrecht, Syllabusuitgave* (2e uitgave), Mechelen: Kluwer 2008, pp. 321-322.

[1101] A misdemeanour is considered to be more severe than a violation, but not as severe as a crime.

11.4.1.2 Political rights

Disputes related to political rights in the subjective litigation that fall under the jurisdiction of Administrative Courts installed by the Legislature in accordance with Article 146 CA are dealt with by the various Administrative Courts. Normally, there is always the possibility of higher appeal, although this is arranged for in very different ways. With regard to cases concerning political rights, the Council of State is the highest Court and has similar authorities compared to the Court of Cassation (Article 14 § 2 Council of State Act). One may only appeal to the Council of State when all other legal remedies are exhausted. The Legislature occasionally decides that the Court of Cassation serves as highest Court in political right issues instead.

One example of an Administrative Court that is established by the Legislature ex Article 146 CA is the Council for Aliens Disputes, installed by Statute of 15 September 2006 CA.[1102] The Council is an Administrative Court with an exclusive jurisdiction over complaints against individual administrative decisions regarding the access to, residence in and the removal from Belgium. The Council's jurisdiction basically encompasses two procedures. In the first, the Council functions as a Court of appeal regarding individual decisions made by the Commissioner-General for refugees and stateless persons. In these cases, the Council re-examines the presented facts, and may confirm or reform the disputed decision, or – when the Council is unable to do so based on the presented facts – the Council may annul the decision.[1103] In the second procedure, the Council may annul *ergo omnes* an administrative action that violates formal requirements, or exceeding or diversion of power, that are related to the access to, residence in and the removal from Belgium. The Council only examines the conformity with the law, and the verdict does technically not concern subjective rights. However, as will appear in Section 12.2, the differences between the legal effect of both procedures seems to be very hypothetical, while in practice, also the verdicts in annulment procedures have a strong (indirect) effect *inter pares.*[1104]

[1102] Wet van 15 september 2006 tot hervorming van de Raad van State en tot oprichting van een Raad voor Vreemdelingenbetwistingen, Belgian State Gazette, 6 October 2006, 53.468. In Dutch: 'Raad voor Vreemdelingenbetwistingen.' The Council replaced the Permanent Appeals Commission for Refugees (in Dutch: Vaste Commissie voor Vluchtelingen).

[1103] Note that before the instalment of the Council for Aliens Disputes, an appeal against the decision of the Commissioner-General was directly brought before the Council of State.

[1104] According to the Council of Aliens Disputes (see its Annual Report 2008-2009, pp. 20-21, available at www.rvv-cce.be), this second procedure mostly concerns decisions that are summed up in the attachments of the Royal Decree on the access to the area, the residence, the establishment, and the removal of Aliens, BS 27 October 1981, 13.740, attachments (Koninklijk Besluit betreffende de toegang tot het grondgebied, verblijf, de vestiging en de verwijdering van vreemdelingen).

11.4.2 The objective litigation

Two Courts operate in the objective litigation: the Council of State (Article 160 CA) and the Constitutional Court (Article 142 CA), which is the former Court of Arbitration.

11.4.2.1 The Council of State

As demonstrated above, the Council of State operates in the subjective litigation as Court of Last Instance in cases concerning political rights. With regard to the objective litigation, the Council of State is entitled to nullify all administrative acts, such as Royal Decrees, Ministerial Decisions, and Decisions by decentralised authorities, if they are in conflict with any higher legal standard, including international legal standards that are directly applicable, and general legal principles, such as the general principles of good governance. Every person or group of persons that demonstrates to have an interest in the disputed action may invoke an annulment (Article 19 Council of State Act). Interest appears when the complainer experiences a disadvantage due to the disputed action, and the disadvantage will cease to exist when the action is nullified.[1105] The Council of State then will take into consideration whether the disputed administrative act is lawful and opportune. The latter is possible due to the fact that the appeal is technically an Administrative Appeal, which is not a part of the competences of the Judiciary (that will usually solely rule on the lawfulness of an act), but rather part of the competences of the Administration in itself.[1106] If the Council of State nullifies an administrative act, the ruling applies *ergo omnes*. As a consequence, the ruling is binding for the administrative body whose act has been nullified, who is therefore not allowed to reestablish a similar action. The ruling is also binding for the Courts, which are obliged to award compensation in case of the existence of a causal relation between the nullified action and damage to the parties involved. Finally, the ruling is also binding for the Legislature, who is not allowed to validate a disputed or nullified administrative act.[1107] Since 1989, an interim injunction proceeding can be raised before the Council of State. The complainer must base his/her claim on serious legal remedies, and demonstrate that the administrative act when executed will cause serious damage that is difficult to repair (Articles 17 and 18 Council of State Act).

[1105] G. Maes, *De afdwingbaarheid van sociale grondrechten*, Antwerp: Intersentia, 2003, p. 17.

[1106] A. Alen and K. Muylle, *Compendium van het Belgisch staatsrecht, Syllabusuitgave* (2ᵉ uitgave), Mechelen: Kluwer 2008, p. 335.

[1107] A. Alen and K. Muylle, *Compendium van het Belgisch staatsrecht, Syllabusuitgave* (2ᵉ uitgave), Mechelen: Kluwer 2008, pp. 341-342.

11.4.2.2 The Constitutional Court

As will appear in Chapter 12, the Constitutional Court rules on most cases in which international economic, social and cultural rights are invoked. During the Court's relatively short history – the Court was only established in 1980 – its authorities shifted from a Court that was authorised to judge on conflicts of competences (a Court of Arbitration) to a Court with the full competences of a Constitutional Court. During this period, the competences of the Court were broadened at three stages. To fully understand the Court's verdicts, it is therefore necessary to examine these stages before analysing its verdicts.

- The first stage: 1980-1989

  Due to the state reforms that took place since 1970,[1108] the division of powers was considerably altered by the recognition of the Regions and Communities that now have exclusive and autonomous legislative powers equal to the legislative powers of the Federation. Not surprisingly, there was a need for an institution that could rule on conflicts of these powers. Therefore, the Court of Arbitration was founded in 1980.[1109] Its original authorities were limited to settle conflicts of power that could arise between the three legislative powers in Belgium. Because the Court would have to rule on conflicts between (democratically chosen) Legislators, the choice was made to compose the judges of 6 jurists and 6 politicians (of whom 6 were Dutch-speaking and 6 were French-speaking). During this first period, the Court's work was limited to a few rulings in which the general principles of division of powers were developed (the so-called '*krachtlijnen*').[1110] The powers of the Communities and Regions were interpreted broadly.

- The second stage: 1989-2003

  In 1989, the Belgian Constitution was reviewed. The old Article 107ter was replaced by the new Article 140, broadening the competences of the Constitutional Court. Cause for this review was the necessity to solve problems in the sphere of the protection of minorities after the transfer of powers regarding educational matters to the Communities. In Belgium, the freedom of education has always been a sensitive issue, especially concerning the freedom of choice for non-confessional education. After a century of political struggle, in which a balance between confessional and non-confessional education was sought, the 'school-pact' of 1958 was concluded, whose principles are now laid down in Article 24 of the Constitution that stipulates the recognition of equal treatment of different forms of education (unless a different treatment can be justified

---

[1108] Especially the first two state reforms. During the first reform (1970-1971), three culture communities, four language areas and three regions were recognised. During the second reform (1980), the culture communities became communities, and the legislative powers of the communities and regions were recognised. See also: Karel Rimanque, *De grondwet toegelicht, gewikt en gewogen*, Antwerp: Intersentia, 2005, pp. 6-8.

[1109] Article 107ter Constitutional Act (old), the current Article 142 Constitutional Act.

[1110] Freely translated: 'power principles'.

based on objective differences).[1111] By broadening the competences of the Court of Arbitration in such way that decrees from the communities concerning educational matters could now be reviewed against Article 24 Constitution, the 'school-pact' and its forthcoming 'school-peace' should be preserved.

The Court's competences were also broadened by allowing it to review legislation against Articles 10 and 11 of the Constitution. Both Articles are often mentioned together and stipulate the equality of all Belgian citizens (Article 10) and the principle of non-discrimination (Article 11). Especially the review of legislation against Articles 10 and 11 of the Constitution resulted in remarkable and extensive case law, in which the Court appeared to have interpreted its competences in a very broad manner.[1112] The Court reviewed against both Constitutional Articles *in conjunction* with other legal standards. Via Articles 10 and 11 Constitution, the Court reviewed legislation indirectly against the other rights that are recognised in Chapter II of the Constitution,[1113] against Provisions from international treaties and Covenants that are directly applicable,[1114] and against treaties and Covenants in general that are binding to Belgium, regardless of the fact whether its Provisions are directly applicable or not.[1115] Also, the Court reviewed against general legal principles.[1116]

During the first stage, only the three autonomous legislative powers had access to the Court of Arbitration. Since the Constitutional reform of 1989 however, every natural legal entity that demonstrates interest may request the annulment of legal standards as well.

- The third stage: since 2003

  The reviewed Constitution also stipulated the authority of the Federal Legislator to further broaden the competences of the Court with regard to the legal standards it may review against by adopting a Statute with a 2/3rd majority. On 21 April 2003, during the Fifth Belgian State Reform (2001-2003), the Federal Legislator used this authority to further broaden the competences of the Court of Arbitration by the coming into effect of the adopted special Act of 9 March 2003.[1117] This Act basically acknowledges the established practice of the Court to review legislation indirectly (via Articles 10 and 11 Constitution) against all rights that are recognised in Chapter II of the Constitution. In addition, the

---

[1111] See: K. Rimanque, *De grondwet toegelicht, gewikt en gewogen*, Antwerp: Intersentia, 2005, pp. 79-82, 314-315; A. Alen (ed.), *20 jaar Arbitragehof*, Mechelen: Kluwer, 2005, p. 7.

[1112] A. Alen, *Twintig jaar grondwettigheidstoezicht op wetgevende normen, Krachtlijnen van de rechtspraak van het Arbitragehof van 1985 tot en met 2004*, in: *Twintig Jaar Arbitragehof*, Mechelen: Wolters Kluwer België, 2005, p. 7-10.

[1113] Constitutional Court, 23/89, 13 October 1989.

[1114] Constitutional Court, 18/90, 23 May 1990; 22/94, 8 March 1994, consideration B.1.

[1115] Constitutional Court, 41/2002, 20 February 2002; later reaffirmed in Constitutional Court, 75/2003, 28 May 2003; Constitutional Court 106/2003, 22 June 2003; and Constitutional Court 189/2004, 24 November 2004.

[1116] Constitutional Court, 72/92, 18 November 1992.

[1117] Bijzondere wet van 9 maart 2003 tot wijziging van de bijzondere wet van 6 februari 1989 op het Arbitragehof.

Court could also review legislation against Articles 170, 172 (equality regarding taxes) and 191 (equality between aliens and Belgians) of the Constitution. The jurisprudence of the Court did not change considerably, for the broadening of its competences was merely a way to avoid unnecessarily complicated legal constructions regarding indirect violations of rights via Articles 10 and 11 Constitution. The Court could now directly review against the rights embedded in Chapter II of the Constitution. Therefore, the case law of the second stage was basically continued: the Court would in a majority of the cases keep on adding Articles 10 and 11 as a standard that would be reviewed against, in addition to the legal standard of the second Chapter of the Constitution. It is the view of the Court that a violation of a fundamental right is *ipse facto* a violation of Articles 10 and 11 Constitution.[1118]

With regard to international human rights law, it is the opinion of the Court that when an international Provision, legally binding to Belgium, has a similar scope compared to one of the Constitutional Provisions the Court may review against, the content of this international Provision is inextricably linked to its national counterpart. Therefore, this international Provision must also be taken into account when legislation is reviewed against a Constitutional norm.[1119] This does not mean that legal standards are reviewed against international legal standards directly (for this would still fall outside the competence of the Court), but the content of the international standard certainly plays a role concerning the interpretation of the analogous Constitutional standard. Valaers distinguishes three possible situations in this context.[1120] The first is that the analogous international Provision offers no broader protection compared to the national Constitutional norm, or the complaining party does not base substantially different arguments on the international standard compared to the arguments based on the national Constitutional standard. In such a case, the Court will not feel the necessity to involve the international standard in its review. In the second case, the international standard may offer somehow a broader protection compared to the analogous national Constitutional standard. In such a case, the Court will strive to offer the broadest protection to the parties involved, and the protection that is offered by national Constitutional Provision will thus be broadened by the additional protection offered by the international standard. A third possibility is that an international standard has no analogous national Constitutional standard. In this case, the Court will most likely turn to the method of reviewing against the international indirectly, via Articles 10 and 11 Constitution.

[1118] Constitutional Court, no. 136/2004, 22 July 2004, Section B.5.3.; reconfirmed in no. 158/2004, 20 October 2004, Section B.5.2.; no. 202/2004, 21 December 2004, Section B.2.2.

[1119] Constitutional Court, no. 136/2004, 22 July 2004, Section B.5.3.; reconfirmed in no. 158/2004, 20 October 2004, Section B.5.2.; no. 202/2004, 21 December 2004, Section B.2.2; no. 101/2005, 1 June 2005, consideration B.2.3.

[1120] See: J. Valaers, *Samenloop van grondrechten: het Arbitragehof, titel II van de Grondwet en de international mensenrechtenverdragen*, in: A. Alen (ed.), *20 jaar Arbitragehof*, Mechelen: Kluwer, 2005.

The Constitutional Court has not only the authority to nullify Statutes, decrees and ordinances, but is also entitled to receive preliminary questions. To promote coherent case law, the Courts, the Administrative Courts and the Council of State are obliged to ask a preliminary question when in their case the question arises whether Statutes, decrees or ordinances are in conflict with one of the standards the Constitutional Court may review against. The Constitutional Court may then rule whether the disputed standard is in conflict with one of these standards or not. The effect of such a preliminary ruling is that the Court that asked the question has to implement the view of the Constitutional Court in its actual decision on the case. The effect of a preliminary ruling is supposed to have *res judicata*, although, as will be demonstrated below, the Courts do not always willingly apply the verdicts of the Constitutional Court.[1121]

11.4.3 The subtle difference between the subjective and the objective litigation

The effect of the difference between the subjective and objective litigation however, must not be overestimated. It clarifies on the one hand the structure and functioning of the Belgian Judiciary, and, could be seen as one – among other – reasons why the reasoning patterns of the Constitutional Court are so different compared to the Court of Cassation or the Council of State. Of course, the fact that one Court rules on the legality of national legislation, and specifically reviews that legislation against Constitutional Provisions, leads to a different approach in case law compared to disputes *inter pares*.

However, on the other hand, the difference has only a limited effect on the actual nature and effect of the verdicts for the involved parties, due to some developments in legal practice. Gunter Maes, who dedicated his PhD thesis to the matter of enforceability of social fundamental rights in Belgium,[1122] thoroughly explained why the difference between the subjective and the objective litigation is a very subtle one, with convincing arguments.

He argues in the first place that the Council of State, in its annulment cases, may indirectly judge in matters concerning subjective rights. The Council of State interpreted its competences broadly, and will therefore tend to receive complaints, also when the disputed administrative act has effect on a very limited group of individuals. The complainers will hardly act for the greater good or general interest, but rather act for their own benefit, defending their subjective interest. Nullifying such an act will therefore indirectly protect subjective rights as a result,

---

[1121] See for instance below, Section 12. 5.3.7.
[1122] G. Maes, *De afdwingbaarheid van sociale grondrechten*, Antwerp, Grondingen, Oxford: Intersentia, 2003, no. 28-49. He also uses a fourth argument that relates to the European Court of Justice. Since the argument has a very limited link with this research, it was excluded. See no. 50-54.

for the annulment will technically have effect *ergo omnes*, but has practically only an effect on the interest of a specific group of individuals. Only in matters in which the Administration refuses to execute a specific legal obligation while the complainer is entitled to its execution, based on unambiguous legislation with no margin of appreciation for the Administration, the Council of State will decide the complaint to be inadmissible, due to the fact that it concerns a clear subjective right. Perhaps in addition to Maes, a same argumentation may apply to the annulment procedures of the Council for Aliens Disputes. The Council may nullify an Administrative Act *ergo omnes*, but practically, the annulment will have almost exclusively consequences for the subjective rights of the complainer. This may clarify why there seems to be hardly any difference in approach or reasoning patters of those Courts which operate in both the subjective and the objective litigation between the two correlating procedures.[1123]

Maes argues in the second place, that the Courts in preliminary injunction procedures, being part of the subjective litigation, also interpret their competences in a broad way, leading to judgments that are different from rulings of (Administrative) Courts on standard cases, showing some characteristics of the objective litigation. In preliminary injunction procedures, the Courts appear to restrict their considerations to a balancing of interest rather than ruling on subjective rights. This can be explained by the function of a preliminary injunction procedure: the task of a Court is then to judge a case preliminary, and on the surface.

Thirdly, Maes observes that especially in the case law of the Council of State, and also by special legislation, the direct interest of the plaintiff is understood rather broadly, for in some proceedings within the subjective litigation, interest groups are admitted as requesting party/plaintiff in cases that relate to their specific statutory purposes. While an interest group will primarily be interested in the effect of a verdict for the common good, the procedure in itself was designed to have effect *inter pares*.

Fourthly, Maes argues that the Constitutional Court considers that cases, especially regarding immigration law in which the Administration has a certain margin of appreciation, falls within the subjective litigation since they concern political rights, thereby using the earlier mentioned '*organic criterion.*' Therefore, it is possible that the Council of State, in its capacity as Court of Last Instance, has jurisdiction in these cases, while when applying a classic approach, this would rather fall under the objective litigation.

In addition to the arguments of Maes, developments in case law of the Constitutional Court might be added. As discussed in Section 11.4.2.2, the Constitutional Court gradually developed from a Court of Arbitration between Legislators to a full

---

[1123] The Court of Cassation, but also the Council for Aliens Disputes.

Constitutional Court. Now, also natural legal entities whose interest is affected can start an annulment procedure before the Court, and the Court's reviewing competences are broadened to all rights embedded in the Constitution. As a result, the cases brought before the Court, as will be demonstrated in Section 12.5, concern very specific groups of persons who claim that their fundamental rights are violated by Legislative Acts, and consequently, the verdict has a strong indirect effect on this group (especially in the cases concerning the entitlements to social benefits). Furthermore, the Constitutional Court seems to gradually move beyond merely ruling whether or not a national Provision is in violation with a Constitutional Provision. Especially in preliminary rulings, in which the Court is asked by a Court operating in the subjective contentious (in this research, mostly the Labour Courts) whether a certain application of a law is in violation with a Constitutional Provision, the Constitutional Court has the tendency to prescribe a specific interpretation to a legal rule in a particular case.[1124] Due to the enhanced *res judicata*, although its effect is disputed, those interpretations of the Constitutional Court may have a profound effect in the subjective litigation.

To conclude, the legal effect of verdicts in the subjective and an objective litigation for the parties involved seems to be increasingly similar, which is exactly the point Maes tries to emphasise. He writes in the context of the issue of applicability of ECOSOC standards. Traditionally, it can be considered that rights (whether recognised nationally or internationally) are considered to be enforceable rights when they are specific and unambiguous enough to be applied without the need for further clarification by a Legislator, and addressed to citizens. According to Maes, this approach is also supported by the Court of Cassation and the Council of State (see also Sections 12.3 and 12.4), which generally consider that ECOSOC standards are no subjective rights as stipulated in Articles 144 and 145 CA, and thus the Courts operating in the subjective litigation are not competent in reviewing against ECOSOC Provisions.[1125] By nuancing the difference between the subjective and objective he finds an argument against this traditional approach. He emphasises that the applicability of ECOSOC standards, due to this subtle difference, should not necessarily be excluded from the subjective litigation. Gunter Maes basically writes about the effect of verdicts in relation to the invoking parties. Indeed, as discussed above, the difference between the subjective and objective contentious should not be overestimated in that light. On the other hand, it should not be marginalised as well, for the differences explain the organisational structure of Belgian Judiciary, and also greatly influence the reasoning patterns used by the Constitutional Court, operating in the objective litigation only.

[1124] J. Smets, *De verhouding van het Arbitragehof tot de verwijzende rechter in het prejudiciële contentieux*, in: *Twintig Jaar Arbitragehof*, Mechelen: Wolters Kluwer België, 2005.

[1125] G. Maes, *De afdwingbaarheid van sociale grondrechten*, Antwerp, Grondingen, Oxford: Intersentia, 2003, no. 21-27.

11.5 Belgian monism

It is generally accepted that Belgium has a monistic system.[1126] However, the Constitution does not regulate the relation between international standards and national standards. Instead, Belgian monism appears from case law. In the famous Franco-Suisse Le Ski ruling, the Court of Cassation considered that an international rule is superior to a contradicting (anterior and posterior) national rule, as long as the international rule has direct effect.[1127] The Constitution only regulates the formalities for a treaty to become effective in the internal legal order. Firstly, a treaty needs Parliamentary approval by the relevant legislative body (Federal Parliament, Communities or Region, as stipulated in Article 167 CA. Secondly, the treaty must be published (Article 190 CA) – normally this is done in the Belgian State Gazette.[1128] Usually, these formalities have no influence on the effect of the treaty involved. It is the notion of 'direct effect' that primarily determines whether or not a treaty Provision can be invoked by a person or legal entity. In the case law of the subjective litigation, the Courts seem to judge whether an international Provision has direct effect by applying a subjective criterion and/or an objective criterion. In applying a subjective criterion, the question is asked what the intention was of the State Parties regarding the direct effect of a standard. It is often considered that this is demonstrated by the addressee of the Provision: when the standard is addressed to citizens, the State Parties intended the Provision to have direct effect, while if the standard is addressed to the Contracting States, it was not intended that the standard would have direct effect. In applying an objective criterion, the question is raised whether a standard is indeed self-executing, meaning that an individual can derive a specific right from the standard, without the need for further clarification in national legislation. It is then often accepted that if a standard leaves a considerable margin of appreciation for the national Legislature or Administration, the standard has no direct effect. The Courts in the subjective litigation seem to apply both criteria randomly, while it must be noted that usually the 'direct effect' of a Provision is not thoroughly examined.[1129] Of course, the idea of *'direct effect'* has a very limited meaning in

---

[1126] D. Van Eeckhoutte and A. Vandaele, *Doorwerking van internationale normen in de Belgische rechtsorde*, Instituut voor Internationaal Recht K.U. Leuven, Working Paper no. 33, October 2002.

[1127] Court of Cassation, 27 May 1971, arresten van het Hof van Cassatie, 1971(p. 959). Before this ruling, it was assumed that Belgian had a monistic system, in the sense that international law could have effect in the national legal order without the interference of a national Legislator. However, international law was not necessarily superior to national law. The principle of *'lex posterior derogate legi priori'* generally applied, and therefore, later national law would be superior to contradicting earlier international law. See: G. Maes, *De afdwingbaarheid van sociale grondrechten*, Antwerp, Grondingen, Oxford: Intersentia, 2003, no. 105-108.

[1128] In Dutch: Belgisch Staatsblad.

[1129] See: G. Maes, *De afdwingbaarheid van sociale grondrechten*, Antwerp, Grondingen, Oxford: Intersentia, 2003, no. 112-146; D. Van Eeckhoutte and A. Vandaele, *Doorwerking van internationale normen in de Belgische rechtsorde*, Instituut voor Internationaal Recht K.U. Leuven, Working Paper No. 33, October 2002. See also the Belgian Government in its reports on the implementation of UN treaties, for instance: E/C.12/BEL/4, 18 June 2012, Section 9; CRC/C/SR.222, 6 June 1995, Sections 8-10.

the objective litigation, especially regarding the Constitutional Court since – as discussed above – the concept of 'direct effect' has no real meaning anymore since its ruling in 2002.[1130]

11.6 Concluding remarks

In this Chapter, the Belgian Constitutional System was discussed, particularly in view of the Belgian Judiciary. It can be concluded that there exists a certain complexity that is partly related,[1131] in the organisation of both the Belgian Legislature and Judiciary. The complexity of the Legislature may have found its origin in the profound differences amongst the Belgians in language and economics. The complexity of the Judiciary may be caused by the fact that the organisation is based on purpose and functionality rather than pragmatic considerations. It is this complexity that leads to a very vivid and extensive case law amongst the different Courts, as will be analysed in the next Chapter. Remarkable aspects of the organisation of the Judiciary are firstly, the existence of two Courts of Last Instance. Both have an exclusive jurisdiction, and therefore, opposing case law is not to be expected. However, it is their relation with the Constitutional Court that also shows many characteristics of a Court of Last Instance, which may lead to a 'battle of the forums.'[1132] Secondly, the development of the Constitutional Court, as discussed in Section 11.4.2.2, is remarkable and leads to case law that is of a different nature and with different reasoning patterns compared to all other Courts, as a result of its function. It is to be expected, as will be discussed in Chapter 12.5, that due to the fact that the Constitutional Court is not bound by the principle of 'direct effect' in ruling in matters in which international human rights Provisions are invoked, there is more room for effect in case law of such internationally recognised Articles. Thirdly, the Belgian Judiciary is organised from the perspective of two different litigations. Whereas this difference thus is important to explain the organisational structure, and also the nature and reasoning patterns of the Courts, the differences in result for invoking parties, as discussed in Section 11.4.3, should not be overestimated, for it seems that there is a trend in Belgian case law in which the objective and subjective litigation increasingly blend.

[1130] Constitutional Court, 41/2002, 20 February 2002.

[1131] The installment of the Constitutional Court for instance, is a direct result of the organisational structure of the legislative branche.

[1132] See for instance the matter of the regularisation request and the entitlements to social benefits as discussed in Section 12.5.3.7.

12. The Belgian case law on the enforceability of the right to adequate food

12.1 Introduction

The structure of the Belgian Judiciary is explained in the previous Chapter, in Section 11.4. In this section, case law regarding the right to adequate food will be analysed or, if there is no significant case law on the right to food, case law regarding the UN Treaties in which the right to food is stipulated will be discussed more generally.

As will appear below, most cases in which the right to food is invoked, or ECOSOC rights in general, concern immigration issues, and especially the issue of illegally residing aliens. In this light, two judicial proceedings are commonly used. Firstly, a legal dispute regarding administrative decisions (mostly regarding the residence status of an alien) starts with an appeal to the Council for Aliens Disputes, and may be continued before the Council of State. Secondly, a dispute on social benefits is usually brought before a Labour Court, continued in appeal before a Labour Court of Appeal, and may eventually be brought before the Court of Cassation. However, the Labour Courts appear to frequently submit preliminary questions to the Constitutional Court on the interpretation and legality of legislation.

In this Chapter, there is a strong focus on the extensive case law of the Courts of last instance and the Constitutional Court. However, it will appear that both the Council of State (Section 12.3) and the Court of Cassation (Section 12.4) hardly deal with cases in which the right to food is explicitly invoked. Therefore, the case law analysis of both Courts of Last Instance is extended to the direct applicability of ICESCR, ICRC and CEDAW Provisions in general. Instead, the Constitutional Court appears to have ruled frequently on cases in which the right to food and related ECOSOC rights are invoked, resulting in interesting case on the matter throughout the years. This development will be discussed in detail in Section 12.5.

To keep this Chapter readable, the case law of the lower Courts is only included when significantly relevant. However, an analysis of the case law of the Council of Aliens Disputes is included, for this relatively new administrative body appears to frequently rule in individual and annulation cases in which the direct applicability of especially ICRC Provisions is disputed (Section 12.2). Furthermore, when relevant, case law of the Labour Courts is included, mostly in the context of the subsequent rulings of last instance, or to demonstrate the effect of a ruling of a Court of Last Instance. This case law is not discussed in a separate Chapter, but included where necessary throughout the Sections. This was done especially in Section 12.5, in relation to the preliminary rulings of the Constitutional Court.

12.2 The Council for Aliens Disputes

As explained in Section 11.4.1.2, the Council for Aliens Disputes has exclusive jurisdiction over complaints against individual administrative decisions regarding the access to, residence in and the removal from Belgium.

Because the Council summarises its case law briefly in its annual reports, this will be explored first. Secondly, an analysis will be made of all relevant case law regarding the direct applicability of ECOSOC rights. The Council for Aliens Disputes rules in both the subjective and the objective litigation. In practice however, also the rulings *ergo omnes* have a strong influence on the legal position of the parties involved. Here thus the difference between verdicts in the subjective and objective litigation appeas to be more of theoretical than of practical relevance. Therefore, no deliberate distinction is made between the two types of procedures in the analysis below, in line with the annual reports of the Council.

12.2.1 The annual reports of the Council for Aliens Disputes

In its ruling of 17 October 2007, the Council ruled that Article 10 ICRC has no direct effect.[1133] This was understood in the Annual report 2007-2008 as a confirmation of the case law of the Council of State, in which it was ruled that most treaty Provisions were not suitable for direct effect.[1134] In several subsequent rulings, the Council reaffirmed that the ICRC Provisions would not be suitable for direct applicability. In its annual report 2008-2009, the Council stated that due to its nature, content and wordings, Articles 2, 3, 9, 28, and 29 as individual rights are not directly applicable. Further legislation would be required to specify and complement the meaning of these Articles. With regard to Articles 3 (1), 5, 7, 8, and 16, the Council reported that complaining parties appeared to have invoked these Articles by merely referring to them, without any further substantiation. Therefore, no answer to the question of direct applicability was given. With regard to Articles 9 and 10, the Council reported that no answer was given on the matter of direct applicability, due to the fact that the complainer failed to adequately specify why these rights were violated. The Council reported furthermore that Article 3 has a very general scope, and could therefore not be invoked against procedural rules embedded in the Aliens Act. The Council also reported that the Covenant could only be invoked on behalf of children, and certainly not by adults, and annulment procedures must always concern administrative actions.[1135] In its third annual report (2009-2010), the Council reaffirmed that in its case law

---

[1133] Case no. 2760, 17 October 2007, published on www.rvv-cce.be.

[1134] Council for Aliens Disputes, Annual report 2007-2008, p. 45.

[1135] Council for Aliens Disputes, Annual report 2008-2009, pp. 70-71. Reference was made to case no. 28476, 9 June 2009, available at: www.rvv-cce.be. See also (available at: www.rvv-cce.be): case no. 77468, 19 March 2012; case no. 79734, 20 April 2012; case no. 81048, 11 May 2012; case no. 82696, 11 June 2012; case no. 82697, 11 June 2012; case no. 82790, 11 June 2012; case no. 83484, 22 June 2012.

Articles 2, 3, 8, 9, 10, and 16 had no direct effect, in line with the case law of the Council of State. Besides that, the Council observes that – regardless of the fact whether a ICRC Provision is directly applicable or not – the parties requesting for the annulment of a certain administrative action often fail to substantiate how the disputed administrative decision contradicts the well-being of the child *in concreto.*[1136]

12.2.2 Case law of the Council for Aliens Disputes

As it appears from the case law of the Council as well, the opinion of the Council of State is greatly valued, in line with one of the goals of the Council which is to promote the unity of law.[1137] Indeed the Council seems to consider that the aforementioned ICRC Provisions, and the ICRC in general, cannot be directly applicable,[1138] or the matter of direct applicability remains unanswered, for the complaining party failed to substantiate why the disputed administrative action would be a violation of the rights of the child.[1139] The exact considerations leading to the decision that ICRC Provisions are not directly applicable however show some variety.

[1136] Council for Aliens Disputes, Annual report 2009-2010, p. 61.

[1137] See for instance: Council for Aliens Disputes, Annual report 2008-2009, Section 2.3.

[1138] Referred to by the Council in its annual reports (available at: www.rvv-cce.be): case no. 15780, 11 September 2008 (Articles 3 and 9); case no. 24487, 13 March 2009 (Article 3); case no. 20139, 9 December 2008 (Articles 2, 3, 6, 7, and 9); case no. 27657, 26 May 2009 (Article 3); case no. 27.951, 28 May 2009 (ICRC in general, and Articles 9 and 10); case no. 37979, 29 January 2010 (Articles 2, 8, 9 10, and 16); case no. 40246 15 March 2010 (ICRC in general, and Article 28); furthermore, in 2012: case no. 72810, 6 January 2012 (Article 3); case no. 74507, 31 January 2012 (Article 3); case no. 75732, 24 February 2012 (Articles 3, 7, and 9); case no. 75953, 28 February 2012 (ICRC in general); case no. 76154, 29 February 2012 (Article 7, 9, and 16); case no. 76891, 9 March 2012 (Articles 3 and 28); case no. 77339, 15 March 2012 (Article 3); case no. 77338, 15 March 2012 (Article 3); case no. 77700, 21 March 2012 (Articles 8 and 28); case no. 78203, 28 March 2012 (ICRC in general, and Articles 3 and 9); case no. 79852, 20 April 2012 (Article 3); case no. 42394, 27 April 2010 (Article 9); case no. 45588, 29 June 2010 (Articles 3, 9, and 10); case no. 80534, 27 April 2012 (Article 3); case no. 81048, 11 May 2012 (Articles 7, 9, and 16); case no. 81092, 14 May 2012 (Article 9); case no. 81503, 22 May 2012 (Articles 3 and 9); case no. 82143, 31 May 2012 (Article 3); case no. 82145, 31 May 2012 (Article 3); case no. 82696, 11 June 2012 (Article 9); case no. 82697, 11 June 2012 (Article 9); case no. 82790, 11 June 2012 (Article 28); case no. 83585, 25 June 2012 (Article 9); case no. 83716, 26 June 2012 (Article 3); case no. 84296, 6 July 2012 (Article 9); case no. 85594, 3 August 2012 (Articles 2, 8, 9, 10, and 16); case no. 91043, 6 November 2012 (Articles 3 and 9); case no. 93190, 10 December 2012 (Article 9);case no. 93287, 11 December 2012 (Article 3); case no. 93272, 11 December 2012 (Articles 3 and 9); case no. 93863 18 December 2012 (Article 3). Considering the extensive amount of cases, only the case law referred to in the annual reports, and the cases covering 2012 – at the time of writing the most recent calendar year – were included in the footnote.

[1139] Referred to by the Council in its annual reports (available at: www.rvv-cce.be): case no. 18.341, 4 November 2008 (Articles 3, 7, and 9); case no. 24320, 11 March 2009 (Articles 9 and 10); case no. 24487, 13 March 2009 (Article 28); case no. 25703, 7 April 2009 (Articles 8 and 16); case no. 27657, 26 May 2009 (Article 2); In 2012: case no. 73575, 19 January 2012 (Article 3); case no. 76427, 1 March 2012 (Article 3); case no. 77918, 23 March 2012 (Article 6); case no. 80146, 25 April 2012 (Articles 3 and 9); case no. 82145, 31 May 2012 (Article 1); case no. 84816, 18 July 2012 (Article 9); case no. 88126, 25 September 2012 (Article 28).

For instance, in its ruling on 17 October 2008,[1140] the Council stated that Articles 27, 28 ICRC and 13 ICESCR could not be invoked, for both the UN Convention on the Rights of the Child and the International Covenant on Economic, Social and Cultural Rights were not directly applicable in general. No further explanation was given.

A more subtle approach to the matter was expressed in a ruling on 31 August 2009, in which a complainer invoked Articles 3, 6, 12, and 28 ICRC. The Council ruled that it appears from the Council of State's (legislation department) advice on the Ratification Act of the ICRC that the ICRC stipulates different types of Provisions.

> The legal department of the Council of State makes a distinction between Provisions that generally meet accepted criteria to determine that international treaties are 'self-executing' and as such have an direct effect in the domestic legal order of the signatory States, Provisions that establish rights whose principles are recognised in the Convention and whose implementation requires a positive legislative action, and Provisions that stipulate rights that belong to the category of economic, social and cultural rights leaving a large margin of discretion to the States regarding their progressive implementation.[1141]

In this consideration thus the ICRC Provisions are divided into three groups: (1) self-executing Provisions; (2) Provisions that require national legislation to be effective; and (3) ECOSOC rights that leave a large margin of appreciation to the Member States with regard to their progressive realisation. In its verdict, the Council held that Articles 3, 6, and 28 resort under the second category and are therefore not directly applicable.[1142] The disputed decision is however reviewed against Article 12 ICRC (the right to be heard in any judicial and administrative proceedings affecting the child), but according to the Council, this standard was not violated by the decision.[1143] The consideration quoted above was later used in other cases as well, each time by the same Magistrate.[1144]

[1140] Case no. 17293, 17 October 2008, available at: www.rvv-cce.be.

[1141] Case no. 30858, 31 August 2009, available at: www.rvv-cce.be, consideration 2.3.2. Original text in Dutch: *'De afdeling wetgeving van de Raad van State maakt namelijk een onderscheid tussen bepalingen die beantwoorden aan de algemeen aanvaarde criteria om te bepalen dat de internationale verdragen 'self-executing' zijn en die als zodanig rechtstreekse gevolgen hebben in de interne rechtsorde van de ondertekende staten, bepalingen die rechten vaststellen waarvan het principe door het Verdrag wordt erkend en waarvan de uitvoering een positief optreden van de wetgever vereist en bepalingen die rechten vaststellen die behoren tot de categorie van economische, sociale en culturele rechten en die aan de Staten een grote beoordelingsvrijheid laten wat de geleidelijke uitvoering ervan betreft.'*

[1142] Case no. 30858, 31 August 2009, available at: www.rvv-cce.be, consideration 2.3.2.

[1143] Case no. 30858, 31 August 2009, available at: www.rvv-cce.be, consideration 2.3.2.

[1144] See for instance (available at: www.rvv-cce.be): case no. 72810, 6 January 2012; case no. 77339, 15 March 2012; case no. 77338, 15 March 2012; case no. 80534, 27 April 2012, consideration 2.7.2, and case no. 83716, 26 June 2012; case no. 37862, 29 January 2010 all ruled by the same Magistrate: M.C. Goethals. In one case brought before the Council of State, it was disputed whether the Council for Aliens disputes would have considered that the ICRC in general was not directly applicable. The complainer argued

With regard to the international right to food, reference was made already above to a case in which *inter alia* Article 27 (includes the right to food) was invoked.[1145] In its ruling on 11 May 2009,[1146] the Council reaffirmed that Article 27 had no direct effect, due to the fact that this Article is not a clear or legally complete Article that would impose clear duties on a Member state.

It is remarkable that ICESCR or CEDAW Provisions are rarely invoked by the complaining parties. No verdicts of significance can be found in which the Council rules on the direct applicability of CEDAW Provisions.[1147] Regarding ICESCR Provisions, no Article 11 ICESCR case could be found during the analysis. Besides some cases in which the Council ruled that an invoked ICESCR Article was clearly not violated, or the complaining party did not substantiate why the invoked Article would be violated,[1148] the Council ruled that Articles 10[1149] and 13[1150] ICESCR had no direct effect.

To conclude, there appears to be a very extensive case law on the direct applicability of ICRC Provisions. While the reasoning to reject direct applicability of ICRC Provisions shows some variety, ICRC Provisions in general (both civil and political rights as well as economic, social and cultural rights) are considered not to be directly applicable. Furthermore, there is so little case law on the ICESCR and the CEDAW, that no general conclusions on the direct effect of those Provisions can be drawn, although also here in the few cases that exist, not a single time the direct applicability of a Provision was recognised. Regarding the right to food in particular, it can be concluded that, according to the Council, Article 27 ICRC has no direct effect, while no rulings could be found on the direct applicability of other international Provisions stipulating the same right. The Council however has the tendency – apart from some exceptions – to only answer the question of

---

that while the Council of Aliens Disputed did not dispute an earlier ruling of the Council of State, in which the Council had ruled that the ICRC had no direct effect, the Council of Aliens Disputes had also earlier ruled that Article 12 could have direct effect, which would violate the principles of legal unity. The Council of State then somewhat confusingly considered that the fact that the Council of Aliens Disputes did not dispute the said ruling of the Council of State would not automatically imply that it fully agreed, and therefore the legal remedy of the complainer was factually incorrect. See: Council of State, 25 March 2010, case no. 202356, available at: www.rvv-cce.be.

[1145] Case no. 17293, 17 October 2008, available at: www.rvv-cce.be.

[1146] Case no. 27146, 11 May 2009, available at: www.rvv-cce.be. *In casu*, also Article 6 ICRC was invoked.

[1147] In case no. 53965, 28 December 2010, available at: www.rvv-cce.be, several CEDAW Provisions were invoked by the complaining party. The Council however ruled that the party involved referred to facts that did not match the administrative files, and therefore, it would not be plausible that any CEDAW Provisions would have been violated.

[1148] Case no. 59130, 31 March 2011, available at: www.rvv-cce.be (Article. 12); case no. 59131, 31 March 2011, available at: www.rvv-cce.be (Article 12); case no. 67931, 5 October 2011, available at: www.rvv-cce.be (Aricle 6); case no. 92302, 27 November 2012, available at: www.rvv-cce.be (Aricle 6); case no. 23968, 27 February 2009, available at: www.rvv-cce.be (Article 13).

[1149] Case no. 28928, 22 June 2009, available at: www.rvv-cce.be.

[1150] Case no. 17293, 17 October 2008, available at: www.rvv-cce.be; case no. 34792, 25 November 2009, available at: www.rvv-cce.be.

direct applicability when it is strictly relevant. In case of inadequate arguments or apparently non-violation of the invoked norm, the matter often remains unanswered. In general, the Council appears to greatly value the opinion of the Council if State (both the legislation department and the judicial department), and seeks to judge in line with the opinion of the Council of State. This is not surprising, as the Council of State is the appellate body for cases ruled by the Council.

12.3 The Council of State

As will appear below, there are hardly any verdicts in which the Council clearly ruled on the direct applicability of an international Provision stipulating the right to food. Only two cases could be found in which the Council denied direct effect to Article 27 ICRC.[1151] More generally, only a limited amount of rulings exist in which the direct applicability of ECOSOC rights is dealt with.[1152] Therefore, in the analysis below, the effect of ICESCR, ICRC and CEDAW Provisions in general will be analysed as well. As discussed in Section 11.4.2.1, the Council of State rules in both the subjective and the objective litigation. No evidence could be found however, that regarding the effect of international Provisions there is a different approach in each function of the Council. Therefore, in the analysis below, no distinction is made between the objective litigation and the subjective litigation, but rather between the analysed treaties.[1153]

12.3.1 The Council of State on ICESCR Provisions

With regard to the direct applicability of ICESCR Provisions, the Council seems to prefer to circumvent the issue, by ruling that the invoking party did not adequately substantiate why the invoked standard is violated,[1154] or, while an ICESCR Provision

---

[1151] Council of State, 12 February 1996, case no. 581221, available at: www.raadvst-consetat.be.

[1152] For the Court case analysis, the author primarily used the search engine installed on the official website of the Council of State (www.raadvst-consetat.be). The database of this search engine only includes rulings since 1994. A case review of the period 1994-2012 appeared to be rather inconclusive. Therefore, only when of significant importance, rulings in the period before 1994 are included in the analysis.

[1153] Since the database of the Council of State stores all case law since 1994, the analysis does not go further back in time, unless a verdict is of significant relevance.

[1154] Council of State, 2 December 1996, case no. 63387, available at: www.raadvst-consetat.be (Article 6 (1)); Council of State, 21 December 1999, case no. 84310, available at: www.raadvst-consetat.be (Article 7); Council of State, 9 December 2004, case no. 138290, available at: www.raadvst-consetat.be (Article 26); Council of State, 12 July 2005, case no. 147579, available at: www.raadvst-consetat.be (Article 13); Council of State, 6 April 2006, case no. 157423, available at: www.raadvst-consetat.be (Article 26); Council of State, 30 August 2006, case no. 162085, available at: www.raadvst-consetat.be (Article 26); Council of State, 7 November 2006, case no. 164420, available at: www.raadvst-consetat.be (Articles 1, 2, and 6); Council of State, 30 October 2008, case no. 187467, available at: www.raadvst-consetat.be (Article 13 (2) (C)); Council of State, 21 February 2011, case no. 211392, available at: www.raadvst-consetat.be (Articles 6 and 8 ICESCR).

was invoked by the plaintiff, the Council chose not to significantly include the Provision in its considerations.[1155] However, the Council ruled in several cases rather clearly that the ICESCR in general would have no direct effect.[1156] The denial of direct effect is not always explained further. However, in a case in 1993 the Council offers a short explanation. *In casu*, a plaintiff argued that a Royal Decree, that stipulated that a prison warder should have the same sex as the prisoners detained, would violate *inter alia* the ICESCR, for it would constitute discrimination between men and women. The Council considered that '*first it should be determined whether the applicant can indeed invoke the violation of international Conventions mentioned by her as legal remedy for annulation before the Council of State, that consequently it should be determined whether or not the by her invoked standards of international law have direct effect in the internal Belgian legal order, that a rule of international or supranational law has direct effect when it can be applied in the legal order in which this rule is in force without any substantial internal implementation measure, that a rule of international or supranational law, however has no direct effect, when it imposes an obligation to the State to act or not to act according to the principles embedded in the rule, that the treaty Provisions that have no direct effect do not possess a normative character towards the individual, do not constitute subjective rights by their nature, but only impose obligations on the contracting parties; that is, that the individual citizen cannot derive rights from it, and therefore are not subject to any obligations.*'[1157] In a later case, the Council explained that this '*appears from Article 2 of the treaty that only obligations to act accordance with the principles*

[1155] Council of State, 17 October 1994, case no. 49701, available at: www.raadvst-consetat.be (Article 7); Council of State, 23 December 1994, case no. 50972, available at: www.raadvst-consetat.be (Article 7); Council of State, 5 November 1997, case no. 69470, available at: www.raadvst-consetat.be (Article 7); Council of State, 1 April 1998, case no. 72893, available at: www.raadvst-consetat.be (Article 8); Council of State, 20 November 2000, case no. 90902, available at: www.raadvst-consetat.be (Article 11); Council of State, 28 January 2002, case no. 102971, available at: www.raadvst-consetat.be (Article 7); Council of State, 10 June 2002, case no. 107611, available at: www.raadvst-consetat.be (Article 7); Council of State, 1 October 2002, case no. 110821, available at: www.raadvst-consetat.be (Article 10); Council of State, 28 February 2005, case no. 141340, available at: www.raadvst-consetat.be (Article 26); Council of State, 29 March 2005, case no. 142691, available at: www.raadvst-consetat.be (Article 26); Council of State, 4 May 2005, case no. 144111, available at: www.raadvst-consetat.be (Articles 1, 2, 6, and 12); Council of State, 8 December 2009, case no. 198664, available at: www.raadvst-consetat.be (Article 12).

[1156] Council of State, 30 December 1993, case no. 45552, S.K. 1994, 6 (ICECR in general was invoked); Council of State, 21 December 2005 case no. 153073, available at: www.raadvst-consetat.be (Article 26 was invoked); Council of State, 29 September 2009, case no. 196475, available at: www.raadvst-consetat. be (Article 6 was invoked).

[1157] Council of State, 30 December 1993, case no. 45552, Sociaalrechtelijke Kronieken 1994, 6. Original text in Dutch: '*Overwegende dat vooreerst dient nagegaan te worden of verzoekster wel degelijk de schending van de door haar vermelde internationale akten als annulatiemiddel voor de Raad van State kan inroepen; dat bijgevolg dient onderzocht te worden of de door ingeroepen normen van internationaal recht al dan niet directe werking hebben in de interne Belgische rechtsorde; dat een regel van internationaal of supranationaal recht directe werking bezit indien hij, zonder enige substantiële interne uitvoeringsmaatregel, kan toegepast worden in de rechtsorde waar deze regel van kracht is; dat een regel van internationaal of supranationaal recht daarentegen geen directe werking bezit, wanneer hij aan de Staat de verplichting oplegt te handelen, of niet te handelen, volgens de principes die vervat zijn in de regel; dat de verdragsverplichtingen die geen directe werking hebben geen normerend karakter bezitten ten opzichte van de individuen, geen subjectieve rechten*

embedded in the Convention are imposed on the States.'[1158] There seem to be some exceptions to that rule however, in which the Council cautiously accepts some effect of ICESCR Provisions in the national legal order.

Firstly, the Council seems to have reviewed against Article 7 ICESCR. In one case, a disappointed civil servant, belonging to the Dutch language group, started legal proceedings against the State of Belgium in a case in which someone else was promoted to Deputy Commissioner-General in one of the Belgian Ministries, due to the fact that the promoted person belonged to the French speaking language group, and the Belgian Government strived for numeric equality on language background amongst the personnel of the State. *In casu*, the Council ruled that Articles 10 and 11 CA, whether or not read in conjunction with Articles 2 and 7 (c) ICESCR were not violated, referring to and in line with the answer to the prejudicial question the Council asked from the Constitutional Court.[1159] Also earlier, in 2002, the Council had reviewed against Article 7 ICESC, ruling that the Provision was not violated, in a case in which a new Ministerial Decree concerning the mode of calculating working hours of the national gendarmerie was contested in an annulment procedure.[1160]

Furthermore, the Council appears to have granted at least some effect to Article 13 ICESCR on several occasions. For instance, in one case, the Council distilled from a combination of Articles, including Articles 23 and 24 CA, Article 2 of the Additional Protocol to the ICCPR, and Article 13 ICESCR, the general principle of freedom of choice for parents regarding the education of their children. *In casu*, a teacher was less positively assessed during an appraisal interview compared to the previous year, due to the fact that he had enrolled his children in a different (competing) school. The Council ruled that this was a violation of this principle.[1161] In another case, in 1995, the Council ruled that the denial of the Ministry of the Interior to grant permission to a not formally recognised and unsubsidised school for higher education to issue certificates to foreign students necessary for a residence permit longer than three months was in violation with *inter alia* Article 13 ICESCR, for it would establish a non-justifiable discrimination between educational institutions,

---

in hunnen hoofde scheppen, maar alleen verplichtingen opleggen aan de verdragsluitende partijen; dat met andere woorden de individuele burger er geen rechten kan aan ontlenen en erdoor aan geen verplichtingen onderworpen wordt.'

[1158] Council of State, 29 September 2009, case no. 196475, available at: www.raadvst-consetat.be. Original text in Dutch: *'blijkt uit artikel 2 van het verdrag dat alleen aan de staten de verplichting wordt opgelegd te handelen overeenkomstig de principes vervat in het verdrag.'*

[1159] Council of State, 27 January 2003, case no. 115050, available at: www.raadvst-consetat.be. Prejudicial question asked in: Council of State, 15 December 1997, case no. 70202, available at: www.raadvst-consetat.be. Prejudicial question answered by the Constitutional Court in: Constitutional Court, 2/99 13 January 1999.

[1160] Council of State, 14 January 2002, case no. 102510, available at: www.raadvst-consetat.be (based on other legal grounds however, the disputed Article of the decree was eventually quashed).

[1161] Council of State, 13 December 2000, case no. 91625, available at: www.raadvst-consetat.be.

but also a non-justifiable discrimination between foreign students who want to study at educational institutions that are allowed to grant the said certificates, and foreign students that want to study at educational institutions that are not.[1162]

Also the right to strike, as embedded in Article 8 ICESCR, seems to be granted some effect. In a case in 1995, the Council ruled that certain disciplinary measures against a strike undertaken by employees of the postal service were against the right to strike, as embedded in Article 6 ESC and Article 8 ICESCR.[1163] The Council had a similar reasoning in other cases, in 2002 and 2006.[1164]

In one case, the Council referred to *inter alia* Article 6 ICESCR to underline that the right to work is a basic principle, albeit that *in casu* this principle was not violated.[1165]

12.3.2 The Council of State on ICRC Provisions

Also regarding the ICRC, a large amount of cases can be found in which the Council circumvented to decide on the direct applicability of ICRC Provisions, by ruling that the invoking parties did not adequately substantiate why the invoked standard was violated.[1166] In other cases, the invoked Articles were not used further – for

---

[1162] Council of State, 30 October 1995, case no. 56106, available at: www.raadvst-consetat.be.

[1163] Council of State, 22 March 1995, case no. 52424, available at: www.raadvst-consetat.be.

[1164] Council of State, 3 December 2002, case no. 113168, available at: www.raadvst-consetat.be; Council of State, 13 February 2006, case no. 154824, 154825, 154826, 154827, 154828, 154829, 154830, 154831 and 154832, available at: www.raadvst-consetat.be. See also: Council of State, 1 April, 1998, case no. 72893, available at: www.raadvst-consetat.be. Also here, disciplinary measures against an employee of the National Railways in case of strike were under dispute. The employee invoked *inter alia* Article 6 ESC, and successfully appealed against these disciplinary measures, although the Council did not implicitly base its verdict on the invoked ESC standard.

[1165] Council of State, 3 July 1998, case no. 74948, available at: www.raadvst-consetat.be.

[1166] See for instance (available at: www.raadvst-consetat.be): Council of State, 24 January 1996, case no. 57793 (ICRC in general); Council of State, 19 March 1997, case no. 65343 (Article 28); Council of State, 7 December 1999, case no. 83940 (Article 15); Council of State, 16 January 2001, case no. 92281 (ICRC in general); Council of State, 16 January 2001, case no. 92285 (ICRC in general); Council of State, 27 March 2002, case no. 105233 (ICRC in general); Council of State, 4 July 2002, case no. 108866 (Article 9); Council of State, 2 March 2004, case no. 128688 (ICRC in general); Council of State, 13 October 2004, case no. 135968 (Article 22); Council of State, 31 March 2005, case no. 142746 (Article 22); Council of State, 12 July 2005, case no. 147579 (Articles 28 and 29); Council of State, 5 December 2005, case no. 52209 (Article 3); Council of State, 14 December 2005, case no. 152699 (the requesting party derived a non-refoulement principle from the ICRC in general); Council of State, 27 December 2005, case no. 153187 (Articles 7, 9, and 10); Council of State, 26 January 2006, case no. 154152 (Article 28, continuation of case 65343); Council of State, 13 December 2006, case no. 165924 (ICRC in general); Council of State, 17 January 2007, case no. 166839 (Article 24); Council of State, 4 May 2007, case no. 170798 (ICRC in general); Council of State, 9 April 2008, case no. 181862 (Articles 9 and 10); Council of State, 29 June 2009, case no. 194819 (ICRC in general); Council of State, 29 June 2009, case no. 194820 (Articles 9 and 10); Council of State, 9 June 2011, case no. 213778 (Article 28); Council of State, 13 December 2011, case no. 216839 (ICRC in general).

several reasons – in the considerations of the Council,[1167] the invoking party would not fall under the scope of the ICRC,[1168] the situation would not fall under the scope of the ICRC and therefore the invoked Provisions are not violated,[1169] or the invoking party could not adequately prove its interest *in casu*.[1170]

Only occasionally, the Council specifically ruled on the matter. In some cases, the direct effect of the ICRC in general was denied '*...because the Convention on the Rights of the Child, ratified by Belgium on 20 November 1989, at first sight, has not direct effect.*'[1171] In another case, the Council considered that '*...to the extent that violation of the Convention on the Rights of the Child is included in the legal remedy it cannot be invoked, for the convention has no direct effect.*'[1172] In other cases, the direct effect of particular Provisions was denied. For instance, in two cases in 1996, the Council stated that Articles 6, 24, 27 (1), and 28 ICRC would have no direct effect, due to the fact that the Provisions were not specific enough to distill specific

---

[1167] Council of State, 26 January 1996, case no. 57840, available at: www.raadvst-consetat.be (Articles 2 and 8); Council of State, 22 March 1996, case no. 58746, available at: www.raadvst-consetat.be (Article 37); Council of State, 4 July 2002, case no. 108862, available at: www.raadvst-consetat.be (Articles 9 and 28); Council of State, 14 September 2004, case no. 134863, available at: www.raadvst-consetat.be (Articles 12, 13, and 16); Council of State, 24 August 2005, case no. 148314, available at: www.raadvst-consetat.be (Articles 7 and 14); Council of State, 9 January 2006, case no. 153367, available at: www.raadvst-consetat.be (Article 3); Council of State, 17 March 2007, case no. 156541, available at: www.raadvst-consetat.be (Article 3); Council of State, 9 September 2010, case no. 207266, available at: www.raadvst-consetat.be (ICRC in general).

[1168] Council of State, 16 October 2002, case no. 111607, available at: www.raadvst-consetat.be (ICRC in general); Council of State, 8 December 2003, case no. 126132, available at: www.raadvst-consetat.be (Article 29); Council of State, 25 October 2005, case no. 150620, available at: www.raadvst-consetat.be (in particular, Articles 1-4); Council of State, 15 December 2009, case no. 198958, available at: www.raadvst-consetat.be (Article 3); Council of State, 23 November 2011, case no. 216416, available at: www.raadvst-consetat.be (Article 3); Council of State, 14 March 2012, case no. 218469, available at: www.raadvst-consetat.be (Article 3); Council of State, 14 March 2012, case no. 218470, available at: www.raadvst-consetat.be (Article 3); Council of State, 14 March 2012, case no. 218474, available at: www.raadvst-consetat.be (Article 3).

[1169] Council of State, 10 October 2002, case no. 111411, available at: www.raadvst-consetat.be (Articles 9, 19, and 29); Council of State, 11 February 2004, case no. 128051, available at: www.raadvst-consetat.be (Articles 3 and 39); Council of State, 18 March 2005, case no. 142387, available at: www.raadvst-consetat.be (Article 22); Council of State, 25 May 2005, case no. 144960, available at: www.raadvst-consetat.be (ICRC in general); Council of State, 31 August 2005, case no. 148485, available at: www.raadvst-consetat.be (Article 3); Council of State, 9 September 2005, case no. 148700, available at: www.raadvst-consetat.be (Articles 2, 3, and 6); Council of State, 28 August 2008, case no. 185919, available at: www.raadvst-consetat.be (Articles 9 and 10); Council of State, 8 May 2009, case no. 193105, available at: www.raadvst-consetat.be (Articles 9 and 10).

[1170] Council of State, 12 March 2001, case no. 93859, available at: www.raadvst-consetat.be (Article 36).

[1171] Council of State, 26 April 2000, case no. 86914, available at: www.raadvst-consetat.be. Original text in Dutch: '*...omdat het Verdrag inzake de rechten van het kind van 20 november 1989, geratificeerd door België, op het eerste gezicht geen rechtstreekse werking heeft.*'

[1172] Council of State, 16 December 2002, case no. 113723, available at: www.raadvst-consetat.be. Original text in Dutch: '*...in de mate dat de schending van het Verdrag inzake de rechten van het kind in het middel vervat zit, dit niet op ontvankelijke wijze kan worden aangevoerd nu voormeld verdrag geen directe werking heeft.*'

rights from their wordings.[1173] In another case, the Court ruled that *'Considering that Article 9.1 of the Convention on the Rights of the Child must be denied direct effect since this treaty Provision is not a clear and legally complete Provision that imposes the States Parties either a duty to refrain or a duty to act in a particular manner; that therefore, the applicant cannot effectively invoke a violation thereof.'*[1174]

A large number of cases referred to above concern disputes concerning the refusal to foreigners of a permit to legally reside on Belgian territories. It is interesting to note here that it seems to be not unusual for the Commissioner-General for refugees and stateless persons to emphasise that the ICRC is applicable to the minor applicant while rejecting the request for a residence permit. Usually, a standard consideration is used: *'I draw the attention of the Minister of the Interior on the fact that under your national law you are considered to be a minor, and that consequently the Convention on the Rights of the Child of November 20, 1989, ratified by Belgium, can be applied.'*[1175] In none of these cases however, this notion resulted in a more favourable verdict for the applicant, nor could the applicant successfully invoke one of the ICRC Articles in appeal.[1176]

12.3.3 The Council of State on CEDAW Provisions

Regarding the CEDAW, it must be concluded that there is no significant case law on CEDAW treaty Provisions. Articles are rarely invoked, and have no significant effect on the verdict of the Council.[1177] Interesting detail in a case in 2009 however, is that the direct effect of Article 16 CEDAW is explicitly denied, due to an interpretation of the Council of Article 2 in conjunction with Article 11 CEDAW. To this end, the

---

[1173] Council of State, 24 January 1996, case no. 57793 (Articles 6, 24, 27 (1), and 28) available at: www.raadvst-consetat.be; Council of State, 12 February 1996, case no. 58122, available at: www.raadvst-consetat.be (Articles 6, 27 (1), and 28).

[1174] Council of State, 9 July 2003, case no. 121461, available at: www.raadvst-consetat.be (Article 3). Original text in Dutch: *'Overwegende dat aan artikel 9.1 van het Verdrag inzake de Rechten van het Kind een directe werking moet worden ontzegd aangezien deze verdragsbepaling geen duidelijke en juridisch volledige bepaling is die de verdragspartijen of een onthoudingsplicht of een plicht om op een welbepaalde wijze te handelen, oplegt; dat verzoeker de schending daarvan bijgevolg niet dienstig kan aanvoeren.'*

[1175] Original text in Dutch: *'Ik vestig de aandacht van de Minister van Binnenlandse Zaken op het feit dat u krachtens uw nationale wet minderjarig bent en dat bijgevolg het Verdrag inzake de rechten van het kind van 20 november 1989, geratificeerd door België, op u kan worden toegepast.'* In some cases, the word 'must' is used instead of 'can'. See for instance: Council of State, 24 Mai 2005, case no. 144825, available at: www.raadvst-consetat.be.

[1176] See for instance (available at: www.raadvst-consetat.be): Council of State, 2 April 2002, case no. 105351; Council of State, 10 March 2003, case no. 116807; Council of State, 23 April 2003, case no. 118541; Council of State, 8 December 2003, case no. 126132; Council of State, 15 December 2003, case no. 126410; Council of State, 31 March 2005, case no. 142746; Council of State, 24 May 2005, case no. 144825; Council of State, 22 September 2005, case no. 149302; Council of State, 1 March 2006, case no. 155724; Council of State, 25 March 2010, case no. 202357. It must be noted that in cases no. 105351, 118541, 126410, 144825 and 202357, in appeal, no ICRC Articles were invoked by the applicant.

[1177] See for instance (available at: www.raadvst-consetat.be): Council of State, 7 March 2002, case no. 104430 (Article 15); Council of State, 23 December 2009, case no. 199259 (Article16).

Council rather selectively quoted two elements of the two Articles: '*States Parties (...) agree to pursue by all appropriate means a policy of eliminating discrimination against women*' and '*States Parties shall take all appropriate measures to eliminate discrimination against women.*' From this, according to the Council, follows that Article 16 CEDAW has no direct effect. A reasoning that is hardly understandable, for there is no relationship with Article 11 and the case, for *in casu*, the applicant appeals against a decision that denies her request to change her family name, while the quoted part of Article 11 could also be taken from Article 16 CEDAW, whose first sentence has the same wordings.[1178]

12.3.3.4 Concluding remarks

It can be concluded that the case law of the Council of State regarding the direct applicability of ECOSOC rights is rather inconsistent. Especially regarding the ICESCR, the Council varies as it seems randomly between denying the direct effect of the entire treaty, to granting some effect to certain Articles, whereas the ICRC seems to have no direct effect at all in the case law of the Council. Regarding the CEDAW, no significant case law could be found in which the Council ruled on the possibility of direct effect. No evidence could be found that the difference between civil and political rights on the one hand, and economic, social and cultural rights on the other, plays a significant role in the considerations of the Council in determining whether or not a treaty Provision has direct effect.

12.4 The Court of Cassation

The Court of Cassation did not rule on matters in which the international right to food was invoked, and only occasionally on international ECOSOC Provisions. In this Section, therefore, rulings in which ICESCR, ICRC, and CEDAW Articles are invoked will be discussed in general.

12.4.1 The Court of Cassation on ICESCR Provisions

The Court of Cassation did not rule very often in cases in which ICESCR Provisions were invoked, and rather exceptionally decided on the direct effect of the Provisions.

For instance, on 20 December 1990,[1179] a plaintiff invoked Article 13 ICESCR in a case in which certain school fees were under dispute that due to Article 2.1. ICESCR, stipulating the obligation to progressively realise the rights in the Covenant, Article 13 therefore cannot have any immediate legal effect, and does not constitute any subjective right for an individual. In a ruling on 25/09/2003,

---

[1178] Council of State, 23 December 2009, case no. 199259, available at: www.raadvst-consetat.be (Article 16).
[1179] Court of Cassation, 20 December 1990, Arresten van het Hof van Cassatie, 1990(91)(p. 445).

the Court had a similar reasoning with regard to Article 15.[1180] *In casu*, a plaintiff invoked Article 15 ICESCR to justify her publication of summaries of scientific articles, without the approval of the authors of these articles. On 26 May 2008, the Court ruled that *inter alia* Articles 9 and 10 ICESCR had no direct effect. Also here, the Court referred to Article 2 ICESCR, and furthermore considered that from the wordings of the invoked Provisions, it followed that the Articles had no direct effect. *In casu*, the amount of child allowance was under dispute. The plaintiff claimed that the difference of child allowance in case of the death of a parent between entrepreneurs on the one hand, and 'regular' employees and civil servants on the other, was discriminative. The Court ruled that the different groups of beneficiaries were not comparable, and therefore the disputed Royal Decree was not discriminative.[1181]

On 21 November 1996, the Court ruled that the use of coercion tools as a security measure for hospitalised detainees could only be an exceptional measure, and not a normal safety rule. In this case, the plaintiff invoked Article 12 ICESCR in a cassation plea against a Court ruling, in which the Court of First Instance had decided that certain straps could lawfully serve as a replacement for several necessary safety equipment that would normally be present in a prison, in case of a hospitalised detainee. The Court decided with a confusing reasoning that Article 12 ICESCR could not be violated without the violation of several national Provisions, together with Articles 2 and 12 ICESCR. The most probable interpretation seems that in this case, the national Provisions offer at least a similar, or possibly a broader protection than Article 12 ICESCR read alone.[1182]

12.4.2 The Court of Cassation on ICRC Provisions

Regarding the ICRC, the Court frequently seems to consider the invoked ICRC Provision, but rules that *in casu*, the invoked Provision is not applicable or it has not been violated.[1183] Only occasionally, the Court decided on the direct

---

[1180] Court of Cassation, 25 September 2003, no. C030026N, published on: http://jure.juridat.just.fgov.be.

[1181] Court of Cassation, 26 May 2008, no. S060105F, published on: http://jure.juridat.just.fgov.be.

[1182] Court of Cassation, 21 November 1996, Arresten van het Hof van Cassatie, 1996 (446).

[1183] For instance: Court of Cassation, 31 October 2006, P.06.0890.N, published on: http://jure.juridat.just.fgov.be (Article 40, due to an interpretative declaration made by Belgium to this Article: '*(a) this Provision shall not apply to minors who, under Belgian law, are declared guilty and are sentenced in a higher Court following an appeal against their acquittal in a Court of the first instance; (b) this Provision shall not apply to minors who, under Belgian law, are referred directly to a higher Court such as the Court of Assize.*'). See also: Court of Cassation, 4 November 1993, Arresten van het Hof van Cassatie, 1993 (p.919) (Article 21); Court of Cassation, 1 October 1997, Arresten van het Hof van Cassatie, 1997(378) (Article 16); Court of Cassation, 14 October 2003, no. P.03.0591.N, published on: http://jure.juridat.just.fgov.be (Articles 3, 9, 12, and 18); Court of Cassation, 22 March 2005, no. P050340N, published on: http://jure.juridat.just.fgov.be (Article 37, 38 and 40); Court of Cassation, 21 March 2006, no. P060211N, published on: http://jure.juridat.just.fgov.be (Article 16); Court of Cassation, 27 January 2010, no. P091686F, published

applicability of ICRC Provisions. In some cases, the Court was rather clear in explicitly denying direct effect of an Article.

On 31 March 1999, the Court ruled in a case in which a child was placed into Court custody in order to resume proper education and restore the relationship with its father. *In casu*, the father did not agree with the Court custody, and invoked *inter alia* Article 25 ICRC. He argued that in Article 25 ICRC, the restoration of the relationship with the father was not recognised as a legal ground for placing a child into the Court's custody. The Court considered among other things that from the wordings of Articles 4 and 24 ICRC it appears that they only stipulate obligations to State Parties, and therefore have no direct effect, and consequently cannot be invoked directly before the national Courts.[1184]

In the ruling of 26 May 2008 discussed above, not only Articles 9 and 10 ICESCR, but also Articles 2.1 and 26.1 ICRC were invoked. The Court considered also, regarding the invoked ICRC Provisions, that from the wordings of the invoked Provisions, it followed that the Articles had no direct effect.[1185]

On 11 March 1994, the Court decided that the Court of Appeal rightfully could decide that the visiting rights of a father could not be enforced by implementing measures or coercive measures against the children involved. The Court considered that the rights embedded in the Convention on the rights of the child (including the visiting rights of parents) should always be considered in view of the best interest of the child. Furthermore, children have the right to express their opinion on matters that concern their interest, and this opinion needs to be taken into careful consideration (as stipulated in Articles 9 (2) and 12 ICRC). *In casu*, there was a deeply disturbed relationship between the father and his daughters, and his daughters had clearly no desire to be in contact with their father.[1186]

The Court decided on 4 November 1999 in two cases[1187] that Articles 3 (1) and 3 (2) could be useful for the interpretation of other legal Provisions, but are of insufficient clarity to have direct effect due to the fact that these Provisions recognise a large margin of appreciation for Member States to, *in casu*, decide on how the interests of the child are best protected in relation to the manner in which the biological origin of the father is determined. Therefore, no substantive

---

on: http://jure.juridat.just.fgov.be (Article12); Court of Cassation, 15 September 2010, no. P101218F, published on: http://jure.juridat.just.fgov.be (Article 12); Court of Cassation, 20 October 2010, no. P090529F, published on: http://jure.juridat.just.fgov.be (Article 3).

[1184] Court of Cassation, 31 March 1999, Arresten van het Hof van Cassatie, 1999(195).

[1185] Court of Cassation, 26 May 2008, no. S060105F, published on: http://jure.juridat.just.fgov.be.

[1186] Court of Cassation, 11 March 1994, Arresten van het Hof van Cassatie, 1994(I, p. 243).

[1187] Court of Cassation, 4 November 1999, Arresten van het Hof van Cassatie, 1999 (588); Court of Cassation, 4 November 1999, Arresten van het Hof van Cassatie, 1999 (589).

rights can be distilled from Article 3. On 10 November 1999,[1188] the Court was less subtle, by simply stating that Articles 3 and 13 ICRC were not directly applicable in criminal proceedings. This was reaffirmed by the Court with regard to Article 3 on 4 March 2008[1189] and 20 October 2010.[1190]

On 23 October 2006,[1191] The Court of Cassation ruled in a case in which the plaintiff invoked Articles 2, 3, 24 (1), 26, and 27 ICRC against a verdict of the Labour Court in higher appeal. The Labour Court denied social benefits to a woman with three children, not legally residing in Belgium. The Labour Court had referred to the Constitutional Court's verdict of case 106/2003,[1192] in which the Constitutional Court ruled that Article 57 § 2 '*Organic Law of 8 July 1976 on public centres for social welfare*' was in violation with Articles 2, 3, 24 (1), 26, and 27 ICRC if it would not allow Public Centres for Social Welfare to provide material help that is necessary for the development of the child. *In casu* however, the plaintiff demanded financial assistance, and not material benefits, which was according to the Labour Court not in accordance with the reasoning of the Constitutional Court, and the resulting revision of the relevant national law. The Court of Cassation however, ruled otherwise, stating that the adapted Article 57 § 2 '*Organic Law of 8 July 1976 on public centres for social welfare*' did not necessarily exclude financial benefits, and therefore could not be a valid ground to deny the financial assistance demanded by the plaintiff. Based on the above, the Court already concluded to annul the disputed ruling, and therefore did not feel the need to consider the cassation plea in which the ICRC Provisions were invoked. *In casu*, the Court appears to be willing to broaden the scope of the verdict of the Constitutional Court.

On 24 October 2012, the Court ruled that Article 8 ECHR needs to be read in conjunction with Articles 7 and 9 ICRC, meaning that '*Thus, the right to respect private and family life, implies the right of the child to be cared for by his parents, and the right of a mother not to be separated from her child against her will, unless such separation is necessary in the interest of the child and to the extent that the measure, which has been taken in accordance with applicable law and applicable procedures, is subject to juridical review on behalf of, in particular, the holders of parental authority, which has thus been compromised.*'[1193] *In casu*, a minor was placed in foster care for an unlimited period, until the age of majority, and the parents had very

---

[1188] Court of Cassation, 10 November 1999, Arresten van het Hof van Cassatie, 1999 (599).

[1189] Court of Cassation, 4 March 2008, no. P071541N, published on: http://jure.juridat.just.fgov.be.

[1190] Court of Cassation, 20 October 2010, no. P090529F, published on: http://jure.juridat.just.fgov.be.

[1191] Court of Cassation, 23 October 2006, no. S050042F, published on: http://jure.juridat.just.fgov.be.

[1192] Constitutional Court, 106/2003, 22 June 2003.

[1193] Court of Cassation, 24 October 2012, no. P121333F, published on: http://jure.juridat.just.fgov.be. Original text in Dutch: '*Aldus houdt het recht op eerbiediging van het familie- en gezinsleven het recht in, voor het kind, om door zijn ouders te worden verzorgd en, voor een moeder, om niet tegen haar wil te worden gescheiden van haar kind, tenzij deze scheiding noodzakelijk is in het belang van het kind en voor zover de maatregel, die in overeenstemming met het toepasselijke recht en de toepasselijke procedures is genomen, door de rechter getoetst kan worden op verzoek van met name de houders van het ouderlijk gezag waaraan aldus afbreuk is gedaan.*'

limited access to legal remedies. The Court ruled that this was an unnecessary infringement to the right to private and family life.

12.4.3 The Court of Cassation on CEDAW Provisions

Also concerning the case law of the Court of Cassation, no rulings of any significance could be found regarding the direct applicability of CEDAW Provisions.[1194]

12.4.4 Concluding remarks

In general, it can be concluded that the Court of Cassation only occasionally rules on the possibility of direct effect of ICESCR and ICRC Provisions. Regarding the ICESCR Articles, no evidence could be found that the Court granted direct effect – or any effect at all – to one of the Provisions. Regarding the ICRC Provisions, there appears to be – rather exceptionally – some room for effect of certain Provisions, although no direct effect was ever explicitly granted, and the Provisions concerned do not stipulate typical ECOSOC Provisions.

One remarkable case is the ruling of 23 October 2006, in which the Court appears to open the door for financial social benefits for illegally residing aliens with children, while this is, as will appear below, clearly a broader scope of benefits than the Constitutional Court, and the subsequently altered Organic Law seem to recognise. The effect *in casu* of the invoked ICRC Provisions however remains unclear.

12.5 The Constitutional Court

The direct applicability of economic, social and cultural rights is in fact a matter in which the Constitutional Court cannot decide, since the Constitutional Court only reviews legislation against the Constitution, and not against international standards. Indirectly however, as demonstrated in Section 11.4.2.2, international standards can play a role in the interpretation of Constitutional standards for which the Court is authorised to review against. As mentioned above, the Court ruled that it was authorised to review legislation indirectly against Provisions

---

[1194] As it seems, in proceedings that concern discrimination against women, Provisions of EU legislation or the ECHR are invoked, rather than CEDAW Provisions. See for instance: Court of Cassation, 17 January 1994, Arresten van het Hof van Cassatie, 1994(I, p.54); Court of Cassation, 5 December 1994, Arresten van het Hof van Cassatie, 1994(II, p.1052); Court of Cassation, 5 December 1994, Arresten van het Hof van Cassatie, 1994(II, p.1055); Court of Cassation, 4 November 1996, Arresten van het Hof van Cassatie, 1996 (411); Court of Cassation, 4 November 1996, Arresten van het Hof van Cassatie, 1996 (412); Court of Cassation, 13 January 1997, Arresten van het Hof van Cassatie, 1997 (28); Court of Cassation, 13 February 1997, Arresten van het Hof van Cassatie, 1997 (83); Court of Cassation, 29 May 1998, Arresten van het Hof van Cassatie, 1998 (279); Court of Cassation, 8 March 1999, Arresten van het Hof van Cassatie, 1999 (136); Court of Cassation, 29 April 2002, no. S010137F, published on: http://jure.juridat.just.fgov.be.

from international treaties and Covenants that are directly applicable, but also against international Provisions that are not directly applicable, as long as they are binding to Belgium.

12.5.1 The first reviews against Article 13 ICESCR

In 1992, the Constitutional Court seemed to review a decree from the French Community that would lead to an increase of certain school fees, against Article 13 ICESCR. The Court decided that the increase of school fees was not a violation of the standstill obligation with regard to 13 ICESCR, due to the fact that such fees were no severe or significant obstacle for the access to education, considering the average income of Belgian citizens.[1195] This ruling however was understood by the requesting party of a case in 1994,[1196] that Article 17 CA – the predecessor of Article 24 CA – should be understood in conjunction with Article 13 ICESCR.[1197] *In casu*, a decree of the French Community was under dispute. The decree introduced school fees for arts education that did not yet exist at the moment on which the ICESCR was ratified by Belgium, and was therefore perceived to be a violation of the standstill obligation. The Court held that Article 13 ICESCR had no direct effect in Belgium, and could not create any subjective rights, but held also that any national Provision that would introduce school fees for primary, secondary and higher education that was accessible free of charge at the time of ratification of the ICESCR would be in violation with Article 13 ICESCR.[1198] Furthermore, perhaps somewhat in contradiction to the previous reasoning, the Court considered that any school fee that would not severely hinder the access to education would not be considered to be a step back compared to the previous situation in which no school fees existed yet.[1199] Moreover, the Court held that the arts education in question did not fall under the scope of primary, secondary or higher education, and the French Community was therefore not obliged to keep this education free of any charge.[1200] The Court finally ruled that the disputed decree was not in violation with Article 24 Constitution in conjunction with Article 13 ICESCR.[1201]

Since then, the Court repeatedly reviewed national legislation against Article 24 Constitution in conjunction with (the standstill effect of) Article 13 ICESCR, but mostly ruled that the national Provision in question did not violate these standards.[1202] However, the Court did not only reject nullification requests in

---

[1195] Constitutional Court, 33/92, 7 May 1992, especially consideration B.8.2.

[1196] Constitutional Court, 40/94, 19 May 1994.

[1197] Constitutional Court, 40/94, 19 May 1994, A.10.1.

[1198] Constitutional Court, 40/94, 19 May 1994, consideration B.2.3.

[1199] Constitutional Court, 40/94, 19 May 1994, consideration B.2.7.

[1200] Constitutional Court, 40/94, 19 May 1994, consideration B.2.8.

[1201] Constitutional Court, 40/94, 19 May 1994. consideration B.2.9.

[1202] For instance, a decree of the French Community refusing finances for students who repeatedly failed to pass their exams (Constitutional Court, 35/98, 1 April 1998, consideration B.5.1.-B.5.2.); a decree of the French Community indirectly allowing Universities to impose under circumstances

which Article 13 was invoked. In a case on 4 June 1998,[1203] the Court ruled that a decree of the French Community, practically resulting in the fact that a quota was established for the acceptation of foreign students due to altered subsidy rules, and also the prohibition to oblige student candidates to do an entrance examination before attending a specific type of arts education, would lead to serious and irreparable harm to the school. The requesting party invoked *inter alia* Articles 10 and 11 Constitution read in conjunction with Article 13 ICESCR. The Court however did not specify on which Articles the verdict was based.[1204] The ruling was later reaffirmed on 15 July 1998.[1205]

In a case in 2001,[1206] another decree from the French Community was under dispute. The decree introduced certain reference systems that explained basic competences in education. According to the requesting party, this system was detailed to such extent that it would impose a particular pedagogic view on schools, which would be in violation with Article 24 CA in conjunction with *inter alia* Article 13 ICESC. The defending party – the French Community Government – held that the invoked international Provisions were inadmissible whereas the requesting party did invoke these Provisions directly (thereby suggesting that

---

additional registration fees (Constitutional Court, 44/98, 22 April 1998, consideration B.4); a decree from the French Community that obliged students with French as mother tongue but attending secondary education in Dutch to do a language exam in order to attend education on a university in the French language (Constitutional Court, 48/98, 22 April 1998, consideration B.5); a decree from the French Community that would lead to less favourable working conditions for education personnel compared to the other communities (Constitutional Court, 134/98, 16 December 1998, consideration B.4); a decree from the French Community allowing rarely visited religious education to coincide with other educational activities (Constitutional Court, 42/99, 30 March 1999, and 90/99, 15 July 1999: in both cases Article 13 ICESCR was invoked, but the Court only reviewed against Constitutional Provisions); a decree from the French Community on the equal granting of social benefits to schools installed by the Government and schools not installed by the Government, in which an exhaustive list of categories of social benefits for schools that are subsidised under strict conditions is stipulated (Constitutional Court, 56/2003, 14 May 2003); a decree from the French Community allowing higher educational institutions that are not a university to raise additional school fees (Constitutional Court, 28/2007, 21 February 2007; 56/2008, 10 March 2008); a decree of the Flemish Community on the fees for teacher education, in which the fees were raised (Constitutional Court, 105/2008, 17 July 2008); a decree of the French Community blocking the possibility in primary education to change school during a college year (Constitutional Court, 119/2008, 31 July 2008; a decree of the French Community altering the enrolment procedures for the admission to secondary education (Constitutional Court, 121/2009, 16 July 2009); a national Law excluding non-subsidised and recognised educational institutions from legal grounds for foreign students to be granted a residence permit (Constitutional Court, 145/2010, 16 December 2010); a national Law to prevent the abuse of asylum procedures in order to gain access to social benefits (Constitutional Court, 135/2011, 27 July 2011); a decree of the Flemish Community allowing schools to give priority to the enrolment of students with the Dutch language as mother tongue in the Brussels-Capital Language Area (Constitutional Court 7/2012, 18 January 2012), were considered not to be in violation with Article 24 Constitution, read in conjunction with Article 13 ICESCR.

[1203] Constitutional Court, 62/98, 4 June 1998.

[1204] Constitutional Court, 62/98, 4 June 1998, consideration B.13.

[1205] Constitutional Court, 91/98, 15 July 1998. The requesting party invokes Articles 10 and 11 Constitution in conjunction with *inter alia* Article 13 ICESCR in the first part of the appeal.

[1206] Constitutional Court, 49/2001, 18 April 2001.

the Provision was not invoked in conjunction with a Constitutional Provision) and did not specify in what manner the disputed decree would violate these standards.[1207] According to the Court, the international standard was invoked in conjunction with a Constitutional standard. Moreover, the Constitutional standard in itself (Article 24 CA) clearly recognises the respect for fundamental rights and freedoms. According to the Court, Article 13 ICESCR is one of these fundamental rights and freedoms, and could therefore be rightfully invoked by the requesting party.[1208] Finally, the Court ruled that the decree was indeed in violation with the freedom of education as recognised in Article 24 CA. Despite the recognition that Article 13 ICESCR could rightfully be invoked by the requesting party, and thus apparently is a standard that needs to be taken into consideration, no reference was made to the earlier invoked international standards in the final verdict.[1209]

12.5.2 Reviews against ECOSOC rights in general

Since the aforementioned early rulings in which the Court indirectly reviewed against Article 13 ICESCR, it also started to review national Provisions, indirectly, against other ICESCR Provisions.

The Court, similar to the Court of Cassation and the Constitutional Court, however does not always choose to review against an invoked international standard when it does not seem necessary. There are plenty of cases in which an international ECOSOC right was somehow invoked or referred to by a requesting or defending party, or referred to in a preliminary question, but not considered further by the Court. In some cases the Court ruled that the invoked international Articles were not violated due to the fact that the invoked protection did not fall under the scope of the invoked Article.[1210] In some cases the Court ruled that the invoked international Provision would not lead to a broader protection, or that no additional arguments were derived from these Provisions by the requesting party.[1211] In all these cases, the Court did not review indirectly against the international Provision. In one case, the requesting party did invoke Article 6 ICESCR, but not in conjunction with a Constitutional Provision. The Court therefore ruled that it was not competent

---

[1207] Constitutional Court, 49/2001, 18 April 2001, A.4.2.

[1208] Constitutional Court, 49/2001, 18 April 2001, considerations B.5.1. and B.5.2.

[1209] Constitutional Court, 49/2001, 18 April 2001, consideration B.12.

[1210] For instance: Constitutional Court, 41/2002, 20 February 2002, considerations B.5-B.6 and verdict (Article 6); 110/2005, 22 June 2005, consideration B.3.2 (Article 13); 147/2005, 28 September 2005, consideration 10.5 (Article 8); 28/2007, 21 February 2007, Consideration B.5.2 (Article 13); 64/2008, 17 April 2008, B.S. 9 May 2008 (9 ICESCR and 12 ESC); Constitutional Court, 64/2009, 2 April 2009 (Article 8).

[1211] For instance: Constitutional Court, 49/95, 15 June 1995, consideration B.6.2 (Article 2); Constitutional Court, 1/95, 12 January 1995, consideration B.9 (Article 9); Constitutional Court, 85/98, 15 July 1998, consideration B.5.9 (Article 13); Constitutional Court, 15/2001, 14 February 2001, consideration B.9 (Article 7); Constitutional Court, 109/2001, 20 September 2001, consideration B.12 (Article 6); Constitutional Court, 137/2011, 27 July 2011 (Article 6).

reviewing the disputed national standard only against an international Provision, hereby drawing a line between direct and indirect reviewing competences.[1212]

On the other hand, in other cases the Court did review (indirectly) against the invoked treaty Provisions. While in some cases the parties involved seem to quarrel over the direct applicability of treaty Provisions,[1213] the Constitutional Court appears not to engage in that debate[1214] but instead seems to consider rather frequently whether or not a restriction on an invoked human right would be necessary in a democratic society, not unreasonable, and not incompatible with its purpose. This way, the Court reviewed indirectly against Articles 6,[1215] 8,[1216] and 13[1217] ICESCR, although such a review seldom resulted in a verdict in which a national rule was considered to be indirectly in violation with an international standard.

[1212] Constitutional Court, 107/98, 21 October 1998, consideration B.7.2.

[1213] For instance, in Constitutional Court, 110/98, 4 November 1998, the Court was asked to review a decree of the Flemish Community against Article 24 Constitution in conjunction with *inter alia* Articles 5 and 7 CEDAW, Articles 2 and 10 ICESCR. According to the Flemish Community Government, this Court could not review against these Articles due to the fact that these Provisions could not be applied in the internal legal order without implementing measures. The requesting party also invoked Article 2 ICRC, but according to the Flemish Community Government, this Provision had no direct effect for the Article recognised merely a modality of the rights recognised in the Convention, but had no independent significance. The Court did not directly respond to the arguments of the Government, but instead quoted all international Provisions invoked, and did review the disputed decree against those Provisions. *In casu*, the Court considered that the Articles were not violated by the decree. The direct applicability of Article 7 ICESCR was also disputed in Constitutional Court, 2/99, 13 January 1999. A preliminary question was raised concerning the compliance of a Board Language Law that obliged a complex but equal division in language groups of certain civil servants with regard to bilingual matters with Articles 10 and 11 Constitution in conjunction with Articles 2 and 7 ICESCR. The Council of Ministers questioned the direct applicability of Article 7 ICESCR. The Court however reviewed the disputed Statute against Articles 10 and 11 Constitution in conjunction with Articles 2 and 7 ICESCR, and decided that the Statute was not in violation with the Constitutional and international standards.

[1214] Although in the period 1990-2002, the Court considered itself to be only competent in indirectly reviewing against international, directly applicable Provisions. During that period – although very rarely – the Court decided on the direct applicability of international Articles. For instance, the Court ruled that the Provisions stipulated in the UDHR had no direct effect. See: Constitutional Court, 22/94, 8 March 1994, Consideration B.1.

[1215] Constitutional Court, 22/94, 8 March 1994, consideration B.9.2; 109/2001, 20 September 2001, consideration B.10.2 (although no specific reference was made to the criteria, they appear to have been considered implicitly).

[1216] Constitutional Court, 62/93, 15 July 1993, consideration B.3.8-B.3.12. 42/2000, 06 April 2000, consideration B.7.4.

[1217] As already referred to in Section 12.5.1. Constitutional Court, 47/97, 14 July 1997, consideration B.3.2; Constitutional Court, 28/2007, 21 February 2007, Consideration B.4.11-B.4.12; Constitutional Court, 56/2008, 10 March 2008; Constitutional Court, 105/2008, 17 July 2008; Constitutional Court, 145/2010, 16 December 2010, consideration B.5-B.6.

12.5.3 Proceedings regarding the right to food as embedded in Article 11 ICESCR

International Provisions recognising the right to adequate food were almost exclusively invoked (with varying degrees of success) in annulment procedures regarding national Provisions on the basis of which aliens may request social benefits, or in cases in which another Court asked a preliminary question, mostly in the same context. Aliens seeking protection by invoking fundamental human rights have led to extensive case law that is explained in detail in the PhD thesis of Steven Bouckaert, a book that I gratefully consulted as an important source of inspiration for this Section.[1218] This Chapter will instead focus on the cases in which the international right to food played a role. However, in some cases this right was not explicitly invoked, while it is important to fully understand the general lines in the case law of the Constitutional Court. Therefore, where necessary, these cases are also discussed below.

12.5.3.1 The first review of Article 57 § 2 O.C.M.W: Case 51/94[1219]

On 29 June 1994, the Court made its first landmark ruling in a series of cases concerning the review of Article 57 § 2 of the '*Organic Law of 8 July 1976 on public centres for social welfare*'[1220] against several human rights Provisions, including Articles stipulating the right to food. The disputed Article was altered by the Federal Government with the aim of restricting immigration. Before, the public centres for social welfare were bound to provide for basic needs to aliens illegally residing on Belgian territory by granting them benefits in kind. After the Amendment of 30 December 1992, Article 57 § 2 stated that '*The Provision of social benefits shall end from the date of the execution of the order to leave the territory, and at the latest, from the date of expiration of the period of the final order to leave the territory. Deviation from the Provisions of the preceding Section shall be allowed during the period that is strictly necessary to enable the person to effectively leave the territory: that period shall in no case exceed one month. Also, deviation shall be allowed in case of urgent medical care.*'[1221]

[1218] S. Bouckaert, *Documentloze vreemdelingen, grondrechtenbescherming doorheen de Belgische en internationale rechtspraak vanaf 1985*, Antwerp-Apeldoorn: Maklu, 2007.

[1219] Constitutional Court, 51/94, 29 June 1994.

[1220] In Dutch: *artikel 57, § 2, van de organieke wet van 8 juli 1976 betreffende de openbare centra voor maatschappelijk welzijn*, abbreviated as: Article 57 § 2 O.C.M.W.

[1221] Constitutional Court, 51/94, 29 June 1994, III. Original text in Dutch: '*Aan de maatschappelijke dienstverlening wordt een einde gemaakt vanaf de datum van de uitvoering van het bevel om het grondgebied te verlaten, en ten laatste, vanaf de datum van het verstrijken van de termijn van het definitieve bevel om het grondgebied te verlaten. Van het bepaalde in het voorgaande lid wordt afgeweken, gedurende de strikt noodzakelijke termijn, om de betrokkene in staat te stellen het grondgebied effectief te verlaten: die termijn mag in geen geval een maand overschrijden. Er wordt eveneens afgeweken ingeval van dringende medische hulp.*'

In casu, the requesting party – a variety of interest groups – argued that this would result in discrimination between illegally residing aliens or aliens whose request for a residence permit was denied on the one hand and legally residing aliens and Belgian citizens on the other hand. The requesting parties thereby invoked Articles 10 and 11 CA in conjunction with Articles 3 ECHR, 7 ICCPR, 11 ICESCR, and 13 ESC.

The Court stated that the disputed Statute did not unlawfully treat persons differently, because in a situation in which a State wishes to restrict immigration, while all measures taken seem to be ineffective, it is not unreasonable to accept different obligations regarding the needs of legally residing persons in Belgium on the one hand and illegally residing persons on the other hand. In addition, the illegally residing person could know beforehand what the consequences would be when he does not leave the territory on time. The Court ruled that the disputed measure was not disproportionate with its purpose, as long as it would guarantee to foreigners the social benefits that are necessary to leave the country for one month, and emergency medical assistance for an unlimited period.[1222]

The Court then affirmed that with regard to Articles 10 and 11 *'the Constitutional Provisions on equality and non-discrimination are applicable with regard to all rights and freedoms, including those resulting from international treaties that are binding to Belgium, made applicable by ratification in the internal legal order and have direct effect.'*[1223]

In this light, the Court firstly ruled that restrictions on the right to social benefits were not in violation with Articles 3 ECHR and 7 ICCPR. The Court considered that the restrictions were not an act of torture or inhuman treatment, for there was no act in which severe pain or suffering of a physical or psychological nature was deliberately inflicted on the victim with the purpose of obtaining information or confessions, nor was there an act of punishing the victim, or putting pressure on or intimidating the victim or third parties. Also, the restrictions would not result in any kind of degrading treatment, for there was no situation in which a victim was hurt in front of other persons, or the human dignity of the victim was seriously affected.[1224]

Secondly, the Court quoted Article 11 ICESCR, and considered that this Article must be read in conjunction with the standstill Provision of Article 2.1 ICESCR. The Court ruled with remarkable consideration that despite the fact that the

---

[1222] Constitutional Court, 51/94, 29 June 1994, consideration B.4.3.

[1223] Constitutional Court, 51/94, 29 June 1994, consideration B.5.2. Original text in Dutch: *'de grondwettelijke regels van de gelijkheid en van de niet-discriminatie zijn toepasselijk ten aanzien van alle rechten en alle vrijheden, met inbegrip van die welke voortvloeien uit internationale verdragen die België binden, die door een instemmingsakte in de interne rechtsorde toepasselijk zijn gemaakt en die directe werking hebben.'*

[1224] Constitutional Court, 51/94, 29 June 1994, consideration B.5.4.

ICESCR stipulates the right to an adequate standard of living for everyone, this right cannot reasonably be understood without limitations. The Court argued that Article 11 ICESCR could only apply to persons for whom the State is responsible, and that the State is not responsible for aliens residing in Belgium who were ordered to leave the territory due to the fact that they did not fulfil (anymore) the conditions to get a residence permit.[1225]

Thirdly, the Court held that it was not necessary to decide whether Article 13 ESC was directly applicable, for the Provision would only add to the previous that the Member States of the ESC were obliged to grant on the basis of equality social benefits to nationals and citizens from other Member States that legally reside on their territory. According to the Court, this Article was not violated, since the different treatment was not based on nationality, but on the legal status of residence.[1226] The Court thus rejected the request.

Interesting however in this case is that the Court ruled that Articles 10 and 11 CA are applicable with regard to all rights and freedoms, including those embedded in directly applicable international Provisions. The Court reviewed the disputed national Provision against Article 3 ECHR and 7 ICCR, and also against Article 11 ICESCR. Is the Court telling us here, that in their view, Article 11 ICESCR is directly applicable? It will probably never be known, for some years later, the Court would explicitly consider being competent to also review against international Provisions that are not directly applicable. As a matter of fact, the Court seems already to hint at a broader understanding of its reviewing capacities in this case implicitly, by considering that Articles 10 and 11 CA could be read in conjunction with '*all rights and freedoms, including those resulting from international treaties that are binding to Belgium, made applicable by ratification in the internal legal order and have direct effect.*'[1227]

Another interesting observation can be made. The Court avoids reviewing against Article 13 ESC, based on the *ratione personae* of this Provision. In fact, as Bouckaert points out, the Court's considerations regarding Article 11 ICESCR concern the *ratione personae* of the Provision as well.[1228] The difference however is that in Article 13 ESC the *ratione personae* is clearly described (all citizens of all Member States, legally residing on the territory of one of the Contracting Parties), and therefore cannot be invoked by the plaintiff, while Article 11 ICESCR recognises the right to an adequate standard of living *for everyone*, without a limitation regarding the *ratione personae*. The Court then ruled that an unlimited *ratione personae* is unreasonable, and therefore decides that the Provision is not violated,

---

[1225] Constitutional Court, 51/94, 29 June 1994, consideration B.5.5.
[1226] Constitutional Court, 51/94, 29 June 1994, consideration B.5.6.
[1227] Constitutional Court, 51/94, 29 June 1994, consideration B.5.2.
[1228] S. Bouckaert, *Documentloze vreemdelingen, grondrechtenbescherming doorheen de Belgische en internationale rechtspraak vanaf 1985*, Antwerp-Apeldoorn: Maklu, 2007, part II, no. 553.

without reviewing against the standstill effect of 11 ICESCR, or the substantive right in itself.

This Court ruling can be considered as a landmark ruling, introducing the basic ideas and direction of the Constitutional Court regarding the restricting of fundamental human rights of illegally residing aliens. For the first time, and certainly not for the last time, the Constitutional Court reviewed Article 57 § 2 *Organic Law of 8 July 1976 on public centres for social welfare'*, representing an important part of the Belgian aliens rules and policy, against several basic human rights.

12.5.3.2 *The period after Case 51/94 and the meaning of 'urgent medical care' and 'final order to leave'*

Although Article 57 § 2 *'Organic Law of 8 July 1976 on public centres for social welfare'* was thus not considered to be a violation of fundamental human rights, the application of the amended Provision led to some confusion regarding the exact meaning of the terms *'urgent medical care'* and *'the final order to leave the territory.'*

Some Labour Courts understood *'urgent medical care'* in a very broad sense, and even considered that the assistance would include the right to adequate food and housing.[1229] As an example to this, Bouckaert refers to the Labour Court of Verviers, that held that *'refugees who are denied asylum retain the right to a minimum of dignity; that is not to die from hunger or cold'*[1230] and to the Labour Court of Appeal of Liège, that ruled that *'this urgent medical assistance is not limited to a medical consultation and the provision of medicines. It should also be understood as granting supplies, including food essential for the survival of a person, and the enjoyment of decent housing.'*[1231] In appeal to the last verdict however, the Court of Cassation ruled otherwise, and considered that adequate housing and food did not fall under the scope of emergency medical assistance.[1232] In response to these developments in case law, the Federal Government amended Article 57 § 2 *'Organic Law of 8 July 1976 on public centres for social welfare'* by authorising the King to determine in more detail the meaning of 'emergency medical assistance.' As a result, in the

---

[1229] S. Bouckaert, *Documentloze vreemdelingen, grondrechtenbescherming doorheen de Belgische en internationale rechtspraak vanaf 1985*, Antwerp-Apeldoorn: Maklu, 2007, part II, no. 571-572.

[1230] Bouckaert referred to: Labour Court of Verviers, 12 August, 1993, S.K. 1993, part 10, 471-473 and Labour Court of Verviers, 22 March 1994, S.K. 1995, part 2, 60-61. Original text in French: *'les candidates réfugiés non admis à l ásile conservent le droit à la dignité minimale de ne mourir de faim, ni de froid.'*

[1231] Labour Court Liège, 13 February 1996, S.K. 1996, part 11, 568. Original text in French: *'cette aide médicale urgente ne se limite pas à une consultation médicale et l'octroi des medicaments. Elle doit aussi s'entendre comme permettant la fourniture, notamment des vivres indispensables à le sauvegarde de la personne humaine ainsi de la jouissance ce d'un logement decent.'*

[1232] Court of Cassation, 17 February 1997, Arresten van het Hof van Cassatie 1997 (91).

Royal Decree of 12 December 1996 (Article 1),[1233] emergency medical assistance was understood as assistance of a pure medical nature, of which the urgency is demonstrated by a medical certificate. The assistance may not consist of financial support, housing or other social services in kind.

A more complex question appeared to be the exact meaning of *'the final order to leave the territory.'* After some consideration,[1234] the Federal Government understood this phrase restrictively, and stated that an order to leave the territory is final when it can be executed. This would practically mean that when an alien would be involved in a legal procedure that does not automatically suspend the execution of the order (as in the case of an annulment procedure before the Council of State) he would have no access to any social benefits, besides urgent medical care, for the duration of this procedure. Therefore, this approach was not supported by the Labour Courts, which unanimously understood *'the final order to leave the territory'* in a broader sense. Most Labour Courts at least understood *'final order'* in the sense that an order is final when the alien could no longer appeal to the order to leave the territory, whether or not this appeal would suspend the execution of the order. Some Courts considered an even broader interpretation.[1235] However, in several verdicts, the Court of Cassation ruled that the interpretation of the Federal Government was the correct one,[1236] and consequently quashed the verdicts of the Labour Courts. However, not all Labour Courts were willing to change their view, and some of them persisted in their broader interpretation.[1237]

The Federal Legislator solved the matter by amending Article 57 § 2 *'Organic Law of 8 July 1976 on public centres for social welfare'* once more, on 15 July 1996.[1238] The granting of social benefits to aliens who were given a final order to leave Belgian territory would be ended on the day of their departure, but at the least from the day of the expiration of the order to leave the country. Deviation from the above would only be allowed for the duration that was strictly necessary to be able to leave the territory, but only when the alien signed a declaration of intent to leave, and only for the maximum period of one month. In fact, this amendment

---

[1233] Royal Decree on the emergency medical assistance provided for by the public centres of social welfare to illegally residing aliens in Belgium, 12 December 1996. Original Dutch name: Koninklijk Besluit van 12 december 1996, betreffende de dringende medische hulp die door openbare centra voor maatschappelijk welzijn wordt verstrekt aan de vreemdelingen die onwettig in het Rijk verblijven.

[1234] S. Bouckaert, *Documentloze vreemdelingen, grondrechtenbescherming doorheen de Belgische en internationale rechtspraak vanaf 1985*, Antwerp-Apeldoorn: Maklu, 2007, part II, no. 577.

[1235] S. Bouckaert, *Documentloze vreemdelingen, grondrechtenbescherming doorheen de Belgische en internationale rechtspraak vanaf 1985*, Antwerp-Apeldoorn: Maklu, 2007, part II, no. 578-581.

[1236] For instance, the earlier mentioned Court of Cassation, 17 February 1997, Arresten van het Hof van Cassatie 1997 (90).

[1237] S. Bouckaert, *Documentloze vreemdelingen, grondrechtenbescherming doorheen de Belgische en internationale rechtspraak vanaf 1985*, Antwerp-Apeldoorn: Maklu, 2007, part II, no. 583.

[1238] Statute of 15 July 1996 amending Article 57 § 2 *'Organic Law of 8 July 1976 on public centres for social welfare'*.

reaffirmed the original understanding of the Government of the phrase 'final order', meaning that it is understood as an 'executable order'. However, compared to the previous version of Article 57 § 2 *'Organic Law of 8 July 1976 on public centres for social welfare'* on 15 July 1996, the amended version made it possible to prolong the full right to social benefits with the period necessary to leave the territory (with a maximum of one month), instead of granting only the right to the aid that was necessary to leave the country.[1239]

12.5.3.3 Article 57 § 2 O.C.M.W. and the appeal to an order to leave: Case 43/98

On 22 April,[1240] some interest groups requested nullification of the Statute that amended Article 57 § 2 *'Organic Law of 8 July 1976 on public centres for social welfare.'*[1241] They claimed that the Article amending the Organic Law was in violation with Articles 10, 11, 23, and 141 Constitution, because the Statute creates a difference in treatment between on the one hand aliens that receive social benefits at the time when they are ordered to leave the Belgian territory and appeal against that decision, and on the other hand aliens that appeal to the order to leave while they reside illegally in Belgium. Both categories are awaiting a decision of the Council of State regarding their appeal against the order to leave, while only the first category of aliens will receive social benefits for another period, until the last day of the expiration of the order to leave the country, while the last category receives no social benefits except for urgent medical care. Also, the requesting parties held that Article 23 CA was violated, for the enjoyment of the right to a dignified life could not legally be depending on the residence status. According to the interest groups, this would be at least a violation of the standstill effect of this basic human right, for it would result in a situation in which a group of aliens is deprived of its right, which is to be considered as a step backwards. Besides that, this would be a disproportionate means in order to achieve the restriction of immigration. In addition, the interest groups argued that this would undermine the right to appeal to an order to leave the territory, as recognised in national legislation (by Statute of 15 December 1980). Furthermore, the requesting parties argued that another difference in treatment was installed, based on financial power, between aliens who could afford to live without social services for the period of the appeal, and those who could not.[1242]

The Council of Ministers basically claimed that in case 51/94 the Court adopted a principle of restricting social benefits, regardless of the fact whether an order to

[1239] S. Bouckaert, *Documentloze vreemdelingen, grondrechtenbescherming doorheen de Belgische en internationale rechtspraak vanaf 1985*, Antwerp-Apeldoorn: Maklu, 2007, part II, no. 590.
[1240] Court of Cassation, 43/98, 22 April 1998, B.S. (04) (83, pp.13348-13357).
[1241] Article 65, Statute of 15 July 1996 amending Article 57 § 2 *'Organic Law of 8 July 1976 on public centres for social welfare'*.
[1242] Constitutional Court, 43/98, 22 April 1998, A.13.

leave the territory is final or merely executable. Furthermore, the Council argued that both categories of aliens referred to by the requesting parties were in the same uncertain situation awaiting a decision on their appeal against the order to leave, and therefore, there was no difference in treatment.[1243]

The Court ruled that indeed the principles forthcoming from their ruling in case 51/94 were still applicable, and reaffirmed that the basic idea of restricting social services to aliens who were ordered to leave Belgian territory was a suitable means for the purpose of restricting immigration. However, the Legislature chose to restrictively interpret '*final order*' which was expressed in the amended law, in the sense that in case of an executable order to leave Belgian territory, except for emergency medical assistance, the Provision of all social services would be stopped. According to the Court, this measure was proportionate in a situation in which the alien wishes to leave the territory, and explicitly expresses the intention to leave. However, this is clearly not the case when the alien appeals against the decision to leave and (in addition) asks for suspension of the execution of the order to leave. According to the Court, the Council of State may, by an accelerated procedure, reject appeals that are manifestly inadmissible or unfounded. This would limit the chance that asylum-seekers would abuse an appeal procedure in order to enjoy social benefits for a longer period. The Court considered that it is unnecessary to deprive the appellants of social benefits during appeal procedures. The amended legislation would thus disproportionately restrict the exercise of fundamental human rights, in particular the right to social services and the right to effective exercise of a judicial appeal.[1244] The Court therefore nullified the word '*executable*' in the phrase '*The Provision of social services to an alien who was making actual use of these services at the time he received an executable order to leave the territory will be ended, with the exception of urgent medical care, on the day the alien actually leaves the territory, and at the latest, on the day of the expiration of the order to leave the territory.*'[1245] Furthermore, the Court ruled that this should be understood as that during the period of the appeal against the order to leave before the Council of State, Article 57 § 2 '*Organic Law of 8 July 1976 on public centres for social welfare*' shall not be applied, practically meaning that the alien has normal access to social benefits during these procedures.[1246]

[1243] Constitutional Court, 43/98, 22 April 1998, A.14

[1244] Constitutional Court, 43/98, 22 April 1998, consideration B.33-36.

[1245] The original text in Dutch: '*De maatschappelijke dienstverlening aan een vreemdeling die werkelijk steuntrekkende was op het ogenblik dat hem een uitvoerbaar bevel om het grondgebied te verlaten werd betekend, wordt, met uitzondering van de dringende medische hulpverlening, stopgezet de dag dat de vreemdeling daadwerkelijk het grondgebied verlaat, en ten laatste de dag van het verstrijken van het bevel om het grondgebied te verlaten*' (Article 65, Statute of 15 July 1996 amending Article 57 § 2 Organic Law of 8 July 1976 on public centres for social welfare).

[1246] Constitutional Court, 43/98, 22 April 1998, consideration B.37 and verdict.

In this case, no international Provisions were invoked, but only domestic Provisions. The Court reaffirmed this verdict in two preliminary rulings concerning this matter.[1247] Although in both cases Article 11 ICESCR – read in conjunction with Articles 10 and 11 Constitution – was included in the prejudicial question, the Court did not explicitly review against Article 11 ICESCR, but referred to its previous verdict (43/98).

12.5.3.4 Article 57 § 2 O.C.M.W. and the request for a prolonged stay: Case 25/99

Ruling 43/98 did not solve all the matters regarding the entitlements to social benefits during legal proceedings in relation to the order to leave. Whereas now it was made clear that during an appeal to the refusal of a residence permit the alien would have entitlements to social benefits, there appeared to be alternative procedures as well that were accessible for aliens, as will appear in this section, but also more extensively in Sections 12.5.3.6-12.5.3.7.

The Labour Court of Liège asked a preliminary question with regard to a procedure of the Statute of 15 December 1980 (Article 9). In this procedure, an alien may request permission to reside in Belgium for a period longer than the normal maximum period of three months. This request must be reviewed by the competent Minister or an authorised representative. The Labour Court asked whether this procedure would discriminate between asylum-seekers whose request for a residence permit was denied, but requested permission for a prolonged stay under the procedure of Article 9 (Statute of 15 December 1980) before they received an order to leave the territory on the one hand, and asylum-seekers whose request for a residence permit was denied, and who did not submit a request for a prolonged stay before they received an order to leave the territory on the other hand. The reason for this question was that the first category of aliens would still have entitlements to social benefits during the procedure stipulated in the Statute of 15 December 1980, while the second category of Asylum-seekers would not have any entitlements to social benefits during this procedure. Therefore, the Labour Court asked in addition whether this would be in violation with Articles 10, 11, and 23 Constitution in conjunction with Articles 2 and 11 ICESCR. The Constitutional Court concluded that there was no difference in treatment due to the disputed Articles, but instead due to a circular of the Minister, ordering the municipalities not to send the order to leave the territory when it was determined that the alien submitted a request to a prolonged stay, based on Article 9 Statute of 15 December

---

[1247] Constitutional Court, 46/98, 22 April 1998, consideration B.2-B.4; Constitutional Court, 108/98, 21 October 1998, consideration B.3.2-B.3.5 and verdict.

1980.[1248] The consequence would be that an alien that requested a prolonged stay would still receive social benefits, for the order to leave was not sent, while other aliens, who did receive an order to leave would not receive social benefits. A circular however is an administrative act, and no legislation. Therefore, The Court ruled that it was incompetent in reviewing the circular against the invoked legislation. Remarkable in this case is that the Court determined in detail that this circular resulted in a difference of treatment, while such a consideration was not necessary to come to the verdict, and it might even be questioned whether such considerations would fall outside the scope of the Court's competences. On the other hand, the Court was well aware of the fact that the discriminative wordings were already omitted in a later circular.[1249]

12.5.3.5 The medical impossibility of Case 80/99[1250]

The Labour Court of Gent asked in a preliminary procedure whether the amended Article 57 § 2 'Organic Law of 8 July 1976 on public centres for social welfare' was in violation with Articles 10 and 11 Constitution in conjunction with Articles 11 ICESCR and 13 ECHR, because it would institute an unjustified difference in treatment regarding the entitlement to social benefits between aliens who were ordered to leave the Belgian territory who can be removed and those who cannot be removed due to medical impossibility.[1251] The Court confirmed in a short consideration that if Article 57 § 2 'Organic Law of 8 July 1976 on public centres for social welfare' would be applied to persons who are in an absolute incapacity due to medical reasons to leave the Belgian territory, this was indeed in violation with Articles 10 and 11 Constitution. In its considerations,[1252] the Court referred to the invoked international Provisions, while in its verdict only a reference was made to the violation of Articles 10 and 11.[1253] According to the Court thus an illegally residing alien, regardless the fact whether he/she appealed against the decision

---

[1248] Circular on the application of Article 9, third Section, of the Act of 15 December 1980 on the access to the territory, residence, establishment and removal of aliens, 9 October 1997, B.S. 14 November 1997. Original title in Dutch: *Omzendbrief betreffende de toepassing van artikel 9, derde lid, van de wet van 15 december 1980 betreffende de toegang tot het grondgebied, het verblijf, de vestiging en de verwijdering van vreemdelingen.*

[1249] Circular on the application of Article 9, third Section, of the Act of 15 December 1980 on the access to the territory, residence, establishment and removal of aliens, 15 December 1998, B.S. 19 December 1998. Original title in Dutch: *Omzendbrief betreffende de toepassing van artikel 9, derde lid, van de wet van 15 december 1980 betreffende de toegang tot het grondgebied, het verblijf, de vestiging en de verwijdering van vreemdelingen.*

[1250] Constitutional Court, 80/99, 30 June 1999.

[1251] Another question was raised by the Labour Court, regarding the difference of treatment between Belgian citizens and legally residing aliens on the one hand, and aliens whose request for a residence permit was denied and who had received an order to leave the territory on the other hand. The Constitutional Court referred to its earlier case law on this matter, especially cases 51/94 and 43/98, and reaffirmed this case law.

[1252] Constitutional Court, 80/99, 30 June 1999, consideration B.5.1.

[1253] Constitutional Court, 80/99, 30 June 1999, verdict.

ordering him/her to leave, who is in a situation in which it is impossible to obey the order to leave due to medical reasons, should receive full social services.

Due to the short consideration of the Constitutional Court, the principle that medical impossibility should result in not applying Article 57 § 2 'Organic Law of 8 July 1976 on public centres for social welfare', Bouckaert observed that therefore the Courts, in their later rulings, further specified (1) whether invoking medical impossibility was only possible when a regularisation request was submitted[1254]; (2) what medical impossibility means;[1255] and (3) what the substance should be of the social benefits when the medical impossibility was invoked.[1256]

12.5.3.6 Artice 57 § 2 O.C.M.W. and the application for recognition of statelessness

In two preliminary rulings[1257], the Constitutional Court was asked whether Article 57 § 2 'Organic Law of 8 July 1976 on public centres for social welfare' would institute an unjustified difference in treatment regarding entitlements to social benefits between Belgian citizens, legally residing aliens and aliens who appealed to a refusal of their request to a residence permit on the one hand, and aliens

---

[1254] Due to the extensive case law on this, and the limited link with the main issue of this thesis, this debate is excluded from this Chapter. For more information see: S. Bouckaert, *Documentloze vreemdelingen, grondrechtenbescherming doorheen de Belgische en internationale rechtspraak vanaf 1985*, Antwerp-Apeldoorn: Maklu, 2007, part II, no. 759-765.

[1255] In case law, certain rules of thumb could be distilled. Four categories seem to have been developed amongst the Labour Courts and Courts of Appeal: (1) most importantly, the intrinsic severity of the condition of the person; but also (2) the impossibility to travel; (3) the impossibility to receive adequate medical care in the country of origin; and (4) medical attests that can prove what the applicant claims. See: S. Bouckaert, *Documentloze vreemdelingen, grondrechtenbescherming doorheen de Belgische en internationale rechtspraak vanaf 1985*, Antwerp-Apeldoorn: Maklu, 2007, part II, no. 766-778. However, not all the Courts seem to have adopted a similar approach towards the principle of medical impossibility. This can be demonstrated by one poignant decision of the Labour Court of Brussels. *In casu*, a refugee who was infected with the HIV virus and therefore under heavy treatment was ordered to leave Belgian territory. It was rather clear in the case that in the country of origin the treatment could not be continued. The Court ruled that the refugee was not entitled to full social services, for the impossibility to obey the order to leave due to medical reasons could not be proven. The refugee also submitted a regularisation request. The Labour Court referred to case 89/2002, in which the Constitutional Court ruled that restricting social services to an illegally residing alien who submitted a regularisation request was not in violation with Constitutional standards. This case seems to reflect a broader discussion, in which the question was raised whether an illegally residing alien that invokes a medical impossibility to obey an order to leave should submit a regularisation request in order to be entitled to full social services. It is doubtful however, whether in this particular case the Court uses case 89/2002 as an argument to deny social services in case of medical impossibility, for both issues were dealt with separately. See: Labour Court of Appeal of Brussels, 16 December 2004, S.K. 2005 (pp. 264-265).

[1256] Basically, a minimum substance of existence will be granted, and where necessary, additional necessary medical aid. See: S. Bouckaert, *Documentloze vreemdelingen, grondrechtenbescherming doorheen de Belgische en internationale rechtspraak vanaf 1985*, Antwerp-Apeldoorn: Maklu, 2007, part II, no. 779-780.

[1257] Constitutional Court, 17/2001, 14 February 2001; Constitutional Court, 89/2002, 5 June 2002.

who are ordered to leave the Belgian territory and submitted an application for recognition of statelessness on the other hand, aliens whose order to leave the territory is definite, and did not appeal or whose legal remedies are exhausted and submitted an application for recognition of statelessness on the other hand (case 17/2001), or, between aliens who appealed to the order to leave the territory in their capacity as refugee on the one hand, and aliens who were ordered to leave the territory in refugee procedures and now submitted an application for recognition of statelessness on the other hand (case 89/2002).[1258] In the first situation of both questions, the alien would receive social services, as was ruled in case 43/98,[1259] while in the second case the alien would receive only emergency medical assistance. The question was whether the principle of an effective legal remedy (and therefore the right to social services) of case 43/98 would analogously apply to an application for recognition of statelessness. In both cases, the Court ruled that this was not an unjustified difference of treatment. Basically, the Court reasoned that there would be a significant risk that in judging otherwise, the procedure of applying for recognition of statelessness might be abused to receive social benefits.[1260] In addition, the Court underlined in both cases that the fact that the alien in a previous procedure submitted a request for obtaining the status of refugee, based on his nationality, was not a convincing fact in a procedure for applying for recognition of statelessness.[1261] Only in case 89/2002, Article 11 ICESCR was invoked in the preliminary question. In its considerations however, the Court ruled that the invoked international Provisions would not lead to a different conclusion.[1262]

12.5.3.7 Article 57 § 2 O.C.M.W. and the regularisation request

The Statute of 22 December 1999 *'on the regularisation of the residence of certain categories of Aliens residing on Belgian Territory*'[1263] instituted a one-time opportunity to submit a regularisation request for certain categories of aliens (1) whose request for a refugee status was not considered within a period of four years (or in cases of families with minors three years); (2) who could not return to the country from which they arrived before entering Belgian territory, due to reasons beyond their control; (3) due to serious illness; or (4) due to humanitarian reasons while

---

[1258] In case 89/2002, a second question was raised, whether an unjustified difference in treatment was installed between aliens who are ordered to leave Belgian territory but appealed to that decision on the one hand and aliens who are ordered to leave Belgian territory and submitted a regularisation request on the other hand.

[1259] Constitutional Court, 43/98, 22 April 1998, consideration B.5.3.

[1260] Constitutional Court, 17/2001, 14 February 2001, consideration B.5.2; 89/2002, 5 June 2002, consideration B.8.

[1261] Constitutional Court, 17/2001, 14 February 2001, consideration B.5.3; Constitutional Court, 89/2002, 5 June 2002, consideration B.8.

[1262] Constitutional Court, 89/2002, 5 June 2002, consideration B.9.

[1263] Original title in Dutch: *Wet van 22 december 1999 'betreffende de regularisatie van het verblijf van bepaalde categorieën van vreemdelingen verblijvend op het grondgebied van het Rijk.'*

having developed considerable lasting social bonds with their country, illegally resided in Belgian territory on 1 October 1999 (Article 2). During the procedure, the alien would not be removed from Belgian territory (Article 14). The question was raised whether 57 § 2 '*Organic Law of 8 July 1976 on public centres for social welfare*' should be applied to those aliens who submitted such a request and were awaiting a decision.[1264] Bouckaert observed that the Council of State advised that during this period, Article 57 § 2 of the Organic Law should not be applied, so that these aliens were entitled to social benefits during the period of the procedure. The Federal Government however expressed in a Ministerial circular an opposite view, and stated that the Statute of 22 December 1999 did not change the illegal residence status and therefore the alien would have no entitlements to social services. A majority of the Labour Courts however, followed the reasoning of the Council of State, and ruled that during the process, Article 57 § 2 '*Organic Law of 8 July 1976 on public centres for social welfare*' would not apply, and as a result, the alien would be entitled to social services. Apparently, the situation was rather unclear, which led to a complex chain of legal proceedings.

In two annulment procedures, interest groups requested nullification of Article 14 of the Statute of 22 December 1999 '*on the regularisation of the residence of certain categories of Aliens residing on Belgian Territory*' for the Legislator, based on the principles of equality and non-discrimination (Articles 10 and 11 CA), should have excluded the application of Article 57 § 2 '*Organic Law of 8 July 1976 on public centres for social welfare*' to aliens who submitted a regularisation request. The Constitutional Court referred to the Parliamentary History of the Statute of 22 December 1999,[1265] and stated that it was obviously not the intention of the Legislature to grant full social benefits to these aliens. However, due to the fact that the requesting party basically argued that the Legislature refused to alter a standard that was not under dispute in the annulment request, the Court considered the request to be inadmissible.[1266]

In five preliminary questions the Labour Courts of Antwerp, Liège and Brussels asked whether Article 57 § 2 '*Organic Law of 8 July 1976 on public centres for social welfare*' was in violation with Articles 10 and 11 Constitution (in two cases read in conjunction with Articles 11 ICESCR and 3 ECHR) when the Provision of social benefits would be limited to emergency medical aid to illegally residing aliens who submitted a regularisation request.[1267] According to the Constitutional Court, it was the competence of the Legislature to adopt a policy regarding the

---

[1264] See: S. Bouckaert, *Documentloze vreemdelingen, grondrechtenbescherming doorheen de Belgische en internationale rechtspraak vanaf 1985*, Antwerp-Apeldoorn: Maklu, 2007, part II, no. 649-659.
[1265] Parliamentary Documents, House of Representatives, 1999-2000, doc. 50 0234/005, p. 60; Parliamentary Documents, Senate, 1999-2000, no. 2-202/3, p. 23; House of Representatives, 1999-2000, meeting of 24 November 1999, HA 50 plen. 017, pp. 7, 8, 18 and 31-32.
[1266] Constitutional Court, 106/2000, 25 October 2000; Constitutional Court, 32/2001, 1 March 2001.
[1267] Constitutional Court, 131/2001, 30 December 2001.

access to, residence in and the removal from Belgium of aliens, and to make regulations to define the conditions of legal residence in Belgium. The Court considered also here that it could be deduced from the Parliamentary Documents that the Legislature by adopting the regularisation procedure did explicitly not intend to alter the juridical status of the illegally residing aliens, and thus did not intend to create a right to social services for those who submitted a regularisation request, but only intended to postpone the actual removal of the territory. The Legislature was clearly looking for a solution to deal in a humane manner with a large group of illegally residing aliens on the one hand, while on the other hand such a procedure needs to be controllable and workable. In addition, it was the intention of the Legislature to discourage illegally residing aliens to abuse such a procedure only to gain access to social benefits.[1268] The Court then considered that it was the competence of the Constitutional Court to judge whether such policy measures were not in violation with the principle of non-discrimination.[1269] The Court underlined that illegally residing aliens had no entitlements to social benefits, except for urgent medical assistance. The Court then argued, in line with the Legislature, that the juridical status of the illegal alien who submitted a regularisation request was not different from the status of illegally residing aliens who did not submit such a request.[1270] In addition, the Court considered that the illegality of the alien was a result of his/her own actions, for apparently he/she did not act in accordance with the relevant regulations.[1271] The Court also stated that the regularisation procedure was not an international obligation, such as the recognition of the status of refugee, but a national choice.[1272] Furthermore, it seemed unreasonable not to guarantee the postponement of removal from the territory when illegally residing aliens were asked to identify themselves, while it seems also unreasonable to then assume that postponing the removal would imply access to social benefits.[1273] Therefore, the Court ruled that the restriction of social services to illegally residing aliens who submitted a regularisation request was not an unreasonable measure. Thus, in its verdict the Court ruled that the application of Article 57 § 2 '*Organic Law of 8 July 1976 on public centres for social welfare*' would not be in violation with the invoked standards in the interpretation that social benefits are restricted to urgent medical care for illegally residing

---

[1268] Constitutional Court, 131/2001, 30 December 2001, consideration B.3.5. The Court referred to Parliamentary Documents, House of Representatives, 1999-2000, doc. 50 0234/001, p. 5. and doc. 50 0234/005, p. 60; Parliamentary Documents, Senate, 1999-2000, no. 2-202/3, pp. 23, 36, and 58. House of Representatives, 1999-2000, meeting of 24 November 1999, HA 50 plen. 017, pp. 7, 8, 18, 31, and 32.
[1269] Constitutional Court, 131/2001, 30 December 2001, consideration B.3.4.
[1270] Constitutional Court, 131/2001, 30 December 2001, consideration B.3.5.
[1271] Constitutional Court, 131/2001, 30 December 2001, consideration B.4.3.
[1272] Constitutional Court, 131/2001, 30 December 2001, consideration B.3.7.
[1273] Constitutional Court, 131/2001, 30 December 2001, consideration B.4.3.

aliens that submit a regularisation request. The Constitutional Court reaffirmed this verdict in four preliminary rulings on 17 January 2002.[1274]

Nevertheless, the matter appeared not to be completely settled. Bouckaert observed that case 131/2001 did not lead to a uniform approach amongst the Labour Courts toward the application of Article 57 § 2 '*Organic Law of 8 July 1976 on public centres for social welfare*' to aliens who submitted a regularisation request.[1275] Since a majority of the Labour Courts had consistently granted social benefits to applicants of a regularisation request, following the ruling of the Constitutional Court would result in an inconsistency in their own case law. Some Labour Courts did alter their original view due to the enhanced *res judicata* of the Constitutional Court's ruling, and while explicitly considering that there are convincing arguments to rule otherwise, they ruled in compliance with case 131/2001.[1276] Other Labour Courts however maintained their original view.[1277] The Labour Court of Brugge for instance,[1278] among others, appeared to be struggling with the legal effect of the Constitutional Court's ruling. The Labour Court recognised the *res judicata* of this verdict, but added that such an effect applied to all rulings, and not only to case 131/2001. The Court considered that when the Constitutional Court ruled that social services were granted to an alien that could not be removed due to medical reasons (case 80/99),[1279] it was unclear why this should not be the case for a group of aliens that the Legislature did not even want to remove. The Court furthermore stated that it remained unclear whether the Constitutional Court decided to change their approach by case 131/2001 in view of their earlier rulings on comparable matters, such as the case of medical impossibility. The Labour Court considered that changing approaches in case law as such was not automatically unjustifiable, but underlined that such a radical change needed to be considered with caution. In the past, the Labour Court had systematically and with detailed motivation ruled that an alien who submitted a regularisation request was entitled to full social service during the period of that procedure. *In casu*, the alien involved belonged to one of the last submitters of a regularisation request. The ruling of the Constitutional Court was published near the end of

---

[1274] Constitutional Court, 14/2002, 17 January 2002, B.6, verdict; Constitutional Court, 15/2002, 17 January 2002, verdict. Constitutional Court, 16/2002, 17 January 2002, consideration B.5, verdict; Constitutional Court, 17/2002, 17 January 2002, consideration B.6, verdict. Only in case 15/2002 Article 11 ICESCR was not invoked.
[1275] S. Bouckaert, *Documentloze vreemdelingen, grondrechtenbescherming doorheen de Belgische en internationale rechtspraak vanaf 1985*, Antwerp-Apeldoorn: Maklu, 2007, part II, no. 665.
[1276] Bouckaert refers for instance to the Labour Court of Leuven, 12 December 2001, published on http://jure.juridat.just.fgov.be.
[1277] Bouckaert refers for instance to the Labour Court of Appeal of Liège, 5 March 2002, published on: http://jure.juridat.just.fgov.be; Labour Court of Appeal of Brussels, 17 April 2002, published on: http://jure.juridat.just.fgov.be; Labour Court of Brugge, 28 January 2002 published on: http://jure.juridat.just.fgov.be.
[1278] Labour Court of Brugge, 28 January 2002, published on: http://jure.juridat.just.fgov.be.
[1279] Constitutional Court, 80/99, 30 June 1999.

the period in which the regularisation procedure was valid. The Court was thus confronted with a dilemma: on the one hand, there was the enhanced *res judicata* of case 131/2001, while on the other hand, by recognising this *res judicata*, the Court would violate the principles of equality and legitimate expectations due to its previous rulings. The Labour Court of Brugge eventually decided not to change its approach in line with case 131/2001, and firmly rejected the ruling of the Constitutional Court.

Not surprisingly, the Court of Cassation, being the Court of Last Instance for Labour Court cases, was also confronted with appeals to the rulings of Labour Courts concerning this issue. On 17 June 2002, the Court ruled that Article 23 Constitution recognised the right to a decent existence, including the right to social benefits. The Court considered that entitlements to social benefits could be restricted to urgent medical care when an alien illegally resides on Belgian territory, by applying Article 57 § 2 *'Organic Law of 8 July 1976 on public centres for social welfare'*, in order to encourage the alien to leave the territory. The Court considered furthermore that the Statute of 22 December 1999 *'on the regularisation of the residence of certain categories of aliens residing on Belgian territory'* opened a possibility for aliens who stayed illegally on Belgian territory to submit a special regularisation request. During this procedure, the aliens could not be removed from Belgian territory. Therefore, the Court ruled that Article 57 § 2 *'Organic Law of 8 July 1976 on public centres for social welfare'* could not be applied to aliens who were allowed to reside on Belgian territory during this procedure.[1280] In other words: the illegally residing alien submitting a regularisation request did not reside illegally on Belgian territory during the regularisation procedure, and was therefore fully entitled to social benefits. The Court reaffirmed this on 7 October 2002[1281] and 7 June 2004.[1282] The ruling of the Court of Cassation does not necessarily contradict the earlier rulings of the Constitutional Court. This is due to the fact that the Constitutional Court only discussed the entitlements to social benefits of illegally residing aliens, while the Court of Cassation considers the aliens who submitted a regularisation request not to be illegally residing.[1283] The two approaches do not excel in consistency either.

Therefore, again, prejudicial questions were raised concerning the legality of applying Article 57 § 2 *'Organic Law of 8 July 1976 on public centres for social welfare'*. The Constitutional Court confirmed its previous case law, using remarkably frequent references to Parliamentary History. Furthermore, the Court considered that there are also differences between the aliens who requested regularisation, for some applicants would already have full access to social benefits based on other

---

[1280] Court of Cassation, 17 June 2002, no. S010148F, published on: http://jure.juridat.just.fgov.be.

[1281] Court of Cassation, 7 October 2002, no. S000165F, published on: http://jure.juridat.just.fgov.be.

[1282] Court of Cassation, 7 June 2004, no. S030008N, published on: http://jure.juridat.just.fgov.be.

[1283] Steven Bouckaert, *Documentloze vreemdelingen, grondrechtenbescherming doorheen de Belgische en internationale rechtspraak vanaf 1985*, Maklu: Antwerp-Apeldoorn, 2007, part II, no. 667.

procedures. Those who had no access to these social benefits did then clearly not act in accordance with the Belgian rules on immigration, and are therefore responsible for their own illegality.[1284] Furthermore, the Court underlined that due to the fact that the regularisation procedure did not include automatic regularisation, but rather a procedure in which each request was judged individually, Article 57 § 2 of the Organic Law could not be interpreted in such way that by merely submitting a regularisation request an entitlement to full social services would be created.[1285] The Court then explicitly ruled that Article 14 of the Statute of 22 December 1999 on the regularisation of the residence of certain categories of aliens residing on Belgian Territory (the alien would not be removed from Belgian territory during the regularisation procedure) should be interpreted in such a way that until the residence status of the alien is regularised, the entitlements to social benefits of the alien are restricted to emergency medical aid only.[1286]

Bouckart observed that the above situation led to an impasse amongst the Labour Courts: rulings of the Constitutional Court have an enhanced *res judicata*, while on the other hand the Court of appeal – the Court of Cassation – may nullify a ruling of the Labour Court that follows the interpretation of the Constitutional Court.[1287] Of course, in this matter the regularisation procedure was only a temporary measure, and the Constitutional Court's verdicts came at the end of this period so that actual legal inconsistencies due to the opposing rulings of the Constitutional Court and the Court of Cassation are of a hypothetical nature. This matter however does reflect a debate in Belgium concerning the competences of the Constitutional Court.[1288] The Court is authorised to review national legislation against Constitutional standards. However, especially in case of a preliminary ruling, the Court may need to review a standard against Constitutional standards in a certain interpretation of that national standard, usually given by the Court that raises the preliminary question. The Constitutional Court may perhaps suggest that it does not approve of that interpretation, or may even suggest one or two other possible interpretations in its verdict. It is however questionable whether the Constitutional Court's competences go this far, that it may impose a specific interpretation on law on a Labour Court. Smets analysed the preliminary rulings of the Court throughout the years, and concludes that the Constitutional Court

---

[1284] Constitutional Court, 203/2004, 21 December 2004, consideration B.13.1.

[1285] Constitutional Court, 203/2004, 21 December 2004, B.S. 25 February 2005, consideration B.14; 204/2004, 21 December 2004, B.S. 03 March 2004, consideration B.15; 205/2004, 21 December 2004, B.S. 04 March 2005, consideration B.14.

[1286] Constitutional Court, 203/2004, 21 December 2004, consideration B.15; Constitutional Court, 204/2004, 21 December 2004, consideration B.16; Constitutional Court, 205/2004, 21 December 2004, consideration B.15.

[1287] S. Bouckaert, *Documentloze vreemdelingen, grondrechtenbescherming doorheen de Belgische en internationale rechtspraak vanaf 1985*, Antwerp-Apeldoorn: Maklu, 2007, part II, no. 673.

[1288] See especially: J. Smets, *De verhouding van het Arbitragehof tot de verwijzende rechter in het prejudiciële litigation*, in: *Twintig Jaar Arbitragehof*, Mechelen: Wolters Kluwer België, 2005.

increasingly interferes with the interpretation that should be given to a national standard in a preliminary procedure.[1289]

12.5.3.8 Article 57 § 2 O.C.M.W. and an appeal to the Council of State against a second or third rejection of a request for asylum

The Aliens Act recognises the possibility for a refugee to request for asylum a second time when he/she is of the opinion that there is new evidence to support the request. The Belgian Immigration Department may in such a second or third request refuse to consider the request when it is doubtful whether there is indeed such new evidence.[1290] The alien can only appeal to this decision before the Council of State.

In three prejudicial cases, the Constitutional Court was asked whether the alien who appealed against such a decision was entitled to full social benefits, in an analogue interpretation of case 43/98, in which *inter alia* the principle of an effective exercise of a judicial appeal was recognised.[1291] The Constitutional Court ruled that the application of Article 57 § 2 *'Organic Law of 8 July 1976 on public centres for social welfare'* to these aliens, and thus restricting their right to social services to emergency medical aid, was not in violation with the invoked Constitutional and international standards. The Court considered in case 21/2001 that in ruling otherwise, the procedure to apply for asylum a second or third time might easily be abused to get entitlements to social benefits only.[1292] The Court furthermore ruled that an alien, who submitted a second request and had already exhausted all legal remedies during the first request, was in a significantly different position compared to the alien who requested asylum for the first time.[1293] The last argument was also (although perhaps not very convincingly) used in case 148/2001, in a situation in which the alien submitted a second request for asylum, but clearly did not exhaust all legal remedies during the first procedure.[1294] The Court however added in case 50/2002 that the alien who did not exhaust or did not use the juridical remedies during the first procedure, and asked for asylum a second time, was in a different situation compared to the alien who asked for asylum for the first time.[1295] Therefore, applying Article 57 § 2 of the Organic

---

[1289] J. Smets, *De verhouding van het Arbitragehof tot de verwijzende rechter in het prejudiciële litigation*, in: *Twintig Jaar Arbitragehof*, Mechelen: Wolters Kluwer België, 2005.

[1290] See for more information regarding the procedure: https://dofi.ibz.be, the official website of the Belgian Immigration Department.

[1291] Constitutional Court, 43/98, 22 April 1998, considerations B. 33-36.

[1292] Constitutional Court, 21/2001, consideration B.4.

[1293] Constitutional Court, 21/2001, 1 March 2001, consideration B.6.

[1294] Constitutional Court, 148/2001, 20 November 2001, consideration B.7.

[1295] Constitutional Court, 50/2002, 13 March 2002, consideration B.6.2. Bouckaert argued that in a case in which the alien did not exhaust all legal remedies during the first procedure, full social services should be granted during the second, when it is apparent that the alien is not abusing the procedure. He

Law was not in violation with the principles of equality and non-discrimination, and not an unreasonable means to prevent the abuse of the procedure.[1296]

In cases 21/2001 and 148/2001 the Court only reviewed against Articles 10 and 11 Constitution, while in case 50/2002 the Court reviewed against Articles 10 and 11 Constitution read in conjunction with Article 11 ICESCR.

12.5.4 Proceedings regarding the International Convention on the Rights of the Child

The ICRC was ratified by Belgium by the decree of the Flemish Community on 15 May 1991, the decree of the German language Community of 9 August 1991, the decree of the French Community of 25 November, and the Federal Statute of 15 January 1992. Therefore, the Convention came into force in the Belgian legal order on 15 January 1992. Since 2003, Article 57 § 2 *'Organic Law of 8 July 1976 on public centres for social welfare'* was also reviewed against national Constitutional Provisions in conjunction with ICRC Articles.

12.5.4.1 Case 106/2003: social benefits for the child of illegally residing aliens

The Labour Court of Brussels asked in a prejudicial question whether Article 57 § 2 *'Organic Law of 8 July 1976 on public centres for social welfare'* was in violation with Articles 10 and 11 CA, separately or read in conjunction with Articles 23 and 191 Constitution, Articles 2, 3, 24, 26, and 27 ICCR (separately or read in conjunction with Article 4 ICCR), Article 11 (1) ICESCR (separately or read in conjunction with Article 2 (1) ICESCR), and Article 3 ECHR, when it restricts the entitlements to social benefits to urgent medical care to an illegally residing minor alien, and consequently institutes a difference in treatment between (1) illegally residing minor aliens on the one hand, and minor Belgian and legally residing minor aliens on the other hand, when this difference of treatment is instituted with the purpose to encourage illegally residing aliens to voluntarily leave Belgian territory; and (2) illegally residing major aliens, who could leave the territory voluntarily on the one hand, and illegally residing minor aliens, who due to their young age could not on the other hand. The Labour Court case concerned a minor that was accompanied by its parents.[1297]

---

supports a more practical approach towards the principle of an effective legal remedy. See: S. Bouckaert, *Documentloze vreemdelingen, grondrechtenbescherming doorheen de Belgische en internationale rechtspraak vanaf 1985*, Antwerp-Apeldoorn: Maklu, 2007, part II, no. 721.

[1296] The Court referred to a verdict of 1994, in which the Constitutional Court had ruled that a Legislator may take measures to prevent the abuse of procedures as long as these measures are not unreasonable or disproportionate. See: Constitutional Court, 83/94, 1 December 1994.

[1297] Constitutional Court, 106/2003, 22 July 2003, B.S. 4 November 2003.

The Council of Ministers argued firstly that the Provisions of the Convention on the Rights of the Child would have no direct applicability, and that the Convention was only applicable to children under the jurisdiction of the contracting Parties, which would not include illegally residing children.[1298] The Council argued furthermore that there was no reason why a child could not, due to his/her age, leave the territory voluntarily, and made a distinction between accompanied and unaccompanied illegally residing minor aliens. In the first situation, the child was accompanied by the parents, who were the main applicant of the residence permit. If the parents would be ordered to leave the territory, this would automatically include the child. In the second situation, the child is unaccompanied, and could therefore request for a residence permit on its own behalf. If in such a case a residence permit is denied, Article 118 of the Royal Decree of 8 October 1981[1299] will be applied, which means that an order of refoulement will be issued against the child, ordering the receiver to take the necessary measures to return the child to the country of origin. Therefore, the Council concluded that the situation of an illegally residing child is similar to the situation of illegally residing adults, and thus the restriction of entitlements to social services is not unreasonable.[1300]

The Court in response clarified its competences, by underlining that it was not bound to decide on the direct applicability of international Provisions, but instead was authorised to assess whether the national Legislature violated international obligations of Belgium in a discriminative manner.

The Court emphasised that Belgium made a reservation to Article 2 (1):'*With regard to Article 2, Section 1, according to the interpretation of the Belgian Government non-discrimination on grounds of national origin does not necessarily imply the obligation for States automatically to guarantee foreigners the same rights as their nationals. This concept should be understood as designed to rule out all arbitrary conduct but not differences in treatment based on objective and reasonable considerations, in accordance with the principles prevailing in democratic societies.*'[1301] The Court however underlined that this declaration should be read in conjunction with Article 141 Constitution, which recognises equal rights between aliens and Belgian citizens, apart from exceptions prescribed by Law. The Court considered that unequal treatment thus can only be legal when it was enacted by law, while the Legislature

---

[1298] Constitutional Court, 106/2003, 22 July 2003, B.S. 4 November 2003, A.3.2.

[1299] Royal Decree of 8 October 1981 on access to the territory, residence, establishment and the removal of aliens. Original title in Dutch: Koninklijk Besluit van 8 oktober 1981, betreffende de toegang tot het grondgebied, het verblijf, de vestiging en de verwijdering van vreemdelingen.

[1300] Constitutional Court, 106/2003, 22 July 2003, A.3.3.

[1301] Constitutional Court, 106/2003, 22 July 2003, consideration B.5.2. Text in Dutch: '*In verband met artikel 2, eerste lid, legt de Belgische Regering niet-discriminatie op grond van nationale afkomst uit als niet noodzakelijk de verplichting voor de Staten inhouden om aan vreemdelingen dezelfde rechten te waarborgen als aan de eigen onderdanen. Dit begrip moet worden verstaan als ertoe strekkende iedere willekeurige gedraging uit te bannen, doch niet verschillen in behandeling, stoelend op objectieve en redelijke overwegingen, overeenstemmend met de beginselen die in democratische samenlevingen gelden.*'

thereby had to take into consideration the fundamental principles embedded in the Constitution.[1302]

The Court also considered that Article 2 ICRC, stipulating that States Parties shall respect and ensure the rights set forth in the Convention to each child within their jurisdiction, implies that there must be a certain connection between the child that invokes the Convention and the Member State that is accused of violating it.[1303] According to the Court, Article 2 ICRC had to be understood in view of the entire Convention, while also the nature of the difference in treatment that is under dispute must be taken into consideration. Therefore, the Court reasoned that the answer to the question whether illegally residing children would fall under the jurisdiction of Belgium according to the Convention, coincided with the assessment of the invoked discrimination.[1304] The Court thus did not directly answer the question whether in this context illegally residing children would fall under Belgian jurisdiction. The fact however – as will be demonstrated below – that the Court reviews against the invoked Articles suggests that the *ratione personae* of the Convention on the Rights of the Child includes illegally residing alien children.[1305]

The Court furthermore considered that the parents had the primary responsibility for the maintenance of the child, as embedded in Article 27 (2) ICRC.[1306] However, the parents *in casu* could not fulfil this obligation due to the illegality of their stay.[1307] The Court underlined that parents who did not obey the order to leave the territory had no rights to social benefits. It would be unreasonable when illegal aliens with children would have indirect access to social benefits by invoking the needs of their child, for this would result in a difference of treatment based on whether or not illegally residing adults are accompanied by children. In addition, granting the parents social benefits, even when limited to the amount that is necessary for the maintenance of the child, would be contrary to the aim of the Belgian Legislator to encourage illegally residing aliens to leave the territory (case 51/94[1308]).[1309] However, the Court considered that this would not justify complete refusal of any aid to the child, when this would lead to a situation in which the child would live in circumstances that are harmful for his/her health and development, while there would be a certainty that the parents cannot enjoy these social benefits. The Court referred in this light to Article 2 (2) ICRC, stipulating

---

[1302] Constitutional Court, 106/2003, 22 July 2003, considerations B.5.2.-B.5.3.
[1303] Constitutional Court, 106/2003, 22 July 2003, consideration B.6.2.
[1304] Constitutional Court, 106/2003, 22 July 2003, consideration B.6.3.
[1305] See also: Steven Bouckaert, *Documentloze vreemdelingen, grondrechtenbescherming doorheen de Belgische en internationale rechtspraak vanaf 1985,* Maklu: Antwerp-Apeldoorn, 2007, part III, no. 32 and 438.
[1306] Constitutional Court, 106/2003, 22 July 2003, consideration B.7.2.
[1307] Constitutional Court, 106/2003, 22 July 2003, consideration B.7.1.
[1308] Constitutional Court, 51/94, 29 June 1994.
[1309] Constitutional Court, 106/2003, 22 July 2003, B.S. 4 November 2003, consideration B.7.3.-B.7.4.

that States Parties shall take all appropriate measures to ensure that the child is protected against all forms of discrimination or punishment on the basis of the status of the parents.[1310] Also, the Court considered that the purpose of Articles 2, 3, 24 (1), 26, and 27 ICRC had to be reconciled with the goal of encouraging illegally residing adults to leave.[1311]

Therefore, the Court ruled that social benefits may be granted under strict conditions. Firstly, the competent authorities had to establish that the parents do not or cannot fulfil their duty to maintain the child. Secondly, the requested social benefits must only cover the necessary expenses for the development of the child. Thirdly, the Centre of Social Welfare involved must be absolutely sure that the social services are indeed used to cover those necessary expenses. To avoid any abuse of these benefits by the parents, the Centres of Social Welfare may only provide for social benefits for the particular need of the child, and the benefits must be provided in kind (or payment to third parties that provide for services in kind).[1312] The Court ruled that the refusal of these social benefits would indeed be a violation of Articles 10 and 11 Constitution, read in conjunction with Articles 2, 3, 24 (1), 26, and 27 ICRC. Remarkably, Articles 11 ICESCR and 3 ECHR were not included in the verdict.

Ruling 106/2003 was later reaffirmed by the Court in case 129/2003[1313] and 189/2004.[1314] Both cases concerned accompanied alien children.

12.5.4.2 Case 131/2005: social services and the right to family life

Cases 106/2003, 129/2003, and 189/2004 resulted in the adaption of a new law: Article 496 of the Programme Act of 22 December 2003[1315] authorised the Centres for Social Welfare to determine the needs of a minor alien illegally residing on Belgian territory whose parents cannot fulfil the duties towards the child. In such circumstances, the social benefits are limited to those that are necessary for the development of the child, and only provided for in Federal Centres for Social Welfare, according to the conditions and modalities established by the King. The restriction that these social benefits could only be provided for in a Federal reception centre could however result in the fact that the child would be separated from his/her parents, who are not entitled to any social benefits except

---

[1310] Constitutional Court, 106/2003, 22 July 2003, B.S. 4 November 2003, consideration B.7.5.

[1311] Constitutional Court, 106/2003, 22 July 2003, B.S. 4 November 2003, consideration B.7.6.

[1312] Constitutional Court, 106/2003, 22 July 2003, consideration B.7.7. But see in this light also the interpretation of the Court of Cassation of this verdict, 23 October 2006, no. S050042F, published on: http://jure.juridat.just.fgov.be. In this verdict, the Court ruled that benefits in kind does not necessarily exclude financial benefits.

[1313] Constitutional Court, 129/2003, 1 November 2003.

[1314] Constitutional Court, 189/2004, 24 November 2004.

[1315] Programme Act 22 December 2003, B.S. 31 December 2003.

for urgent medical assistance. The Labour Courts were confronted with many cases in which the question was raised whether separating the family due to the fact that the child was housed in a reception centre receiving social benefits, while the parents were not, was in violation with human rights.[1316]

On 19 July 2005,[1317] some interest groups requested the nullification of Article 496 of the Programme Act. To support their request, they firstly invoked Articles 22 and 23 Constitution in conjunction with Articles 191 Constitution, 8 ECHR, 17 and 23.1 ICCPR, 2 (1), 10 (1), and 10 (3). ICESCR, and 3 and 16 ICRC. The requesting parties argued that separating the child from his/her family would be an unreasonable infringement of a the right to privacy and family life '...*to the extent that law implicitly but with certainty requires that the limited material social benefits, which are essential for the development of the child, are exclusively granted to the child, without taking into account the situation of all Members of the family.*'[1318] In response, the Court considered that on the one hand granting social benefits to the parents of the illegally residing child would be contrary to the principles introduced in case 51/94, that is to encourage the illegally residing parents to voluntarily leave the Belgian territory. On the other hand, the Court stated that it had to consider whether the disputed Programme Act would make the existence of a family life impossible.[1319] The Court considered that a restriction to family life could only be allowed if it was enacted by law and foreseeable, with a legitimate goal and proportionate, as stipulated in Articles 22 CA and 8 ECHR. The Court then referred to Parliamentary Documents, a Ministerial circular and case law of the European Court of Human Rights, in which the importance of the presence of the parents near to the child is stressed.[1320] Based on this, the Court ruled that by restricting the Provision of the social benefits necessary for the child's development to only the location of Federal Centres for Social Welfare, the right to privacy and family life was violated.[1321]

Secondly, the Court did not consider a legal remedy, in which the requesting parties invoked *inter alia* Articles 23 Constitution, 11 and 13 ICESCR, 27, 28, and 29 ICRC, 16 and 17 ESC (the treaty Provisions in conjunction with Articles 10 and 11 CA), due to the fact that a violation of those Provisions could not lead to a broader nullification than the already established violation of privacy and family life.[1322]

[1316] S. Bouckaert, *Documentloze vreemdelingen, grondrechtenbescherming doorheen de Belgische en internationale rechtspraak vanaf 1985*, Antwerp-Apeldoorn: Maklu, 2007, part III, no. 488-522.

[1317] Constitutional Court, 131/2005, 19 July 2005.

[1318] Constitutional Court, 131/2005, 19 July 2005, A.3.1.1. Original text in Dutch: '...*in zoverre die wet impliciet doch met zekerheid bepaalt dat de tot de materiële hulp beperkte maatschappelijke dienstverlening, die absoluut noodzakelijk is voor de ontwikkeling van het kind, hulp is die uitsluitend aan het kind wordt voorbehouden, zonder rekening te houden met de situatie van alle leden van het gezin.*'

[1319] Constitutional Court, 131/2005, 19 July 2005, considerations B.3.2.-B.4.

[1320] Constitutional Court, 131/2005, 19 July 2005, considerations B.5.1.-B.5.5.

[1321] Constitutional Court, 131/2005, 19 July 2005, considerations B.5.5.-B.6.

[1322] Constitutional Court, 131/2005, 19 July 2005, considerations B.9.1.-B.9.2. and B.10.1-B.10.2.

Thirdly, the requesting parties argued that Article 496 of the Programme Act was in violation with Articles 10 and 11 CA, for it would institute a difference in treatment between illegally residing children or their families on the one hand, and legally residing children or their families on the other hand, since the residence status would determine how the interest of the child is understood. The Court ruled that the application of Article 496 of the Programme Act is not discriminative, for both categories of children would have entitlements to social benefits. The fact that the kind of social benefits might differ was also not discriminative, due to the fact that it was the purpose of the Act to reconcile the principles of the Convention on the Rights of the Child with the purpose to encourage illegally residing aliens to leave Belgian territory.

Another difference in treatment was invoked, based on Articles 10 and 11 CA, whether or not in conjunction with Articles 2 (2) ICESCR and 2 ICRC. The interest groups argued that the Programme Act made an unjust difference in treatment regarding the Provision of social benefits between illegally residing children that are accompanied by their parents on the one hand, and illegally residing children that are not accompanied by parents on the other hand. The Court however ruled that this difference in treatment was not discriminating and it could be reasonably justified that children accompanied by their parents would receive different social services compared to minors who are not accompanied by their parents.[1323]

Finally, the Court ruled that Article 496 of the Programme Act of 22 December 2003 violates Article 23 CA and equivalent international Provisions, but only to the extent that the Provision does not guarantee that parents can accompany their child to the centre for social welfare, where the child receives the social assistance.[1324]

As a consequence, the Federal Legislator once again changed its legislation by adopting Article 22 of the Programme Act of 27 December 2005,[1325] recognising that an illegally residing child that receives social benefits necessary for his/her development in a Federal Centre for Social Welfare can enjoy the presence of the parents or persons who exercise parental authority.

12.5.4.3 Case 194/2005: social services, the right to family life, medical impossibility to leave the territory

The course set by the Constitutional Court in case 131/2005 was later continued in other cases with different circumstances. In case 194/2005, the Labour Court of Brussels asked in a preliminary question whether Article 57 § 2 '*Organic Law*

[1323] Constitutional Court, 131/2005, 19 July 2005, consideration B.11.2.
[1324] Constitutional Court, 131/2005, 19 July 2005, consideration B. 12.2. and verdict.
[1325] Programme Act, 9 January 2006, B.S. 30 December 2005.

of 8 July 1976 on public centres for social welfare' was in violation with Articles 10 and 11 Constitution, whether or not read in conjunction with Articles 22, 23, and 191 Constitution and 2, 3, 24 (1), 26, and 27 ICRC, when it was applied to an illegally residing mother of a severely handicapped child, while the medical situation of the child made it absolutely impossible to return to the country of origin. The Court referred to its ruling in case 80/99[1326] in which it recognised the principle that an illegally residing alien who cannot be removed from Belgian territory due to medical reasons has entitlements to full social services.[1327] The Court considered furthermore that it is a fundamental element of family life that parents and children live together, as embedded in Articles 22 CA and 8 ECHR.[1328] Therefore, the Court finally ruled that Article 57 § 2 *'Organic Law of 8 July 1976 on public centres for social welfare'* was in violation with Articles 10, 11, and 22 CA, because it treated persons equally who were in fundamentally different position, without a reasonable justification: on the one hand those aliens who could be removed, and on the other hand, those aliens who cannot leave the Belgian territory due to the fact that they can prove to be the parent of a severely handicapped child that cannot possibly leave the territory due to medical reasons. *In casu*, the Court considered that it was impossible for the child to leave the territory due to these medical reasons, because it was in need for adequate medical healthcare that cannot be offered in the country of origin or another country that should receive the family.[1329]

12.5.4.4 Social services to Belgian children with illegally residing alien parent(s)

The Labour Court of Brussels asked in a preliminary question whether Article 57 § 2 *'Organic Law of 8 July 1976 on public centres for social welfare'* was in violation with Articles 10 and 11 Constitution, alone or in conjunction with Articles 2 (2), 3.2, 9, 10, and 27 ICRC, for it would institute an unequal treatment between children born to Belgian parents, minor aliens with alien parents who were allowed residence on Belgian territory, minor aliens born to minor parents who were illegally residing on Belgian territory on the one hand, and an illegally residing alien who is the parent of a Belgian child on the other hand. In case of the first category, due to the case law mentioned above, the parents would be entitled to social benefits, while in the second category, the child has the Belgian nationality and therefore, Article 57 § 2 *'Organic Law of 8 July 1976 on public centres for social welfare'* is not applicable, while in case of an illegally residing child, the parents may receive social services via the child based on this Article (case 131/2005).

[1326] Constitutional Court, 80/99, 30 June 1999.
[1327] Constitutional Court 194/2005, 21 December 2005, consideration B.4.2.
[1328] Constitutional Court 194/2005, 21 December 2005, consideration B.5.1.
[1329] Constitutional Court 194/2005, 21 December 2005, consideration B.5.2. and verdict.

Consequently, the entitlements to social services of the illegally residing parent of a Belgian child are limited to urgent medical care.[1330]

The Court ruled that Articles 10 and 11 Constitution were not violated, thereby referring to the legitimate policy of the State to encourage illegally residing aliens to leave the territory.[1331] In the opinion of the Court the child had, due to his nationality, a right to social benefits of his own, and therefore Articles 2 (2) and 3 (2) ICRC were not violated. However, the Court considered that the fact that an illegally residing parent of a Belgian child that has no individual entitlements to social benefits does not automatically imply that the Public Centres for Social Welfare do not have to take into account the specific family situation. These centres were free to select the appropriate instrument to meet the medical and development needs of the child. The Court ruled that in determining the amount of the social benefits to the child, the Centres for Social Welfare had to take into account the fact that the social services of the mother were restricted to emergency medical aid, and the fact that the (Belgian) father had a legal duty to maintain the child.[1332] In this interpretation, the Court ruled that Article 57 § 2 of the Organic Law was not in violation with the invoked Provisions.

Interesting detail in this case is that the Court explicitly considered that it was not asked to determine whether the illegally residing mother should be granted a residence permit. The Court underlined that therefore a review against Articles 9 and 10 ICRC would be necessary, which was not part of the prejudicial question.[1333] Most probably the Court is giving a hint to the Labour Court on how to settle the issue.

In cases 35/2006[1334] and 44/2006[1335] the Court was asked the same preliminary question as in 32/2006, although now the cases concerned two illegally residing parents of a Belgian child. The Court reaffirmed its ruling 32/2006 also to this case, and ruled that Article 57 § 2 of the Organic Law did not violate the invoked Provisions, although the Centres for Social Welfare were free to determine the appropriate social benefits, taking into consideration the particularities of the case.

In case 110/2006[1336] not the social benefits of the parents, but the benefits of the child were under dispute. The Labour Court of Brussels asked in a preliminary question whether the fact that a Belgian child with illegally residing parents was

---

[1330] Constitutional Court, 32/2006, 1 March 2006.

[1331] Constitutional Court, 32/2006, 1 March 2006, considerations B.6.1.-B.6.4. and B.10.

[1332] Constitutional Court, 32/2006, 1 March 2006, consideration B.10.

[1333] Constitutional Court, 32/2006, 1 March 2006, consideration B.9.

[1334] Constitutional Court, 35/2006, 1 March 2006.

[1335] Constitutional Court, 44/2006, 15 March 2006.

[1336] Constitutional Court 110/2006, 28 June 2006.

not entitled to certain family benefits[1337] was in violation with Articles 10 and 11 Constitution, whether or not read in conjunction with Articles 2 and 26.1 ICRC. Also here, the Court ruled that the disputed Articles were not in violation with the invoked Constitutional and International Provisions, but despite that, the Centres of Social Welfare were free to select the appropriate instrument to meet the medical and development needs of the child, thereby referring to Articles 2 (1) and 26 ICRC.[1338]

12.5.4.5 Case 43/2006: the principle of legality in Articles 22 and 23 Constitution

The Labour Court of Brussels asked in a preliminary question whether Article 57 § 2 *Organic Law of 8 July 1976 on public centres for social welfare* violated Articles 22 and 23 (2) (3) Constitution (alone or together) whether or not in conjunction with Article 191 Constitution when authorising the King to set the conditions and modalities for the establishment of social benefits in kind that are provided for in Federal Centres for Social Welfare to minor aliens and their parents who illegally reside on Belgian territory, instead of establishing those conditions by law. In addition, the Court asked the same question with regard to the fact that the disputed legal standard entrusted the King with establishing the minimum guarantees regarding the right to privacy and family life as recognised in Article 22 Constitution, and the prerequisites for entitlements to social benefits and adequate housing in kind in Federal Social Welfare Centres as recognised in Article 23 (2) and (3) Constitution. In Articles 22 and 23 the Constitution clearly (and exclusively) authorises the Legislature (by Statute, decree or ordinance) to establish the aforementioned conditions. Therefore, the Labour Court asked whether these competences of the King are in accordance with the principle of legality. The Court ruled that the Legislature sufficiently met its obligations in Articles 22 and 23 Constitution, and merely entrusted the King with the specific execution of the legal standards.[1339] The Court added however, that this does not mean that the King has unlimited authority, for He is also obliged to take into account the Constitution and the ICRC. In this light, the Court especially referred to the right to health (24 ICRC), the right to an adequate standard of living (27 ICRC) and the right to education (28 ICRC and 24 Constitution). The Court finally concluded that both the Administrative Courts and the regular Courts had the obligation to nullify any measures taken by the Crown regarding the granting of social benefits that were in violation with these human rights (159 Constitution).[1340]

[1337] *In casu*, based on Article 1 (6). Statute of 20 July 1971 on the establishment of guaranteed family benefits. Original title in Dutch: wet van 20 juli 1971 tot de instelling van gewaarborgde gezinsbijslag.
[1338] Constitutional Court 110/2006, 28 June 2006, B.S. 15 September 2006, consideration B.6.
[1339] Constitutional Court, 43/2006, 15 March 2006, considerations B.20.-B.21.
[1340] Constitutional Court, 43/2006, 15 March 2006, consideration B.22.

12.6 Concluding remarks

From the analysis of case law, some conclusions can be drawn. Firstly, in general there appears to be hardly any case law in Belgium on the effect of CEDAW Provisions. Furthermore, there seems to be a difference between the Council of State and the Court of Cassation on the one hand, who are very hesitant in applying ICESCR and ICRC Provisions directly in their case law, and on the other hand the Constitutional Court, which does not need to rule on the direct effect of those Articles and therefore – not surprisingly – seems to be more willing to indirectly review against international Provisions, including the right to food. Thirdly, there is occasionally a 'battle of the forums' in Belgium. The Courts of Last Instance do not always seem to agree with the Constitutional Court,[1341] and the Labour Courts seem to have some difficulties in recognising the *res judicata* in prejudicial rulings of the Constitutional Court.[1342] This does not contribute to the principles of legal certainty.

12.6.1 The Council of State and the Court of Cassation

It can be concluded that both the Council of State and the Court of Cassation hardly ruled on matters in which international Provisions stipulating the right to food were invoked. The only two cases were rulings of the Council of State in which *inter alia* Article 27 ICRC was invoked. The Council ruled, without further explanation, that the Provision had no direct effect.[1343] In general, the Council of State and the Court of Cassation have the tendency to circumvent the issue of direct effect, and only rule on the matter when it is absolutely necessary.

When ruling on the direct effect of a Provision, the case law of the Council of State does not excel in consistency. Where in a majority of the cases, the direct effect of ICESCR and ICRC Provisions is denied, and even in some verdicts the direct effect of the entire Covenant or Convention is rejected, there appear to be – almost randomly – some cases in which the Council is willing to review against some of the ICESCR Provisions, or at least to take into consideration the principles stipulated in those Articles. Furthermore, no evidence could be found that the Council ever granted direct effect to an ICRC Provision. The Council of Aliens Disputes seems to greatly value the case law of the Council of State and – in line with this case law – consistently decides that ICRC Provisions invoked have no direct effect. Also, direct effect is denied in the few cases in which ICESCR Provisions were invoked.

[1341] See for instance Constitutional Court, 106/2003, 22 July 200 and the Court of Cassation, 23 October 2006, no. S050042F, published on: http://jure.juridat.just.fgov.be, on the interpretation of the concept *'benefits in kind.'*

[1342] See for instance the aftermath of Constitutional Court, 131/2001, 30 December 2001.

[1343] Council of State, 12 February 1996, case no. 58122l, available at: www.raadvst-consetat.be.

The Court of Cassation, in the scarce cases in which ICESCR Provisions are invoked, generally rules that ICESCR Articles have no direct effect. Regarding the ICRC, the tendency appears to be that most Provisions have no direct effect either, although in some cases, without explicitly recognising direct effect, there seems to be some room for application of ICRC Provisions (albeit mostly Articles stipulating civil and political rights).

Perhaps the occasional (although very rare) exceptions to the principle that ECOSOC standards have no direct effect, might be clarified by the observations made by Gunter Maes, as discussed in Section 11.4.3, who underlined that the difference between the subjective and objective litigation is gradually diminishing. Especially in cases in which the Courts, operating in the subjective litigation, are weighing interests rather than ruling on subjective rights, the matter of 'direct effect' becomes less relevant, so there might be some more effect for internationally recognised ECOSOC rights.[1344]

12.6.2 The Constitutional Court

The right to food however, was often invoked in proceedings brought before the Constitutional Court. Since 1989, the Court was competent in reviewing national legislation against Constitutional Provisions, and throughout the years the Court gradually started to review indirectly against international human rights Provisions. Since 1992, the Court reviewed against the right to (equal treatment in) education, as stipulated in Articles 24 CA and 13 ICESCR. This opened the door for the review against other ICESCR Provisions via Articles 10 and 11 CA. It seems that due to indirect reviewing, there is more room for effect of international Provisions stipulating human rights.

Since 1994, the right to food as embedded in Articles 11 ICESCR and 27 ICRC was frequently invoked in cases concerning illegally residing aliens. In these cases, Article 57 § 2 of the 'Organic Law of 8 July 1976 on public centres for social welfare' was usually reviewed against Articles 10 and 11 CA, whether or not read in conjunction with other Constitutional and international human rights Provisions. Usually, in this review against the non-discrimination principle, the Court seems to balance the interest of the State's policies to restrict immigration with the individual interest of the illegally residing foreigner.[1345] In case 51/94, the principle was introduced that applying Article 57 § 2 of the Organic Law to

---

[1344] See for instance the aforementioned Court of Cassation, 11 March 1994, Arresten van het Hof van Cassatie, 1994 (I, p.243), and Council of State, 13 December 2000, case no. 91625, available at: www.raadvst-consetat.be. In both cases there seems to be a weighing-up of principles rather than an assessment on whether or not a subjective right has been violated. In the first case, some effect was granted to Articles 9 and 12 ICRC, and in the latter to Article 13 ICESCR.

[1345] This was quite accurately described by Belgium in the second periodic report on the implementation of the ICRC: CRC/C/83/Add.2, 25 October 2000, Sections 28-29.

illegally residing asylum-seekers, resulting in the fact that their entitlements to social benefits are restricted to urgent medical care only, was not in violation with the invoked Provisions. This principle was further specified and nuanced in a period of approximately 12 years. Generally, in both annulment and preliminary procedures, the Constitutional Court was confronted with particular groups of aliens that claimed to be exempted from this principle due to several reasons, as discussed into more detail above. Article 11 ICESCR was mainly invoked in cases concerning illegally residing aliens, ordered to leave the territory, while they were awaiting the outcome of a legal procedure (an appeal to the order to leave the territory, a request to a prolonged stay, the recognition of statelessness, a regularisation request, and the appeal to a rejection to a second or third request for asylum), or could not leave the territory due to a medical impossibility. Only the medical impossibility and the appeal to the (first) order to leave the territory resulted in a deviation of the principle of case 51/94. The ICRC, including Article 27, was later invoked by parents of children in several situations, also claiming to be exempted from the basic principle of case 51/94. This led to a chain of case law in which firstly the child of illegally residing parents was granted the necessary social benefits in kind, whilst later the Provision of these social benefits was broadened to the benefit of the parents too, based on the right to family life, as well as to the parents whose child could not leave the territory due to a medical impossibility. Although the illegally residing parents of a Belgian child were not exempted from the main principle, the Court ruled that the Centres for Social Welfare should, when establishing the social benefits for the child, take the family situation into account.

12.6.3 To what extent has the right to food legal effect in the domestic legal order of Belgium?

As it seems, the right to food and other ECOSOC rights as embedded in the ICESCR and ICRC, have some (indirect) effect in the case law of the Constitutional Court. Purely considered from a technical juridical point of view this is interesting, for it appears that indeed the right to food, alongside with other internationally recognised human rights, can be invoked before the Constitutional Court – and successfully so – since the Court seems to consider that the standard has been violated more than once. However, due to the very strong interrelationship with domestic Provisions, and perhaps due to the indirectness of the review, it is doubtful whether invoking those Articles truly makes a difference. This is demonstrated by the fact that firstly, in some cases, the Court, despite the fact that during the procedure international Provisions were invoked, only refers to national Articles in its verdict;[1346] secondly, the fact that the Court occasionally made principle decisions in a series of similar cases, while in some of those cases only national Provisions were invoked, and in other cases those national Provisions were read

---

[1346] For instance, Constitutional Court, 80/99, 30 June 1999.

in conjunction with international Articles;[1347] thirdly, that in the chain of case law analysed above, in which the cases are strongly interconnected, it is not self-evident that international Provisions are invoked or reviewed against. To understand the complete course of events, it is also necessary to analyse case law in which only national Provisions are discussed. Therefore, no particular evidence could be found that the international Articles led to a significantly broader protection compared to the national Provisions invoked in the various cases. The case law seems to be focussed rather 'inwards' and the international Provisions are occasionally listed, even quoted, and analysed, but do not appear to play a role of any significance. On the other hand, this does not mean that there are no rights for those who need it most (usually illegally residing asylum-seekers), but those rights are rather determined in balancing between on the one hand the well-being and human dignity of the requesting part, and on the other hand the justified policies of the Belgian (Federal) Government.

[1347] See for instance: Constitutional Court, 21/2001 and 148/2001, in which the Court only reviewed against Articles 10 and 11 Constitution, while in 50/2002, the Court reviewed against Articles 10 and 11 CA read in conjunction with Article 11 ICESCR.

13. The Belgian periodic country reports

13.1 Introduction

As stated in Section 2.6.3, this Chapter will focus on the question: *What does the Government of Belgium communicate in its reports in the United Nations arena regarding the enforceability of the right to food in its domestic legal order?*

To this end, the Belgian reports submitted on the implementation of the ICESCR (Section 13.2), the ICRC (Section 13.3) and the CEDAW (Section 13.4) are analysed. Furthermore, when relevant, the reporting cycle of Belgium in view of the Universal Periodic Review will be discussed (Section 13.5). Belgium submitted four periodic reports on the implementation of the ICESCR.[1348] On the implementation of the ICRC[1349] Belgium submitted four periodic reports as well, from which the third and fourth were submitted as combined reports.[1350] Six periodic reports were submitted on the implementation of the CEDAW, from which the third and fourth as well as the fifth and sixth were combined reports.[1351] In the analysis special attention is drawn to the federal structure, as this appears to have raised the concern of the different treaty bodies in relation to a coherent and uniform implementation of the rights stipulated in the treaties. Furthermore, there is a focus on the Belgian Aliens policies and legislation, for they are strongly interrelated with the right to food and ECOSOC rights in general, and also in the reports often discussed separately. Of course the right to food and the direct applicability of the treaties are the primary focus.

13.2 The ICESCR reports

13.2.1 Introduction

In general, it must be concluded that Belgium expresses little self-criticism in its reports. The ICESCR reports mainly consist of a long list with legal and policy

---

[1348] Initial report: E/1990/5/Add.15, 13 May 1993; second periodic report: E/1990/6/Add.18, 5 March 1998; third periodic report: E/C.12/BEL/3, 21 September 2006; fourth periodic report: E/C.12/BEL/4, 18 June 2012. Please note that the Belgian Delegation replied to the Committee's List of Issues on the initial report during the sessions and not in a written reply, contrary to all later reporting procedures.
[1349] Initial report: CRC/C/11/Add.4, 6 September 1994; second periodic report: CRC/C/83/Add.2, 25 October 2000; combined third and fourth periodic report: CRC/C/BEL/3-4, 4 December 2009.
[1350] CRC/C/BEL/3-4, 4 December 2009, Section 1. As recommended by the Committee on the Rights of the Child, see: CRC/C/114, 14 May 2002, p. 5. This report encompasses mainly information on the period 2002 – July 2008, instead of the official period 1999-2004 (third report) and 2004-2009 (fourth report), due to the fact that the second report was already updated in written replies in 2002, in preparation of the sessions with the Committee, and because Belgium was asked to hand in the combined report 18 months before the deadline set in 2009.
[1351] Initial report: CEDAW/C/5/Add.53, 25 September 1987; second periodic report CEDAW/C/BEL/2, 8 April 1993; combined third and fourth report: CEDAW/C/BEL/3-4, 29 September 1998; combined fifth and sixth report: CEDAW/C/BEL/6, 22 June 2007.

initiatives in a certain field that is related to the rights stipulated in the ICESCR. However, points for improvement are hardly included, which was also noted by the Committee. For instance, in its Concluding Observations on the third periodic report, *'The Committee notes the absence of any factors or difficulties preventing the effective implementation of the Covenant in Belgium.'*[1352]

When discussing the right to adequate food, also the reporting on Article 9 was taken into consideration, due to the fact that Belgium considers Article 23 CA to be equivalent to Article 11,[1353] although discusses many aspects of the implementation of Article 23 CA in view of Article 9 ICESCR.

Please note that at the time of writing only a fourth report was submitted, while the follow up of that procedure was scheduled for 2013. Therefore, only the content of the fourth report could be taken into consideration in this Section, and not the List of Issues, replies, summary records and Concluding Observations.

13.2.2 Federal structure

The initial report on the implementation of the ICESCR was submitted far too late. This was regretted by Belgium, and it was explained that this was partially caused by the State reforms that took place simultaneously.[1354] In response, the Committee stated in its Concluding Observations that it *'regretted, however, that the report was submitted nearly 10 years late'*, and was quite firm in its opinion on Constitutional reform as a cause for this delay: *'The Committee wishes to emphasise that those explanations should not be considered by the Belgian Government as justification. States Parties must comply with the reporting obligations they have freely assumed under the Covenant.'*[1355]

During the sessions on the second periodic report, Belgium needed to explain its complex federal structure, especially in the light of overlapping powers regarding international instruments. The Delegation explained that it could occur that both the Federal Government and the Governments of federated areas had to ratify the ICESCR, due to the different aspects of the instrument, covering the authorities of both the Federal Government and the Governments of the federated areas. It underlined that this was indeed a complex system, but also very democratic, and therefore *'the degree of bonding of the State to the international instrument is only higher.'*[1356] The Delegation had to recognise that the Belgian authorities had not developed a national plan on human rights, but stressed that *'at the*

[1352] E/C.12/BEL/CO/3, 4 January 2008, Section 9.
[1353] As reported by Belgium, see: E/1990/6/Add.18, 5 March 1998, Sections 151-152.
[1354] E/1990/5/Add.15, 13 May 1993, introduction; E/C.12/1994/SR.15, 17 May 1994, Sections 1 and 3.
[1355] E/1995/22, 20 May 1994, Section 145.
[1356] Mr Noirfalisse and Mr Nayer, E/C/12/2000/SR.64, 27 November 2000, especially Sections 12-13. Original text in French: *'le degré d'adhésion de l'État à l'instrument international n'en est que plus élevé.'*

federal level and in the regions and communities, many bodies ensure the promotion and enforcement of human rights. There can be no doubt about the unquestionable commitment of different powers to human rights and one might even say that there is some competition between them on the matter.[1357] The Delegation added that the Ministry of Foreign Affairs normally is the first to receive the Committees Observations, and ensured the implementation of the Committee's *'Jurisprudence'*. The Federal Government, the Communities and the Regions are responsible for applying their international commitments. The Delegation pointed out that it cannot be said that there was disparity in the application of human rights across the Regions and the Communities, and explained that there was in fact some cooperation. The Delegation explained that if there were differences in policies in the field of economic rights, there was one legal system ensuring equal rights to all citizens, and the highest Courts ensured a uniform case law regarding human rights throughout the country.[1358] However, in its Concluding Observations, the Committee noted with concern *'that there are no sufficient mechanisms to coordinate and ensure uniformity of compliance, at both the federal and regional levels, with the State Party's international human rights obligations.'*[1359]

In the third periodic report, Belgium again referred to the autonomous powers of the Communities and the Regions,[1360] but assured that *'The Constitution contains Provisions designed, on the one hand, to ensure that the internal autonomy of the Communities and Regions is as extensive as possible at the international level and, on the other hand, to guarantee, through appropriate mechanisms, consistency and unity in the country's external relations.'*[1361] The Committee was not convinced however, and repeated its concern *'relating to the lack of appropriate and effective mechanisms to ensure compliance, at the federal, regional and community levels, with the State Party's obligations under the Covenant.'*[1362]

In the fourth periodic report, Belgium once more clarified its Constitutional system, in a request of the Committee to update its Core Document in accordance with the 2006 harmonised guidelines on a common core document.[1363]

[1357] E/C/12/2000/SR.64, 27 November 2000, Section 35. Original text in French: *'au niveau fédéral que dans les régions et les communautés, de nombreuses instances veillent à la promotion et à l'application des droits de l'homme. On ne saurait mettre en doute l'attachement incontestable des différents pouvoirs aux droits de l'homme et on pourrait même dire qu'il y a une certaine concurrence entre eux à ce sujet.'*
[1358] Mr Vandamme, E/C/12/2000/SR.64, 27 November 2000, Section 36.
[1359] E/C.12/1/Add.54, 1 December 2000, Section 5.
[1360] E/C.12/BEL/3, 21 September 2006, Section 6.
[1361] E/C.12/BEL/3, 21 September 2006, Section 7.
[1362] E/C.12/BEL/CO/3, 4 January 2008, Section 11.
[1363] As requested in E/C.12/BEL/CO/3, 4 January 2008, Section 42. The harmonised guidelines can be found in: HRI/GEN/2/Rev.4, 21 May 2007. See the update in the fourth report: E/C.12/BEL/4, 18 June 2012, Sections 15-26.

13.2.3 The right to food in the ICESCR reports

13.2.3.1 The initial report

In the first initial report, no food-related issues were discussed in particular. On the implementation of both Articles 9 and 11 ICESCR, Belgium clarified the functioning of its social security system. Noteworthy is the reference to Article 1 of the Act of 8 July 1976,[1364] stipulating that *'everyone is entitled to social welfare. The aim of social welfare is to permit everyone to live in a manner befitting the dignity of the human person.'* Belgium stated that this Provision had two important cornerstones. The first was the universal scope, for entitlements to social benefits applied to everyone, and the second was *'the explicit reference to human dignity.'*[1365] Particularities regarding the social security system in relation to foreigners will be discussed below in Section 14.2.3. During the sessions with the Committee on Economic, Social and Cultural Rights however, the Delegation had to admit that Belgium's social security was quite expensive, and therefore the goal was set to streamline the system in a sustainable way, not with the aim to restrict the rights but to avoid *'wastage that may in the long run undermine the system.'*[1366] The Delegation furthermore stated that while Belgium had a high standard of living, there was still poverty and growing unemployment. However, *'the social security system covers a large part of the population.'*[1367] Also, the Delegation noted that the Belgian food industry was highly developed, hygiene was strictly controlled, and a an extensive regulatory framework Law on the matter was in force.[1368]

The Committee invited the Delegation to provide an answer to one of the questions that were included in the List of Issues, regarding the implementation of Article 11 ICESCR: *'Please describe the system adopted in Belgium for determining the standard of living of the population and in this connection give information on the most vulnerable groups – the unemployed, pensioners, migrant workers, etc.'*[1369] The Delegates' reply was short: *'there was no single system in Belgium for determining the standard of living of the population. Various income guarantee systems existed, including the unemployment benefit, pensions, and collective agreements on minimum wages in the various occupational sectors. The final safety net was the public welfare centre, which provided financial, housing, psycho-social and legal assistance that*

[1364] Organic Law of 8 July 1976 on the Public Centres for Social Welfare. Original title in Dutch: *Organieke wet van 8 juli 1976 betreffende de openbare centra voor maatschappelijk welzijn*, B.S. 5.VIII.1976 – err. B.S. 26.XI.1976). At the time of publishing, Article 1 as referred to in the initial report was still applicable.

[1365] E/1990/5/Add.15, 13 May 1993, Sections 143-145.

[1366] Mr Vandamme, E/C.12/1994/SR.15, 17 May 1994, Section 15. Original text in French: *'les gaspillages qui risquent à la longue de saper le système.'*

[1367] Mr Vandamme, E/C.12/1994/SR.15, 17 May 1994, para 17. Original text in French: *'le système de sécurité sociale couvre une grande partie de la population'.*

[1368] Mr Vandamme, E/C.12/1994/SR.15, 17 May 1994, Section 21.

[1369] Mrs Vysokajova, in her capacity as chairperson, E/C.12/1994/SR.16/Ad.1, 18 May 1994, Section 41.

was indexed.'[1370] No further questions were asked by the Commission regarding the issue. However, in its Concluding Observations, the Committee noted with concern in this light *'that the most vulnerable groups of society in Belgium are not always adequately protected.'*[1371]

13.2.3.2 The second periodic report

Belgium reported on Article 9 that no significant changes had been made regarding Belgium's social welfare system. Renewed legislation was merely aimed at the modernisation of the existing system.[1372]

Regarding Article 11, Belgium reported the adoption of Article 23 in its Constitution, including the right to *'lead a life consistent with human dignity'*, and the *'right to decent accommodation'*: fundamental rights, similar to the rights embedded in Article 11 ICESCR. Belgium added however that the *'comments made in connection with the discussion of Article 2 of the Covenant naturally apply to them as well.'*[1373] Hereby thus suggesting that the rights embedded in Article 11 ICESCR, but also Article 23 CA, are of a programmatic nature and can therefore not easily be invoked directly before a Court.[1374] To demonstrate this, Belgium made reference to Constitutional Court case of 26 June 1994, in which the Court ruled that *'The scope of the fundamental right to a satisfactory standard of living is at present subject of debate in Belgium. In its judgment of 26 June 1994,*[1375] *the Court of Arbitration ruled that a fundamental right of this kind could be subject to certain limitations in the case of well-defined general policy objectives. The judgment concerned a case of limitation of the right to welfare of illegal foreigners who had been ordered to leave the country.'*[1376] Furthermore, Belgium reported the draft of two reports on the combat of social exclusion and the prevention of poverty: the General Report on Poverty, and Habitat II.[1377] Also, the legislation setting the conditions for applicants for the guaranteed minimum income was discussed, as well as several initiatives on the improvement of social integration.[1378] On federated level, the Regions reported on their housing policies.[1379] Right to food issues were not discussed as a separate issue.

[1370] Mr Vandamme, E/C.12/1994/SR.16/Ad.1, 18 May 1994, Section 42.

[1371] E/1995/22, 20 May 1994.

[1372] E/1990/6/Add.18, 5 March 1998, Sections 107-134.

[1373] E/1990/6/Add.18, 5 March 1998, Sections 151-152.

[1374] E/1990/6/Add.18, 5 March 1998, Sections 2-3.

[1375] Constitutional Court, 51/94, 29 June 1994. The case referred to is exhaustively discussed in Section 12.5.3.1., and is the first in line of a series determining the limits of Belgium's Asylum Legislation, basically by reviewing Article 57 § 2 *'Organic Law of 8 July 1976 on public centres for social welfare'* against human rights.

[1376] E/1990/6/Add.18, 5 March 1998, Sections 151-152.

[1377] E/1990/6/Add.18, 5 March 1998, Sections 148-149.

[1378] E/1990/6/Add.18, 5 March 1998, Sections 148-156.

[1379] E/1990/6/Add.18, 5 March 1998, Sections 157-167.

The Committee's questions in the List of Issues on the implementation of Article 11 basically concerned housing legislation and policy.[1380] One question however was food-related, and concerned information on food contaminated by dioxine, the measures to prevent this, and the compensation to victims.[1381]

During the sessions with the Committee on Economic, Social and Cultural Rights, the Belgian Delegation explained the wording of the recently adopted Article 23 of the Constitution, and clarified that the Article was actually a compromise between two views. On the one hand, some Parliamentarians wanted to incorporate a directly applicable right in the Constitution, while others favoured a solemn declaration independent from the Constitution. The result was Article 23, which stipulates the right to lead a life consistent with human dignity, but also leaves a large margin of appreciation. The Provision is therefore indirectly applicable, requiring the federal Parliament and the Communities to enact legislation to effectuate the right.[1382] On the other hand, the Delegation stated that even before the adoption of Article 23 in the Constitution, Belgium already had legislation that gave effect to the ICESCR, basically by upholding a *'social security system of a very high quality.'*[1383] However, it was unclear to the Committee what the precise effect of Article 23 then would be, and it therefore invited Belgium to further explain this in its next report.[1384] Also, the question was raised what the general criteria were under which Article 23 would have direct effect instead of having only a declarative character.[1385] The Committee furthermore expressed its concern on the fact that it appeared that individuals did not know the principles of the application of Article 23, and also that Courts were reluctant to fully implement the Article. Belgium was urged to solve that matter as quickly as possible.[1386] The Delegation replied that Article 23 Constitution was of a programmatic nature and consequently was never directly invoked before a Court. However, the Provision was occasionally invoked indirectly before the Constitutional Court, via Articles 10 and 11 Constitution.[1387] In response, the Committee stated in its Concluding Observations that *'Article 23 of the Constitution represents a step forward in that it incorporates a number of economic, social and cultural rights, leaving the guarantee of such rights to statutes and royal decrees. However, such legislation has so far not been adopted. While Article 23, read in conjunction with other fundamental rights guarantees of the Belgian Constitution, could be interpreted to be applicable directly in the domestic legal order, such interpretation still depends on the exercise of discretion by*

[1380] E/C/12/Q/BELG/1, 13 December 1999, questions 27-29.
[1381] E/C/12/Q/BELG/1, 13 December 1999, question 30.
[1382] Mr De Neve, E/C/12/2000/SR.64, 27 November 2000, Section 19.
[1383] E/C/12/2000/SR.64, 27 November 2000, Section 20. Original text in French: '...*régime de sécurité sociale de très grande qualité'*.
[1384] Mr Riedel, E/C/12/2000/SR.64, 27 November 2000, Section 27.
[1385] Mr Wimer Zambrano, E/C/12/2000/SR.64, 27 November 2000, Section 31
[1386] Mr Antanovich, E/C/12/2000/SR.64, 27 November 2000, Section 30.
[1387] Mr Vandamme, E/C/12/2000/SR.64, 27 November 2000, Section 37.

the national Courts.'[1388] Clearly, the Committee did not attach much value to the Delegation's opinion that there was already legislation that gave effect to Article 23 Constitution by upholding a social security system of a very good quality. It is remarkable here, that the Commission also considers the discretion of the Courts in relation to the direct effect of Article 23 CA a reason for concern. Furthermore, the Committee expressed its concern that *'the most vulnerable groups of society in Belgium are not always adequately protected. In this regard, it is particularly concerned about the reductions made in 1993 with regard to subsidies and certain social benefits, such as some categories of reimbursable medicines. The Committee regrets that this negative trend is developing in Belgium as well as in other European countries.'*[1389]

In general, the Committee criticised the second periodic report of Belgium for the fact that *'the report did not adequately deal with the situation of economic, social and cultural rights in Belgium. In future there must be better coordination among the various departments participating in the drafting of Belgium's report.'*[1390] The Belgian Delegation promised that *'the next report would be fuller and better balanced.'*[1391]

13.2.3.3 The third periodic report

The concerns of the Committee on the effect of Article 23 CA seem hardly to be taken into consideration in drafting the third periodic report: Belgium sufficed with merely referring to the existence of the Provision in reporting on the (legislative) measures taken on Federal level to fulfil its obligations under Article 2 ICESCR, while no further explanation or clarification was given on the working of this Provision.[1392]

On Article 11 ICESCR, Belgium reported at the federal level *inter alia* on national action plans for social inclusion and certain changes regarding the legislation on minimum incomes in which the focus shifted from income grants to enabling persons to generate their own income. Also, due to market liberalisation of the gas and power market, Belgium introduced legislation for social benefits for households that cannot pay their energy bills.[1393] In that light, as a temporary measure, households could be provided with heating allowances to meet winter heating requirements.[1394] On federated level, the Regions and Communities reported on

---

[1388] E/C.12/1/Add.54, 1 December 2000, Section 6.

[1389] E/1995/22, 20 May 1994, Section 153.

[1390] Mr Grissa, E/C.12/2000/SR.66, 14 March 2002, Section 37.

[1391] Mr Vandamme, E/C.12/2000/SR.66, 14 March 2002, Section 38.

[1392] E/C.12/BEL/3, 21 September 2006, Section 14.

[1393] Reference was made to the Statute of 4 September 2002, on the assignment to the public centres on social welfare of the guidance and financial social assistance to the most deprived on the supply of energy. Original title in Dutch: *Wet van 4 september 2002, houdende toewijzing van een opdracht aan de openbare centra voor maatschappelijk welzijn inzake de begeleiding en de financiële maatschappelijke steunverlening aan de meest hulpbehoevenden inzake energielevering*, B.S. 28 September 2002.

[1394] E/C.12/BEL/3, 21 September 2006, Sections 423-424.

their programmes and activities on social assistance[1395] and housing.[1396] Regarding food related issues, both the French-speaking and Flemish Communities reported on several initiatives on awareness-raising programmes for healthy nutrition.[1397]

Since food security was reported to be a focus of Belgian development cooperation,[1398] Belgium reported that a number of bilateral cooperation projects were undertaken, and that the Belgian Parliament had established the Survival Fund (EBS) *'which supports integrated programmes for arid and depressed areas in sub-Saharan Africa.'* In addition, another more general fund was established for the Third World. The goal of the EBS was to *'improve food security of the most vulnerable population groups in the least developed countries.'*[1399] The EBS was the predecessor of the later established Belgian Fund for Food Security.[1400]

Regarding food safety, Belgium reported on the establishment in 2000 of the Belgian Federal Agency for the Safety of the Food Chain (FASFC) *'in the wake of controversies regarding food crisis management in the late 1990s.'* This agency, in line with EU law, was reported to be an autonomous federal organisation under supervision of the Ministry of Health that supervises the entire food production process.[1401]

In its List of Issues, right to food issues were not addressed directly. Regarding Article 11 ICESCR, the Committee mainly asked questions concerning the situation of social housing (and to what extent Belgium implemented the Committee's previous recommendations on the matter), as well as the State's policy regarding poverty acquisition and its impact on social housing.[1402] However, in response to the question on what measures Belgium had adopted to achieve its objective to devote 0.7% of the GDP to international cooperation for development and how Belgium's policy on development cooperation contributed to the realisation of

---

[1395] E/C.12/BEL/3, 21 September 2006, Sections 426-492.

[1396] E/C.12/BEL/3, 21 September 2006, Sections 493-539.

[1397] E/C.12/BEL/3, 21 September 2006, Sections 541-550.

[1398] Belgium reported that in its development cooperation, as regulated by the Act of 25 May 1999 (amended on 19 July 2005), there was a focus on five specific sectors, and four cross-cutting themes. One of the sectors was agriculture and food safety, and one of the cross-cutting themes was the respect for the rights of the child. Also, Belgium reported to have subscribed to the Millennium Goals: *'Although they do not include in detail all rights stipulated in the Covenant, those goals are not attainable unless economic, social and cultural rights are exercised in the developing countries. Accordingly, pursuit of the the MDGs may increase pressure on Governments to ensure respect for those rights.'* See: E/C.12/BEL/3, 21 September 2006, Section 50. In the report, reference was made to: Statute of 25 May 1999 on Belgian international cooperation. Original title in Dutch: *Wet van 25 mei 1999 betreffende de Belgische internationale samenwerking,* B.S. 1 July 1999.

[1399] E/C.12/BEL/3, 21 September 2006, Sections 551-555.

[1400] See their current website: http://diplomatie.belgium.be/en/policy/development_cooperation/ partnerships/special_programmes/belgian_fund_food_security.

[1401] E/C.12/BEL/3, 21 September 2006, Sections 556-558.

[1402] E/C.12/BEL/Q/3, 10 April 2007, Sections 18-19.

economic, social and cultural rights, right to food-related issues were discussed.[1403] Belgium responded by underlining several initiatives on development aid and improvements in the field of development cooperation.[1404] Belgium stressed '*to work tirelessly to achieve the Millennium Development Goals.*'[1405] Regarding goal no. 1 on poverty and hunger, Belgium stated that '*the eradication of poverty has a central place in Belgium's cooperation activities. One of the ways in which Belgium is working to eradicate poverty is by supporting the drafting and implementation of Poverty Reduction Strategy Papers (in particular, through the Belgian Poverty Reduction Partnership, a multilateral World Bank programme). The eradication of poverty is also the main objective of Governmental cooperation. Belgium places particular importance on reducing hunger, including through the Belgian Survival Fund (FSF).*'[1406] Regarding goal no. 7 on the environment, access to water and sanitation, Belgium underlined that it '*supports the United Nations Environment Programme and a World Bank programme designed to provide access to water and sanitation for the poorest peri-urban populations in the Democratic Republic of the Congo and Rwanda.*'[1407]

The fact that right to food issues were not discussed in the report directly was also noticed by one of the Committee Members: '*Since the term "right to adequate food" was not included in the report, she wished to know what had been done to guarantee that right, and ensure that the poor had access to food. Travelling communities were particularly vulnerable to deprivation of their right to food, and she wondered what measures were taken to overcome that problem.*'[1408] The question was however never answered by the Delegation, although one Delegate admitted that '*his Delegation had taken note of the Committee's concerns, particularly regarding improvements in the method of drafting periodic reports.*'[1409]

In its Concluding Observation, the right to food was not discussed further. However, a compliment was made on the Belgian system for social welfare: '*The Committee notes with satisfaction the high quality, comprehensiveness and almost universal coverage of the social security and healthcare systems existing in the State Party.*'[1410]

13.2.3.4 The fourth periodic report

Regarding the implementation of Article 9 ICESCR, Belgium reported that '*the right to social security is guaranteed by Article 23 of the Belgian Constitution.*'[1411] An

---

[1403] E/C.12/BEL/Q/3, 10 April 2007, Section 2.
[1404] E/C.12/BEL/Q/3/Add.1, 1 November 2007, Sections 24-49.
[1405] E/C.12/BEL/Q/3/Add.1, 1 November 2007, Section 30.
[1406] E/C.12/BEL/Q/3/Add.1, 1 November 2007, Section 41.
[1407] E/C.12/BEL/Q/3/Add.1, 1 November 2007, Section 46.
[1408] Mrs Bonoan-Dandan, E/C.12/2007/SR.42, 10 December 2007, Section 39.
[1409] Mr Vandamme, E/C.12/2007/SR.42, 10 December 2007, Section 40.
[1410] E/C.12/BEL/CO/3, 4 January 2008, Section 5.
[1411] E/C.12/BEL/4, 18 June 2012, Section 149.

interesting choice of words, for the Article is considered, as discussed above, to have no direct effect, or at most non-direct effect before the Constitutional Court. However – not surprisingly – Belgium recalls in its fourth periodic report the compliments made by the Committee in itsr Concluding Observations concerning the Belgian social security system,[1412] and furthermore updates on the latest developments in legislation in the field.[1413]

Also regarding the implementation of Article 11 ICESCR, an update was provided on newly adopted legislative measures: '*although Belgium already had a successful mechanism for ensuring a decent income in place, many new elements have been introduced in recent years.*'[1414] This included a federal anti-poverty plan, the publication of a report on poverty every two years, a national plan for social inclusion and against poverty, and updates on the legislation on social benefits.[1415] Also, an update was provided on the latest policies and legislation on housing.[1416]

This time, indeed and as requested by the Committee during the sessions on the third periodic report, the right to food was included as a separate reporting topic. A reference was made to a national nutrition and health plan, health promotion targets of which one was food in the Flemish Community, and the establishment of a health promotion advisory board in the German-speaking Community.[1417] Also, the right to water was addressed, for Belgium had ratified the Protocol on Water and Health to the Convention on the Protection and Use of Transboundary Watercourses and International Lakes in 1999.[1418] As part of the implementation, Belgium reported to have '*submitted its first report on the measures taken to ensure access to safe water in sufficient quantities.*'[1419] Furthermore, a reference was made to regional initiatives to help people who were living in poverty or precarity paying the waterbills.[1420]

13.2.4 The Belgian asylum policies in the ICESCR reports

13.2.4.1 The initial report

In its initial report on the implementation of the ICESCR, Belgium reported several times on its system of social benefits, and repeatedly stated that this system was

---

[1412] E/C.12/BEL/4, 18 June 2012, Section 151.
[1413] E/C.12/BEL/4, 18 June 2012, Sections 149-169.
[1414] E/C.12/BEL/4, 18 June 2012, Section 237.
[1415] E/C.12/BEL/4, 18 June 2012, Sections 235-243.
[1416] E/C.12/BEL/4, 18 June 2012, Sections 249-262.
[1417] E/C.12/BEL/4, 18 June 2012, Sections 244-246.
[1418] See: The Protocol on Water and Health to the 1992 Convention on the Protection and Use of Transboundary Watercourses and International Lakes, 18 October 1999, MP.WAT/2001/1, EUR/ICP/EHCO 020205/8Fin.
[1419] E/C.12/BEL/4, 18 June 2012. Section 247.
[1420] E/C.12/BEL/4, 18 June 2012. Section 248.

equally accessible to all persons, regardless of their nationality or race. With regard to Article 9, Belgium reported that *'the Belgian social security system applies to every Belgian or foreign salaried or wage-earning worker who is employed in Belgium by an employer established in Belgium or who is attached to an establishment located there'.*[1421] In view of Article 11, Belgium firmly stated that the relevant legislation on social welfare does not restrict welfare to Belgians, for *'the Act excludes any requirement as to the existence and duration of residence in Belgium. The only consideration is the presence of a destitute person in Belgium, regardless whether he or she has the status of a resident. The universality of the terms of the Act makes it impossible to restrict welfare to foreigners who are duly registered in a municipal population or aliens' register, and ensures that the requirement of lawful residence in Belgium territory may in no case constitute a condition sine qua non.'*[1422] Belgium reported that *'anyone who has reached the age of majority is entitled to subsistence provided he or she actually resides in Belgium, does not have sufficient means of subsistence and is not capable of obtaining them.'*[1423] Regarding the demand to reside in Belgium, the report explained that *'anyone who customarily and permanently lives in Belgium territory is considered as being actually resident in Belgium.'*[1424]

13.2.4.2 The second periodic report

While in the second periodic report hardly any reference was made to legislation regarding (the entitlements to social benefits of) aliens, the Committee asked some critical questions concerning changes in legislation on the position of aliens. In the first place, the Committee referred to a change in Belgian legislation on entitlements to social benefits of refugees. Before, refugees were aided financially, but as a result of abuse of these regulations, the aid provided for had been replaced by aid in kind *'(in particular, food and clothing)'*. The Committee asked why Belgium, despite the criticism of Amnesty International and other NGOs regarding *'the disastrous effect of the latter policy on the human dignity of refugees'* did not consider a compromise solution. Now, all refugees were punished for the abuse of the rules of a few.[1425] The Delegation replied that *'Belgium had to deal with ever-increasing influxes of refugees. As the capacities of the two centres in Liège and Antwerp were no longer sufficient, new centres, some of which were run by the Red Cross and others by local public welfare offices, had been opened. In a statement to Parliament in October 2000, the Prime Minister had said that the best strategy for combatting trafficking in human beings by criminal organisations, was to give refugees aid in kind rather than financial support. Such aid was thus being provided by decision of the Government and had not yet been established in law. The matter was now before Parliament. In any event, the*

[1421] E/1990/5/Add.15, 13 May 1993, Section 109.
[1422] E/1990/5/Add.15, 13 May 1993, Section 150
[1423] E/1990/5/Add.15, 13 May 1993, Section 153.
[1424] E/1990/5/Add.15, 13 May 1993, especially Section 154.
[1425] Mr Riedel, E/C.12/2000/SR.66, 14 March 2002, Section 7.

State would cover medical expenses incurred by sick refugees.'[1426] Secondly, questions were asked about the regularisation procedure. Illegally residing foreigners could *'regularise their situation but, if they did so, would lose their job and their right to a pension under that job'.* The Commission asked whether all the consequences of such a procedure were taken into account.[1427] The Delegation replied that *'only periods of legal employment subject to the payment of social security contributions were taken into account in calculating pensions. That appeared to be fair and he* (the Delegate) *did not think that more favourable measures would be adopted in the near future.*'[1428] Thirdly, the Committee, recalling the concerns expressed by the Committee on the Rights of the Child[1429] *'about the fact that although unaccompanied minors whose asylum requests had been rejected by the Belgian authorities were permitted to remain in the country until they were 18, they incurred the risk of being denied their identity and certain rights, such as the right to education.*'[1430] The Delegation replied that unaccompanied asylum-seekers were not *'simply abandoned'* but housed in a federal reception centre. Also, a series of measures was adopted, *'amending the procedure for handling asylum requests and providing for improved treatment of the asylum-seeker'.* Furthermore, the Centre for Equal Opportunity and Action to Combat Racism was competent to *'receive complaints from persons who believed they had been the victim of discriminatory treatment, assisted persons who wished to undertake legal action.*'[1431]

Comparing the concerns of the Committee with issues dealt with in the case law analysed in Chapter 12, it is remarkable that they are of a different nature. Where the case law of especially the Constitutional Court was mainly about the access of illegally residing aliens to social benefits in itself, the concerns of the Committee are about the nature of the social benefits. Indeed, in 2000 Article 57 § 2 *'Organic Law of 8 July 1976 on public centres for social welfare'* was amended, and social benefits for refugees awaiting their asylum procedure were only to be provided with social benefits in kind.[1432] Secondly, where in case law the access to social benefits of illegally residing aliens during the regularisation procedure was under dispute the Committee was concerned about the long-term consequences on the pensions of the requesting parties. Thirdly, where in case law the access of unaccompanied and illegally residing minors to social benefits was disputed,

---

[1426] Mr Donis, E/C.12/2000/SR.66, 14 March 2002, Section 9.

[1427] Mr Riedel, E/C.12/2000/SR.66, 14 March 2002, Section 7.

[1428] Mr Deneve, E/C.12/2000/SR.66, 14 March 2002, Section 10.

[1429] Probably, the Committee referred to the following section of the concluding observations on Belgium's initial report on the ICRC: CRC/C/15/Ad.38, 20 June 1995, Section 9.

[1430] Mr Ahmed, E/C.12/2000/SR.66, 14 March 2002, Section 13.

[1431] Mr Vandamme, E/C.12/2000/SR.66, 14 March 2002, Section 16.

[1432] See for an overview of the social benefits and a historical overview of the Provision of social benefits: http://www.kruispuntmi.be, an expert centre on migration, integration and cultural diversity. See in particular: http://www.kruispuntmi.be/vreemdelingenrecht/wegwijs.aspx?id=14882. See Article 57 § 1 *'Organic Law of 8 July 1976 on the Public Centres for Social Welfare.'*

the Committee was concerned about the long-term consequences for their stay in Belgium.

13.2.4.3 The third periodic report

The Belgian approach in its aliens' policies and legislation had changed compared to the period of the initial report. Belgium reported in its third periodic report on the legislative measures taken to fulfil its obligations under Article 2 ICESCR. A reference was made to Article 191, which stipulates that *'all foreigners on Belgian soil shall benefit from the protection provided to persons and property, save for the exceptions specified by the law.'* Thus, exceptions by law could *'provide for a partial exception to these rights.'* As an example, it was explained that for the exercise of certain rights embedded in Article 23 Constitution, foreigners had to meet *'a number of conditions related to their status.'*[1433] On the implementation of Article 11 ICESCR, Belgium reported however that *'any person residing legally or illegally in the national territory is entitled to emergency medical care, the need for which must be attested by a physician.'* Reference was made to Articles 1 and 57 of the *'Organic Law of 8 July 1976 on public centres for social welfare'.*[1434]

In their shadow report, NGOs expressed profound criticism on Belgium's aliens policies. The NGOs argued that there were two main grounds which allow foreigners to legally reside on Belgian territory. The first is on humanitarian grounds, and the second on socio-economic grounds, such as employment (only for EU and EEA citizens, and exceptionally for non-EU nationals), family reunification, education and tourism. The NGOs argued that only a small number of people could invoke such a legal ground, leading to far more people remaining illegally on Belgian territory. The NGOs referred to the rights enshrined in the International Convention on the Protection of the Rights of All Migrant Workers and Members of Their Families, stating that *'independent of their legal (documented) or illegal (undocumented) status, migrants are entitled to the full respect, protection and fulfilment of their fundamental rights, including economic, social and cultural rights.'*

Indeed, the Commission asked for some clarification regarding the entitlements to social benefits of aliens, and therefore *'asked whether asylum-seekers whose cases had not yet been decided were entitled to emergency medical assistance in the same way as persons in an irregular situation.'*[1435] In response, the Delegation *'explained that there was a difference between procedures for asylum-seekers and persons in the country illegally who did not have a pending application.'*[1436] Furthermore, the Delegation emphasised *'that asylum-seekers were entitled to material assistance at*

[1433] E/C.12/BEL/3, 21 September 2006, Sections 15-16.
[1434] E/C.12/BEL/3, 21 September 2006, Section 425.
[1435] Mr Riedel, E/C.12/2007/SR.41, 29 February 2008, Section 52.
[1436] Mr Vulder, E/C.12/2007/SR.42, 10 December 2007, Section 20.

the federal reception centres. However, persons in the country illegally who did not have an application pending were entitled only to emergency healthcare, and it was for the person's doctor to decide what constituted an emergency. The principle of the best interests of the child applied, however, and a family with children in the country illegally, without any possibility of legalising its status, could receive material assistance, including accommodation and healthcare, at the federal reception centres.[1437]

The Committee however, seemed to be not convinced that this policy was in conformity with the obligations stipulated in the ICESCR, for in its Concluding Observations it noted *'with concern that access to healthcare facilities, goods and services for persons belonging to vulnerable and disadvantaged groups, such as undocumented migrant workers and members of their families, is limited to access to urgent medical care.'*[1438] And therefore, *'Taking into account general comment No. 14 (2000) on the right to the highest attainable standard of health, the Committee urges the State Party to adopt all appropriate measures to ensure that persons belonging to vulnerable and disadvantaged groups, such as undocumented migrant workers and members of their families, have access to adequate health-care facilities, goods and services, on an equal basis with legal residents of the State Party.'*[1439]

13.2.4.4 The fourth periodic report

In its fourth periodic report, Belgium gives an update on its policy and legislation on asylum proceedings and the entitlements to social benefits of aliens during and after those procedures. Belgium explains that an asylum-seeker has entitlements to material social benefits from the moment of the *'lodging of an asylum application'* to *'any appeal lodged in either in plenary jurisdiction before the Conceil du Litigation des Étrangers,*[1440] *an Administrative Court of appeal, or before the Council of State (...) including during the time limits for lodging appeals.'*[1441] Then, there are three possibilities. First, the application is rejected, and once the deadline of the order to leave the territory expires, the alien is illegally residing, and therefore only entitled to urgent medical care. Second, the application is granted, and the alien is considered a refugee, is issued a residence permit of unlimited duration, and is therefore entitled to social integration assistance. Third, the application is rejected, however, an application for subsidiary assistance[1442] is granted, meaning that the alien is issued a registration certificate that is valid for one year, which may be

---

[1437] Mrs Proumen, E/C.12/2007/SR.42, 10 December 2007, Section 21.

[1438] E/C.12/BEL/CO/3, 4 January 2008, Section 21.

[1439] E/C.12/BEL/CO/3, 4 January 2008, Section 35.

[1440] The Council for Aliens Disputes.

[1441] E/C.12/BEL/4, 18 June 2012, Section 217.

[1442] Subsidiairy protection of persons is an EU concept and is stipulated in: Council Directive 2004/83/EC, 29 April 2004, on minimum standards for the qualification and status of third-country nationals or stateless persons as refugees or as persons who otherwise need international protection and the content of the protection granted. See for the definition in particular Articles 2 (e) and (f).

extended or renewed for a maximum period of five years. After this period, the alien is granted a residence permit for an unlimited period. During the period of subsidiary protection, the alien is entitled to (normal) welfare assistance.[1443] Regarding the distribution of the social benefits during the asylum procedure, Belgium reported that *'asylum-seekers are entitled to reception in the form of material assistance provided by Fedasil or one of its reception partners. When asylum proceedings end, the entitlement to material assistance may be extended in the following situations: to safeguard family unity, to allow a child to complete the school year, in the event of pregnancy, when returning to the person's country of origin is impossible, if the person is the parent of a Belgian child, if the person has signed a commitment to leave voluntarily, or for medical reasons.'*[1444] Regarding the nature of those benefits, Belgium stated that *'the material assistance provided in open collective or individual reception structures must be adapted to individual needs and enable the asylum-seeker to live decently. Such assistance comprises the following elements: housing, food, social support, free legal aid, medical and psychological support, right to education for minors, training for adults and access to the services of translators and interpreters.'*[1445]

13.2.5 Direct effect of the ICESCR rights in the ICESCR reports

13.2.5.1 The initial report

Belgium explained in its initial report that Provisions of international treaties would have direct effect when the Provision was *'sufficiently precise and binding for a national judge to be able to apply them'.*[1446] In that light, Belgium argued that the ICESCR Provisions did not have direct effect, and underlined that, Article 2 ICESCR *'provides that the implementation of the "rights enunciated in the (...) Covenant" depend on "the available resources" of the State and "the adoption of legislative measures." The programmatic nature of this requirement prevents the Provisions of the Covenant from being directly invoked by complainants before Belgian Courts and tribunals.'*[1447] However, Belgium pointed out that it was nevertheless bound by the standstill effect of the Covenant's Provisions, *'because of the requirement that treaties be performed in good faith, rules of domestic law by which rights embodied in the Covenant were already secured at the time it entered into force in Belgium cannot be challenged at a later date.'*[1448] A principle that was recognised in a Court of Cassation ruling on

---

[1443] E/C.12/BEL/4, 18 June 2012, Section 218.
[1444] E/C.12/BEL/4, 18 June 2012, Section 219.
[1445] E/C.12/BEL/4, 18 June 2012, Section 220.
[1446] E/1990/5/Add.15, 13 May 1993, Section 1.
[1447] E/1990/5/Add.15, 13 May 1993, Section 2.
[1448] E/1990/5/Add.15, 13 May 1993, Section 3.

20 December 1990.[1449] Furthermore, Belgium stated that the Covenant may be used as a *'basis for interpretation of the law applied by judges (...)'*.[1450]

During the sessions with the Committee on Economic, Social and Cultural Rights, the Delegation underlined that Belgium *'particularly values the concept of the indivisibility of human right.'*[1451] Reference was made to a statement of the Minister of Foreign Affairs during the World Conference on Human Rights held in Vienna 1993, in which he emphasised that *'a true democracy can only be based on an inclusive society, socially just and economically stable.'*[1452] The Committee asked to explain whether the Covenant and various human rights instruments may be invoked before and directly applied by the Belgian Courts or administrative tribunals.[1453] The Delegation replied that some of the ICCPR Articles were directly applicable, but the ICESCR had no direct effect except for its standstill-clause. The Delegation explained that the Court of Cassation applied two main criteria to determine whether a Provision has direct effect. The first criterion was whether the Provision was formulated in clear and unambiguous terms. The second criterion was whether the Provision was not dependant on further national legislation for its application. The Delegation referred to the Court rulings of the Court of Cassation concerning Article 13 ICESCR, in which the Court ruled that the Provision had no direct effect. Also, the Delegation underlined that the Courts could use the Covenant as a tool for interpretation of domestic legislation or general principles of law.[1454] Not surprisingly, the Committee did not share the vision of the Delegation, and referred *inter alia* to its General Comment 4, in which *'one may find a careful analysis of certain Provisions of the Covenant from which it does not appear that one can say that they have absolutely no direct effect.'*[1455] As an example, the Committee referred to Article 13 ICESCR, which stipulates that primary education shall be free and compulsory for all. The Committee stated that this sub-part of Article 13 should have direct effect. The Committee therefore *'regrets to some extent the*

[1449] E/1990/5/Add.15, 13 May 1993, Section 3. Reference was made to: Court of Cassation, 20 December 1990, Arresten van het Hof van Cassatie, 1990(91)(P.445); as discussed in Section 12.4.1., although the standstill effect was recognised, the invoked Article 13 ICESCR did not have any immediate consequences, and did not create any subjective rights that should be safeguarded by the Courts.
[1450] E/1990/5/Add.15, 13 May 1993, Section 4.
[1451] Mr Reyen, E/C.12/1994/SR.15, 17 May 1994, Section 4. Original text in French: *'attache une importance particulière à la notion de l'indivisibilité des droits de l'homme'*.
[1452] Mr Reyen, E/C.12/1994/SR.15, 17 May 1994, Section 4. Original text in French: *'une véritable démocratie ne peut être fondée que sur une société solidaire, socialement juste et économiquement stable.'*
[1453] Mrs Bonoan-Dandan, in her capacity as chairperson, E/C.12/1994/SR.15, 17 May 1994
[1454] E/C.12/1994/SR.15, 17 May 1994, Section 29.
[1455] Mr Simma, E/C.12/1994/SR.15, 17 May 1994, Section 33. Original text in French: *'on y trouve une analyse minutieuse de certaines dispositions du Pacte dont il ne semble pas que l'on puisse dire qu'elles n'ont absolument aucun effet direct'*.

attitude of the Belgian Courts which, together with Courts of many other countries, insist in considering that the Covenant as a whole is not directly applicable.'[1456]

Furthermore, referring to the report, the Committee underlined that the implementation of the rights embedded in the Covenant does not by definition depend on financial resources, and therefore such dependence, as mentioned in Article 2 ICESCR, could not always be used as an argument to deny direct applicability.[1457] This was repeated in the Concluding Observations of the Committee.[1458] The Delegation replied that the Court of Cassation '*recognises that it appears that there are Provisions in the Covenant that are clear and unambiguous – including the Provision of Article 8 on the right to strike – which should in principle be directly applicable if not Article 2 of the Covenant stipulated the principle of progressive implementation.'*[1459] The Delegation added that the Council of State had a more subtle approach, by recognising some effect of Article 13 ICESCR, in particular that '*if the compulsory and free primary education is established, it is not possible to derogate from that later and such derogation would be null and void.'*[1460] Regarding the interpretation of the standstill effect, the Delegation referred to the Committee of Independent Experts for the implementation of the ESC, which stated in a Dutch case that a standstill effect does not mean the obligation to maintain the social benefits as they were at the moment of ratification of the relevant international obligations, but rather to keep the same level of social benefits comparable to the moment of ratification.[1461]

No further questions were asked by the Committee on the matter of direct applicability, neither was the issue addressed in its Concluding Observations.

In addition, the Delegation was asked what the Belgian position was regarding the draft of an Optional Protocol to the ICESCR.[1462] The Belgian Delegation replied with some caution that '*since the protocol in question does not yet exist, it is difficult for the Belgian Delegation to express an opinion on the matter. From a theoretical point of view, however, the position of the Belgian Government is as follows: such a protocol*

[1456] Mr Simma, E/C.12/1994/SR.15, 17 May 1994, Section 33. Original text in French: '*déplore dans une certaine mesure l'attitude des tribunaux belges comme des tribunaux de beaucoup d'autres pays consistant à considérer que le Pacte dans son ensemble n'est pas d'application automatique'*.

[1457] E/C.12/1994/SR.15, 17 May 1994, Section 41.

[1458] E/1995/22, 20 May 1994, Section 146.

[1459] Mr Deneve, E/C.12/1994/SR.15, 17 May 1994, Section 43. Original text in French: '*reconnaît semble-t-il qu'il y a dans le Pacte des dispositions claires et nettes – notamment une disposition de l'Article 8 relative au droit de grève – qui devraient en principe être directement applicables si l'Article 2 du Pacte ne prévoyait pas une application progressive'*

[1460] Mr Deneve, E/C.12/1994/SR.15, 17 May 1994, Section 43. Original text in French: '*si l'obligation et la gratuité de l'enseignement primaire sont établies, il n'est plus possible d'y déroger ultérieurement et que toute dérogation serait nulle et non avenue.'*

[1461] Mr Deneve, E/C.12/1994/SR.15, 17 May 1994, Sections 44 and 52.

[1462] E/C.12/1994/SR.15, 17 May 1994, Section 51.

is possible, provided that Belgian citizens can not apply directly to the Committee after exhausting all their legal remedies.'[1463]

In General, the Delegation regretted that the Covenant was not well-known amongst the judges, and promised future measures to make the Covenant better known amongst them.[1464]

13.2.5.2 The second periodic report

In its second periodic report, Belgium almost literally repeated its view on the working of Article 2 ICESCR regarding its programmatic nature as stated in its initial report.[1465] However, Belgium added that the Court of Cassation and the Council of State had ruled that Belgium is nevertheless bound by the 'standstill' effect of the ICESCR Provisions, which means that *'rules of domestic law by which rights embodied in the Covenant were already secured at the time it entered into force in Belgium cannot be challenged at a later date.'*[1466] Since then, judgments of the Constitutional Court (at that time the Arbitration Court) show a *'trend towards recognition of direct effect of certain Provisions.'*[1467] As an example, Belgium discussed two cases, from 15 July 1993,[1468] recognising direct effect of Article 8 ICESCR, and 8 March 1994,[1469] recognising direct effect of Article 6 ICESCR.[1470] However, Belgium rightfully added that *'A distinction must be made (...) between the decisions of the Court of Arbitration, which exercises indirect control since it assesses the degree to which laws in conformity with Articles 10 (equal treatment), 11 (discrimination), and 24 (freedom of instruction) of the Constitution (which incorporates the texts of duly ratified international treaties) and the decisions of the Courts and tribunals, in particular the Court of Cassation and the Council of State, which exercise direct control. The latter have not yet recognised any direct effect of the Covenant.'*[1471]

During the sessions with the Committee on Economic, Social and Cultural Rights, the Committee referred to its previous concerns that it was not possible to invoke the Covenant before the Courts, and asked whether the situation had improved,

---

[1463] Mr Van Craen, E/C.12/1994/SR.15, 17 May 1994, Section 53. Original text in French: *'puisque le protocole en question n'existe pas encore, il est difficile pour la délégation belge d'exprimer une opinion en la matière. D'un point de vue théorique, cependant, le principe qui inspire la position du Gouvernement belge est le suivant: un tel protocole est envisageable, pour autant que les citoyens belges ne puissent s'adresser directement au Comité qu'après extinction de l'ensemble des voies de recours internes.'*
[1464] Mr Deneve, E/C.12/1994/SR.15, 17 May 1994, Section 45.
[1465] E/1990/6/Add.18, 5 March 1998, Section 2.
[1466] E/1990/6/Add.18, 5 March 1998, Section 3.
[1467] E/1990/6/Add.18, 5 March 1998, Section 3.
[1468] Constitutional Court, 62/93, 15 July 1993, considerations B.3.8-B.3.12.
[1469] Constitutional Court, 22/94, 8 March 1994, consideration B.9.2.
[1470] E/1990/6/Add.18, 5 March 1998, Sections 4-7.
[1471] E/1990/6/Add.18, 5 March 1998, Section 8.

and whether international law would now prevail over national law.[1472] Regarding the possibility to invoke ICESCR Articles before the Courts, the Delegation basically repeated the Belgian view on Article 2 (1) ICESCR – the ICESCR is of a programmatic nature, and therefore its Provisions have no direct effect before the Courts – and the recognition of the 'standstill' effect.[1473] The view of the Courts regarding the matter had not changed: '*The three main Belgian Courts, namely the Court of Arbitration, the Court of Cassation and the Council of State, apply on a regular basis the "standstill" effect, while the other Courts are bound by their case law.*'[1474] Once again, a reference was made to the two Constitutional Court cases referred to in the second periodic report, to demonstrate that the Court had granted a certain effect to Articles 6 and 8 ICESCR, and therefore the Court seems to move towards the acceptance of direct effect of certain ICESCR Provisions. Remarkable however is the observation of the Delegation that '*since then, the Court of Arbitration and other Courts such as the Council of State, once again recognised that certain Provisions of the Covenant are directly applicable.*'[1475] It is unclear what Cases of the Council of State is referred to, and it seems inappropriate to refer to rulings of the Constitutional Court as proof for the possibility of direct effect, due to its indirect reviewing competences. Only later, during the sessions, the Delegation explained the indirect application of the Constitutional Court of international Provisions via Articles 10 and 11 Constitution.[1476]

Regarding the question whether international law prevails over national law in Belgium, the Delegation replied that '*Belgium has legal monism, which means that international law and national law are part of a single legal order. Therefore, the Provisions of the Covenant have precedence over domestic laws to the extent that they are directly applicable Provisions.*'[1477] In addition, the Delegation underlined that the Court of Cassation recognised the primacy of international law over domestic law in 1971 (this must be a reference to the Franco-Suisse Le Ski case[1478]), and this doctrine was never altered since. It must be noted here that in the Franco-Suisse Le Ski case, the Court of Cassation indeed recognised the primacy of international

---

[1472] Mr Kouznetsov, E/C/12/2000/SR.64, 27 November 2000, Section 9.

[1473] Mr Deneve, E/C/12/2000/SR.64, 27 November 2000, Section 21.

[1474] Mr Deneve, E/C/12/2000/SR.64, 27 November 2000, Section 22. Original text in French: '*Les trois Courts principales de Belgique, à savoir la Cour d'arbitrage, la Cour de cassation et le Conseil d'État, appliquent depuis, de manière régulière, l'effet "standstill", et les autres tribunaux sont liés par leur jurisprudence.*'

[1475] Mr Deneve, E/C/12/2000/SR.64, 27 November 2000, Section 23. Original text in French: '*Par la suite, la Cour d'arbitrage et d'autres tribunaux comme le Conseil d'État, ont reconnu à nouveau que certaines dispositions du Pacte sont directement applicables.*'

[1476] Mr Vandamme, E/C/12/2000/SR.64, 27 November 2000, Section 37.

[1477] Mr Deneve, E/C/12/2000/SR.64, 27 November 2000, Section 24. Original text in French: '*la Belgique connaît le monisme juridique, c'est-à-dire que le droit international et le droit national font partie d'un seul ordre juridique. Il s'ensuit que les dispositions du Pacte priment sur les dispositions internes dans la mesure où il s'agit des dispositions directement applicables.*'

[1478] Court of Cassation, 27 May 1971, arresten van het Hof van Cassatie, 1971 (p. 959).

Provisions over national Provisions but, as discussed in Section 11.5, only insofar these international Provisions have an direct effect.

According to the Committee, the interpretation of Belgium on Article 2 ICESCR was in contradiction with the authoritative interpretation of the Committee, for *'The possibility to implement the rights gradually at the level of available resources are not obstacles to the direct application of the Covenant.'*[1479] Also, the Committee considered that the precedence of international law over domestic law was in contradiction with the fact that the Courts decided whether a Provisions was directly applicable or not. In addition, the Committee referred to General Comment 9, Section 11,[1480] in which the Committee expressed the view that *'the Covenant does not negate the possibility that the rights it contains may be considered self-executing in systems where that option is provided for.'*[1481]

An interesting question was raised regarding the Concluding Observations of the Committee on the Rights of the Child, which noted with satisfaction that in Belgium, the ICRC is self-executing and that its Provisions may be invoked before the Court.[1482] The Committee asked what made the ICRC directly applicable, and the ICESCR not.[1483] The Delegation replied that the ICRC did not contain an equivalent Provision to Article 2 (1) ICESCR, in which the implementation of the Covenant was made dependent on the resources of the state. For this reason, the Belgian Courts first considered that all Provisions of the ICESCR were not directly applicable, but this view was later corrected, *'in particular in view of the concluding observations of the Committee after consideration of the initial report of Belgium.'*[1484] Again, reference was made of the Court cases of the Constitutional Court, in which Articles 6 and 8 ICESCR were invoked.[1485] A curious response since firstly, as will appear below, the Concluding Observation of the Committee on the Rights of the Child was a misunderstanding: most ICRC Provisions have no direct effect in Belgium (see also Chapter 12). It seems unlikely that a Delegation does not know this. Secondly, the Constitutional Court did not refer to the any Concluding Observations in its verdict, and therefore, the suggestion that there is a causal link between the Committee's Observations and the alleged change of view of the Belgian Judiciary seems to be inaccurate.

[1479] Mr Dasi, E/C/12/2000/SR.64, 27 November 2000, Section 28. Original text in French: *'La possibilité de mettre en œuvre des droits progressivement et le niveau des ressources disponibles ne sont pas des obstacles à l'application directe des dispositions du Pacte.'*

[1480] Mr Ceville, E/C/12/2000/SR.64, 27 November 2000, Section 29.

[1481] See: /C.12/1998/24, 3 December 1998, CESCR, *General Comment 9, The Domestic Application of the Covenant*, Section 11.

[1482] Reference was made to document CRC/C/15/Add.38, 20 June 1995. See Section 6.

[1483] Mr Ahmed, E/C/12/2000/SR.64, 27 November 2000, Section 34.

[1484] Mr Deneve, E/C/12/2000/SR.64, 27 November 2000, Section 38. Original text in French: *'notamment à la lumière des observations finales émises par le Comité après l'examen du rapport initial de la Belgique.'*

[1485] Mr Deneve, E/C/12/2000/SR.64, 27 November 2000, Section 38.

To conclude, in its Concluding Observations, the Committee recommended Belgium to '*take appropriate steps to fully guarantee the direct applicability of the Covenant in the domestic legal order.*'[1486]

13.2.5.3 The third periodic report

In its third report, Belgium did not address the issue of direct applicability of the Covenant at all. Perhaps unexpectedly, especially after the extensive discussion on the matter during the sessions with the Committee on the second periodic report, the NGOs did not address the issue either.[1487] The Committee however, asked in its List of Issues '*whether Covenant Provisions have been invoked before, or directly enforced by, the Courts, other tribunals or administrative authorities.*'[1488] To this Belgium replied that the Constitution and the ECHR were in fact enforced by Belgian Courts on a regular basis in cases involving ECOSOC rights. Also, in specific cases, the social law of the EU may be enforced as well, especially regarding the freedom of movement of workers and fender equality.[1489] Furthermore, Belgium replied that the ICRC was invoked in '*very specific cases*'.[1490] As an example, Belgium referred to the Constitutional Court case 106/2003 '*which affirms the right of underage irregular migrants to receive social assistance, provided that the authorities have ascertained that the parents do not or cannot discharge their duty to support them, that the claim manifestly concerns expenses necessary for the minor's development and, lastly, that the assistance is used solely for those expenses.*'[1491] This indeed summarises the core content of the Court's verdict. Regarding the ICESCR, however, Belgium reported that the Covenant, '*just like the European Social Charter*', was seldom enforced by the Courts. The reason for this was reported to be the fact that '*a significant majority of the Provisions of these texts have no direct effect in domestic law.*'[1492] However, Belgium provided for a non-exclusive list of examples in which ICESCR Articles were somehow involved, although '*never considered on its own but rather always in conjunction with other instruments.*'[1493] The reported Constitutional Court cases are: case 17/2002, 89/2002, ('*no discrimination between foreigners who have filed for recognition or refugee status (who are entitled to social assistance) and those who have filed for recognition of stateless persons' status ('who are not entitled to social assistance'),* 49/2002 (violation of the right to education by a French decree establishing a '*platform of skills in French as a basis for granting subsidies*'), 131/2001, 14/2002, 16/2002, 17/2002, 205/2004 ('*The lack of a specific status for irregular migrants who have applied for regularisation is not discriminatory*'),

---

[1486] E/C.12/1/Add.54, 1 December 2000, Section 20.

[1487] E/C.12/BEL/NGO/3, 23 October 2007.

[1488] E/C.12/BEL/Q/3, 10 April 2007, Section 3.

[1489] E/C.12/BEL/Q/3/Add.1, 1 November 2007, Section 50 and footnote 7.

[1490] E/C.12/BEL/Q/3/Add.1, 1 November 2007, Section 51.

[1491] E/C.12/BEL/Q/3/Add.1, 1 November 2007, footnote 8.

[1492] E/C.12/BEL/Q/3/Add.1, 1 November 2007, Section 51.

[1493] E/C.12/BEL/Q/3/Add.1, 1 November 2007, Section 51.

50/2002 (*'It is not discriminatory to deny'* the right to social assistance *'to foreigners who are applying for refugee status for the second or third time'*), 89/2002 (*'It is not discriminatory to accord foreigners who have submitted an application for regularisation on grounds of exceptional circumstances the right to emergency medical treatment only'*), 106/2003 (Although the Court ruled that the ICRC was violated, it was *'unnecessary to consider a violation of the Covenant'*), 5/2004 (although an Act was rescinded, the ICESCR was not violated), 107/2004 (*'Rescission of a decree by the French Community restricting access to postgraduate studies in health sciences'*, violating the right to freedom of education) and 131/2005 (Although the ICESCR was not violated, *'the holding of a minor irregular migrant in a reception centre violates the right to privacy and to family life, if the presence of the parents in the same centre is not guaranteed'*).[1494] Indeed, as was also observed in Section 12.5, the ICESCR Provisions were never considered on their own merits and had only a limited effect in the quoted verdicts. Furthermore, reference was made to four Court cases of other Belgian tribunals in which the standstill effect of Article 13 ICESCR was acknowledged (Council of State, 6 September 1989, Court of Cassation, 20 December 1990), the direct effect of the right to strike enshrined in the ESC was recognised (Council of State, 3 December 2002), and the denial of the direct effect of the right of an individual to benefit from the protection of his scientific, literary or artistic performance (Court of Cassation, 25 September 2003).[1495] No further explanation was given.

The matter was also discussed during the sessions with the Committee, although no new arguments vice versa seem to have been invoked. The Delegation, on request on the Committee, underlined that *(...) not all the Provisions were directly applicable, because direct application was regarded as a matter for the Courts; it was up to the Courts to decide whether a treaty was sufficiently clear and explicit. Hence, most of the Provisions of the Covenant had had to be given concrete form in a legislative and regulatory framework. The Provisions of the Covenant were not directly enforceable in the Courts, and references to it could only be made in support of an interpretation of Belgian law. A number of rights enunciated in the Covenant (right to work, right to equitable working conditions and right to housing, among others) had been included as such in the latest version of the Constitution in order to make them more effective.*[1496]

Once more, the Committee expressed in the Concluding Observations its concern regarding the fact that most ICESCR Provisions, as well as some Provisions of Article 23 CA, have no direct effect and, as a result, are *'rarely invoked separately before, and directly enforced by, national Courts and other tribunals or administrative authorities.'*[1497] Therefore, *'The Committee draws the attention of the State Party to*

[1494] E/C.12/BEL/Q/3/Add.1, 1 November 2007, Section 51.

[1495] E/C.12/BEL/Q/3/Add.1, 1 November 2007, Section 51.

[1496] Mr Vandamme, E/C.12/2007/SR.41, 29 February 2008, Section 17.

[1497] E/C.12/BEL/CO/3, 4 January 2008, Section 12.

its general comment No. 9 (1998) on the domestic application of the Covenant, and recommends, as already mentioned in Section 20 of its previous concluding observations, that the State Party take all appropriate steps in accordance with Article 2, Section 1 of the Covenant, to guarantee the direct applicability of the Covenant Provisions in its domestic legal order. The Committee also requests that the State Party provide detailed information about the measures adopted in its next periodic report.[1498]

13.2.5.4 The fourth periodic report

In the fourth periodic report, in response to the previous Concluding Observations, Belgium only briefly addresses the issue of direct applicability. Firstly, a reference is made to the two conditions that must be met in order for an international Provision to be directly applicable, that is, (a) it must be the intention of the State Parties to establish individual rights, and (b) the Provision must be *'sufficiently precise and comprehensive'.* Furthermore, Belgium reported again that the Belgian Courts rarely apply the ICESCR Provisions. *'In the absence of any case law and in view of doctrinal disagreements, it is difficult to assess whether Covenant Provisions are directly applicable under Belgian law. Those Provisions are in fact formulated in a rather programmatic way: they commit States to taking measures but do not directly declare subjective individual rights.'*[1499] It is unclear what exactly those doctrinal disagreements are, although it is most likely a reference to a debate at a national level on the direct applicability of international standards amongst scholars.[1500]

In this light, it is interesting to note that Belgium proudly reported to be one of the first countries to have signed the Optional Protocol to the ICESCR, on 24 September 2009.[1501]

13.3 The ICRC reports

13.3.1 Introduction

In accordance with the harmonised treaty-specific reporting guidelines,[1502] the Belgian reports on the ICRC are not structured in line with the order of appearance of the Provisions in the Convention, but rather in categories of rights and sub-categories of substantive issues, whereby each time reference is made to the related Articles. In addition, due to the Belgian Constitutional structure, the issues

---

[1498] E/C.12/BEL/CO/3, 4 January 2008, Section 25.
[1499] E/C.12/BEL/4, 18 June 2012, Sections 9 and 10.
[1500] See for instance: G. Maes, *De afdwingbaarheid van sociale grondrechten*, Antwerp: Intersentia 2003.
[1501] E/C.12/BEL/4, 18 June 2012, Section 7.
[1502] CRC/C/58/Rev.2, 23 November 2010, treaty-specific guidelines regarding the form and content of periodic reports to be submitted by parties under Article 4, Section 1(b) of the Convention on the Rights of the Child.

are discussed on the Federal level, Community level and Region level. See also Section 9.3.1.3, on the Dutch reporting obligations.

Issues on direct applicability are often discussed in Section 1: *'General measures of implementation'*, in particular in sub-Section 1: *'measures taken to harmonise Belgian law and policy with the Provisions of the Convention'.* The right to food, or related issues, is basically discussed in Section V on *'family environment and alternative care'*, more specifically in sub-Section V, *'recovery of maintenance of the child'*, Article 27 Section 4, Section IV, *'basic health and welfare'* (more specifically in sub-Section IV, *'social security and child-care services and facilities'*, Articles 26 and 18, Section 3, and sub-Section V, *'standard of living'*, Article 27, Sections 1-3). The issues relating the rights of foreign children (especially unaccompanied foreign children) in Belgium are mostly discussed in Section VII, *'special protection measures'* (mostly sub-Section I, *'children in emergency situations'*, under A: *'refugee children'*, Article 22). Since the right to food is very often invoked in the extensive case law on the rights of foreigners in Belgium, special attention to the Belgian reporting on the matter will be given below.

In general, it can be said that the reporting behaviour of Belgium was appreciated by the Committee. No significant delays – constructive and in compliance with the guidelines. For instance, in its Concluding Observations, the Committee noted with appreciation *'that the report followed the guidelines for reporting. It notes that the report was timely, comprehensive, and self-critical in nature, as were the written replies to the List of Issues (...) and welcomes the additional information provided in the Annexes. The discussion in the report on the follow-up to the Committee's earlier recommendations was especially appreciated. The Committee also notes with appreciation the presence of a high-level Delegation, which contributed to an open dialogue and a better understanding of the implementation of the Convention in Belgium.'*[1503]

13.3.2 Federal structure

On various occasions during the reporting processes, also here the complex Belgian Constitution led to discussions on the possibility of an effective and uniform approach towards human rights in Belgium. More than once, it was necessary to further explain the Belgian Constitutional structure.[1504]

In its initial report, Belgium explains existing or planned mechanisms for coordinating policies regarding children and for monitoring the implementation of the convention, basically on the Community and Regional level. It appears that on a federal level no such mechanisms existed yet. However, Belgium reported to

---

[1503] CRC/C/15/Add.178, 13 June 2002, Section 2.
[1504] For instance during the sessions with the Committee on the Rights of the Child, Mr Willems, CRC/C/SR.222, 6 June 1995, Section 4.

plan the establishment of a group of experts that should be responsible for '*following up the implementation of the Convention in Belgium and supervising its execution, but also for coordinating the various initiatives taken at the Federal, Community, Regional or even local level relating to the rights of the child.*'[1505] Indeed, Belgium reports a large variety of initiatives undertaken in the federated areas that relate to the rights of the child, mostly coordinated by the French Community's Birth and Children Office (ONE), the Flemish Community's *Kind en Gezin*, and the German-Language Community's *Kind und Familie* (DKF).[1506] In its List of Issues, the Committee on the Rights of the Child asked whether Belgium was '*planning to establish a mechanism at the national level to coordinate and evaluate measures for implementing the Convention.*'[1507] In its written replies, Belgium responded that several working groups were installed to investigate such a national approach. The working group set up by the King Baudouin Foundation specifically focussed on the monitoring in the context of the ICRC.[1508] During the sessions with the Committee on the Rights of the Child, the Committee asked on what manner a universal approach towards the rights embedded in the Convention could be guaranteed in such a complex Governmental system.[1509] In response, the Delegation basically referred to the many initiatives to coordinate and monitor matters that relate to the Convention, but also discussed the different approaches between the Federation and the Communities.[1510] The Flemish representative however explained that the cooperation between the Flemish Ombudsmen (falling under the scope of *Kind en Gezin*) and the Delegate General of the French-speaking Community (a post outside the normal administration, with a duty to ensure that the interests of the child were safeguarded in the French-Speaking Community) was '*politically speaking, a thorny issue. "Personisable matters" from one region to another, differed, and the situation was likely to detoriate still further. It was therefore difficult to devise a body for overall cooperation.*'[1511]

In its List of Issues on the second periodic report, the Committee wondered how '*in the light of the different competences of the different autonomous Governments (...) the co-ordination and co-operation on child rights*' was achieved.[1512]

During the sessions with the Committee, the first meeting consisted for the larger part of a discussion on the Belgian Constitutional structure and the difficulties in coordinating the rights of the child nationally.[1513] In this light, the Chairperson

---

[1505] CRC/C/11/Add.4, 6 September 1994, Section 21.

[1506] CRC/C/11/Add.4, 6 September 1994, Sections 22-33.

[1507] CRC/C.9/WP.4, 17 February 1995, issue 7.

[1508] Belg/1, 04 May 1995, reply to question 7.

[1509] Mr Hammarberg, CRC/C/SR.222, 6 June 1995, Section 21.

[1510] See in particular: CRC/C/SR.222, 6 June 1995, Sections 33-48.

[1511] Mr Van Keymeulen, CRC/C/SR.222, 6 June 1995, Section 46.

[1512] CRC/C/Q/BELG/2, 8 February 2002, part I, B.2.

[1513] CRC/C/SR.782, 24 July 2002.

noted that the debate had been '*stimulating and informative*', and found the '*occasional disagreement among Delegation Members themselves on certain points (...) encouraging.*'[1514] In its Concluding Observations, the Committee made several recommendations on the uniform and nationwide implementation and monitoring of the rights of the child, including the establishment, as already agreed upon by Belgium,[1515] of a national commission for the rights of the child.[1516]

In its combined third and fourth periodic report, the Belgian Constitutional structure was once more clarified.[1517] As recommended by the Committee, the establishment of a National Commission on the Rights of the Child was reported, based on a cooperation agreement between the State and all federated areas. The main purpose of this Commission is the '*more extensive monitoring of the Convention's implementation in Belgium and the effective coordination of measures for the benefit of the child*',[1518] and its principle task is the draft of the reports on the implementation of the ICRC.[1519] The Belgian Delegation proudly commented that Belgium was the first ICRC State Party with such a national body.[1520] In the NGO shadow-report however, the coordination of children's rights issues was critically reviewed. The NGOs stated that there is lack of coordination amongst the different policy areas concerning minors, and therefore recommended the establishment of a coordinating Minister on children's rights on the federal level. In addition, the work of the National Commission on the Rights of the Child was criticised, due to the fact that usually Government representatives would take the most important decisions in the body, resulting in lack of autonomy and power to influence political decisions on children's rights-related issues.[1521]

During the sessions, the Committee expressed its concerns '*about the possibility that the federal structure of the State Party could hamper the implementation of the Convention*' and pointed out that since the Federal Government had signed the Convention it was solely responsible for implementing international law and therefore, national coordination was of vital importance. The Country Rapporteur

---

[1514] CRC/C/SR.783, 29 May 2002, Section 64.
[1515] CRC/C/83/Add.2, 25 October 2000, Sections 158-164.
[1516] CRC/C/15/Add.178, 13 June 2002, especially Sections 10 and 11.
[1517] CRC/C/BEL/3-4, 4 December 2009, Sections 1-13.
[1518] CRC/C/BEL/3-4, 4 December 2009, Section 17.
[1519] CRC/C/BEL/3-4, 4 December 2009, Section 18. Besides the drafting of the country report, its duties were '*to contribute to drafting other documents*', '*to coordinate the collection, analysis and processing of data intended for the Committee on the Rights of the Child*', '*to encourage partnerships and an ongoing exchange of information between the authorities and bodies dealing wit children's rights*', '*to monitor and review the implementation measures needed to respond to the suggestions and recommendations of the Committee on the Rights of the Child*', and '*to give advice.*'
[1520] Mrs D'hondt, CRC/C/SR.1521, 10 June 2010, Section 4.
[1521] Shadow-report of the NGOs on the implementation of the International Convention on the Rights of the Child. Original title in Dutch: *Alternatief Rapport van de NGOs over de toepassing van het Internationaal Verdrag inzake de Rechten van het Kind in België*, 2010, available at: www.kinderrechtencoalitie.be, Chapter 1, Section 1.

(advising the Committee) stated that in the combined third and fourth periodic report a wealth of information was given on coordinating measures, suggesting that *'the problem lay not just in coordination, but in ensuring that children throughout the State Party could enjoy their rights equally.'*[1522] The Belgian Delegation replied that *'all Regions and Communities within Belgium had equal status and were competent to implement the Convention. The federal authorities provided the necessary guarantees and safeguards to ensure compliance with the Convention nationwide but, given the autonomy of the regions and communities, priorities and financing could vary in line with their different requirements and situations.'*[1523] The Country Rapporteur found it difficult *'to gain an accurate understanding of the level of enjoyment of those rights by all children nationwide given that coordination between different regions, bodies and authorities was extremely complex.'*[1524]

Therefore, in its Concluding Observations, the Committee recommended to ensure that all legislation, nationwide, was fully in conformity with the Convention, because of the varying legislation in the Communities.[1525] Furthermore, the Committee recommended the establishment of *'an effective system of coordination of the Convention and ensure cooperation of the coordination mechanisms established at the federal and Community level so as to achieve a comprehensive and coherent child rights policy.'*[1526] Also, the Committee regretted that the previously recommended[1527] national plan of action for children was not implemented.[1528]

13.3.3 The right to food in the ICRC reports

13.3.3.1 The initial report

In its initial report the right to food is not discussed as a separate issue, and the right to an adequate standard of living is mostly discussed in the context of social benefits. Noteworthy is that on Community level, the report informed about the adoption by the French Community of the Young Children's charter, a *'declaration of intent constituting an intermediate step towards the adoption of a Covenant on young children.'*[1529] Article 1 of this Charter stipulated the *'right of the child to an adequate standard of living to permit its physical, intellectual, emotional and social development'*, and Article 4 recognised the *'right of the child to benefit from social security.'*[1530]

[1522] Mr Citarella, CRC/C/SR.1521, 10 June 2010, Section 16.

[1523] Mrs D'hondt, CRC/C/SR.1521, 10 June 2010, Section 31

[1524] Mr Citarella, CRC/C/SR.1523, 11 June 2011, Section 48.

[1525] CRC/C/BEL/CO/3-4, 18 June 2010, Sections 11-12.

[1526] CRC/C/BEL/CO/3-4, 18 June 2010, Section 14.

[1527] CRC/C/OPAC/BEL/CO/1, 9 June 2006, Section 9, Concluding Observations on the initial report on the ICRC-OP on the rights of the child on the involvement of children in armed conflict.

[1528] CRC/C/BEL/CO/3-4, 18 June 2010, Sections 15-16.

[1529] CRC/C/11/Add.4, 6 September 1994, Section 13.

[1530] CRC/C/11/Add.4, 6 September 1994, Section 13.

Belgium reported extensively on its system of social benefits, which appears to be quite complex due to the fact that both the Federal Government and the Communities have certain responsibilities towards the allowances of these benefits. Belgium reported *inter alia* that all children are equal under Belgian law, regardless of their filiation (Article 10 CA, leading to amendments in the Civil Code by Statute of 31 March 1987),[1531] that in case a child is aided by a Public Centre for Social Welfare, naturally the best interests of the child prevails (a reference was made to several Articles of the *'Organic Law of 8 July 1976 on public centres for social welfare'*),[1532] and that family allowance is increased when a child is disabled (Royal Decree of 19 December 1993).[1533]

In the Report, reference was made to several plans and initiatives for the near future to improve the systems of social benefits. For instance, the goal was set *'of detaching the family allowances from the social-occupational status of the parents'* for *'family allowances should no longer be a right stemming from the parents' socio-occupational situation, but an inherent right of the child as such.'*[1534]

Belgium furthermore explained that a principal rule in the Belgian Civil Code is that *'Minors, like everyone, are entitled to social assistance. But the law does not explain whether a minor may also file an application for assistance and receive it, or, for that matter, who should file the application when the beneficiary is a minor. It is a fundamental rule of civil law that a minor is legally incapable of taking action. Case law, however, has observed that minors are in fact capable of taking action in respect of acts of everyday life. Acts necessary for acquiring absolutely vital resources to enable himself to lead a life in keeping with human dignity can be seen as representing acts of everyday life.'*[1535] The child has thus entitlements of its own regarding social benefits that are necessary for an adequate standard of living.

13.3.3.2 Second periodic report

In the second report, the right to food is nowhere addressed as a separate issue either. Instead, Belgium further explains how its social security system works.

With regard to Article 27 (4) ICRC, Belgium reported that legislation on collective settlements of debts includes Provisions that protect the rights of the 'maintenance

---

[1531] CRC/C/11/Add.4, 6 September 1994, Section 74. See: Statute of 31 March amending certain Provisions relating to the origin. Original title in Dutch: *Wet van 31 maart 1987 tot wijziging van een aantal bepalingen betreffende de afstamming.*

[1532] CRC/C/11/Add.4, 6 September 1994, Sections 88-89. See: *'Organic Law of 8 July 1976 on the Public Centres for Social Welfare'.* Original title in Dutch: *Organieke Wet van 8 juli 1976 betreffende de openbare centra voor maatschappelijk welzijn,* B.S. 5.VIII.1976 – err. B.S. 26.XI.1976.

[1533] CRC/C/11/Add.4, 6 September 1994, Section 297.

[1534] CRC/C/11/Add.4, 6 September 1994, Section 320.

[1535] CRC/C/11/Add.4, 6 September 1994, Section 350.

creditor.' Thus, responsibilities towards the child as maintenance creditors can still be fulfilled, also in case of extensive debts or bankruptcy.[1536] With regard to Article 23 ICRC, special forms of care were introduced for disabled children, including increased insurance benefits and family allowance.[1537] The report furthermore highlighted that access to healthcare (Article 24 ICRC) was guaranteed for children, regardless whether their parents were insurable or not, and discussed several initiatives on healthcare issues, including a number of initiatives on the cot-death syndrome, support systems for diabetic women wishing to become pregnant and pregnant diabetes patients, additional help for diabetic children, the creation of reference centres for children suffering from metabolic illness, mucoviscidosis and neuromuscular ailments and AIDS.[1538] On a federated level, action programmes were discussed, mainly initiated by the French Community's Birth and Children Office (ONE), and Flemish Community's *Kind en Gezin*. The French Community for instance introduced a five-year programme consisting of a health promotion and prevention policy. Certain elements of this policy concerned food issues, such as the promotion of healthy eating habits, combatting malnutrition and the Provision of information on child health and nutrition.[1539]

With regard to the right to social security, Belgium reported on specific regulations regarding family allowances for children in complex family situations. Also, benefits that are paid directly to children (and not indirectly via its legal representatives) are discussed: '*The range of circumstances in which allowance may be paid to the child himself is being extended in order to prevent loss of the right to family allowances. Such allowances are now paid when children attain the age of 16 (previously 18), when they have a separate principal residence or when they are themselves beneficiaries for their own children.*'[1540]

On the standard of living (Article 27 (1-3) ICRC), Belgium reported that at the federal level, the Civil Code was amended, setting forth the role of fathers and mothers, which included '*the obligation to assume, according to their means, maintenance, monitoring, raising and education of their children.*'[1541] On the federated level, the report mainly highlighted initiatives on the support of disadvantaged children and families.[1542]

[1536] CRC/C/83/Add.2, 25 October 2000, Section 402.
[1537] CRC/C/83/Add.2, 25 October 2000, Sections 500-507.
[1538] CRC/C/83/Add.2, 25 October 2000, Section 535.
[1539] CRC/C/83/Add.2, 25 October 2000, Sections 537-541.
[1540] CRC/C/83/Add.2, 25 October 2000, Sections 552-561.
[1541] CRC/C/83/Add.2, 25 October 2000, Section 575.
[1542] CRC/C/83/Add.2, 25 October 2000, Sections 576-586.

13.3.3.3 The combined third and fourth periodic report

Again, in the combined third and fourth report, no direct reference was made to the right to food, although certain aspects of social benefits to the child were discussed in detail, including occasionally food-related issues.

In general, Belgium expressed its intention to ratify the Convention on the Rights of Persons with Disabilities several times,[1543] and confirmed the ratification during the sessions with the Committee on the Rights of the Child.[1544]

Furthermore, Belgium reported on Article 3 that *'the interests of the child are taken into account on the coordinated legislation on family allowances for employees.'* Measures were adopted to assure that the child benefits from all its entitlements.[1545]

On Article 27 (4) ICRC Belgium referred to the Act of February 2003,[1546] installing a Maintenance Recovery Service, collecting and/or recovering maintenance payments on behalf of the beneficiaries and – when necessary – making advance payments when the maintenance money was not properly paid.[1547] As a future objective, the relatively vulnerable single parent families will be more intensively aided by this Maintenance Recovery Service.[1548]

Regarding Article 24 ICRC, Belgium reported on new federal legislation *'designed to guarantee the quality of care for children in hospital.'*[1549] Special care for children is provided for, as well as arrangements for parents to accompany the child in the care process. In this light *'resources are provided to ensure the quality of food given to patients, the psychological and social support of child and family, and for the organisation of leisure and educational activities.'* Furthermore, the care is both internally and externally assessed.[1550] Also, reference was made to a pilot to develop the WHO-UNICEF driven 'Baby Friendly Hospital Initiative', aiming

---

[1543] CRC/C/BEL/3-4, 4 December 2009, Sections 53, 449, and 595.

[1544] Ms. D'hondt, CRC/C/SR.1521, 10 June 2010, Section 6.

[1545] CRC/C/BEL/3-4, 4 December 2009, Section 208.

[1546] See: the Statute of 21 February 2003, establishing a service for maintenance claims at the Ministry of Finance. Original title in Dutch: *Wet van 21 februari 2003 tot oprichting van een Dienst voor alimentatievorderingen bij de FOD Financiën.*

[1547] CRC/C/BEL/3-4, 4 December 2009, Sections 366-369.

[1548] CRC/C/BEL/3-4, 4 December 2009, Section 417.

[1549] Especially the Royal Decree of 13 July 2006, establishing the standards that care for children must comply with in order to be recognised, and amending the Royal Decree of 25 November 1997 establishing the standards for the 'surgical day treatment' function must comply with in order to be recognised. Original title in Dutch: *Koninklijk Besluit van 13 juli 2006, houdende vaststelling van de normen waaraan het zorgprogramma voor kinderen moet voldoen om erkend te worden en tot wijziging van het Koninklijk Besluit van 25 november 1997 houdende vaststelling van de normen waaraan de functie 'chirurgische daghospitalisatie' moet voldoen om te worden erkend.*

[1550] CRC/C/BEL/3-4, 4 December 2009, Section 492.

at *inter alia* the encouragement of breastfeeding.[1551] Furthermore a national advertising campaign by the national plan for nutrition was reported, consisting of a television commercial, five guides to nutrition (of which three guides were especially aimed at children and young persons), and a website. Also, future actions were mentioned, such as the establishment of nutrition work groups, the stimulation of breastfeeding, and education on healthy nutrition.[1552] In this context, the Flemish and French Community Governments and the Walloon Region also reported on several initiatives on the stimulation of healthy eating habits.[1553]

With regard to Articles 26 and 18 (3) ICRC, basically the general social security system of Belgium was explained.[1554]

On Article 27 (1-3) ICRC, Belgium mainly reported that childcare costs are tax-deductible, and changes had been made to extend those costs arrangements, resulting in a positive effect for the standard of living of children.[1555] Furthermore, Belgium reported improvements on the system of social benefits: firstly, the social benefits will increase when a recipient of the guaranteed minimum income has a child, regardless of the further family situation, which was not the case before. Secondly, reference was made to the improvement regarding the entitlements to social benefits of children of illegally residing parents in Belgium, which will be discussed in Section 13.3.4. Belgium also discussed some difficulties and future objectives related to poverty and the standard of living. As a future objective, Belgium reported to work towards reducing the poverty rate of 5%, as recommended by the UNICEF's Innocenti Research Centre, although has to admit that this is a very ambitious level.[1556] A *'multi-faceted approach'* would be required to *'cover all aspects and all types of difficulties encountered by families living in insecurity'*. Furthermore, Belgium set the goal to improve the access to information on the assistance to vulnerable persons.[1557] Also, it was planned to carry out a study on the possible measures of assistance that can be adopted to ensure access to energy, such as gas and electricity, for it appeared that families could be cut off from energy supplies (except during the winter from December until March), also when children were involved.[1558] More in general, Belgium reported that people living in financial insecurity are – as a result – often living an unhealthy life, for instance because they *'defer medical treatment because they do not have the*

[1551] CRC/C/BEL/3-4, 4 December 2009, Section 494.

[1552] CRC/C/BEL/3-4, 4 December 2009, Sections 503-504.

[1553] CRC/C/BEL/3-4, 4 December 2009, Sections 516 and 543-544.

[1554] CRC/C/BEL/3-4, 4 December 2009, especially Sections 555-566.

[1555] CRC/C/BEL/3-4, 4 December 2009, Sections 586-588.

[1556] CRC/C/BEL/3-4, 4 December 2009, Section 607.

[1557] CRC/C/BEL/3-4, 4 December 2009, Section 610.

[1558] CRC/C/BEL/3-4, 4 December 2009, Sections 612-615. See also with regard to the Flemish Community: CRC/C/SR.1523, 11 June 2011, Section 26.

means to pay'. Combatting poverty thus should also lead to a reduction of health problems.[1559]

During the sessions with the Committee on the Rights of the Child, the problem of (child) poverty was addressed on several occasions.[1560]

In its Concluding Observations, the Committee on the Rights of the Child urged the State Party to take *'measures to monitor the state of health of children from the most disadvantaged families in their first year of life, ensuring access to health services to all children and encourage parents to seek the health services that are available for their children. The Committee also recommends that the State Party review health insurance systems in order to lower the cost of health services for the most disadvantaged families. The Committee further recommends the State Party to strengthen enforcement of the International Code of Marketing of Breast-milk Substitutes in all parts of the State.'*[1561] Furthermore, the Committee expressed its concern about the fact that 16.9% of all children in Belgium live below the poverty line.[1562]

13.3.4 The rights of foreign children in the ICRC reports

13.3.4.1 The initial report

Belgium made an interpretative declaration on Article 2 of the ICRC: *'With regard to Article 2, Section 11, according to the interpretation of the Belgian Government non-discrimination on grounds of national origins does not necessarily imply the obligation for States to automatically grant foreigners the same rights as their nationals. This concept should be understood as designed to rule all arbitrary conduct but not differences in treatment based on objective and reasonable considerations, in accordance with the principles prevailing in democratic societies.'*[1563] This declaration was discussed during the sessions with the Committee on the Rights of the Child. The Committee, which was not surprisingly concerned about the declaration, asked whether *'that reservation was really necessary given that non-discrimination was one of the major principles of the Convention',* and recommended reconsideration of the declaration.[1564] The Belgian Delegation responded that the declaration was *'precautionary since the Convention affected various areas.'* Furthermore, it stated that the Belgian Government *'already adopted a number of principles and sought to ensure equality of treatment for Belgians*

[1559] CRC/C/BEL/3-4, 4 December 2009, Section 616.
[1560] Mr Citarella and Mrs D'hondt, CRC/C/SR.1521, 10 June 2010, Sections 17, and 33.
[1561] CRC/C/BEL/C/3-4, 18 June 2010, Section 57.
[1562] CRC/C/BEL/C/3-4, 18 June 2010, Section 64.
[1563] See for an overview of all reservations and interpretative declarations the UN official treaty database: http://treaties.un.org.
[1564] MrsSanto Pais, CRC/C/SR.222, 6 June 1995, Section 26.

and aliens'.[1565] Somehow, the above discussion led the Committee to believe that the Delegation was willing to consider the withdrawal of the declaration.[1566]

Indeed, in the initial report, several measures to ensure equality between Belgian citizens and foreigners were discussed. With regard to foreign children, reference was made to Article 1, Section 1 of the '*Organic Law of 8 July 1976 on public centres for social welfare*': '*Everyone is entitled to social assistance. The goal of social assistance is to enable everyone to live a life in keeping wit human dignity.*'[1567] This Article has two components, according to the report: '*on the one hand, the universal scope of the right to social assistance, and on the other hand, the explicit reference to human dignity. The right to social assistance is considered to be an inalienable right founded on recognition of the humanity of everyone.*'[1568] This has thus the consequence that '*No criteria of nationality or race are needed in order to avail oneself of this right. The only criterion is that of leading a life that is not (or is no longer) compatible with human dignity.*'[1569] Furthermore, the report held that '*Social assistance is not reserved for nationals alone, and the law also prohibits any requirement of prior residence for a particular number of years. The only element taken into consideration is the presence in Belgium of a destitute person, regardless of that person's residence status.*'[1570] And: '*For example, the general terms which Article 1 of the Organisational Act make it impossible to reserve social assistance to foreigners properly entered in the town population or aliens registers, or to require legitimate residence in Belgium as an exclusive condition.*' Furthermore, Belgium reported that a minimum income was guaranteed to persons holding Belgian nationality, nationals of Member Countries of the EEC who are subject to Council Regulation 1612/68 of 15 October 1968 (on the free movement of workers within the Community), stateless persons and recognised political refugees.[1571]

In its List of Issues, the Committee on the Rights of the Child asked whether there were '*specific measures or procedures in force or envisaged with regard to unaccompanied minors in order to guarantee their full enjoyment of the rights recognised by the convention'.*[1572] Belgium responded in its written replies that there was no such specific procedure: unaccompanied minor foreigners could follow the

---

[1565] Mr Debrulle, CRC/C/SR.222, 6 June 1995, Section 50.

[1566] CRC/C/15/Add.38, 20 June 1995, Section 4. See also the responses of the Committee Members during the sessions, especially CRC/C/SR.224, 9 June 1995, Sections 72 (Mrs Badran) and 73 (Mrs Santos Pais).

[1567] CRC/C/11/Add.4, 6 September 1994, Section 341.

[1568] CRC/C/11/Add.4, 6 September 1994, Section 341. See also the initial report on the implementation of the ICESCR: E/1990/5/Add.15, 13 May 1993, Sections 143-145. See furthermore: Article 1, § 1, '*Organic Law of 8 July 1976 on the Public Centres for Social Welfare'.* Original title in Dutch: *Organieke Wet van 8 juli 1976 betreffende de openbare centra voor maatschappelijk welzijn*, B.S. 5.VIII.1976 – err. B.S. 26.XI.1976).

[1569] CRC/C/11/Add.4, 6 September 1994, Section 342.

[1570] CRC/C/11/Add.4, 6 September 1994, Section 348.

[1571] CRC/C/11/Add.4, 6 September 1994, Section 354.

[1572] CRC/C.9/WP.4, 17 February 1995, question 23.

same asylum procedures as adult asylum-seekers. The only difference was that an unaccompanied minor could not be removed from Belgian territory in case of refusal of a residence permit, for 'no order to leave the territory may be delivered to a foreigner who is less than 18 years old or whose personal status is that of a minor'.[1573] Instead, a reconduction order will be delivered, and a specified person will be enjoined to, if possible, return the minor to the place where he/she came from. However, Belgium admitted that the fact that unaccompanied children have no specific status in Belgian law gave rise to problems, especially with regard to social assistance. A commission was installed to determine such status for unaccompanied minors.[1574]

It may therefore come as no surprise that in its Concluding Observations, the Committee was 'concerned about the application of the law and policy concerning children seeking asylum, including unaccompanied children. It is particularly concerned that unaccompanied minors who have had their asylum request rejected, but who can remain in the country until they are 18 years old, may be deprived of an identity and denied the full enjoyment of their rights, including healthcare and education. Such a situation, in the view of the Committee, raises concern as to its compatibility with Articles 2 and 3 of the Convention.'[1575]

13.3.4.2 The second periodic report

Unaccompanied minor foreigners

In response to the Concluding Observations on the initial report, Belgium had to admit in its second periodic report that meanwhile still no specific regulations related to the care of unaccompanied minor asylum-seekers were adopted. Belgium stated that 'social assistance centres are responsible for such care but, in practice, they do not provide it. Nor are there any regulations regarding minors whose application for asylum has been rejected but who are "allowed" to stay on Belgian territory.'[1576] And furthermore: 'In certain circumstances there are no specific regulations concerning guardianship of unaccompanied minors seeking asylum, so that protection of their interests is not guaranteed. The Legislature has entrusted the task to the Public Social Assistance Centres (CPAS), but, in practice, it is seldom implemented.'[1577] Basically, the normal procedure that is applicable to adults thus also applies to minors. However, Belgium reported that when a child seeks asylum but is not accompanied by both parents, the wishes of the child are also examined.[1578] Furthermore, in Belgium, unaccompanied minors whose asylum request has been rejected may not be forced

---

[1573] Belg/1, 04 May 1995, reply to question 23.
[1574] Belg/1, 04 May 1995, reply to question 23.
[1575] CRC/C/15/Ad.38, 20 June 1995, Section 9.
[1576] CRC/C/83/Add.2, 25 October 2000, Section 13.
[1577] CRC/C/83/Add.2, 25 October 2000, Section 657.
[1578] CRC/C/83/Add.2, 25 October 2000, Section 655.

to leave Belgian territory. However, Belgium had to admit that '*a number of minors, mainly aged 16-17 and supposedly sufficiently "mature", are expelled in the same way as foreign adults.*'[1579] Again, Belgium promised improvement, and underlined to have taken part in several international symposiums on the matter.[1580] Also, it was emphasised that illegally residing children in both the French-language and the Flemish Communities are allowed to participate in education.[1581]

The Committee on the Rights of the Child already announced in its List of Issues that the matter of unaccompanied minor, asylum-seeking and refugee children would be addressed during the sessions,[1582] and asked for further data on the issue on the number of unaccompanied children that '*(a) have applied for asylum; (b) have pending cases; and (c) were granted residency.*'[1583] In response, Belgium provided for this data.[1584] With regard to the federated areas, the Flemish Community Government reported that '*the situation of foreign unaccompanied minors is being carefully monitored by the Flemish Community*', and informed on several initiatives taken on the matter.[1585]

Unfortunately, the Committee hardly addressed the issue during the sessions, as it had promised in its List of Issues. When the meetings were almost at an end, the chairperson stated in the context of '*adoption reform, discrimination, abuse and neglect and tax measures for poor families*'[1586], that it '*seemed that current draft legislation provided that unaccompanied asylum-seekers could be refused entry at the frontier and sent to a centre to be regarded as extraterritorial. He wondered what law would apply here.*' Furthermore, the chairperson noted that in Belgium a Provision applied for minors to be sent back to their country of origin or another country for family reunification. He noted that families were not always willing to be reunited.[1587] The Belgian Delegation however did not answer all those questions, but instead suggested that it would be better to answer them later by means of written replies.[1588] The Delegation only underlined the fact that also unaccompanied foreign minors had access to free education, and that due to

---

[1579] CRC/C/83/Add.2, 25 October 2000, Section 659.

[1580] CRC/C/83/Add.2, 25 October 2000, Section 14.

[1581] CRC/C/83/Add.2, 25 October 2005, Sections 15-16.

[1582] CRC/C/Q/BELG/2, 8 February 2002, part IV, 6.

[1583] CRC/C/Q/BELG/2, 8 February 2002, part 1, A.6.

[1584] However, it must be noted that major parts of the data (such as the number of unaccompanied asylum-seekers granted residency) were only available since January 2002 (the written replies were received by the Committee on the Rights of the Child on 2 May 2002).CRC/C/RESP/7, received on 3 May 2002, Section 1, A.6.

[1585] In particular, schools were informed on the right to education of illegally residing foreign minors, an Association ('*t Huis*) was founded that aimed at the reception of these children, and under certain conditions, facilities for disabled children were opened up for foreign unaccompanied minors. See: CRC/C/RESP/7, received on 3 May 2002, Flemish Community, part VIII.

[1586] Mr Doek, in his capacity as chairperson, CRC/C/SR.783, 29 May 2002, Section 45.

[1587] Mr Doek, in his capacity as chairperson, CRC/C/SR.783, 29 May 2002, Section 52.

[1588] Mr Debrulle, CRC/C/SR.783, 29 May 2002, Section 53.

new legislation, all disabled children had access to facilities, regardless of their nationality.[1589] Unfortunately, the written replies as suggested by the Delegation were never submitted.

In its Concluding Observations, the Committee regretted that some of the recommendations after the consideration of the initial report were *'insufficiently addressed.'* The Committee referred to *inter alia* Section 9 of its Concluding Observations on the initial report[1590] regarding its concerns about the position of unaccompanied foreign minors in Belgium. Therefore, the issue was reiterated in the second Concluding Observations,[1591] and thus *'The Committee urges the State Party to make every effort to address the previous recommendations that have not yet been implemented, and the list of concerns contained in the previous concluding observations.'*[1592] In general, the Committee welcomed the adoption of Article 22*bis* of the Belgian Constitution on the protection of children.[1593] Furthermore, the Committee recommended to ensure that all laws concerning children *'are right-based and in conformity with international human rights standards, including the Convention'*, especially in view of the draft legislation on unaccompanied minors that was addressed by the chairperson of the Committee during the sessions.[1594] The Committee also recommended to *'take all necessary measures to ensure that all children within its jurisdiction enjoy all the rights set out in the Convention without discrimination.'*[1595] Regarding the issue of unaccompanied foreign minors, the Committee welcomed a number of initiatives, such as a special bureau on unaccompanied minors in the Aliens Office for handling their request to stay, as well as some other activities, but also stressed that *'there are not yet, as the Government acknowledges, specific regulations for unaccompanied minors, whether seeking asylum or not.'*[1596] The Committee therefore made the following, rather detailed, recommendations:

> In accordance with the principles and Provisions of the Convention, especially Articles 2, 3, and 22, and with respect to unaccompanied persons under 18 years of age, the Committee recommends that the State Party:
> a. Expedite efforts to establish special reception centres for unaccompanied minors, with special attention to those who are victims of trafficking and/or sexual exploitation.

[1589] Mr Parmentier, CRC/C/SR.783, 29 May 2002, Sections 54-56.
[1590] CRC/C/15/Add.38, 20 June 1995, Section 9.
[1591] CRC/C/15/Add.178, 13 June 2002, Section 4.
[1592] CRC/C/15/Add.178, 13 June 2002, Section 5.
[1593] CRC/C/15/Add.178, 13 June 2002, Section 3.
[1594] CRC/C/15/Add.178, 13 June 2002, Section 9.
[1595] CRC/C/15/Add.178, 13 June 2002, Section 19(b).
[1596] CRC/C/15/Add.178, 13 June 2002, Section 27.

b. Ensure that the stay in those centres is for the shortest time possible and that access to education and health is guaranteed during and after the stay in the reception centres.

c. Approve as soon as possible the draft law on the creation of a guardianship service, in order to ensure the appointment of a guardian for an unaccompanied minor from the beginning of the asylum process and thereafter as long as necessary, and make sure that this service is fully independent, allowing it to take any action it considers to be in the best interests of this minor.

d. Ensure unaccompanied minors are informed of their rights and have access to legal representation in the asylum process.

e. Improve cooperation and exchange of information among all actors involved, including the Aliens Office and other relevant authorities, police services, tribunals, reception centres and NGOs.

f. Ensure that, if family reunification is carried out, it is done in the best interests of the child.

g. Expand and improve follow-up of returned unaccompanied minors.'[1597]

The Belgian declaration on Article 2 ICRC

With regard to the interpretative declaration on Article 2 ICRC, Belgium reported, contrary to the understanding of the Committee in its Concluding Observations to the initial report,[1598] that '*the Belgian Government considers it necessary to retain this interpretative declaration. In fact, the purpose of the Provision of Article 2 is to rule out arbitrary conduct and not differences in treatment based on legitimate considerations. The given interpretation of the concept "discrimination" is in accordance with the now universally accepted doctrine regarding the term, in other words, the criterion used to constitute a breach of the principle of equality of treatment is the absence of an objective and reasonable justification for distinction. Such a justification is more likely if the purpose is to achieve a legitimate goal and there is a reasonable proportional link between the means used and the goal to be achieved.*'[1599] Furthermore, Belgium reported that the '*Belgian institutions constantly endeavour to ensure the non-discriminatory treatment of persons on Belgian territory*'.[1600] To underline this, reference was made to Constitutional Court ruling 43/98[1601]. It was proudly reported that in its verdict, the Constitutional Court ruled that social assistance should be provided to an illegally residing foreigner whose request for asylum had been rejected and was ordered to leave Belgian territory. No further explanation was given, suggesting that this ruling is standard case law in Belgium. *In casu* however, as demonstrated in Section 12.5.3.3, the asylum-seeker in question was awaiting a higher appeal

---

[1597] CRC/C/15/Add.178, 13 June 2002, Section 28.
[1598] CRC/C/15/Add.38, 20 June 1995, Section 4.
[1599] CRC/C/83/Add.2, 25 October 2000, Sections 28-29.
[1600] CRC/C/83/Add.2, 25 October 2000, Section 30.
[1601] Constitutional Court, 43/98, 22 April 1998, B.S. (04)(83, pp.13348-13357).

procedure before the Council of State, and therefore, as an exception to the earlier case law of the Court, was entitled to social benefits for the period of the appeal. In fact, in case 43/98, the Court confirmed the doctrine of case 51/94[1602] in which it had ruled that restricting the entitlements to social services was a legitimate means to restrict immigration. In this light, referring to this case to demonstrate that foreigners are not discriminated seems to be rather absurd. In addition, a technical juridical note could be made here, although it is only a detail: strictly considered, the Court only nullified the word *'executable'* in the phrase *'The social services to an alien (...) shall be stopped at the moment that an executable order to leave is addressed to him (...).'*, of a Statute amending Article 57 § 2 *'Organic Law of 8 July 1976 on public centres for social welfare'*, and did not (even could not) rule that social assistance must be granted to an illegally residing foreigner. The explanation in the report therefore seems to contradict the technical juridical working of the verdicts of the Constitutional Court. On the other hand, indeed the Legislature did amend the applicable legislation in line with the Court's ruling, and since then, consequently, foreigners awaiting an appeal procedure against the refusal of their request for asylum were entitled to social benefits for this period.

In its List of Issues, the Committee on the Rights of the Child confirmed that Article 2 ICRC did not prohibit differences in treatment, *'but only those which are based on grounds that are arbitrary and objectively unjustifiable, such as those enumerated in Article 2 (1), including nationality.'* The Committee therefore asked what the declaration *'means in practice with respect to non-national children in Belgium.'*[1603] In its reply, Belgium repeated that Belgian law did not *'grant the same rights to foreigners as to nationals or to different categories of foreigners'*, but argued that this was not discriminate in the meaning of Article 2 ICRC *'whenever the differences are based on grounds that are objective and reasonable.'*[1604] Belgium held that States decided for themselves under what conditions a foreigner was allowed to reside on their territory, *'within the limits allowed by international law'*. The fact that Belgium was also a Member State of the Fourth Protocol to the European Convention on Human Rights, recognising the freedom to leave any country, including ones own, did not change that principle, for *'this right does not imply that a person may enter the territory of another country and remain there without permission.'* However, Belgium admitted that *'the discretion allowed to States with regard to establishing rules governing admission and residence has for a long time been restricted by international conventions.'* Belgium then explained that the normal regulation on the admission, residence, establishment and expulsion of foreigners was not always applied, for not all foreigners had a so-called 'normal' status. There were categories of foreigners enjoying a more privileged status,

---

[1602] Constitutional Court, 51/94, 29 June 1994, B.S. 1994(07)(pp. 18551-18558); R.W. 1994(95)(pp. 356-359).
[1603] CRC/C/Q/BELG/2, 8 February 2002, part I, B.1.
[1604] CRC/C/RESP/7, received on 3 May 2002, Section 1, B.1.

such as EC and related nationals, nationals covered by agreements concluded between the EU and third countries, refugees, stateless persons, students, and other privileged foreigners. Then, Belgium clarified that, as a Member State of the ICRC, it *'endeavours to avoid any discrimination with regard to children'*, and reported that in case of a minor foreigner involving admission and residence, the ICRC is applied, *'especially Articles 3 (best interest of the child), 8, 9, and 10 (preservation of the child's identity, including nationality and family unity).'* Furthermore, in order to comply with the EU resolution of 26 June 1997,[1605] and thereby protect foreign unaccompanied minors from human trafficking, a service memorandum was drafted by the Aliens Office, granting these children a protected status outside the 'normal' asylum procedures. Also, it was underlined that in asylum procedures, minors were always heard in their mother tongue.[1606]

Interestingly, the Belgian reply mainly involves an explanation why Belgian law is not in contradiction with Article 2 ICRC, and initiatives to reduce discrimination regarding foreign children, or to improve their situation. However, it appears that the question what the practical effect of the interpretative declaration is, was not answered.

In response, naturally, the Committee expressed its concern in the Concluding Observations about the fact that Belgium did not intend to withdraw the declaration on Article 2 ICRC, and repeats that the Convention *'prohibits differences in treatment on grounds that are arbitrary and objectively unjustifiable, including nationality.'* The Committee therefore *'is concerned that the declaration on Article 2 may restrict the enjoyment of non-Belgian children of rights contained in the Convention'*, and emphasised that the non-discrimination principle was applicable to *'each child within (the State Party's) jurisdiction.'*[1607]

13.3.4.3 *The combined third and fourth periodic report*

During the reporting cycle of the combined third and fourth periodic report, the position of minor foreigners was extensively discussed. In the report and the NGO shadow-report, a clear distinction is made between the status of an unaccompanied and an accompanied foreign minor. This section is therefore subdivided into a sub-section on foreign minors in general, unaccompanied foreign minors, and accompanied foreign minors. Thereafter, since the position of foreign minors was discussed altogether into less detail in the List of Issues, during the sessions with the Committee, and the Concluding Observations, those parts of the reporting

---

[1605] Council Resolution 97/C 221/03 of 26 June 1997 on unaccompanied minors who are nationals of third countries.
[1606] CRC/C/RESP/7, received on 3 May 2002, Section 1, B.1.
[1607] CRC/C/15/Add.178, 13 June 2002, Section 6.

process are discussed separately. Furthermore, a separate sub-section will be dedicated to the interpretative declaration on Article 2 ICRC.

Foreign minors in general

Belgium reported in its combined third and fourth periodic report that some alterations were made in the residence procedure of foreigners. A subsidiary protection was introduced, in line with EU law,[1608] for foreigners who are no refugees, but are nevertheless considered to be in danger for serious reasons when a residence permit would be refused.[1609] Also, in line with the case law of the Constitutional Court,[1610] a foreigner cannot be refused residence in case of the medical impossibility of leaving the territory.[1611]

Regarding the position of foreign minors, Belgium reported in view of Article 3 ICRC that '*the interests of the child are always taken into consideration in any decision taken in relation to a foreign minor and pertaining access to the territory, residence, establishment and expulsion*'.[1612] Also, as mentioned in Section 13.3.3.3, improvement was reported regarding the entitlements to social benefits of children of illegally residing foreign parents: when the parents '*are not fulfilling or are unable to fulfil their duty of care in relation to a foreigner under the age of 18 who is residing with them illegally. In those circumstances, the child will be able to benefit from material assistance that will be provided in a federal care centre. The presence in the care of persons who effectively exercise parental authority is also guaranteed.*'[1613] Also here, the effect of the case law of the Constitutional Court is recognisable.[1614]

Unaccompanied foreign minors

Regarding the specific situation of unaccompanied foreign minors, Belgium reported the establishment of the Guardianship Service,[1615] '*providing legal assistance to all unaccompanied minors in Belgium, by appointing to each of them a guardian*'. The Guardian Service then determines '*whether the minor meets the legal requirements*

[1608] Council Directive 2004/83/EC, 29 April 2004, on minimum standards for the qualification and status of third country nationals or stateless persons as refugees or as persons who otherwise need international protection and the content of the protection granted. See for the definition in particular Article 2 (e) and (f).

[1609] CRC/C/BEL/3-4, 4 December 2009, Sections 750-751.

[1610] See especially the verdicts of the Constitutional Court, 90/99, 30 May 1999 and Constitutional Court 194/2005, 21 December 2005.

[1611] CRC/C/BEL/3-4, 4 December 2009, Section 752.

[1612] CRC/C/BEL/3-4, 4 December 2009, Section 203.

[1613] CRC/C/BEL/3-4, 4 December 2009, Section 589.

[1614] See especially the verdicts of the Constitutional Court, 106/2003, 22 July 2003, B.S. 4 November 2003; and Constitutional Court, 131/2005, 19 July 2005, B.S. 8 August 2005, p. 34459.

[1615] Programme Act of 24 December 2004, Title XIII, Chapter IV: Guardianship of unaccompanied foreign minors, B.S. 31 December 2004. Original title in Dutch: *Programmawet van 24 december 2004*; original title of Chapter: *Voogdij over niet-begeleide minderjarige vreemdelingen.*

to benefit from the protection', appoints a guardian to the minor who represents the minor in *'all legal acts and procedures provided for under the legislation on access to the territory, residence, establishment and the removal of foreigners, and coordinates contacts with the authorities responsible for asylum and residence in regard to care and accommodation'*, approves *'guardians to secure the representatives of minors, coordinate and supervise the practical arrangements for guardians'*, and ensures that *'a sustainable solution, in the young persons' best interest, is sought'*.[1616] It is thus the task of a guardian to *'assist the minor throughout all stages of the residence procedure, monitor care and school attendance, ensure that healthcare and psychological support are provided and act on the minor's behalf in dealings with the bodies responsible for asylum-seeking and immigration. The Guardian must also submit to the Aliens Office a proposal for a sustainable solution.'*[1617]

Furthermore, it was reported that when a minor is interviewed somewhere in the asylum procedure, special attention is given to the interviewing method, *'depending on the minor's age, understanding and maturity'*. A special questionnaire was introduced for unaccompanied foreign minors.[1618]

Regarding the reception of unaccompanied minors, Belgium introduced a two-stage reception system. The first stage lasts a maximum of two weeks (which may be extended once), during which the child is received in a general reception centre. The children will be identified *'in order to channel them to the most appropriate reception facility'*, and a Guardian will be appointed. During the second stage, the minor will be received in a collective facility for a period of 6 months to one year where 24 hours a day care is provided. Belgium reported the set up of two centres that take care of all unaccompanied minors. All the above then should put an end to holding unaccompanied minors in closed centres, except for exceptional cases in which the age of the person in question is under dispute.

However, the NGOs reported in their shadow-report that while the special reception centres for unaccompanied minors were indeed established, those centres were not suited for the youngest minors, or minors with health or other serious problems. Also, the capacity of these centres would have been exhausted, resulting in the frequent reception of unaccompanied minors in adult reception centres once again. In their report, the NGOs even stated that in some cases reception to unaccompanied minors is refused, and the minors have no access to any form of assistance. Furthermore, the NGOs express their concerns regarding the juridical status of the residence permit of unaccompanied foreign minors who did not request

---

[1616] CRC/C/BEL/3-4, 4 December 2009, Sections 728-732.

[1617] CRC/C/BEL/3-4, 4 December 2009, Section 733-145. During the sessions with the Committee on the Rights of the Child, concerns regarding the understaffing of the guardian services were raised by the Committee. See: CRC/C/SR.1521, 10 June 2010, Sections 58 (Mr Kotrane) and 59 (Mrs Lee in her capacity as chairperson).

[1618] CRC/C/BEL/3-4, 4 December 2009, Section 747.

for asylum. While such minors are allowed to remain on Belgian territory, such a permit was not based on the Aliens Act, but on a circular of the Minister of the Interior, offering no legal certainty to the minors.[1619] Also, the NGOs recommended the expansion of the reception facilities for minors, the adaption of the reception facilities for unaccompanied minors in line with their specific needs, and the granting of specialised care to some categories of extra vulnerable unaccompanied minors (such as pregnant minors, minors with children or minors with serious psychological problems). Furthermore, the NGOs recommended that a right for unaccompanied minors should be embedded in the Aliens Act, to remain on Belgian territory until a sustainable solution for the minor has been found. Also, decisions taken on the residence status should be taken by a committee of experts rather than by the Aliens Office, considering the best solution for the child. In addition, the NGOs recommended the draft of a cooperation agreement between the relevant authorities.[1620]

Accompanied foreign minors

Accompanied children however, may still be held in closed centres, in order to preserve and maintain the family unit. Belgium reported that since 2002, an effort was done to give the closed centres '*a more humane face*' by organising recreational activities. '*Children may, for example, attend courses and take part in recreational, cultural and sports activities*'.[1621]

The NGOs however stated in their shadow report that children held in closed centres were mostly children accompanied by their parents who do not possess the required documents to access Belgian territory. They underlined that research showed that such reception was a traumatising experience for the child. Furthermore, the NGOs argued that the living conditions in these closed centres were not adapted to the presence of children: the food is often monotonous and inappropriate as a long-term diet to children, there are no meaningful daily schedule and activities for children, healthcare is restricted to a minimum, and the children are not enrolled in school. The NGOs stressed that the situation regarding children who live on the streets with their parents are in even more precarious conditions and their rights are clearly violated: there is no housing, education nor quality access to the asylum procedure. The NGOs report on a distressing practice in which families receiving a negative decision on their asylum request were ordered to leave the reception centres, regardless of their right to social benefits in kind. It is my understanding

---

[1619] *Alternatief Rapport van de NGOs over de toepassing van het Internationaal Verdrag inzake de Rechten van het Kind in België*, 2010, available at: www.kinderrechtencoalitie.be, Chapter 6, Section 2.

[1620] Shadow-report of the NGOs on the implementation of the International Convention on the Rights of the Child. Original title in Dutch: *Alternatief Rapport van de NGOs over de toepassing van het Internationaal Verdrag inzake de Rechten van het Kind in België*, 2010, available at: www.kinderrechtencoalitie.be, Chapter 6, aanbevelingen (recommendations).

[1621] CRC/C/BEL/3-4, 4 December 2009, Section 774.

that here the NGOs refer to the changes in Belgian law[1622] as a result from the case law of the Constitutional Court regarding the matter. That is basically the case law starting from case 106/2003,[1623] in which the Court decided that accompanied illegally residing children were entitled to social services in kind provided for at a reception centre, and case 131/2005,[1624] in which the Court decided that these children had the right to be accompanied by their parents in the reception centre where the social services in kind were received (see more into detail: Sections 12.5.4.1 and 12.5.4.2). In the opinion of the NGOs, the Belgian authorities are violating their own legislation with regard to the reception of aliens.[1625]

The NGOs were positive about a new initiative of the Belgian Government to establish the so-called 'return homes (*terugkeerwoningen*)', in which the families could reside for a period instead of a reception centre, and where the family is guided towards their return by a coach from the Aliens Office. However, the NGOs stated that such alternative for detention unfortunately was not available for all minors, and the coach was only then appointed when the family definitively had to leave the country.[1626]

In their report the NGOs recommended to put an end to the detention of children by legally banning all administrative detention of minors, to adopt more efficient measures to assure that every minor who has a right to social benefits effectively can be housed in a reception centre, and to assure the equal quality of such reception, regardless of the residence status or the family situation of the child. Furthermore, they recommended the Belgian Government to respect the child's privacy and family life, to start with an integrated approach to assist families towards integration in society and guidance in the development of a realistic return perspective outside Belgium if that is in the best interest of the child, to assure that detention and transfer of families are done in a manner that is as humane as possible.[1627] Also, in their Chapter on violence against children, the NGOs recommended to provide for alternate forms of reception to residential placement (in closed centres), especially due to the prevention of violence which is more likely to occur in a residential structure, against vulnerable groups of

---

[1622] Programme Act 9 January 2006, B.S. 30 December 2005.

[1623] Constitutional Court, 106/2003, 22 July 2003.

[1624] Constitutional Court, 131/2005, 19 July 2005.

[1625] Shadow-report of the NGOs on the implementation of the International Convention on the Rights of the Child. Original title in Dutch: *Alternatief Rapport van de NGOs over de toepassing van het Internationaal Verdrag inzake de Rechten van het Kind in België*, 2010, available at: www.kinderrechtencoalitie.be, Chapter 6, Section 1.

[1626] *Alternatief Rapport van de NGOs over de toepassing van het Internationaal Verdrag inzake de Rechten van het Kind in België*, 2010, available at: www.kinderrechtencoalitie.be, Chapter 6, Section 1.

[1627] *Alternatief Rapport van de NGOs over de toepassing van het Internationaal Verdrag inzake de Rechten van het Kind in België*, 2010, available at: www.kinderrechtencoalitie.be, Chapter 6, Section 1.

children (for instance children who are living in poverty, are illegally residing, are handicapped, or are in conflict with the law).[1628]

The written replies to the List of Issues on questions relating foreign minors

In its List of Issues, the Committee on the Rights of the Child asked Belgium to '*indicate which categories of children do not have appropriate social and medical cover.*'[1629] In its reply, Belgium explained that '*as regards coverage of medical costs through the compulsory health insurance system, to our knowledge no particular category of children is excluded from the system. Although children's rights to reimbursement of health costs are most often exercised through their parents (enrolment as dependant), they can also be granted entitlement in their own right, if they are registered with the National Civil Register.*' However, Belgium explained that foreigners who are not authorised to remain on Belgian territory longer than three months would have no entitlements based on entry in the National Civil Register and consequently, some children would be excluded from the coverage of medical costs. Belgium added that with regard to unaccompanied foreign minors, '*have enjoyed access to healthcare in their own right since 2009.*'[1630] However, reading between the lines, accompanied foreign children from illegally residing parents may find themselves excluded from the coverage of medical costs (except for urgent medical care, see Section 12.5.3).

The Flemish Government responded on the same question that its welfare policy focussed on the very young (under 14 years) unaccompanied foreign minors, a coordination agreement between the Federal Government and the Communities regarding their reception was under way, residential and non-residential capacity regarding the target group was expanded, further consultation with the Federal Government and the French Community regarding the reception and support of foreign minors were presumed, and several initiatives on informing unaccompanied foreign minors on their rights were mentioned. In addition, the Flemish Government stated in a rather obscure Section that '*Health insurance now covers all foreign minors (...). To be eligible, they must fulfil the following criteria: have a guardian and attending school.*'[1631] Since the Section was written in a Section on unaccompanied foreign minors, and the guardian services were intended for this specific group of foreign minors, it is unclear whether the intention of the Flemish Government was to report that indeed 'all' foreign minors had access to health insurance.

[1628] *Alternatief Rapport van de NGOs over de toepassing van het Internationaal Verdrag inzake de Rechten van het Kind in België*, 2010, available at: www.kinderrechtencoalitie.be, Chapter 4, Section 4.

[1629] CRC/C/BEL/Q/3-4, 8 March 2010, part I, question 9.

[1630] CRC/C/BEL/Q/3-4/Add.1, 27 April 2010, part I, reply to Section 9 of the List of Issues. See for instance: http://www.kruispuntmi.be/vreemdelingenrecht/wegwijs.aspx?id=3985.

[1631] CRC/C/BEL/Q/3-4/Add.1, 27 April 2010, part 1, reply to Section 9 of the List of Issues.

Furthermore, the Committee asked to '*indicate what problems affecting children the State Party considers to be priorities and in need of urgent attention with regard to the implementation of the Convention.*'[1632] Belgium replied that one of the future projects relating to children would be to '*protect child asylum-seekers, unaccompanied foreigners and abused children*'. Also, '*a resolution on the protection of foreign minors adopted by the Senate on 21 January 2010 requests*, inter alia, *the Federal Government to establish a central database for automatic registration of unaccompanied foreign minors, provide them with better guidance and professional support, take all possible measures to permit the reunification of children with their parents, evaluate the Guardianship Act, and ease the administrative procedures for obtaining a residence permit.*'[1633]

The sessions with the Committee on the Rights of the Child on foreign minors

During the sessions, the Belgian Delegation stated that unaccompanied minors '*had not been held in closed centres since the end of 2007*'. This is contrary to the NGO shadow-report, in which the NGOs held that due to a lack of capacity, unaccompanied minors were still held in centres for adults or were refused any reception at all.[1634] The Delegation underlined furthermore that families with children '*that were in the State Party illegally had not been held in closed centres, but in repatriation centres where they received professional support, and the children could attend school. Thus, foreign children, whether accompanied or not, were no longer held in closed centres.*'[1635] The Committee however was not entirely convinced, and replied that '*according to information before the Committee, there was a crisis within the reception centres and those centres were unable to care for many unaccompanied foreign minors. There were reports that those children were often detained by the police (...)*'[1636] The Belgian Delegation explained that the lack of capacity was due to an increase of asylum request of 40%, but additional budget was granted to the Agency for the Reception of Asylum-seekers. '*With regard to the detention of unaccompanied minors, the information that Mrs Otiz had cited was not correct; minors were no longer detained in closed Centres.*' It is unclear whether this information referred to is the NGO shadow-report. However, it is obvious that the NGO shadow-report paints a very different picture on the matter compared to the official statements of the Belgian Government, in both the report and through their Delegation.

[1632] CRC/C/BEL/Q/3-4, 8 March 2010, part I, question 11.

[1633] CRC/C/BEL/Q/3-4/Add.1, 27 April 2010, part 1, reply to Section 11 of the List of Issues.

[1634] *Alternatief Rapport van de NGOs over de toepassing van het Internationaal Verdrag inzake de Rechten van het Kind in België*, 2010, available at: www.kinderrechtencoalitie.be, Chapter 6, Section 2.

[1635] Mrs D'hondt, CRC/C/SR.1521, 10 June 2010, Section 10.

[1636] Mrs Otiz, CRC/C/SR.1523, 11 June 2011, Section 34.

The Concluding Observations on foreign minors

In its Concluding Observations, the Committee on the Rights of the Child seems to doubt the statements of the Belgian Delegation concerning the type of reception of children. It expressed its concerns about the fact that unaccompanied minors may still be held in adult centres, and, in some cases, are excluded from any type of assistance, and recommended to ensure protection and assistance to all unaccompanied foreign minors, to ensure that all unaccompanied minors are appointed a guardian, to ensure that family reunification is dealt with in a positive, humane and expeditious manner, and to implement and accede to international law concerning stateless persons.[1637] Also regarding accompanied minor aliens, the Committee was concerned about the fact that they are sometimes held in closed centres, and urged Belgium to *'put an end to the detention of children in closed centres, create alternatives to detention for asylum-seeking families and take the necessary measures to urgently find temporary housing solutions for families whose asylum request has been rejected and who live on the streets.'*[1638] After all, the NGO shadow-report seems to have had its effect.

The Belgian Declaration on Article 2 ICRC

It is quite remarkable, especially bearing in mind the Concluding Observations on the second periodic report procedure,[1639] that in the combined third and fourth periodic report no reference was made to the declaration on Article 2. The report mainly highlighted the equal treatment of foreigners and nationals in certain matters of family allowances and healthcare.[1640]

The NGOs reported in their shadow-report their concerns on the matter. The NGOs complained that they had no view on the current situation regarding the Declaration, and expressed their fear that no steps were taken to withdraw the Declaration.[1641] It is remarkable that the Committee on the Rights of the Child did not address this issue in its List of Issues.[1642]

During the sessions with the Committee on the Rights of the Child, the Country Rapporteur Mr Citarella welcomed the decision by the Sate Party to look again at its interpretative declaration on Article 2.[1643] In this light, the Belgian Delegation stated that the declaration was *'in accordance with the modern interpretation of*

[1637] CRC/C/BEL/C/3-4, 18 June 2010, Sections 74-75.
[1638] CRC/C/BEL/C/3-4, 18 June 2010, Sections 76-77.
[1639] CRC/C/15/Add.178, 13 June 2002, Section 6.
[1640] CRC/C/BEL/3-4, 4 December 2009, Sections 152-158.
[1641] *Alternatief Rapport van de NGOs over de toepassing van het Internationaal Verdrag inzake de Rechten van het Kind in België*, 2010, available at: www.kinderrechtencoalitie.be, Chapter 1, Section 3.
[1642] CRC/C/BEL/Q/3-4, 8 March 2010.
[1643] Mr Citarella, CRC/C/SR.1521, 10 June 2010, Section 14.

that Article made by the Constitutional Court of Belgium and the European Court of Human Rights as well as with the jurisprudence of the Committee. The suggestion to withdraw that Declaration would be examined from a legal and political perspective, but it should be noted that a decision to rescind the Declaration could have consequences for the domestic Courts and that the same interpretative Declaration had been made in respect of the International Covenant on Economic, Social and Cultural Rights.'[1644] Again, the Committee on the Rights of the Child recommended in its Concluding Observations to withdraw the Declaration on Article 2.[1645]

13.3.5 Direct applicability of ECOSOC rights in the ICRC reports

13.3.5.1 The initial report

In its first periodic report, Belgium states that since the ICRC entered into force, *'there has been a trend in both the Legislature and the judicial practice towards compliance with the requirements of the convention in respect of, firstly, Article 12, and secondly, legislation on child labour.'*[1646] According to the report, the direct applicability of Article 12 ICRC was recognised by a judgment of the Court of Appeal of Mons (Bergen), resulting in an effective right of a minor to be heard in legal procedures.[1647] This is the only reference to the direct effect of ICRC Provisions in the first report. Therefore, in its List of Issues, the Committee asked to *'indicate to what extent the Provisions of the Convention other than Article 12 can be or have actually been invoked in Court.'*[1648]

In its written replies, Belgium indeed referred to five Court cases in which ICRC Articles were invoked (Articles 3, 7, 9, and 14). Remarkably, one of the cases was the prejudicial ruling of the Constitutional Court on 14 July 1994.[1649] *In casu* however, there were no parties who literally 'invoked' the Provisions of the ICRC. Besides the fact that this is not possible in a prejudicial ruling, in the original prejudicial question, the ICRC Provision was not even mentioned.[1650] It is the Constitutional Court that eventually decided to take into consideration Articles

---

[1644] Mr Brauwers, CRC/C/SR.1521, 10 June 2010, Section 30.

[1645] CRC/BEL/CO/3-4, 18 June 2010, Section 10.

[1646] CRC/C/11/Add.4, 6 September 1994, Section 3.

[1647] CRC/C/11/Add.4, 6 September 1994, Section 5.

[1648] CRC/C.9/WP.4, 17 February 1995, issue 3.

[1649] Constitutional Court, 62/94, 14 July 1994.

[1650] The Prejudicial Question was: *'Does Article 319 § 4 of the Civil Code violate Articles 6 and 6a of the Constitution because, when the mother is in the impossibility to express her will, the Court is authorised, depending on its assessment of the interests of the child, to quash the recognition of a minor, not emancipated child by a man of whom it has not been established that he is not the biological father?'* Original text in Dutch: *'Houdt artikel 319, § 4, van het Burgerlijk Wetboek een schending in van de artikelen 6 en 6bis van de Grondwet doordat het, wanneer de moeder in de onmogelijkheid is haar wil te kennen te geven, aan de rechtbank de bevoegdheid toekent om, afhankelijk van haar beoordeling van de belangen van het kind, de erkenning van een minderjarig, niet ontvoogd kind door een man van wie niet is aangetoond dat hij niet de biologische vader is, te vernietigen?'*

3 and 7 in its ruling, and thus, indirectly, the Provisions of the ICRC had some effect on the outcome of the case. However, on the other hand it seems somewhat excessive to speak of ICRC Provisions that were invoked in Court and considered to be directly applicable. In this light, it is remarkable that the Delegation would later, during the sessions, explain more about the particularities of the cases referred to in the report, except for this Constitutional Court ruling.

During the sessions with the Committee on the Rights of the Child, the Belgian Delegation further explained in its opening statement how it was determined in the Belgian monistic system whether or not an international treaty Provision would have direct effect. As also discussed in Section 11.5, three conditions had to be fulfilled. The first was that the legal instrument is in force in international law. The second was that all national formalities required for legal effect should be fulfilled. In Belgium, this is basically the approval of the competent authorities (in case of the ICRC, this were the Federal Government and the Community Councils), ratification by the King, and promulgation in the official gazette. The third condition was that the content of the international instrument would meet certain criteria: the Provisions should have a legal purpose intended by the parties (a subjective criterion), and the instrument should indeed be self-executing in the sense that additional national legislation is not required (an objective criterion). The Belgian Delegation furthermore added that the Convention did not explicitly state that the Provisions should have direct effect in the national legal order of its Member States, and that therefore the Belgian Courts would have to decide whether a Provision would have direct effect, in the light of the Vienna Convention on the Law of Treaties of 1969.[1651] The Belgian Delegation continued to give more detailed information on the quoted Court cases in the written replies, and indeed demonstrated the direct effect of Articles 3, 12, 9 and 14 by referring to cases ruled by the Court of First Instance of Courtrai, the Ghent Court of Appeal and the Civil Court of Arlon and Liège. The Delegation held that '*the examples cited showed a willingness by Courts to implement the Convention, as well as a high degree of judicial autonomy in applying its Provisions.*'[1652] As mentioned earlier, the ruling of the Constitutional Court (62/94) mentioned in the written replies however was not further discussed by the Delegation.

The Delegation furthermore pointed out that since the Franco-Suisse Le Ski ruling, it was clear that in Belgium international law had primacy over national law.[1653]

A curious thing is that the above was apparently understood by the Committee on the Rights of the Child that the entire Convention had direct effect in Belgium. In its Concluding Observations, the Committee wrote that it '*welcomes the fact that*

[1651] Mr Debrulle, CRC/C/SR.222, 6 June 1995, Section 8-10.
[1652] Mr Debrulle, CRC/C/SR.222, 6 June 1995, Section 11-13.
[1653] CRC/C/SR.222, 6 June 1995, Section 14. See also Section 11.5.

the Convention is self-executing and that its Provisions may be, and in practice have been in several instances, invoked before the Court. It also notes with satisfaction the fact that Belgium applies the principles of the primacy of international human rights standards over national legislation in case of conflict of law.'[1654]

13.3.5.2 The second and combined third and fourth periodic report

Due to the conclusion of the Committee on the Rights of the Child regarding direct applicability of the Convention, it comes as no surprise that in the second and combined third and fourth periodic report, Belgium did not address the issue, nor did the Committee specifically ask questions on the matter in its List of Issues.

However, it was during the sessions with the Committee on the combined third and fourth report that the matter of direct applicability was discussed again. One of the Members of the Committee stated that *'while, in theory, the Convention was directly applicable in Belgium, the Committee had heard of no cases in which participants in a trial, lawyers or judges had directly invoked it.'* The question was raised whether the Convention was sufficiently well known in judicial circles.[1655] The Delegation replied that although judges did not receive special training on the ICRC, training on certain child-related issues were offered to judges, barristers and lawyers. *'With respect to the direct applicability of the Convention, the Court of Cassation, the Court of Arbitration (...) and the Council of State had given their views on several Articles of the Convention, including Articles 3 and 27, and all were considered to be directly applicable in Belgian Courts.'* A remarkable answer for several reasons. Firstly, technically the Constitutional Court is unable to apply international Provisions directly. Although it was demonstrated in Section 12.5 that through indirect application ICRC Articles could have some effect in the case law of the Constitutional Court, the effect is very limited. Secondly, as discussed in Chapter 12 (Sections 12.3 and 12.4), considering the case law of the Council of State and the Court of Cassation, it is the rule that ICRC Articles have no direct effect. Only in some very rare occasions, the Court of Cassation has shown to grant some effect to ICRC Provisions. Thirdly, regarding the direct applicability of Article 3, at most it was recognised (but later also denied) by the Court of Cassation that the Provision could be used as a tool for interpretation of other Articles, but certainly had no direct effect. In addition, the direct applicability of Article 27 was never recognised by any Court in Belgium, and only indirectly applied by the Constitutional Court.

It is regrettable that here the discussion on the direct effect of the ICRC ends. No further questions were asked to the Delegations, and the Committee did not include any concern or recommendation on the matter in its Concluding Observations.

[1654] CRC/C/15/Add.38, 20 June 1995, Section 6.
[1655] Mr Kotrane, CRC/C/SR.1521, 10 June 2010, Section 21.

13.4 The CEDAW reports

13.4.1 Introduction

Also during the reporting cycles on the implementation of the CEDAW, the federal structure (Section 14.4.2), the direct applicability of CEDAW Provisions (Section 14.4.4) and the rights of migrant women (Section 14.4.5) were discussed. The right to food was hardly addressed as a separate issue, which is not a real surprise, considering the wordings of Article 12 CEDAW (Section 14.4.3).

The initial report has not been published in the currently used official databases, neither by the UN nor by the Belgian Government. Considering the quality, briefness and lack of detailed information of the second periodic report, it is not to be expected that any information of significance will be found in the initial report, dating back from 1987.[1656] Therefore, the initial report was not included in the analysis below.

13.4.2 The Federal Structure

Also in the reporting procedures of the CEDAW, the federal structure was more than once food for discussion, and on several occasions Belgium needed to offer clarifications on the matter.[1657] For instance, during the sessions on the combined third and fourth report, the Committee on the Elimination of Discrimination of All Forms of Discrimination against Women stated that '*the complex structure of the federal system, with its different levels of authority each having competence in the separate regions, could appear to be a way of derogating from the implementation.*'[1658] The Delegation answered that '*the federal State and federated entities had different jurisdictions, although there were some split jurisdictions in some areas. The system in place allowed the various authorities to meet the specific needs and priorities of the different communities and regions.*'[1659]

13.4.3 The right to food in the CEDAW reports

The right to food or food-related issues were seldom mentioned as a separate issue during the reporting procedures. Regarding most ECOSOC rights enshrined in the CEDAW, the Belgians mostly held that there was no gender based discrimination concerning these Provisions, in particular Article 12.

[1656] CEDAW/C/5/Add.53, 25 September 1987.
[1657] See for instance: Mrs Abaka, CEDAW/C/SR.560, 20 December 2002 (summary record on second periodic report), Section 31; CEDAW/C/BEL/3-4, 29 September 1998, on Article 3 (combined third and fourth periodic report); CEDAW/C/BEL/6, 22 June 2007, p. 4.
[1658] Mrs Kapalata, CEDAW/C/SR.559, 25 June 2005, Section 51.
[1659] Mrs Paternottre, CEDAW/C/SR.559, 25 June 2005, Section 58.

It must be said that the second periodic report is rather limited, especially concerning ECOSOC rights.[1660] This was also noted by the Committee, which stated that *'the report could, however, have been more specific with regard to the information provided under certain Articles of the Convention.'*[1661] This was also expressed in the Concluding Observations, in which the lack of factual information as occasionally considered to be a principal subject of concern,[1662] and improvement on the matter was recommended.[1663]

The combined third and fourth is indeed much more detailed. On Article 12, no food-related issues were reported. Naturally, considering the wordings of Article 12, there was a focus on initiatives in the field of healthcare, education and awareness-raising on healthcare issues and the combatting of specific diseases, such as specific forms of cancer and osteoporosis. Also, the Belgian regulations on voluntary abortion were discussed.[1664]

In its fifth and sixth combined report, Belgium reports even more extensively and into detail on the implementation of the CEDAW rights. Food-related issues are also here however seldom discussed. On the implementation of Article 12, Belgium mainly reports at the federal level on measures favourable to women regarding healthcare, studies on health, conducted by the Institute for the Equality of Women and Men and provides figures and statistics regarding the health of Belgian women.[1665] Also on the federated level, a variety of initiatives is discussed in the field of healthcare.[1666] Noteworthy detail in the report is the fact that women are considered to be a specific target group concerning food security in development cooperation of Flanders with their partner countries.[1667]

13.4.4 Direct applicability of ECOSOC rights in the CEDAW reports

In its second periodic report, Belgium stated that the Provisions of the CEDAW were for a major part already effectively guaranteed in Belgium, due to already existing international instruments and domestic legislation. Furthermore, Belgium argued that it *'was already implementing the general Provisions of the Convention well before ratifying it and that the national law in force at the time did not require much amendment following the entering into force of the Convention.'*[1668] Then, Belgium underlined that *'this was due to the Belgian legislative system, which gives international*

[1660] CEDAW/C/BEL/2, 8 April 1993, especially Articles 10-14. The shortest explanation was on Article 13, merely stating that: *'there is no discrimination in these fields.'*
[1661] CEDAW/C/SR.301, 6 February 1996, Section 9.
[1662] A/51/38, 9 May 1996, Section 185.
[1663] A/51/38, 9 May 1996, especially Section 189.
[1664] CEDAW/C/BEL/3-4, 29 September 1998, on Article 12.
[1665] CEDAW/C/BEL/6, 22 June 2007, pp. 109-112.
[1666] CEDAW/C/BEL/6, 22 June 2007, pp. 113-117.
[1667] CEDAW/C/BEL/6, 22 June 2007, p. 75.
[1668] CEDAW/C/BEL/2, 8 April 1993, Article 1.

law precedence over national law; indeed, in most of the areas covered by the Convention a large number of international instruments, United Nations conventions, International Labour Organisation conventions, Council of Europe conventions and Directives of the European Economic Community had almost completely eliminated discrimination against women in the country.[1669] A rather obscure Section in which an unclear causal link was made between existing national legislation and directly applicable international law.

Regarding CEDAW case law, the Belgian Delegation explained during the sessions with the Committee on the Elimination of Discrimination Against Women, that the case law on discrimination against women *'mainly concerned rights relating to employment and social security.'* Also, an increase of cases concerning sexual harassment was observed. The Delegation pointed out that Belgian judges increasingly use their authority *'to bring preliminary issues before the Court of Justice of the European Union.'*[1670] No reference was made however to case law in which any of the CEDAW Provisions was invoked. The Delegation furthermore discussed the implementation of non-discrimination Provisions in Belgian law, especially in its Constitution: Articles 10 and 11.[1671] Exceptions to the principle of non-discrimination were only permitted when *'based on objective criteria and precise rules formulated as laws following broad consultations with institutional and other partners. Their purpose was to permit certain territorial adaptations in the organisation of the principles recognised in the Constitution; the principles themselves could not be changed.'* Moreover, they must be adopted by a two-third majority of all linguistic groups in the Parliament and the Senate, *'to ensure that the laws enacted for such purposes reflected the wishes of a large part of the population.'*[1672]

In its Concluding Observations, the Committee on Elimination of Discrimination Against Women did not address the issue.[1673]

In the combined third and fourth periodic report, no reference was made to the issue of direct effect, except for one remark of the Delegation during the sessions with the Committee. The Delegation stated that *'under Belgian law, an international instrument to which Belgium was a party could indeed be invoked in a Court of law. It fell to the judge, however, to determine whether the Provisions were sufficiently precise to allow for their application under domestic law. Although the Convention had been found to be directly applicable, there were other instruments for which this was not the case.'*[1674] A remarkable statement, since this does not appear from the case law analysis of Chapter 12. During the sessions, the Delegation's observation that the

---

[1669] CEDAW/C/BEL/2, 8 April 1993, Article 1.
[1670] Mrs Paternottre, CEDAW/C/SR.300, 1 February 1996, Section 11.
[1671] CEDAW/C/BEL/2, 8 April 1993, Article 1.
[1672] Mrs Paternottre, CEDAW/C/SR.300, 1 February 1996, Section 8.
[1673] A/51/38, 9 May 1996.
[1674] Mrs Paternottre, CEDAW/C/SR.559, 25 June 2005, Section 32.

CEDAW was directly applicable in the domestic legal order was not substantiated with any further arguments or case law. Nor did the Committee pay any attention to the matter.

Also in the combined fifth and sixth periodic report, no reference was made to the issue of direct effect. However, in its List of Issues, the Committee asked whether there are *'instances of any Article of the Convention having been directly invoked before the Courts given that in Belgium, it is generally recognised that an international Provision produces direct effects? Please provide examples of any pertinent case law.'*[1675] In its written replies, Belgium answered that *'Two conditions must be met if a Provision of international law is to have direct effect in Belgian law. Firstly, the intention of the parties must have been to create rights for private persons. Secondly, the Provision in question must be sufficiently precise and complete to be directly applicable in the domestic Courts without the need for an execution measure. Such matters are generally dealt with in case law. Hence, it is generally accepted that the Convention for the Protection of Human Rights and Fundamental Freedoms and the International Covenant on Civil and Political Rights have direct effect.'* Nevertheless, Belgium had to admit that *'a perusal of Belgian case law suggests that litigants have not invoked the CEDAW Convention, but rely more on other Provisions of Belgian law or European law relating to discrimination between women and men.'*[1676] This was reconfirmed by the Delegation during the sessions.[1677] In its Concluding Observations, the Committee expressed its concern about the fact *'that the Convention has not been given central importance as a binding human rights instrument and as a basis for the elimination of all forms of discrimination against women and the advancement of women in the State Party. In this connection, the Committee is concerned about the absence of direct reliance on the Convention by litigants, lawyers and judges, although its Provisions are in principle directly applicable.'*[1678]

13.4.5 The rights of migrant women

In the second periodic report, no significant information was given on the status of foreign women in relation to the rights embedded in the CEDAW. Therefore, in its Concluding Observations, the Committee recommended to include in the next periodic report information on *'Programmes and projects to address the needs of migrant women and other vulnerable women should be made available in the next report.'*

However, in the combined third and fourth periodic report, hardly any reference was made to the position of foreign women either. Briefly, a reference was made

---

[1675] CEDAW/C/BEL/Q/6, 6 March 2008, Section 4.
[1676] CEDAW/C/BEL/Q/6/Add.1, p. 5.
[1677] Mrs Fastre, CEDAW/C/SR.853, 7 January 2009, Section 22; Mrs Grisard, CEDAW/C/SR.853, 26 December 2008, Section 8.
[1678] CEDAW/C/BEL/CO/6, 7 November 2008, Section 17.

in view of Article 6 CEDAW to social measures on behalf of trafficking victims: '*A system of temporary stay permits for victims has been instituted: victims are allowed a stay of 45 days in order to seek counselling and assistance from specialised social agencies before returning to their country of origin, and a renewable three-month stay permit is issued to victims who bring a complaint and cooperate with the justice system in the course of a judicial inquiry. Victims who have a temporary stay permit are entitled to social assistance.*'[1679] Furthermore, it was reported that the number of foreign prostitutes is on the rise.[1680] The absence of detailed information on the status of migrant women was noticed by the Committee, which requested during the sessions, in the next report, for more statistics on migrant women.[1681] The Belgian Delegation replied that '*with regard to the right to stay in Belgium, its Government applied the 1951 Geneva Convention in a broad manner. Refugee women could be granted political asylum on account of their sex, but there had to be reasonable grounds of fear of persecution. In 2000, 45 per cent of the applications for asylum had been granted, out of which 35 per cent were for women. Belgium not only penalised genital mutilation but also granted asylum for that reason. Once asylum-seekers had been granted asylum and had been recognised as refugees, they enjoyed rights on a par with Belgian citizens. Moreover, Belgium applied, by and large, the principle of family reunion.*'[1682] Also, the Delegate of the Flemish Community further explained the several responsibilities of the Community regarding the reception of migrants.[1683] In its Concluding Observations the Committee once again expressed its concern '*that the report provides insufficient information about the situation of migrant and refugee women*', and therefore '*calls on the State Party to provide comprehensive information on these groups of women in its next periodic report.*'[1684]

Finally, in its combined fifth and sixth periodic report, Belgium discussed in more detail the situation of migrant women. In particular, it is reported that when a women is the victim of sexual violence or acts directed against persons by reason of their sex, or against children, this is taken into consideration by the Asylum Courts when deciding on an asylum request.[1685] Furthermore, two coordinators had been designated by the Office for Aliens to pay specific attention to '*those requesting asylum with reference to the problem of gender.*' Also, women are specifically informed on asylum procedures, and are, as much as possible, questioned by female investigators during those procedures.[1686] In addition, an information focus point on gender-related problems was installed at the General Commissariat for Refugees

---

[1679] CEDAW/C/BEL/3-4, 29 September 1998, on Article 6 (in particular, p. 44).

[1680] CEDAW/C/BEL/3-4, 29 September 1998, on Article 6 (in particular, p. 46).

[1681] Mrs Acar, in her capacity as chairperson, CEDAW/C/SR.559, 25 June 2005, Section 53.

[1682] Mrs Verzele, CEDAW/C/SR.559, 25 June 2005, Section 66.

[1683] Mrs Franken, CEDAW/C/SR.560, 20 December 2002, Section 8

[1684] A/57/38 part II, 8 October 2002, Sections 155-156.

[1685] CEDAW/C/BEL/6, 22 June 2007, pp. 23-24.

[1686] This was later discussed during the sessions: Mr Flinterman, CEDAW/C/SR.852, 7 January 2009, Section 18, in response: Mrs Fastre, Section 35. See also the Committee's Concluding Observations: CEDAW/C/BEL/CO/6, 7 November 2008, Sections 36 and 37.

and Stateless Persons.[1687] Belgium provided statistics on asylum requests, and informed (almost proudly) that significantly more women than men were granted a residence permit.[1688] Regarding the reception of asylum-seekers, Belgium stated that '*During the first phase of the procedure for requesting asylum, those requesting asylum receive material assistance only sufficient for meeting their basic needs. Care is provided either in large care centres or in individual residences. In addition to housing and food, those requesting asylum receive social and administrative support as well as medical care. They may take part in a large variety of leisure, educational, and sports activities.*'[1689] Then, Belgium reported on the capacity and gender of the personnel charged with providing care to asylum-seekers, the materials and infrastructure available for female asylum-seekers[1690] and activities they can participate in.[1691] Also on the federated level, the position of migrant women was (briefly) discussed.[1692] During the sessions, the Belgian Delegation further elaborated on its asylum procedures.[1693]

13.5 The Universal Periodic Review

Although the right to adequate food has not been significantly discussed during the reporting cycle of the UPR, the federal structure,[1694] the right to social security,[1695] the Belgian asylum policy,[1696] and the rights of the child[1697] were discussed, touching similar aspects compared to the treaty reports and subsequent procedures.

13.6 Concluding remarks

To conclude, some observations can be made. In general, firstly Belgium seems to prefer to report on initiatives, policies and changes in legislation that lead to a better implementation of the human rights instruments discussed, but is cautious

---

[1687] CEDAW/C/BEL/6, 22 June 2007, pp. 24-25.

[1688] CEDAW/C/BEL/6, 22 June 2007, p. 25.

[1689] CEDAW/C/BEL/6, 22 June 2007, p. 26.

[1690] According to the report (CEDAW/C/BEL/6, 22 June 2007, p. 27): ' *(a) in a standard package of monthly materials, women receive items of feminine hygiene and tampons. Deodorant and razors are furnished upon request; (b) women have access to contraceptives through the medical services at the care centres; (c) restrooms for women are separated from those for men; (d) recreational halls are created for women and their children; (e) in certain centres, women have a special room for watching television; (f) in large care centres, women are housed in special wings; (g) women are questioned regarding their needs with respect to infrastructure and materials, and, whenever possible, their needs are met; (h) certain centres provide a room for ironing that is only accessible to women; similarly, certain Centres provide kitchen stoves, sewing rooms, or beauty salons exclusively for women.*'

[1691] CEDAW/C/BEL/6, 22 June 2007, p. 27.

[1692] CEDAW/C/BEL/6, 22 June 2007, p. 33 (Flemish Region and Community), p. 43 (Walloon Region), p. 46 (German-speaking Community).

[1693] Mrs De Ruyck, CEDAW/C/SR.852, 7 January 2009, Section 36.

[1694] A/HRC/WG.6/11/BEL/1, 16 February 2011, Sections 2-7.

[1695] A/HRC/WG.6/11/BEL/1, 16 February 2011, Sections 34-36.

[1696] A/HRC/WG.6/11/BEL/1, 16 February 2011, Sections 43-53.

[1697] A/HRC/WG.6/11/BEL/1, 16 February 2011, Sections 54-59.

in reporting on points for improvement. Therefore, with the exception of some of the ICRC reports,[1698] the Committee more than once considers the submitted reports as incomplete.[1699]

Secondly, Belgium hardly reported on the right to food as a separate issue, except for once, on the request of the Committee on Economic, Social and Cultural Rights,[1700] in its fourth periodic report on the implementation of the ICESCR.[1701] Issues that relate to the right to adequate food are however discussed more extensively, such as the right to social security, the right to healthcare, and the position of foreigners in relation with entitlements to social benefits.

With regard to the last categories, it must be concluded thirdly, that the Committees are often concerned about the principle not to grant illegally residing aliens adequate social benefits. Of course, especially regarding children, there are plenty of exceptions to that principle, although the Committee on the Rights of the Child seems hardly convinced that the position of illegally residing (unaccompanied or accompanied) children is in accordance with the rights stipulated in the ICRC.

Fourthly, it seems that the complex federal structure of Belgium causes difficulties in the communication between the Government and the treaty bodies. The Committees are often concerned that the federal structure stands in the way of a uniform, coherent approach towards the implementation of the human rights embedded in the treaties. Also, the complexity leads occasionally to misunderstandings, especially regarding the issue of direct applicability. In particular, the function of the Constitutional Court is not always clear amongst the Committees, and the Belgian Government does not hesitate to refer to its case law to demonstrate that certain Provisions are directly applicable in Belgium. Those are questionable statements, considering the fact that the Courts of Last Instance in the subjective litigation usually rule otherwise, and that the Constitutional Court at most grants some indirect effect to international Provisions.

In this light, fifthly, regarding the case law in Belgium, the Delegations seem to be poorly informed on what actually happens with regard to the assessment on direct applicability amongst the Belgian Courts. It is completely unclear how the Belgian Delegation in the sessions on the combined third and fourth report on the implementation of the CEDAW could conclude that the entire CEDAW had direct effect in the national legal order.[1702] It is also unclear how it is possible that both

---

[1698] See for instance: CRC/C/15/Add.178, 13 June 2002, Section 2.
[1699] See for instance: E/C.12/BEL/CO/3, 4 January 2008, Section 9; see also the frequent requests of the Committee on the Elimination of All Forms of Discrimination Against Women for data on the position of migrant women, as discussed in Section 14.4.5.
[1700] Mrs Bonoan-Dandan, E/C.12/2007/SR.42, 10 December 2007, Section 39.
[1701] See in particular: E/C.12/BEL/4, 18 June 2012, Sections 244-246.
[1702] Mrs Paternottre, CEDAW/C/SR.559, 25 June 2005, Section 32.

the Committee on Economic, Social and Cultural Rights and the Committee on the Rights of the Child were led to believe that the entire ICRC had direct effect in Belgium.[1703] It is even more remarkable that Belgium, whether or not through its Delegations, did nothing to correct, or even confirmed this understanding.[1704] Furthermore, the cases referred to are not always portrayed in their proper context, painting an overly positive picture, while this is certainly not a proper reflection of legal reality.[1705]

Sixthly, Belgium made an interpretative Declaration to Article 2 ICRC, declaring that non-discrimination on grounds of national origins does not necessarily imply the obligation for States to automatically guarantee foreigners the same rights as their nationals. Apparently the Committee on the Rights of the Child does not necessarily oppose to such reasoning, and even confirmed that Article 2 ICRC does not prohibit, under strict conditions, a difference in treatment.[1706] While the Committee is wondering what then for Belgium could possibly be the reason for not withdrawing the Declaration, Belgium persists, although without explaining why.[1707] It is furthermore remarkable that a similar interpretative declaration on the ICESCR did not lead to any discussion at all.

[1703] See: CRC/C/15/Add.38, 20 June 1995, Section 6 and Mr Ahmed, E/C/12/2000/SR.64, 27 November 2000, Section 34.
[1704] See for instance during the sessions on the third periodic report on the implementation of the ICESCR: Mr Deneve, E/C/12/2000/SR.64, 27 November 2000, Section 38.
[1705] See: CRC/C/83/Add.2, 25 October 2000, Section 30.
[1706] CRC/C/Q/BELG/2, 8 February 2002, part I, B.1.
[1707] Most obviously: CRC/C/RESP/7, received on 3 May 2002, Section 1, B.1.

14. Evaluation and comparison

14.1 Introduction

In this Chapter, the findings of Chapters 11-13 will be summarised. Furthermore, the legal practice in Belgium will be compared with the reporting behaviour of the Belgian Government on the enforceability of the right to adequate food (Comparison II).

14.2 The legal practice of Belgium regarding the enforceability of the rights of the child

Due to differences in language, culture and economy, Belgium has, since its profound State reforms in 1970, a rather complex *trias politica*, with a Legislature consisting of three legislative powers, equally and exclusively competent. This had an effect on the organisation of the Judiciary, culminating in the instalment of the Constitutional Court. The Belgian Judiciary seems organized, based on purpose and functionality, rather than pragmatic considerations. The organisation is complex. On the one hand there is a subjective contentieux, with two Courts of last instance (the Court of Cassation and the Council of State), and on the other hand an objective contentieux, with two highest tribunals (the Council of State and the Constitutional Court).

While the difference between both litigations may increasingly become less important for the parties involved in a Court case, since the legal effects of the verdicts seems to 'merge' rather than remain of a principally different nature, the judicial organisation has a strong effect on the legal reasoning of the Courts. Belgium is a monistic State, which follows not directly from the Constitution, bur from case law of the Court of Cassation. This means that in principle, an international standard is superior to a contradicting (anterior and posterior) national rule, as long as the international rule has direct effect. Besides some formal prerequisites, an international standard has direct effect when the State Parties involved expressed the intention that the standard should have direct effect, mostly recognisable when the standard is addressed to citizens and not to States (the subjective criterion) and/or when the standard invoked is self-executing (the objective criterion). Neither the Court of Cassation nor the Council of State appear to have recognised the direct effect of the right to adequate food, and as a rule, with some minor, seemingly random, exceptions, consider that ICESCR and ICRC Provisions have no direct effect. There is hardly any case law on (the direct effect of) CEDAW Articles.

However, the Constitutional Court is not bound by the principle of direct effect due to its primary function to review national legislation against the Constitution. As a result, international standards are only indirectly applied, via Articles 10,

11, and 24 CA, and later also via all other rights stipulated in Chapter II CA. The 'school-pact' prompted the Legislature to broaden the reviewing competencies of the Constitutional Court. It is now authorized not only to settle conflicts between the three legislative powers, but also to review national legislation against fundamental rights. This is also reflected in the case law of the Court. The first rulings in which the Court indirectly reviewed against internationally embedded ECOSOC rights concerned Article 13 ICESCR. Only later, the Court also reviewed (mostly indirectly) against other ECOSOC provisions, including articles stipulating the right to adequate food.

Right to food Provisions, and other related ECOSOC rights embedded in UN treaties, mostly play a role in cases concerning the status of illegally residing aliens. In most cases, in both annulment procedures and preliminary rulings, the Court reviewed Article 57 § 2 of the '*Organic Law of 8 July 1976 on public centres for social welfare.*' This Provision restricts social benefits entitlements of illegally residing asylum seekers to urgent medical care only. The Court assessed whether this restriction, in light of the invoked fundamental rights, was objectively and reasonably justified, and proportionate towards a legitimate goal. The basic principle seems to be that it is the legitimate purpose of Belgium to restrict immigration, and that therefore it is not unreasonable to accept different obligations regarding the needs of legally residing persons in Belgium on the one hand and illegally residing persons on the other hand. That is to say that Article 11 ICESCR cannot reasonably be understood without limitations, and therefore, does not apply to persons Belgium is not responsible for, such as illegally residing aliens. In a series of rulings, the Court has further specified and nuanced this notion. Occasionally the Court allows for exceptions from this rule. In particular, the rights of the child usually seem to have more effect than rights stipulated in the ICESCR. In general, in line with its function, the Court shows little hesitation in nullifying legislation, or advising in preliminary proceedings that a Constitutional norm is violated, when it considers the disputed national legislation not objectively and reasonably justified, and/or a disproportionate means to reach a justified goal. Therefore, verdicts of the Constitutional Court often resulted in amendments to legislation, or the adoption of new regulations. However, while international Provisions in this case law are often reviewed against indirectly, no evidence could be found that those provisions made any difference of significance towards the outcome of the rulings on top of the – mostly equivalent – invoked national provisions. This is due to the 'inward' focus of the case law, involving mainly an assessment of national legislation and policies.

14.3 The reporting behaviour of Belgium on the enforceability of the right to adequate food

In its reports, Belgium seems to prefer to communicate initiatives, policies and legal measures that positively contribute to the implementation of human rights.

The reports show little self-evaluation or self-critique. The right to food is seldom addressed as a right in itself. Instead the Government seems to focus on related rights such as the right to healthcare, the right to social security, and more in particular, the position of foreigners in relation with entitlements to social benefits. Regarding the latter, the Committees often express concern. In particular they do not seem to be convinced that withholding rights based on the status of illegally residing foreigners, is compatible with the relevant human rights treaties, especially not in case of accompanied and unaccompanied children.

Furthermore, the complexity of the Belgian federal structure seems to be a reason for concern. The Committees regularly question whether Belgium is capable of a uniform and consistent approach in implementing human rights. It also leads to misunderstandings, especially regarding the issue of direct effect in combination with the function of the Constitutional Court. The delegations underline the direct effect of international ECOSOC provisions by referring to case law of the Constitutional Court. However, they fail to point out that this is at odds with the case law of the Court of Cassation and the Council of State. It appears that more than once, the impression is created that international human rights treaties, such as the ICRC and the CEDAW, have direct effect in the national legal order, even though this is contrary to the legal practice. The Committees seem to be poorly informed on the Belgian constitutional system. They do not really address the issue in-depth. The Belgian Government on the other hand is reluctant in correcting those misunderstandings.

14.4 Comparing the legal practice with the reporting behaviour

Considering the above, it must be concluded that the Belgian reports on the enforceability of the ICESCR, the ICRC and the CEDAW hardly match legal reality. Indeed, the complex federal structure, especially regarding the organisation of the Judiciary, in combination with the Committees' lack of knowledge of Belgian constitutional law, seems to lead to misunderstandings and inaccurate assumptions. Legal practice shows that ECOSOC rights in general, with some rare exceptions, do not have direct effect in the subjective contentieux. Nevertheless, in several reporting cycles it was suggested or assumed that the both ICRC and CEDAW do in their entirety have direct effect. Interestingly, the case law of the Constitutional Court is often invoked by Belgium to demonstrate the direct effect of international human rights provisions, or to paint a positive picture on the implementation of treaty provisions, while such examples seem inaccurate, for there is at most a little indirect effect for the international provisions. In addition, its added value *in casu* should not be overestimated.

Part 4
Conclusions and recommendations

15. Conclusions

15.1 Introduction

In this Chapter, the Dutch and Belgian legal practices will be compared to one another (comparison I), and also with the UN perception of enforceability of the right to food (comparison III). In doing so, an answer will be given to the main research question: *What legal factors explain the differences and similarities regarding the response of the Dutch and Belgian Judiciary and Government to the enforceability of the right to adequate food in view of the UN human rights system?*

Section 15.2 will compare the Dutch and Belgian monism, for it is in this context that the Judiciaries have to operate. Section 15.3 will compare their case law on the right to food, while in Section 15.4 their reporting behaviour will be compared. In Section 15.5 the national legal practice of the Netherlands and Belgium will be compared with the *tertium comparationis*: the UN perception of the enforceability of the right to adequate food. Then, in Section 15.6, the findings will be traced back to the model as discussed in Chapter 2: the functional method based on equivalent functionalism and on an epistemology of constructive functionalism. Finally, in Section 15.7, some concluding remarks will be made.

15.2 Dutch and Belgian monism compared

The Constitutional context in which a Judiciary operates naturally influences the way domestic Courts respond to the enforceability of the right to adequate food. Both the Netherlands and Belgium could be classified as a monistic system regarding the way in which international standards have effect in the domestic legal order. Also, both countries have limited the monistic effect to those standards that fulfil certain formal and substantive requirements. The formal prerequisites in both countries are that the standard, after being ratified by the Legislature, must have been published (Article 93 Dutch CA, Article 190 Belgian CA). In both countries, the formalities usually do not play a decisive role in assessing whether a standard is enforceable in the domestic legal order. Instead, the substantive requirements are of more importance. In the Netherlands and Belgium, a standard must have direct effect to be invoked in Court (the Dutch Constitution refers to this as standards that are 'binding on all persons'). At this point, the similarities mostly end.

In the Netherlands, Article 93 and 94 CA[1708] regulate the supremacy of international standards over national standards under the (strict) condition that they must be

---

[1708] It must be noted here, that it could be debated whether or not Dutch monism follows directly from Articles 93 and 94 CA (see in this light especially the viewpoints expressed in: M.C. Burkens, H.R.B.M. Kummeling, B.P. Vermeulen and R.J.G.M. Widdershoven, *Beginselen van de Democratische*

binding on all persons, while in Belgium, this is not regulated by Constitution, but rather follows from the case law of the Court of Cassation (basically the Franco-Suisse Le Ski ruling).[1709]

In the Netherlands, it seems to be unclear what exactly is regulated by Articles 93 and 94, especially regarding the question who eventually determines by what criterion whether or not an international standard is directly applicable. While both the Legislature and the Judiciary understand the Constitutional Provisions thus that it is the Judiciary who has a final say, especially because this is stipulated in Article 94 CA, there seems to be a profound role for the Legislature on the matter. After analysing literature and Parliamentary History, it can be concluded that the Legislature, throughout the years, explicitly expressed its view on the direct applicability of international human rights Provisions, and expects that the Judiciary takes its view into serious consideration while ruling on the direct applicability of an international standard. However, there seems to be a difference in hierarchy and nature of the criterion used by the Courts as portrayed in on the one hand literature, Parliamentary History, and occasionally the Courts themselves, and on the other hand, legal reality. Strong evidence could be found that in the assessment whether an international Provision stipulating an ECOSOC right has direct effect, the Courts generally follow the lead of the Legislature. Only in the absence of a clear view of the Legislature on the direct effect, the Courts seem to consider the nature, content and wordings of the Provision. This resulted in a rather blunt distinction between on the one hand civil and political rights that may have direct effect, an on the other hand, economic, social and cultural rights, that are not directly applicable.

In Belgium, the matter of direct effect is dealt with only by the Courts and tribunals operating in the subjective litigation.[1710] The Courts seem to randomly apply an objective criterion and/or a subjective criterion to determine whether or not an international Provision has direct effect. The subjective criterion encompasses the question whether or not it was the intention of the State Parties to adopt a standard that has direct effect, which is mostly demonstrated by the addressee of the Provision (either citizens or States). The objective criterion encompasses the

---

rechtstaat, inleiding tot de grondslagen van het Nederlandse staats- en bestuursrecht (5ᵉ druk), Deventer: Tjeenk Willink, 2001). What is obvious however, after analysing the Parliamentary History on the various Constitutional reforms, is that it was the intention of the Legislator to regulate the supremacy of international standards over national standards by Constitution. Since this is the most important effect of monism, the discussion whether monism in itself follows from the Constitution or from earlier case law (for instance, the 'Grenstractaat Aken' ruling; see: Supreme Court, 3 March 1919, NJ 1919, p. 371), seems to be only of theoretical relevance here.

[1709] Court of Cassation, 27 May 1971, arresten van het Hof van Cassatie, 1971 (p. 959).

[1710] Although the Constitutional Court only considers itself competent in reviewing against international Provisions that have no direct effect since 2002 (Constitutional Court, 41/2002, 20 February 2002), and only reviewed against international standards that are directly applicable in the period 1990-2002 (Constitutional Court, 18/90, 23 May 1990).

question whether a standard is self-executing, and thus no further clarification in national legislation is required. The Constitutional Court holds a special position in the Belgian Constitutional context. Installed in the first place to rule in competence matters between the three equal legislative powers, the Court evolved into a true Constitutional Court, and is competent in reviewing national legislation against the Constitution. Firstly, the Court ruled against Articles 10, 11, and 24 CA, but used Articles 10 and 11 CA in a creative way, and consequently reviewed via those Articles, indirectly, against the other fundamental rights enshrined elsewhere in the Constitutional Act, but also against Articles stipulating International Human Right. Later, the Constitutional Court's competences were broadened in line with legal practice, and the Court could now directly review against *inter alia* all rights stipulated in Chapter II of the Belgian Constitutional Act. However, the case law of the Court shows that it tends to continue involving Articles 10 and 11 in its rulings. If the Constitutional structure of Articles 10 and 11 CA are the 'filter' through which international Provisions have effect in the domestic legal order, they have in that sense, for the Constitutional Court only, a similar function compared to Articles 93 and 94 of the Dutch CA.

While in the Netherlands there appears to be a significant role for the Legislature in the assessment of the enforceability of international legal standards in the domestic legal order, the analysis of Belgian case law demonstrates that the Belgian Legislature seems to play no role of any significance. This might be explained by the relatively complex organisation and purpose of the Belgian Legislature and Judiciary. The organisational structure of the Legislature finds its origin in the differences within Belgium in language, culture and economy. Therefore, to guarantee an optimal democratic process, the Legislature consists of three Legislators, whose competences are based on the principles of equality and exclusiveness. It is the original purpose of the Constitutional Court (then referred to as Court of Arbitration) to settle disputes amongst the Legislators regarding their competences, and thus where necessary to nullify legislation that is in contradiction with the Constitution. Gradually, those reviewing competences also implied the indirect review against international Provisions. Comparing this process with the Dutch, it can be concluded that the Constitutional Court has a much more autonomous role compared to the Dutch Courts when it concerns the application of international Provisions. This may also be related to the fact that in the Netherlands, Constitutional review is exclusively a competence of the Legislature (Article 120 CA), while in Belgium this competence is attributed to the Constitutional Court. In general, it can thus be concluded that in Belgium there is a different relation between Legislature and Judiciary, in which the Constitutional Court plays a crucial role compared to the Netherlands. Due to the Dutch, rather obscure Constitutional structure stipulated in Articles 93 and 94 CA, the Dutch Legislature is able to greatly influence the case law of the Courts when it concerns the effect of international Provisions in the domestic legal order.

In general it can thus be concluded that these Constitutional circumstances and developments play a profound role in the enforceability of international standards, and thus not in the first place, the content of the international Provisions that are invoked in Court.

15.3 Dutch and Belgian case law compared

Comparing Dutch case law on the right to adequate food with the Belgian, it can be concluded in general, that in the Netherlands there is a rather clear distinction between civil and political rights, that may be directly applicable, and economic, social and cultural rights, that are not. ECOSOC rights are sometimes even all treated the same way in one instance. The right to food is thus considered not to be directly applicable. Only occasionally, Article 27 is used as an interpretative standard. In Belgium, the situation is somewhat more differentiated. In the subjective litigation there are hardly any cases in which right to food Provisions play a significant role and it seems that in general ECOSOC rights are not directly applicable. However, there appear to be incidental exceptions in which at least some effect is granted to ECOSOC Provisions. In Belgian case law the contrast between civil and political rights on the one hand, and ECOSOC rights on the other hand appears not to be as explicit as in the Dutch.[1711] A contributing factor for this could be that traditionally, the right to education (in particular as embedded in Articles 24 CA and 13 ICESCR), is a sensitive issue in Belgium, and played an important role in the early rulings of the Constitutional Court, while also the Council of State seems to grant some effect to Article 13 ICESCR. In the objective litigation, the matter of direct effect is not an issue, but the Constitutional Court has developed a vivid and extensive case law, involving the indirect review against international Provisions stipulating the right to food.

In both countries, indirect application of the right to food is mostly done in the context of a balancing of interests. The national policies with the aim to restrict immigration are balanced with the individual interests of the foreigner. However, such weighing-up may stand in the way of truly assessing the content of a human right, and makes it subdued to legal uncertainty. For instance, this resulted in the Netherlands in the fact that in case of a legally residing foreign minor, Article 27 ICRC is applied indirectly based on its core content, while in case of an illegally residing foreign minor, the right is (mostly) declared to be not directly applicable, without considering the content. A similar remark could be made about the case law of the Constitutional Court, where the non-discrimination principles are only applied in a balancing of interests. However, in this balancing of interests it seems

---

[1711] However, it is in this light remarkable that also in Belgium, the Council of State considers the right to strike to have direct effect (Article 8 ICESCR), similar to the Dutch case law, on Article 6 ESC. The right to strike is one of the few ECOSOC rights that imply negative obligations of States as their core content, instead of positive duties.

that in the Dutch cases, Article 27 (among others) has a noticeable impact on the rulings, while it can be questioned whether invoked international Provisions in the rulings of the Constitutional Court offer a broader protection, or have any effect of significance on the verdicts besides the invoked national Provisions.

Furthermore, in the case law in both countries in which Provisions are invoked stipulating the right to food, the Courts prefer to solve the matter by applying domestic legislation instead of international standards, and therefore, the case law could be characterised as 'focussed inwards'. No evidence could be found that the authoritative legal interpretations on the invoked standards, as adopted by the treaty bodies, were taken into serious consideration by the Dutch or Belgian Courts.[1712]

15.3.1 ICESCR

In the Netherlands and Belgium, no evidence could be found that Article 11 is enforceable in the Courts, in a way that the Article can be successfully invoked by an individual. Amongst the Dutch Courts, the Provision is also not used as an interpretative norm. In Belgium, the Constitutional Court frequently reviewed indirectly against the Provision.

15.3.2 ICRC

It is remarkable, that in both countries the rights enshrined in the ICRC seem to have more influence compared to other international human rights treaties. In the Netherlands, increasingly, case law seems to emerge in which ICRC standards, including Article 27, are used as an interpretative norm, in some cases via Article 2 or 3 ICRC. While technically the standards invoked are not directly applied, they seem to have a significant effect on the verdict. Here, the Articles appear to have a clear added value on top of domestic legislation. Also in Belgium the Courts seem to be more merciful when the position of children is involved, which is demonstrated by the case law of the Constitutional Court that reviewed occasionally against *inter alia* Article 27 ICRC.

15.3.3 CEDAW

It is interesting to learn that in both countries, there is hardly any case law of significance on the enforceability of CEDAW provisions that stipulate ECOSOC rights. This contrasts with the extensive reporting history of both countries on the implementation of the CEDAW Provisions.

[1712] Exception to that rule however is the Dutch ruling Central Court of Appeal, 8 August 2005, LJN AU0687, in which a brief reference was made to the General Comments of the Committee on the Rights of the Child.

15.3.4 Foreigners and the right to food

Perhaps it is not surprising that in two relatively prosperous countries, both having implemented an extensive system of social security, the right to food has been invoked most often by those whose access to that social security system is not self-evident: foreigners, and more in particular, illegally residing foreigners. In both countries this has led to extensive case law, in which the policies regarding foreigners were often contested and the Courts had to assess whether they are in compliance with the human rights duties of the Netherlands and Belgium. In both countries, the Courts have ruled that the aliens policies, mainly excluding illegally residing foreigners from access to social benefits, are a suitable means to achieve the purpose of restricting immigration. However, the legal reasoning resulting in the adoption of this principle of the Dutch Courts differs from Belgian Constitutional Court, when it concerns the assessment whether or not international human rights are violated. The Dutch Courts merely stated that the invoked Provisions, including Articles 11 ICESCR and 27 ICRC, are not directly applicable due to various reasons as discussed in Chapter 4, and therefore no individual rights can be distilled from the Articles. The Belgian Constitutional Court however reasoned differently and argued that the right to an adequate standard of living could not reasonably be understood without limitations, and excluded the illegally residing alien from the *ratione personae* of the Provision, considering that Belgium was not responsible for them.

However, it are the exceptions to the principle that limiting access to social benefits to illegally residing aliens is a justifiable means to restrict immigration, in which human rights seem to have some effect. In the Netherlands, invoking especially Article 27 ICRC, has led to the exception that foreign children, awaiting a decision on their application for a residence permit, should have access to the means necessary for adequate living conditions. However, this does not apply to illegally residing children, although a development in case law seems to emerge in which the COA is, based indirectly on *inter alia* Article 27 ICRC, not permitted to remove illegally residing children and their families from their facilities. In Belgium, as a result of the case law of the Constitutional Court, there are more exceptions to the principle. In a chain of case law, in which national Provisions, whether or not read in conjunction with international Provisions (including Articles 11 ICESCR and 27 ICRC) were invoked, several exceptions were accepted. Aliens who were in the medical impossibility to leave Belgian territory, or appealed to the (first) order to leave the territory were granted entitlements to social benefits (Article 11 ICESCR cases). Furthermore, the children of illegally residing parents were granted the necessary social benefits in kind. Later, also the Public Centres for Social Welfare had to take into account the fact that when the child was accompanied by parents, it had a right to family life. Therefore, in granting social benefits, this should be taken into consideration (Article 27 ICRC cases).

To conclude, it appears that the Belgian Constitutional Court allows more exceptions to the principle that illegally residing aliens have no entitlements to social benefits, compared to the Dutch Courts. It might be tempting here to conclude that the Constitutional Court's rulings are more humane. However, a thorough comparison of all existing legislation and correlating case law in the field of aliens law would be necessary to determine whether one of the two countries has a more favourable approach towards (illegally residing) aliens. Only an analysis of Court cases on the enforceability of international standards stipulating the right to food cannot possibly lead to such a conclusion. However, of course the basic principle adopted in both countries that illegally residing aliens have no entitlements to social benefits, not even to basic needs to lead a humane life, can be questioned seriously. Especially when the interest of the Country to restrict immigration appears to be more important than the interest of the individual to have access to *inter alia* adequate food.

15.4 The Dutch and Belgian reports compared

Analysis of the various reporting cycles of the Netherlands and Belgium demonstrates that both countries have the tendency to focus primarily on domestic policies, legislation and initiatives that positively contribute to the implementation of the rights stipulated in the related international treaties. It can be seriously questioned whether the reporting duties are taken seriously. Firstly, the reports are very often submitted generously overdue. Secondly, the quality of the reports is questioned more than once by the various Committees, since not all the relevant information seems to be included and especially matters of concern are omitted. Thirdly, more than once improvement, reconsideration or information is promised, but the promises are hardly fulfilled. Furthermore, both Governments seem to refer to the domestic Courts when the matter of direct applicability/direct effect is discussed, and argue that it is their responsibility to assess whether or not an international Provision is directly applicable/has direct effect. While this might be a reasonable reflection of legal reality in Belgium, this is questionable in the case of the Netherlands, considering the significant influence of the Legislature on the enforceability of international standards. Also, the matter of asylum-seekers is discussed in detail, and often a primary point of concern for the Committees. In general, it can be concluded that both countries are painting an overly positive picture of the implementation of human rights Provisions, especially when it concerns the enforceability of these rights. The Committees appear to be poorly informed on the particularities of the Constitutional context in which the enforceability of the human rights are discussed. This leads to occasional misunderstandings about the functioning of Articles 93 and 94 CA in the Netherlands, and about the complex Constitutional organisation in Belgium, while the Delegations, when clarifying national case law appear to be bending legal reality into questionable directions, or even providing the Committees with incorrect information. The NGOs' parallel reports and later assigned country rapporteurs are only partly capable correcting this. In this light, it is also noteworthy that there seems to be insufficient time during the sessions to exhaustively discuss all the relevant issues in-depth.

Also some differences in reporting behaviour could be found. Firstly, in the Belgian reporting cycles the Constitutional structure is a recurring cause for debate. The Committees are concerned that the structure stands in the way of a coherent and uniform approach towards the implementation of human rights. This is hardly the case in the Netherlands, although the structure of Articles 93 and 94 CA often requires some further clarification. Secondly, the Netherlands seem to report more specifically on the right to food compared to Belgium, especially in view of Article 11 ICESCR. Thirdly, the direct effect of the ICRC is hardly discussed in the reporting cycles of the Netherlands, while this was a point of concern during the Belgian reporting cycles. Fourthly, the difference between civil and political rights on the one hand and ECOSOC rights on the other hand, regarding their enforceability, is more emphasised by the Netherlands compared to Belgium. The Netherlands frequently emphasise the different nature of ECOSOC rights, and underline that due to their impreciseness most ECOSOC rights are not directed at States. The Belgian reports are less blunt. In view of the ICRC and CEDAW, a possible distinction between two different types of rights was never discussed during the Belgian reporting cycles. Instead, as discussed in Chapter 13, the direct effect of the treaty in general was debated: a discussion that was mostly led by misunderstandings and miscommunications. Regarding the direct effect of ICESCR, Belgium focussed on the effect of Article 2 and stated that the progressive realisation implied a margin of discretion for the Legislature, resulting in the fact that the ICESCR Articles had no direct effect, although Belgium was bound by the standstill effect of the treaty. Fifthly, perhaps also due to the bluntness of the Dutch in their reports on the distinction between types of rights and the fact that some direct questions of the Committees (especially the CESCR) were only partly answered or ignored, the atmosphere during the reporting cycles becomes more stiff along the way. This development cannot be encountered that clearly when observing the Belgian reporting cycles.

15.5 The enforceability of the right to food in the Netherlands and Belgium in view of the UN human rights system

As discussed in Chapter 3, it is difficult to exactly establish whether the right to food is considered an enforceable right in the UN arena, when considering the viewpoints of all bodies that are related to the matter, including those in which intergovernmental decision-making procedures are followed based on the principle of consensus. However, the bodies that are installed specifically to contribute to the realisation of ECOSOC rights in general, and more particularly the right to food, such as the Committees and the Special Rapporteurs, have frequently addressed the urgency of enforceable rights, and certainly consider a core content of the right to food justiciable, or even self-executing.

In general, it can be concluded that the right to adequate food as embedded in the UN human rights system is not enforceable in the domestic Dutch and Belgian Courts by individuals, in the sense that the right is directly applicable in the domestic legal order, constituting subjective rights. Especially the Dutch Courts, in line with the viewpoints of the Legislature, generally seem to make a clear distinction in this between civil and political rights on the one hand, and ECOSOC rights on the other hand. As discussed in Chapter 3, such an approach is widely criticised in UN circles. However, in both the Netherlands and Belgium, elements of the right to food may have an indirect effect in case law. To put this in the terms used by the CESCR in its General Comments 3 and 9: while both countries technically have a monistic system in which it could be possible, the rights are considered not to be self-executing, but there seems to be some room for justiciability, although these cases seem to be the exception to the rule. As discussed in Chapter 3, the CESCR and the Special Rapporteurs on the right to food, consider that the right has at least a core content, consisting of a minimum substance, and a non-discrimination principle. This then should at least have some effect in the domestic Court (either as justiciable or self-executing rights).

In this light, the Dutch rulings on Article 11 ICESCR, consistently denying the right of any effect at all in the Court seems to be in contradiction with the understanding in the UN arena. Especially where the Courts use the argument that the Provision is not specific enough to distill subjective rights from its nature, content and wordings. The case law on Article 27 ICRC seems to be more in line, indeed guaranteeing a core content of the right to food in some specific situations. However, illegally residing minor asylum-seekers have been mainly excluded from this treatment, which appears to be in contradiction with the non-discrimination principle. On the other hand, new developments in case law also show some mercy for the latter category.

The Belgian rulings in the subjective litigation, in the spare cases that exist on the right to food (two rulings of the Council of State on Article 27 ICRC), also do not seem to consider the core content of the right. They merely rule that the right has no direct effect, also using the argument that the right is not specific enough. Furthermore, the rulings of the Constitutional Court on the one hand seem to violate the principle of non-discrimination when excluding illegally residing aliens from the *ratione personae* of the right to food. On the other hand, in the exceptions to this rule, the Court seems to consider the non-discrimination principle in determining whether the alleged discrimination is indeed against the law, or justified. This reasoning is mainly caused by the habit to review indirectly via Articles 10 and 11 CA, in a balancing of interests. The side effect of this is that there seems to be less focus on the core content of the right.

15.6 Some comparative considerations

As discussed in Chapter 2, in terms of comparative law the commonly shared social problem was defined as *'the right to food and ECOSOC rights in general can be understood to imply immediate state obligations that should be enforceable through domestic Courts. There seems to be at least a core content, consisting of a minimum substance without which the right would be stripped of its raison d'être, and a non-discrimination principle, that should be justiciable, or even self-executing.'* The national social institutions that respond to this problem are the domestic Courts through their case law, and the Governments through their periodic reports. It has already been observed that those responses (the response of the Courts compared to the response of the Government) differ in both countries. The response in itself is heavily influenced by the Constitutional circumstances: in the Netherlands, the structure as embedded in Articles 93 and 94 CA, and in Belgium, the complex organisation and development of their *trias politoca*. Considering this in relation to the analysed case law, in both countries it is thus doubtful whether the content of the internationally embedded human rights play a primary role in their effect in the domestic legal order. Furthermore, as it seems, most cases in which the right to food, or ECOSOC rights in general are invoked, concern foreigners, mostly illegally residing on Dutch or Belgian territory. In both countries, any possible effect of an international Provision stipulating the right to food seems to emerge from a balance of interests, in which the domestic policies to restrict immigration are the counterbalance of the individual interest of the foreigner. This leads to the impression that the Courts through their case law respond to another shared social problem, that is: the restriction of immigration.

15.7 Concluding remarks

This research found its origin and inspiration in the analysis of two Court rulings, one Dutch and one Belgian, in which entitlements based on the right to food were denied to a claimant. This sparked the question which added value such a human right would have in these two countries, which have the means to ensure adequate food to all within their jurisdiction. In line with the nature of human rights, one would expect that such rights basically would have to serve as a safety net, to catch those individuals that somehow fall through the – mostly already very solid and extensive – domestic safety nets. The more so, as both countries communicate in their reports on the implementation of various treaties to fulfill all their duties the right to food implies. In short, the question was raised what the two countries actually do, what they say they do, and what they should do regarding the enforceability of the right to adequate food.

This research has shown that unfortunately it is the main rule that the right to food does not serve as a safety net, and it is hardly possible to invoke the right in Court in both the Netherlands and Belgium. I say 'hardly', because there are some developments in which the Courts do show some compassion for those people who are in most distressing situations, and allow international provisions that stipulate the right to food to have some (indirect) effect in their verdicts. It is those exceptions to the rule that we could be hopeful for. In the Netherlands that is mostly the increasing use of the ICRC as an interpretative standard, whereas in Belgium it is the increasing indirect review against international standards through Constitutional Provisions. The general impression however is that it are the coincidental Constitutional circumstances that mainly determine the enforceability of the right to food, rather than the actual content of the right in itself.

This research has also shown that there is a difference between the legal practice in Belgium and the Netherlands on the one hand, and what is reported on this practice on the other hand. Both countries paint a positive picture of their performances in human rights implementation, while there is little space for self-evaluation or self-criticism regarding the implementation of ECOSOC rights in general, and the right to food in particular. Both countries appear to have hardly any ambition to improve the implementation of the rights enshrined in the treaties they ratified, and certainly not because it is part of their States' duties. In the international arena thus both countries want to show off their good intentions, whilst those differ from the domestic legal practices. An interesting contradiction, to underline this impression, is that the Netherlands and Belgium signed the OP-ICESCR, involving complaints procedures, while the ICESCR Provisions in their domestic legal order have no direct effect.

Lastly, while it might be disputed what countries should do in this light, the treaty bodies and the Special Rapporteurs, with sound arguments consequently defend the position that at least a core substance of the right to food (including a non-discrimination principle) should be justiciable or even self-executing in the Courts. While in both countries developments can be observed that tend towards more justiciability, in the large majority of all cases the right to food is not considered to be a justiciable right, let alone a right that is self-executing.

Therefore, it can be concluded that the countries do not do what they should do, and do not say what they really do.

16. Recommendations

Based on the previous research, the following recommendations can be made

16.1 The Netherlands

Currently, the Constitutional system as embedded in Articles 93 and 94 leads to legal uncertainty that has lasted for decades now. This research shows that the effect of the indistinctiveness in the Articles is more extensive than one might expect at first glance, and truly hinders the Courts in seriously considering the effect of the invoked international standards based on their substance. It might be a welcome development if the Constitutional Provisions would be amended in such way that it is unambiguously clear what the role of the Legislature is with regard to the assessment of the Courts in determining whether or not an international Provision is binding on all persons. Since amendments to the Constitutional Act are not easy to achieve, meanwhile it would be helpful if the Parliamentarians, Senators and Ministers who are responsible for adopting new ratification Bills to international treaties stipulating human rights would study the Constitutional structure of Articles 93 and 94 and the developments in case law in the field of human rights in-depth.

The Courts are reluctant in recognising direct effect of international Provisions stipulating ECOSOC rights, in line with the opinion of the Legislature. The case law on Article 27 ICRC, in which the Provision is used as an interpretative standard, shows that there is more between direct applicability and non-applicability. While it is not perfect and certainly also has its disadvantages, this route could inspire more Courts in their verdicts and open the door to a more differentiated case law that does more justice to the nature, content and wordings of the human rights Provisions. Perhaps the indirect application through non-discrimination Provisions, as used by the Belgian Constitutional Court, might also be a source for inspiration.

In the periodic reports, submitted to the UN treaty bodies, an additional focus on points for improvement as well as accurately reporting on the Dutch case law – including the recent developments in the indirect application of ICRC Provisions – might contribute to a more fruitful dialogue with the different Committees.

16.2 Belgium

It is not always clear what the exact role is of international Provisions when the Constitutional Court reviews against a Constitutional standard in conjunction with an international standard. It might enrich the case law when the Court would reach slightly more 'outwards' and is more precise on what elements of the international Provision are considered in what exact context. Besides the fact that this would do more justice to internationally recognised human rights

Provisions, this might also even further differentiate the balancing of interests in view of the general principle of non-discrimination.

Through the prejudicial procedure, but also through annulment procedures, the effects of the verdicts of the Constitutional Court are increasingly having an impact on subjective rights of individuals. In that light, it seems that nothing stands in the way for the Courts operating in the subjective litigation to simultaneously apply (if need be: indirectly) international Provisions. This might lead to a more coherent approach amongst the Courts, especially the Court of Cassation and the Council of State, in their case law on international Provisions, in particular those that stipulate ECOSOC rights.

In the periodic reports, submitted to the UN treaty bodies, an additional focus on points for improvement might contribute to a more fruitful dialogue with the different Committees. In addition, it might be helpful if this dialogue is based on accurate information on the functioning of the domestic Judiciary. Especially the exact functioning of the Courts of Last Instance (in particular the Constitutional Court), the subdivision in an objective and subjective litigation, its nuances in legal practice, and the effects of the verdicts of the Courts on individuals might be further clarified, as well as the actual case law produced by the Courts.

16.3 The Committees of the UN

While the argument of Courts of the Netherlands or Belgium may often be that the right to food is too imprecise to apply directly, this research shows that it can be questioned whether that is the real argument. The characteristics of the monistic system, mostly regulated in the Constitution or Constitutional law of the country, have an influence that is not to be underestimated. Furthermore, in case of indirect application, policy considerations and the balancing of national interests seem to be the major influence on the effect of the right to food. It is doubtful therefore, whether more attempts to further clarify the right to food in the international arena will be of any help in this case. Instead, a coordinated focus on Constitutional obstacles in the direct effect of international Provisions, or best practices in indirect applications (for instance through non-discrimination Provisions) might be more effective. Furthermore, the question can be asked whether the right to food as a justiciable or self-executing right is a realistic goal to strive for. While there are certainly good arguments to consider that at least a core content of the right to food is a justiciable right, the consistent legal practice of the Netherlands and Belgium shows otherwise. Since human rights naturally stand for entitlements, the final goal might still be a right to food that is (partly) self-executing, but it might be helpful in the communication with Member States to also focus on the indirect solutions. While indirect application of human rights is far from perfect, and as demonstrated in Chapter 15.3 may be blurring the content the right should have, it is a step forward in the right direction.

16.4 Legal comparists

While the current debates on comparative law are extremely vivid, inspiring, friendly and all-inclusive, they did not yet result in a widely accepted method that can be used by researchers, especially not in the field of public law, or more in particular, human rights law. Perhaps the intense debates might shift somewhat to a focus on actually performing a comparison, comparing best practices, and didactics on how to teach students to compare. In other words: it might yield more practical result when considering comparative law slightly more as a means to an end, instead of a goal in itself.

Bibliography

References

Akkermans, P.W.C., Bax, C. and Verhey, L., 2005. Grondrechten, grondrechten en grondrechtsbescherming in Nederland. Kluwer, Heerlen, the Netherlands.

Alen, A. (ed.), 2005. 20 jaar Arbitragehof. Kluwer, Mechelen, Belgium.

Alen, A. and Muylle, K., 2008. Compendium van het Belgisch staatsrecht, syllabusuitgave (2ᵉ uitgave). Kluwer, Mechelen, Belgium.

Alston, P. and Tomasevski, K. (eds.), 1984. The right to food. International Studies in Human Rights, Utrecht, the Netherlands.

Amin Al-Midani, M., Cabanettes, M. (translation) and Akram, S.M. (revision), 2006. Arab charter on human rights 2004. Boston University International Law Journal 24: 147-164.

Bellekom, Th.L., Heringa, A.W., Van der Velde, J., and Verhey, L.F.M., 2007. Compendium van het staatsrecht (10th Ed.). Kluwer, Deventer, the Netherlands.

Bouckaert, S., 2007. Documentloze vreemdelingen, grondrechtenbescherming doorheen de Belgische en internationale rechtspraak vanaf 1985. Maklu, Antwerp-Apeldoorn, Belgium-the Netherlands.

Burkens, M.C., Kummeling, H.R.B.M., Vermeulen, B.P. and Widdershoven, R.J.G.M., 2001. Beginselen van de democratische rechtstaat, inleiding tot de grondslagen van het Nederlandse staats- en bestuursrecht (5th Ed.). Tjeenk Willink, Deventer, the Netherlands.

De Graaf, J.H., Limbeek, M.M.C., Bahadur, N.N. and Van der Mey, N., 2012. De toepassing van het internationaal verdrag inzake de rechten van het kind in de Nederlandse Rechtspraak. Ars Aequi Libri, Nijmegen, the Netherlands.

Devroe, W., 2010. Rechtsvergelijking in een context van europeanisering en globalisering. Uitgeverij Acco, Leuven, Belgium.

Glenn, H.P., 2007. Legal traditions of the world. Oxford University Press, Oxford, UK.

Gorlé, F., Bourgeois, G., Bocken, H., Reyntjens, F., De Bondt, W. and Lemmens, K., 2007. Rechtsvergelijking, Kluwer, Mechelen, Belgium.

Hospes, O. and Hadiprayitno, I. (eds.), 2010. Governing food security, law, policies and the right to food. Wageningen Academic Publishers, Wageningen, the Netherlands.

Hospes, O. and Van der Meulen, B. (eds.), 2009. Fed up with the right to food. Wageningen Academic Publishers, Wageningen, the Netherlands.

Jackson, V.C., 2005. Constitutional comparisons: convergence, resistance, engagement. Harvard Law Review 119: 109.

Kalverboer, M. and Zijlstra, E., 2006. Kinderen uit asielzoekersgezinnen en het recht op ontwikkeling, het belang van het kind in het vreemdelingenrecht. Uitgeverij SWP, Amsterdam, the Netherlands.

Kent, G. (ed.), 2008. Global obligations for the right to food. Rowman and Littlefield Publishers Inc., Lanham, MD, USA.

Kent, G., 2005. Freedom from want: the human right to Food. Georgetown University Press, Washington, DC, USA.

Knuth, L. and Vidar, M., 2011. Constitutional and legal protection of the right to food around the world. Right to food studies. FAO, Rome, Italy.

Bibliography

Knuth, L., 2009. The right to food and indigenous people, how can the right to food help indigenous people? FAO, Rome, Italy.

Koekoek, A.K. (ed.), 2000. De Grondwet, een systematisch en artikelgewijs commentaar (3rd Ed.). Tjeenk Willink, Deventer, the Netherlands.

Kortmann, C.A.J.M., 2008. Constitutioneel recht (6th Ed.). Kluwer, Deventer, the Netherlands.

Law, D.S., 2005. Generic constitutional law. Minnesota Law Review 89: 652.

Maes, G., 2003. De afdwingbaarheid van sociale grondrechten. Intersentia, Antwerp, Belgium.

Mattei, U., 1997. Three patterns of law: taxonomy and change in the world's legal system. The American Journal of Comparative Law 45: 5-44.

McGovern, G., 2001. Third freedom, ending hunger in our time. Rowman and Littlefield Publishers Inc., Lanham, MD, USA.

Moeckli, D., Shah, S., Sivakumaran, S. (eds.) and Harris, D. (cons. ed.), 2010. International human rights law. Oxford University Press, Oxford, UK.

Örücü, E. and Nelken, D. (eds.), 2007. Comparative law, a handbook. Hart Publishing, Portland, OR, USA.

Pintens, W., 1998. Inleiding tot de rechtsvergelijking. Universitaire Pers Leuven, Leuven, Belgium.

Prakke, L. and Kortmann, C. (eds.), 2004. Constitutional law of 15 EU Member States. Kluwer, Deventer, the Netherlands.

Prakke, L. and Kortmann, C. (eds.), 2004. Constitutional law of 15 EU Member States, Kluwer, Deventer, the Netherlands.

Reimann, M. and Zimmermann, R. (eds.), 2006. The Oxford handbook of comparative law. Oxford University Press, Oxford, UK.

Rimanque, K., 2005. De grondwet toegelicht, gewikt en gewogen. Intersentia, Antwerp, Belgium.

Ruitenberg, G.C.A.M., 2003. Het internationaal kinderrechtenverdrag in de Nederlandse rechtspraak. IVRK series 1. Uitgeverij SWP, Amsterdam, the Netherlands.

Runge, C.F., Senauer, B., Pardey, P.G. and Rosegrant, M.W., 2003. Ending hunger in our lifetime, food security and globalization. John Hopkins University Press, Baltimore, MD, USA, London,UK.

Sachs, J.D., 2006. The end of poverty, economic possibilities for our time. Penguin Books, London, UK.

Slingenberg, C.H., 2006. Illegale kinderen en recht op bijstand in het licht van het IVK. In: Migratierecht 2006-02, Forum, Utrecht, the Netherlands, pp. 54-57.

Ssenyonjo, M., 2009. Economic, social and cultural rights in international law. Hart Publishing, Portland, OR, USA.

Van der Pot, C.W., 2006. (adapted by Elzinga, D.J. and De Lange, R.). Handboek van het Nederlandse staatsrecht (15th Ed.). Kluwer, Deventer, the Netherlands.

Van Eeckhoutte, D. and Vandaele, A., 2002. Doorwerking van internationale normen in de Belgische rechtsorde. Working paper no. 33. Instituut voor Internationaal Recht K.U. Leuven, Leuven, Belgium.

Van Laer, C.J.P., 1998. The applicability of comparative concepts. Electronic Journal of Comparative Law 2.2. Available at: http://www.ejcl.org/22/abs22-1.html.

Vande Lanotte, J., Bracke, S. and Goedertier, G., 2010. België voor beginners. Die Keure, Brugge, Belgium.

Vasak, K., 1977. Human rights: a thirty-year struggle: the sustained efforts to give force of law to the universal declaration of human rights. UNESCO Courier 30:11. United Nations Educational, Scientific, and Cultural Organisation, Paris, France.

Vlemminx, F.M.C., 2002a. De autonome rechtstreekse werking van het EVRM, de Belgische en Nederlandse rechtspraak over verzekeringsplichten ingevolge het EVRM. Kluwer, Deventer, the Netherlands.

Vlemminx, F.M.C., 2002b. Een nieuw profiel van de grondrechten, een analyse van de prestatieplichten ingevolge klassieke en sociale grondrechten. Boom Juridische uitgevers, The Hague, the Netherlands.

Vonk, G., 2006. Ongewenste kinderen; opmerkingen naar aanleiding van CRvB 24 januari 2006. Sociaal Maandblad Arbeid: 131-134. Available at: http://rechten.eldoc.ub.rug.nl/FILES/root/2008/ongekiopo/2006_Ongewenste_kinderen.pdf.

Wachira, G.M., 2008. African court on human and peoples' rights: ten years on and still no justice. Minority Rights Group International, London, UK.

Wahlgren, P., 2000. Legal reasoning, a jurisprudential model. Scandinavian Studies in Law 40: 199-282.

Wernaart, B.F.W., 2010. The plural wells of the right to food. In: Hospes, O. and Hadiprayitno, I. (eds.) Governing food security, law, policies and the right to food. Wageningen Academic Publishers, Wageningen, the Netherlands

Whitman, J.Q., 2000. Enforcing civility and respect: three societies. 109 Yale law journal, Faculty Scholarship Series, paper 646.

Ziegler, J., Golay, C., Mahon, C. and Way, S.-A., 2011. The fight for the right to food, lessons learned. The graduate institute, Geneva, Switzerland.

Zweigert, K. and Kötz, H., 1998. An introduction to comparative law. Oxford University Press, Oxford, UK.

International legal documents

217 A (III), 10 December 1948, the Universal Declaration on Human Rights.

A/4354, 20 November 1959, General Assembly Resolution 1386 (XIV), Declaration on the Rights of the Child.

A/56/210, 23 July 2001, Jean Ziegler, report of the Special Rapporteur on the right to food to the General Assembly.

A/57/356, 27 August 2002, Jean Ziegler, report of the Special Rapporteur on the right to food to the General Assembly.

A/58/330, 28 August 2003, Jean Ziegler, report of the Special Rapporteur on the right to food to the General Assembly.

A/59/385, 27 September 2004, Jean Ziegler, report of the Special Rapporteur on the right to food to the General Assembly.

A/60/2005, 12 September 2005, Jean Ziegler, report of the Special Rapporteur on the Right to food to the General Assembly.

A/C.3/SR.1266, 1963.

A/CONF.151/26/Rev.1 (Vol. 1), Rio de Janeiro, 3-14 June 1992, Agenda 21.

Bibliography

A/CONF.157/23, 12 July 1993, the world conference on human rights, Vienna declaration and programme of action.

A/CONF.177/20, Beijing, China, 4-15 September 1995, Report of the fourth world conference of women.

A/HRC/1/L.10, 29 June 2006, Human Rights Council Resolution 2006/3.

A/HRC/6/WG.4/2, 23 April 2007, draft Optional Protocol to the International Covenant on Economic, Social and Cultural Rights.

A/HRC/7/5, 10 January 2008, 10 January 2008, Jean Ziegler, report of the Special Rapporteur on the right to food to the Human Rights Council.

A/HRC/8/WG.4/2, 24 December 2007, revised draft Optional Protocol to the International Covenant on Economic, Social and Cultural Rights.

A/HRC/8/WG.4/3, 25 March 2008, (second) revised draft Optional Protocol to the International Covenant on Economic, Social and Cultural Rights.

A/HRC/9/23, 8 September 2008, Olivier de Schutter, report of the Special Rapporteur on the right to food to the Human Rights Council.

A/HRC/17/36, 16 May 2011, Report of the Open-ended Work Group on an Optional Protocol to the Convention on the Rights of the Child to provide a communications procedure.

A/HRC/RES/5/1, 18 June 2007, Human Rights Council Resolution.

A/HRC/RES/6/2, 27 September 2007, Human Rights Council Resolution.

A/HRC/RES/6/102, 27 September 2007, Human Rights Council Resolution.

A/HRC/RES/7/14, 27 March 2008, Human Rights Council Resolution.

A/HRC/RES/7/22, 28 March 2008, Human Rights Council Resolution.

A/HRC/RES/13/4, 14 April 2010, Human Rights Council Resolution.

A/HRC/RES/16/2, 8 April 2011, Human Rights Council Resolution.

A/HRC/WG.6/1/NLD/1, 7 March 2008, the Netherlands, first Universal Periodic Report.

A/RES/21/2200, 16 December 1966, International Covenant on Economic, Social and Cultural Rights, International Covenant on Civil and Political Rights and Optional Protocol to the International Covenant on Civil and Political Rights.

A/RES/32/130, 16 December 1977, Alternative approaches and ways and means within the United Nations' system for improving the effective enjoyment of human rights and fundamental freedoms.

A/RES/34/180, 18 December 1979, Convention on the Elimination of All Forms of Discrimination Against Women.

A/RES/44/25, 20 November 1989, Convention on the Rights of the Child.

A/RES/45/158, 18 December 1990, The International Convention on the Protection of the Rights of All Migrant Workers and Members of Their Families.

A/RES/54/4, 15 October 1999, Optional Protocol to the Convention on the Elimination of All Forms of Discrimination Against Women.

A/RES/60/251, 3 April 2006, 3 April 2006, Human Rights Council.

A/RES/61/106, 24 January 2007, The Convention on the Rights of Persons with Disabilities.

A/RES/61/295, 2 October 2007, United Nations Declaration on the Rights of Indigenous Peoples.

A/RES/63/117, 10 December 2008, Optional Protocol to the International Covenant on Economic, Social and Cultural Rights, adopted by the General Assembly.

A/RES/66/138, 27 January 2012, Optional Protocol to the Convention on the Rights of the Child on a communications procedure.

A/RES/2106 (XX), 21 December 1965, The International Convention on the Elimination of All Forms of Racial Discrimination.

A/RES/2263 (XXII), 9 November 1967, Declaration on the elimination of discrimination against women.

A/RES/260 (III) (A), 9 December 1948, International Convention on the Prevention and Punishment of the Crime of Genocide.

A/RES/3180 (XXVIII), 17 December 1973, Convening the World Food Conference.

A/RES/3348 (XXIX), 17 December 1974, Endorsing the Universal Declaration on the Eradication of Hunger and Malnutrition.

A/RES/429 (IV), 14 December 1950, Draft Convention relating to the Status of Refugees.

African Charter on Human and Peoples' Rights and related documents

- OAU Doc. CAB/LEG/67/3 rev. 5, 21 I.L.M. 58, June 1981, African Charter on Human and Peoples' Rights.
- Additional Protocol to the African Charter on Human and Peoples' Rights on the rights of Women in Africa, 11 July 2003.
- Additional Protocol to the African Charter on Human and Peoples' Rights on the establishment of an African Court on Human and Peoples' Rights, 10 June 1998.
- African Union Convention for the Protection and Assistance of Internally Displaced Persons in Africa (Kampala Convention), 22 October 2012.
- OAU Doc. CAB/LEG/24.9/49, July 1990, African Charter on the Rights and Welfare of the Child.

The American Convention on Human Rights, 22 November 1969, adopted at the Inter-American Specialised Conference on Human Rights.

Arab Charter on Human Rights, May 22, 2004, League of Arab States.

Asian Human Rights Charter, a peoples' charter, declared in Kwangju, South Korea, 17 May 1998.

Cairo Declaration on Human Rights in Islam, 5 August 1990, adopted at the Nineteenth Islamic Conference of Foreign Ministers.

Charter of the United Nations, 1945.

Commission on Human Rights Resolution 2000/10, 17 April 2000.

Commission on Human Rights Resolution 2001/25, 20 April 2001.

Commission on Human Rights Resolution 2001/30, 20 April 2002.

Commission on Human Rights Resolution 2002/24, 22 April 2002.

Council Directive 2004/83/EC, 29 April 2004, on minimum standards for the qualification and status of third country nationals or stateless persons as refugees or as persons who otherwise need international protection and the content of the protection granted.

Council Resolution 97/C 221/03 of 26 June 1997 on unaccompanied minors who are nationals of third countries.

Declaration of the World Food Summit: five years later, 10-13 June 2002.

E/1989/22, annex III at 87 (1989), 24 February 1989, CESCR, General Comment 1: Reporting by States Parties.

E/1991/23, annex III at 86 (1991), 14 December 1990, CESCR, General Comment 3: The Nature of States Parties' Obligations.

E/1994/104/add.30, 23 August 2005, Third periodic reports submitted by States Parties under Articles 16 and 17 of the Covenant, Addendum, the Netherlands.

E/1996/22, annex IV at 97 (1995), 24 November 1995, CESCR, General Comment 6: the Economic, Social and Cultural Rights of Older Persons.

UN Doc. E/CN.4/1324

E/CN.4/1997/105, 18 December 1996, Draft Optional Protocol to the International Covenant on Economic, Social and Cultural Rights, adopted by the Committee on Economic, Social and Cultural Rights.

E/CN.4/1998/21, 15 January 1998, Report of the High Commissioner for human rights.

E/CN.4/1999/45, 20 January 1999, Report of the High Commissioner for human rights.

E/CN.4/2001/53, 7 February 2001, Jean Ziegler, report of the Special Rapporteur on the right to food.

E/CN.4/2001/148, 30 March 2001, Report of the High Commissioner for human rights.

E/CN.4/2002/58, 10 January 2002, Jean Ziegler, report of the Special Rapporteur on the right to food.

E/CN.4/2003/54, 10 January 2003, Jean Ziegler, report of the Special Rapporteur on the right to food.

E/CN.4/2004/10, 9 February 2004, Jean Ziegler, report of the Special Rapporteur on the right to food.

E/CN.4/2005/47, 24 January 2005, Jean Ziegler, report of the Special Rapporteur on the right to food.

E/CN.4/2006/44, 16 March 2006, Jean Ziegler, report of the Special Rapporteur on the right to food.

E/C.12/2008/2, 24 March 2009, Guidelines on treaty-specific documents to be submitted under articles 16 and 17 of the International Covenant on Economic, Social and Cultural Rights.

E/C.12/2000/6, 7 July 2000, Substantive issues arising in the implementation of the International Covenant on Economic, Social and Cultural Rights.

E/CN.4/Sub.2/1998/7, 10 June 1998, El Hadji Guissé, report of the Special Rapporteur on the right of access of everyone to drinking water supply and sanitation services.

E/CN.4/Sub.2/1999/12, 28 June 1999, Updated study on the right to food, submitted by Mr Asbjørn Eide in accordance with Sub-Commission decision 1998/106.

E/CN.4/Sub.2/2003/12, 26 August 2003: Draft Standards on the Responsibilities of Transnational Corporations and Other Business Enterprises with Regard to Human Rights.

E/CN.4/Sub2/2004/20, 14 July 2004, El Hadji Guissé, report of the Special Rapporteur on the right of access of everyone to drinking water supply and sanitation services.

E/CN.4/Sub.2/2005/25, 11 July 2005; El Hadji Guissé, report of the Special Rapporteur on the right of access of everyone to drinking water supply and sanitation services.

Economic and Social Council Resolution 5 (I), 16 February 1946.

Economic and Social Council Resolution 1985/17, 28 May 1985.

Economic and Social Council Resolution 1988/4, 24 May 1988.

E/C.12/1997/8, 12 December 1997, CESCR, General Comment 8, The relationship between economic sanctions and respect for economic, social and cultural rights.

E/C.12/1998/24, 3 December 1998, CESCR, General Comment 9, The domestic application of the Covenant.

E/C.12/1999/5, 12 May 1999, CESCR, General Comment 12, Right to Adequate Food.

E/C.12/1999/10, 8 December 1999, CESCR, General Comment 13, the Right to Education (Article 13).

E/C.12/2000/4, 11 August 2000, CESCR, General Comment 14, The right to the highest attainable standard of health.

E/C.12/2002/11, 20 January 2003, CESCR, General Comment 15, The Right to Water.

E/C.12/2008/2, 24 March 2009, Guidelines on treaty-specific documents to be submitted under Articles 16 and 17 of the International Covenant on Economic, Social and Cultural Rights.

E/C.12/GC/19, 4 February 2008, CESCR, General Comment 19, The Right to Social Security (Article 9).

E/C.12/GC/20, 10 June 2009, CESCR, General Comment 20, Non-Discrimination in Economic, Social and Cultural Rights (Art. 2, para. 2).

European Convention on Human Rights

- European Convention on Human Rights, 4 November 1950, Rome, in: Council of Europe Treaty Series, No. 5.
- Protocol to the Convention for the Protection of Human Rights and Fundamental Freedoms, 20 March 1952, Paris, in: Council of Europe Treaty Series, No. 9.

European Social Charter

- European Social Charter, 18 October 1961, Turin, in: European Treaty Series 35.
- Amending Protocol of 1991 reforming the supervisory mechanism, 21 October 1991, Turin, in: European Treaty Series 142.
- Additional Protocol to the European Social Charter Providing for a System of Collective Complaints, 9 November 1995, Strassbourg, in: European Treaty Series 158.
- European Social Charter (revised), 3 May 1996, Strasbourg, in: European Treaty Series 163.
- Form for the reports to be submitted in pursuance of the European Social Charter (revised), 31 March 2008, Strassbourg.

FAO Doc. WSFS 2009/2, Rome, 16-18 November 2009, Declaration of the World Food Summit on Food Security.

The Food Assistance Convention, adopted on 12 April 2012, Londen

General Assembly, resolution 42/115, 7 December 1987, The impact of property on the enjoyment of human rights and fundamental freedoms.

Geneva Conventions and Protocols

- First Geneva Convention for the Amelioration of the Condition of the Wounded and Sick in Armed Forces in the Field, 12 August 1949.
- Third Geneva Convention relative to the Treatment of Prisoners of War, 12 August 1949.
- Fourth Geneva Convention relative to the Protection of Civilian Persons in Time of War, 12 August 1949.
- Protocol Additional to the Geneva Conventions of 12 August 1949, and relating to the Protection of Victims of International Armed Conflicts, 8 June 1977.
- Protocol Additional to the Geneva Conventions of 12 August 1949, and relating to the Protection of Victims of Non-International Armed Conflicts, 8 June 1977.

Geneva Declaration of the Rights of the Child, 26 September 1924, adopted by the League of Nations.

Bibliography

HRI/GEN/2/Rev.6, 3 June 2009, Compilation of guidelines on the form and content of reports to be submitted by States Parties to the International Human Rights Treaties.

The Human Rights Committee, 13 March 1984, General Comment 12, the right to self-determination of peoples (Article 1).

ILO Convention 87, 17 June 1948, Freedom of Association and Protection of the Right to Organise Convention.

The Limburg Principles on the Implementation of the International Covenant on Economic, Social and Cultural Rights, published in: E/CN.4/1987/17, Annex and Human Rights Quarterly, Vol. 9 (1989), pp. 122-135.

The Maastricht Guidelines on Violations of Economic, Social and Cultural Rights, Maastricht, January 22-26, 1997.

Office of the United Nations High Commissioner for Human Rights and the World Health Organisation, factsheet no. 31, The right to health, Geneva: UN, 2008.

The Protocol on Water and Health to the 1992 Convention on the Protection and Use of Transboundary Watercourses and International Lakes, 18 October 1999, MP.WAT/2001/1, EUR/ICP/EHCO 020205/8Fin.

Rome Declaration on World Food Security, 13 November 1996.

The San Salvador Additional Protocol to the American Convention on Human Rights, November 1988, adopted by the General Assembly of the Organisation of America States.

The voluntary guidelines to support the progressive realisation of the right to adequate food in the context of national food security, adopted by the 127th session of the FAO Council, November 2004.

Tripartite Declaration of Principles concerning Multinational Enterprises and Social Policy, adopted by the Governing Body of the International Labour Office at its 204th Session (Geneva, November 1977) as amended at its 279th (November 2000) and 295th session (March 2006).

UN Commission on Human Rights, Resolution 1987/18, 10 Mars 1987, On the Impact of Property on the Economic and Social Development of Member States.

Universal Islamic Declaration of Human Rights, 19 September 1981, adopted by the Islamic Council.

OECD, OECD Guidelines for Multinational Enterprises, OECD Publishing, 2011.

UN Human Rights Committee, 30 April 1982, General Comment No. 6: Article 6, Right to Life.

Universal Declaration on the Eradication of Hunger and Malnutrition, adopted on 16 November 1974 by the World Food Conference convened under General Assembly resolution A/RES/3180 (XXVIII) of 17 December 1973, and endorsed by General Assembly resolution A/RES/3348 (XXIX) of 17 December 1974.

Vienna Covenant on the Law of Treaties, adopted on 23 May 1966, entered into force on 27 January 1980, United Nations, Treaty Series, vol. 1155, p. 331.

Court Rulings

International Court Rulings

International Court of Justice, Advisory Opinion of 28 May 1951, Reservations to the Convention on the Prevention and Punishment of the Crime of Genocide.
Court of Justice of the European Union, 05-02-1963, 26/62 (Van Gend en Loos Ruling).

Dutch Court Rulings

Central Appeals Court for Public Service and Social Security Matters, 31 March 1995, JB 1995, 161.
Central Appeals Court for Public Service and Social Security Matters, 22 April 1997, JB 1997, 158.
Central Court of Appeal, 3 July 1986, TAR 1986, 215.
Central Court of Appeal, 16 February 1989, AB 1989, 164.
Central Court of Appeal, 26 June 2001, LJN AB2324.
Central Court of Appeal, 18 June 2004, JB 2004, 303.
Central Court of Appeal, 29 March 2005, LJN AT3468.
Central Court of Appeal, 8 April 2005, LJN AT4112.
Central Court of Appeal, 25 May 2004, USZ 2004, 241.
Central Court of Appeal, 14 June 2005, LJN AT8038.
Central Court of Appeal, 5 July 2005, LJN AT9963.
Central Court of Appeal, 8 July 2005, LJN AT9102.
Central Court of Appeal 25 August 2005, LJN AU1850.
Central Court of Appeal, 1 November 2005, LJN AU5600.
Central Court of Appeal, 24 January 2006, JB 2006, 66.
Central Court of Appeal, 9 May 2006, LJN AX2177.
Central Court of Appeal, 09 October 2006, LJN AY9940.
Central Court of Appeal, 12 June 2007, LJN BA7026.
Central Court of Appeal, 6 September 2007, LJN BB6188.
Central Court of Appeal, 11 October 2007, LJN BB5687.
Central Court of Appeal, 21 November 2007, LJN BB9625.
Central Court of Appeal, 7 April 2008, LJN BD0221.
Central Court of Appeal, 10 July 2008, LJN BD8630.
Central Court of Appeal, 1 October 2008, LJN BF4589.
Central Court of Appeal, 22 December 2008, LJN BG8776.
Central Court of Appeal, 07 April 2009, RSV 2008, 211.
Central Court of Appeal, 21 May 2009, LJN BI8400.
Central Court of Appeal, 11 June 2009, LJN BI9325.
Central Court of Appeal, 18 June 2009, LJN NI9928.
Central Court of Appeal, 06 October 2009, LJN BK0734.
Central Court of Appeal, 26 January 2010, LJN BL1686.
Central Court of Appeal, 30 March 2010, LJN BM1913.
Central Court of Appeal, 14 April 2010, LJN BM3583.
Central Court of Appeal, 19 April 2010, LJN BM0956.

Central Court of Appeal. 30 March 2010, USZ 2010, 166.

Central Court of Appeal, 11 May 2010, LJN BM6748.

Central Court of Appeal, 14 July 2010, LJN BN1274.

Central Court of Appeal, 23 July 2010, LJN BN2492.

Central Court of Appeal; 28 September 2010, LJN BN9571.

Central Court of Appeal, 20 October 2010, LJN BO3581.

Central Court of Appeal, 2 November 2010, LJN BO3025.

Central Court of Appeal, 14 March 2011, NJB 2011, 755.

Central Court of Appeal, 26 April 2011, LJN BQ3795.

Central Court of Appeal, 21 June 2011, LJN BR0385.

Central Court of Appeal, 15 July 2011, JV 2011, 393.

Central Court of Appeal, 5 Augustus 2011, LJN BR4248.

Central Court of Appeal, 5 Augustus 2011, LJN BR4268.

Central Court of Appeal, 5 Augustus 2011, LJN BR4785.

Central Court of Appeal, 30 December 2011, USZ 2012, 58.

Central Court of Appeal, 13 March 2012, USZ 2012, 101.

Central Court of Appeal, 2 May 2012, USZ 2012, 158.

Central Court of Appeal, 13 April 2012, USZ 2012, 161.

Central Court of Appeal, 23 May 2012, www.rechtspraak.nl, 30 May 2012.

Court of Appeal, 's-Gravenhage, 7 September 2005, NJ 2005, 473.

Court of Appeal, 's-Gravenhage, 30 October 2007, NJF 2007, 532.

Court of Appeal, 's-Gravenhage, 20 December 2007, NJ 2008, 133.

Court of Appeal, 's-Gravenhage, 28 April 2009, LJN BI3564.

Court of Appeal, 's-Gravenhage, 27 July 2010, LJN BN2164.

Court of Appeal of 's-Gravenhage, 11 January 2011, JV 2011, 91.

Court of Appeal, 's-Gravenhage, 16 August 2011, LJN BR6656.

Court of Appeal, 's-Hertogenbosch, 2 February 2010, LJN BL6583.

Court of Appeal, Rotterdam, 19 September 2007, LJN BB5715.

Court of First Instance of the Netherlands Antilles, 28 January 2009, EJ 2008/497.

Court of First Instance of the Netherlands Antilles, 30 March 2010, EJ 2009/579.

Council of State, 28 November 2001, LJN AD7067.

Council of State, 26 February 2003, JV 2003, 164.

Council of State, 1 March 2005, JV 2005/176.

Council of State, 13 September 2005, JV 2005, 409.

Council of State, 15 February 2007, LJN AZ9524.

Council of State, 12 April 2007, LJN BA3394.

Council of State, 27 April 2007, LJN BA4654.

Council of State, 13 June 2007, LJN BA7088.

Council of State, 5 December 2007, AB 2008, 35.

Council of State, 26 November 2007, www.rechtspraak.nl, 2 January 2008.

Council of State, 9 April 2008, LJN BC9087.

Council of State, 08 October 2010, LJN BO0685.

Council of State, 22 December 2010, AB 2011, 169.

Council of State, 29 June 2011, AB 2011, 327.

Council of State, 22 February 2012, JV 2012, 200.

Council of State, 11 June 2012, www.rechtspraak.nl, 18 June 2012.

District Court of Alkmaar, 20 July 2005, LJN AT9598.

District Court of Almelo, 28 November 2005, LJN AU7003.

District Court of Amsterdam: 4 August 1999, LJN AA4043.

District Court of Amsterdam, 13 March 2001, LJN AB0942.

District Court of Amsterdam, 4 April 2002, LJN AE3473.

District Court of Amsterdam, 19 December 2005, AWB 04/19508.

District Court of Amsterdam, 19 April 2007, LJN BA6900.

District Court of Amsterdam, 08 November 2007, LJN BF1926.

District Court of Amsterdam, 3 December 2008, LJN BG7017.

District Court of Amsterdam, 12 December 2008, LJN BG6963.

District Court of Amsterdam, 12 December 2008, LJN BG6965.

District Court of Amsterdam, 23 April 2009, LJN BJ1021.

District Court of Amsterdam, 30 January 2009, LJN BH4457.

District Court of Amsterdam, 2 June 2009, LJN BJ3914.

District Court of Amsterdam, 9 February 2010, LJN BL6113.

District Court of Amsterdam, 3 May 2011, LJN BQ9532.

District Court of Amsterdam, 18 February 2011, LJN BQ5256.

District Court of Amsterdam, 1 June 2012, www.rechtspraak.nl, 15 June 2012.

District Court of Arnhem, 25 May 2007, LJN BA6562.

District Court of Arnhem, 29 March 2012, www.rechtspraak.nl, 13 April 2012.

District Court of Arnhem, 13 November 2012, www.rechtspraak.nl, 22 November 2012.

District Court Assen, 10 April 2012, www.rechtspraak.nl, 25 May 2012.

District Court of Dordrecht, 25 September 2008, LJN BG3517.

District Court of Dordrecht, 23 April 2009, LJN BI8643.

District Court of Dordrecht, 06 October 2009, LJN BK0734.

District Court of Dordrecht, 21 December 2009, LJN BL9388.

District Court of Dordrecht, 16 September 2010, LJN BN7990.

District Court of Dordrecht, 5 November 2010, www.rechtspraak.nl, 10 November 2010.

District Court of Dordrecht, 27 May 2011, LJN BR5744.

District Court of 's-Gravenhage, 30 August 2000, LJN AA6959.

District Court of 's-Gravenhage, 6 September 2000, JV 2000, 224.

District Court of 's-Gravenhage, 1 Augustus 2002, AWB 02/54360 and 02/54358.

District Court of 's-Gravenhage, 1 Augustus 2002, AWB 02/54362 and 02/54361.

District Court of 's-Gravenhage, 29 August 2002, LJN AF2534.

District Court of 's-Gravenhage, 14 August 2003, LJN AM3133.

District Court of 's-Gravenhage, 25 March 2004, LJN AO6655.

District Court of 's-Gravenhage, 23 January 2006 LJN AV0548.

District Court of 's-Gravenhage, 22 May 2006, LJN AX4451.

District Court of 's-Gravenhage, 2 October 2006, LJN AY9546.

District Court of 's-Gravenhage, 22 August 2007, LJN BC0745.

District Court of 's-Gravenhage, 08 November 2007, LJN BB9819.

District Court of 's-Gravenhage, 07 December 2007, LJN BC2933.

District Court of 's-Gravenhage, 20 December 2007, LJN BC1047.

District Court of 's-Gravenhage, 24 January 2008, LJN BC2955.

District Court of 's-Gravenhage, 21 March 2008, AWB 07/28996.

District Court of 's-Gravenhage, 10 April 2008, LJN BC9445.

District Court of 's-Gravenhage, 24 July 2008, LJN BF0906.

District Court of 's-Gravenhage, 25 September 2009, LJN BK7090.

District Court of 's-Gravenhage, 7 October 2009, LJN BK3052.

District Court of 's-Gravenhage, 23 December 2009, LJN BL2473.

District Court of 's-Gravenhage, 2 March 2010, LJN BM2383.

District Court of 's-Gravenhage, 20 April 2011, LJN BQ3199.

District Court of Groningen, 14 July 2010, LJN BN2935.

District Court of Haarlem, 17 May 2005, LJN AT6534.

District Court of Haarlem, 23 November 2006, LJN AZ4222.

District Court of Haarlem, 21 June 2007, LJN BA9025.

District Court of Haarlem, 11 July 2007, LJN BB0998.

District Court of Haarlem, 29 August 2007, LJN BB3043.

District Court of Haarlem, 28 February 2008, LJN BC6101.

District Court of Haarlem 8 April 2008, LJN BD3399.

District Court of Haarlem, 29 July 2008 LJN BE9491.

District Court of Haarlem, 21 November 2008, LJN BG6130.

District Court of Haarlem, 6 May 2009, LJN BI3326.

District Court of Haarlem, 14 June 2010, LJN BM9368.

District Court of Haarlem, 29 March 2012, LJN BW2431.

District Court 's-Hertogenbosch, LJN BC4003.

District Court of Leeuwarden 28 November 2005, LJN AU7449.

District Court of Leeuwarden, 05 July 2010, LJN BN0391.

District Court of Maastricht, 25 June 2008, LJN BD5759.

District Court of Rotterdam, 19 September 2007, LJN BB 5715.

District Court of Rotterdam, 24 December 2007, LJN BC0852.

District Court of Zutphen, 12 December 2008, LJN BJ1349.

District Court of Zwolle, 17 February 2003, LJN AF4890.

District Court of Zwolle, 19 March 2003, LJN AF6351.

District Court of Zwolle, 19 March 2003, LJN AF6354.

District Court of Zwolle-Lelystad, 7 June 2006, LJN AY8861.

District Court of Zwolle-Lelystad 16 April 2009, LJNBI1369.

District Court of Zwolle-Lelystad, 29 April 2009, LJN BJ5171.

District Court of Zwolle-Lelystad, 19 April 2011, LJN BQ3967.

District Court of Zwolle-Lelystad, 21 April 2011, LJN BQ9140.

District Court of Zwolle-Lelystad, 3 May 2011, LJN BQ5114.

District Court of Zwolle-Lelystad, 9 June 2011, LJN BR3569.

The Joint Court of Justice of Aruba, Curaçao, Sint Maarten and of Bonaire, Saint Eustatius and Saba, 27 October 2009, EJ 497/08.

President of the judicial Division of the Council of State, 10 May 1979, AB 1979, 472.

Public Service Tribunal, Amsterdam, NJCM-Bulletin 1984, p. 245-252.

Special Council of Cassation, 12 January 1949, NJ 1949, no. 87 ('Rauter' ruling).

Supreme Court, 3 March 1919, NJ 1919, p. 371 ('Grenstractaat Aken' ruling).

Supreme Court, 6 March 1959, NJ 1962, 2 (Nyugat ruling).

Supreme Court, 24 February 1960, NJ 1960, 483.

Supreme Court, 28 November 1961, NJ 1962, 90.

Supreme Court, 25 April 1967, NJ 1968, 63.

Supreme Court, 6 December 1983, NJ 1984, 557.

Supreme Court, 12 februari 1984, NJB 1984, 45.

Supreme Court, 18 February 1986, NJ 1987, 62.

Supreme Court, 30 May 1986, NJ 1986, 668 (railway-trike ruling).

Supreme Court 14 April 1989, AB 1989, 207/NJ 1989, 469 ('Harmonisation Act' ruling).

Supreme Court, 20 April 1990, RvdW 1990, 88, NJ 1992/636.

Supreme Court, 7 May 1993, NJ 1995, 259.

Supreme Court, 30 January 2004, LJN AM2312.

Supreme Court, 3 June 2005, LJN AT3445.

Supreme Court, 12 February 2010, NJB 2010, 405.

Supreme Court, 12 February 2010, LJN BI9729.

Supreme Court, 9 April 2010, LJN BK4549.

Supreme Court, 1 April 2011, JB 2011, 115.

Supreme Court, 21 September 2012, NJ 2013, 22.

Supreme Court, 21 September 2012, NJ 2013, 22.

Supreme Court, 23 November 2012, JV 2013, 115.

Belgian Court Rulings

Constitutional Court

All rulings of the Constitutional Court are published on: http://www.const-court.be.

Constitutional Court, 23/89, 13 October 1989.

Constitutional Court, 18/90, 23 May 1990.

Constitutional Court 33/92, 7 May 1992.

Constitutional Court, 72/92, 18 November 1992.

Constitutional Court, 62/93, 15 July 1993.

Constitutional Court, 22/94, 8 March 1994.

Constitutional Court, 40/94, 19 May 1994.

Constitutional Court, 51/94, 29 June 1994.

Constitutional Court, 62/94, 14 July 1994.

Constitutional Court, 83/94, 1 December 1994.

Constitutional Court, 1/95, 12 January 1995.

Constitutional Court, 49/95, 15 June 1995.

Constitutional Court, 14/97, 18 March 1997.

Constitutional Court, 47/97, 14 July 1997.

Constitutional Court, 35/98, 1 April 1998.

Constitutional Court, 43/98, 22 April 1998.

Bibliography

Constitutional Court, 44/98, 22 April 1998.

Constitutional Court, 46/98, 22 April 1998.

Constitutional Court, 48/98, 22 April 1998.

Constitutional Court, 62/98, 4 June 1998.

Constitutional Court, 85/98, 15 July 1998.

Constitutional Court, 91/98, 15 July 1998.

Constitutional Court, 107/98, 21 October 1998.

Constitutional Court, 108/98, 21 October 1998.

Constitutional Court, 110/98, 4 November 1998.

Constitutional Court, 134/98, 16 December 1998.

Constitutional Court, 2/99 13 January 1999.

Constitutional Court, 42/99, 30 March 1999.

Constitutional Court, 80/99, 30 June 1999.

Constitutional Court, 90/99, 15 July 1999.

Constitutional Court, 42/2000, 06 April 2000.

Constitutional Court, 106/2000, 25 October 2000

Constitutional Court, 15/2001, 14 February 2001.

Constitutional Court, 17/2001, 14 February 2001.

Constitutional Court, 32/2001, 1 March 2001.

Constitutional Court, 49/2001, 18 April 2001.

Constitutional Court, 109/2001, 20 September 2001.

Constitutional Court, 131/2001, 30 December 2001.

Constitutional Court, 148/2001, 20 November 2001.

Constitutional Court, 14/2002, 17 January 2002.

Constitutional Court, 15/2002, 17 January 2002.

Constitutional Court, 16/2002, 17 January 2002.

Constitutional Court, 17/2002, 17 January 2002.

Constitutional Court, 41/2002, 20 February 2002.

Constitutional Court, 50/2002, 13 March 2002.

Constitutional Court, 89/2002, 5 June 2002.

Constitutional Court, 56/2003, 14 May 2003.

Constitutional Court, 75/2003, 28 May 2003.

Constitutional Court, 106/2003, 22 June 2003.

Constitutional Court, 129/2003, 1 November 2003.

Constitutional Court, 136/2004, 22 July 2004.

Constitutional Court, 158/2004 20 October 2004.

Constitutional Court, 189/2004, 24 November 2004.

Constitutional Court, 202/2004, 21 December 2004.

Constitutional Court, 203/2004, 21 December 2004.

Constitutional Court, 204/2004, 21 December 2004.

Constitutional Court, 205/2004, 21 December 2004.

Constitutional Court, 101/2005, 1 June 2005.

Constitutional Court, 110/2005, 22 June 2005.

Constitutional Court, 131/2005, 19 July 2005.

Constitutional Court, 147/2005, 28 September 2005.
Constitutional Court 194/2005, 21 December 2005.
Constitutional Court, 32/2006, 1 March 2006.
Constitutional Court, 35/2006, 1 March 2006.
Constitutional Court, 44/2006, 15 March 2006.
Constitutional Court 110/2006, 28 June 2006.
Constitutional Court, 28/2007, 21 February 2007.
Constitutional Court, 56/2008, 10 March 2008.
Constitutional Court, 64/2008, 17 April 2008.
Constitutional Court, 105/2008, 17 July 2008.
Constitutional Court, 119/2008, 31 July 2008.
Constitutional Court, 64/2009, 2 April 2009.
Constitutional Court, 121/2009, 16 July 2009.
Constitutional Court, 145/2010, 16 December 2010.
Constitutional Court, 135/2011, 27 July 2011.
Constitutional Court, 137/2011, 27 July 2011.
Constitutional Court 7/2012, 18 January 2012.

Council for Aliens Disputes

Council for Aliens Disputes, Annual report 2007-2008.
Council for Aliens Disputes, Annual report 2008-2009.
Council for Aliens Disputes, Annual report 2009-2010.
Case no. 2760, 17 October 2007. Available at: www.rvv-cce.be.
Case no. 15780, 11 September 2008. Available at: www.rvv-cce.be.
Case no. 17293, 17 October 2008. Available at: www.rvv-cce.be.
Case no. 18.341, 4 November 2008. Available at: www.rvv-cce.be.
Case no. 20139, 9 December 2008. Available at: www.rvv-cce.be.
Case no. 23968, 27 February 2009. Available at: www.rvv-cce.be.
Case no. 24320, 11 March 2009. Available at: www.rvv-cce.be.
Case no. 24487, 13 March 2009. Available at: www.rvv-cce.be.
Case no. 25703, 7 April 2009. Available at: www.rvv-cce.be.
Case no. 27657, 26 May 2009. Available at: www.rvv-cce.be.
Case no. 27951, 28 May 2009. Available at: www.rvv-cce.be.
Case no. 28476, 9 June 2009. Available at: www.rvv-cce.be.
Case no. 28928, 22 June 2009. Available at: www.rvv-cce.be.
Case no. 30858, 31 August 2009. Available at: www.rvv-cce.be.
Case no. 34792, 25 November 2009. Available at: www.rvv-cce.be.
Case no. 37979, 29 January 2010. Available at: www.rvv-cce.be.
Case no. 40246 15 March 2010. Available at: www.rvv-cce.be.
Case no. 42394, 27 April 2010. Available at: www.rvv-cce.be.
Case no. 45588, 29 June 2010. Available at: www.rvv-cce.be.
Case no. 53965, 28 December 2010. Available at: www.rvv-cce.be.
Case no. 59130, 31 March 2011. Available at: www.rvv-cce.be.

Case no. 59131, 31 March 2011. Available at: www.rvv-cce.be.

Case no. 67931, 5 October 2011. Available at: www.rvv-cce.be.

Case no. 72810, 6 January 2012. Available at: www.rvv-cce.be.

Case no. 73575, 19 January 2012. Available at: www.rvv-cce.be.

Case no. 74507, 31 January 2012. Available at: www.rvv-cce.be.

Case no. 75732, 24 February 2012. Available at: www.rvv-cce.be.

Case no. 75953, 28 February 2012. Available at: www.rvv-cce.be.

Case no. 76154, 29 February 2012. Available at: www.rvv-cce.be.

Case no. 76427, 1 March 2012. Available at: www.rvv-cce.be.

Case no. 76891, 9 March 2012. Available at: www.rvv-cce.be.

Case no. 77339, 15 March 2012. Available at: www.rvv-cce.be.

Case no. 77338, 15 March 2012. Available at: www.rvv-cce.be.

Case no. 77468, 19 March 2012. Available at: www.rvv-cce.be.

Case no. 77700, 21 March 2012. Available at: www.rvv-cce.be.

Case no. 77918, 23 March 2012. Available at: www.rvv-cce.be.

Case no. 78203, 28 March 2012. Available at: www.rvv-cce.be.

Case no. 79852, 20 April 2012. Available at: www.rvv-cce.be.

Case no. 79734, 20 April 2012. Available at: www.rvv-cce.be.

Case no. 80146, 25 April 2012. Available at: www.rvv-cce.be.

Case no. 80534, 27 April 2012. Available at: www.rvv-cce.be.

Case no. 81048, 11 May 2012. Available at: www.rvv-cce.be.

Case no. 81092, 14 May 2012. Available at: www.rvv-cce.be.

Case no. 81155, 14 May 2012. Available at: www.rvv-cce.be.

Case no. 81503, 22 May 2012. Available at: www.rvv-cce.be.

Case no. 82143, 31 May 2012. Available at: www.rvv-cce.be.

Case no. 82145, 31 May 2012. Available at: www.rvv-cce.be.

Case no. 82696, 11 June 2012. Available at: www.rvv-cce.be.

Case no. 82697, 11 June 2012. Available at: www.rvv-cce.be.

Case no. 82790, 11 June 2012. Available at: www.rvv-cce.be.

Case no. 83484, 22 June 2012. Available at: www.rvv-cce.be.

Case no. 83585, 25 June 2012. Available at: www.rvv-cce.be.

Case no. 83716, 26 June 2012. Available at: www.rvv-cce.be.

Case no. 84296, 6 July 2012. Available at: www.rvv-cce.be.

Case no. 84816, 18 July 2012. Available at: www.rvv-cce.be.

Case no. 85594, 3 August 2012. Available at: www.rvv-cce.be.

Case no. 88126, 25 September 2012. Available at: www.rvv-cce.be.

Case no. 91043, 6 November 2012. Available at: www.rvv-cce.be.

Case no. 92302, 27 November 2012. Available at: www.rvv-cce.be.

Case no. 93190, 10 December 2012. Available at: www.rvv-cce.be.

Case no. 93272, 11 December 2012. Available at: www.rvv-cce.be.

Case no. 93287, 11 December 2012. Available at: www.rvv-cce.be.

Case no. 93863 18 December 2012. Available at: www.rvv-cce.be.

Council of State

Council of State, 30 December 1993, case no. 45552, S.K. 1994, 6.

Council of State, 24 January 1996, case no. 57793. Available at: www.raadvst-consetat.be.

Council of State, 17 October 1994, case no. 49701. Available at: www.raadvst-consetat.be.

Council of State, 23 December 1994, case no. 50972. Available at: www.raadvst-consetat.be.

Council of State, 22 March 1995, case no. 52424. Available at: www.raadvst-consetat.be.

Council of State, 30 October 1995, case no. 56106. Available at: www.raadvst-consetat.be.

Council of State, 24 January 1996, case no. 57793. Available at: www.raadvst-consetat.be.

Council of State, 26 January 1996, case no. 57840. Available at: www.raadvst-consetat.be.

Council of State, 22 March 1996, case no. 58746. Available at: www.raadvst-consetat.be.

Council of State, 2 December 1996, case no. 63387. Available at: www.raadvst-consetat.be.

Council of State, 19 March 1997, case no. 65343. Available at: www.raadvst-consetat.be.

Council of State, 5 November 1997, case no. 69470. Available at: www.raadvst-consetat.be.

Council of State, 1 April 1998, case no. 72893. Available at: www.raadvst-consetat.be.

Council of State, 3 July 1998, case no. 74948. Available at: www.raadvst-consetat.be.

Council of State, 7 December 1999, case no. 83940. Available at: www.raadvst-consetat.be.

Council of State, 21 December 1999, case no. 84310. Available at: www.raadvst-consetat.be.

Council of State, 26 April 2000, case no. 86914. Available at: www.raadvst-consetat.be.

Council of State, 20 November 2000, case no. 90902. Available at: www.raadvst-consetat.be.

Council of State, 13 December 2000, case no. 91625. Available at: www.raadvst-consetat.be.

Council of State, 16 January 2001, case no. 92281. Available at: www.raadvst-consetat.be.

Council of State, 16 January 2001, case no. 92285. Available at: www.raadvst-consetat.be.

Council of State, 12 March 2001, case no. 93859. Available at: www.raadvst-consetat.be.

Council of State, 14 January 2002, case no. 102510. Available at: www.raadvst-consetat.be.

Council of State, 28 January 2002, case no. 102971. Available at: www.raadvst-consetat.be.

Council of State, 7 March 2002, case no. 104430. Available at: www.raadvst-consetat.be.

Council of State, 27 March 2002, case no. 105233. Available at: www.raadvst-consetat.be.

Council of State, 2 April 2002, case no. 105351. Available at: www.raadvst-consetat.be.

Council of State, 10 June 2002, case no. 107611. Available at: www.raadvst-consetat.be.

Council of State, 4 July 2002, case no. 108862. Available at: www.raadvst-consetat.be.

Council of State, 4 July 2002, case no. 108866. Available at: www.raadvst-consetat.be.

Council of State, 1 October 2002, case no. 110821. Available at: www.raadvst-consetat.be.

Council of State, 10 October 2002, case no. 111411. Available at: www.raadvst-consetat.be.

Council of State, 16 October 2002, case no. 111607. Available at: www.raadvst-consetat.be.

Council of State, 3 December 2002, case no. 113168. Available at: www.raadvst-consetat.be.

Council of State, 16 December 2002, case no. 113723. Available at: www.raadvst-consetat.be.

Council of State, 27 January 2003, case no. 115050. Available at: www.raadvst-consetat.be.

Council of State, 10 March 2003, case no. 116807. Available at: www.raadvst-consetat.be.

Council of State, 23 April 2003, case no. 118541. Available at: www.raadvst-consetat.be.

Council of State, 9 July 2003, case no. 121461. Available at: www.raadvst-consetat.be.

Council of State, 8 December 2003, case no. 126132. Available at: www.raadvst-consetat.be.

Council of State, 15 December 2003, case no. 126410. Available at: www.raadvst-consetat.be.

Council of State, 11 February 2004, case no. 128051. Available at: www.raadvst-consetat.be.

Bibliography

Council of State, 2 March 2004, case no. 128688. Available at: www.raadvst-consetat.be.

Council of State, 13 October 2004, case no. 135968. Available at: www.raadvst-consetat.be.

Council of State, 14 September 2004, case no. 134863. Available at: www.raadvst-consetat.be.

Council of State, 9 December 2004, case no. 138290. Available at: www.raadvst-consetat.be.

Council of State, 28 February 2005, case no. 141340. Available at: www.raadvst-consetat.be.

Council of State, 18 March 2005, case no. 142387. Available at: www.raadvst-consetat.be.

Council of State, 29 March 2005, case no. 142691. Available at: www.raadvst-consetat.be.

Council of State, 31 March 2005, case no. 142746. Available at: www.raadvst-consetat.be.

Council of State, 4 May 2005, case no. 144111. Available at: www.raadvst-consetat.be.

Council of State, 24 May 2005, case no. 144825. Available at: www.raadvst-consetat.be.

Council of State, 25 May 2005, case no. 144960. Available at: www.raadvst-consetat.be.

Council of State, 12 July 2005, case no. 147579. Available at: www.raadvst-consetat.be.

Council of State, 24 August 2005, case no. 148314. Available at: www.raadvst-consetat.be.

Council of State, 31 August 2005, case no. 148485. Available at: www.raadvst-consetat.be.

Council of State, 9 September 2005, case no. 148700. Available at: www.raadvst-consetat.be.

Council of State, 22 September 2005, case no. 149302. Available at: www.raadvst-consetat.be.

Council of State, 25 October 2005, case no. 150620. Available at: www.raadvst-consetat.be.

Council of State, 5 December 2005, case no. 52209. Available at: www.raadvst-consetat.be.

Council of State, 14 December 2005, case no. 152699. Available at: www.raadvst-consetat.be.

Council of State, 21 December 2005 case no. 153073. Available at: www.raadvst-consetat.be.

Council of State, 27 December 2005, case no. 153187. Available at: www.raadvst-consetat.be.

Council of State, 9 January 2006, case no. 153367. Available at: www.raadvst-consetat.be.

Council of State, 26 January 2006, case no. 154152. Available at: www.raadvst-consetat.be.

Council of State, 13 February 2006, case no. 154824. Available at: www.raadvst-consetat.be.

Council of State, 13 February 2006, case no. 154825. Available at: www.raadvst-consetat.be.

Council of State, 13 February 2006, case no. 154826. Available at: www.raadvst-consetat.be.

Council of State, 13 February 2006, case no. 154827. Available at: www.raadvst-consetat.be.

Council of State, 13 February 2006, case no. 154828. Available at: www.raadvst-consetat.be.

Council of State, 13 February 2006, case no. 154829. Available at: www.raadvst-consetat.be.

Council of State, 13 February 2006, case no. 154830. Available at: www.raadvst-consetat.be.

Council of State, 13 February 2006, case no. 154831. Available at: www.raadvst-consetat.be.

Council of State, 13 February 2006, case no. 154832. Available at: www.raadvst-consetat.be.

Council of State, 1 March 2006, case no. 155724. Available at: www.raadvst-consetat.be.

Council of State, 6 April 2006, case no. 157423. Available at: www.raadvst-consetat.be.

Council of State, 30 August 2006, case no. 162085. Available at: www.raadvst-consetat.be.

Council of State, 7 November 2006, case no. 164420. Available at: www.raadvst-consetat.be.

Council of State, 13 December 2006, case no. 165924. Available at: www.raadvst-consetat.be.

Council of State, 17 January 2007, case no. 166839. Available at: www.raadvst-consetat.be.

Council of State, 17 March 2007, case no. 156541. Available at: www.raadvst-consetat.be.

Council of State, 4 May 2007, case no. 170798. Available at: www.raadvst-consetat.be.

Council of State, 9 April 2008, case no. 181862. Available at: www.raadvst-consetat.be.

Council of State, 28 August 2008, case no. 185919. Available at: www.raadvst-consetat.be.

Council of State, 30 October 2008, case no. 187467. Available at: www.raadvst-consetat.be.

Council of State, 8 May 2009, case no. 193105. Available at: www.raadvst-consetat.be.

Council of State, 29 June 2009, case no. 194819. Available at: www.raadvst-consetat.be.

Council of State, 29 June 2009, case no. 194820. Available at: www.raadvst-consetat.be.

Council of State, 29 September 2009, case no. 196475. Available at: www.raadvst-consetat.be.

Council of State, 8 December 2009, case no. 198664. Available at: www.raadvst-consetat.be.

Council of State, 15 December 2009, case no. 198958. Available at: www.raadvst-consetat.be.

Council of State, 23 December 2009, case no. 199259. Available at: www.raadvst-consetat.be.

Council of State, 25 March 2010, case no. 202356. Available at: www.raadvst-consetat.be.

Council of State, 25 March 2010, case no. 202357. Available at: www.raadvst-consetat.be.

Council of State, 9 September 2010, case no. 207266. Available at: www.raadvst-consetat.be.

Council of State, 21 February 2011, case no. 211392. Available at: www.raadvst-consetat.be.

Council of State, 9 June 2011, case no. 213778. Available at: www.raadvst-consetat.be.

Council of State, 23 November 2011, case no. 216416. Available at: www.raadvst-consetat.be.

Council of State, 13 December 2011, case no. 216839. Available at: www.raadvst-consetat.be.

Council of State, 14 March 2012, case no. 218469. Available at: www.raadvst-consetat.be.

Council of State, 14 March 2012, case no. 218470. Available at: www.raadvst-consetat.be.

Council of State, 14 March 2012, case no. 218474. Available at: www.raadvst-consetat.be.

Court of Cassation

Court of Cassation, 27 May 1971, Arresten van het Hof van Cassatie, 1971 (p.959).

Court of Cassation, 20 December 1990, Arresten van het Hof van Cassatie, 1990(91) (p.445).

Court of Cassation, 4 November 1993, Arresten van het Hof van Cassatie, 199 (p.919).

Court of Cassation, 17 January 1994, Arresten van het Hof van Cassatie, 1994 (I, p.54).

Court of Cassation, 5 December 1994, Arresten van het Hof van Cassatie, 1994 (II, p.1052).

Court of Cassation, 5 December 1994, Arresten van het Hof van Cassatie, 1994 (II, p.1055).

Court of Cassation, 4 November 1996, Arresten van het Hof van Cassatie, 1996 (411).

Court of Cassation, 4 November 1996, Arresten van het Hof van Cassatie, 1996 (412).

Court of Cassation, 21 November 1996, Arresten van het Hof van Cassatie, 1996 (446).

Court of Cassation, 13 January 1997, Arresten van het Hof van Cassatie, 1997 (28).

Court of Cassation, 17 February 1997, Arresten van het Hof van Cassatie 1997 (91).

Court of Cassation, 1 October 1997, Arresten van het Hof van Cassatie, 1997 (378).

Court of Cassation, 29 May 1998, Arresten van het Hof van Cassatie, 1998 (279).

Court of Cassation, 8 March 1999, Arresten van het Hof van Cassatie, 1999 (136).

Court of Cassation, 31 March 1999, Arresten van het Hof van Cassatie, 1999 (195).

Court of Cassation, 4 November 1999, Arresten van het Hof van Cassatie, 1999 (588).

Court of Cassation, 4 November 1999, Arresten van het Hof van Cassatie, 1999 (589).

Court of Cassation, 10 November 1999, Arresten van het Hof van Cassatie, 1999 (599).

Court of Cassation, 29 April 2002, no. S010137F, published on: http://jure.juridat.just.fgov.be.

Court of Cassation, 17 June 2002, no. S010148F, published on: http://jure.juridat.just.fgov.be.

Court of Cassation, 7 October 2002, no. S000165F, published on: http://jure.juridat.just.fgov.be.

Court of Cassation, 25 September 2003, no. C030026N, published on: http://jure.juridat.just.fgov.be.

Court of Cassation, 14 October 2003, no. P030591N, published on: http://jure.juridat.just.fgov.be.

Court of Cassation, 7 June 2004, no. S030008N, published on: http://jure.juridat.just.fgov.be.

Bibliography

Court of Cassation, 22 March 2005, no. P050340N, published on: http://jure.juridat.just.fgov.be.
Court of Cassation, 21 March 2006, no. P060211N, published on: http://jure.juridat.just.fgov.be.
Court of Cassation, 23 October 2006, no. S050042F, published on: http://jure.juridat.just.fgov.be.
Court of Cassation, 4 March 2008, no. P071541N, published on: http://jure.juridat.just.fgov.be.
Court of Cassation, 26 May 2008, no. S060105F, published on: http://jure.juridat.just.fgov.be.
Court of Cassation, 27 January 2010, no. P091686F, published on: http://jure.juridat.just.fgov.be.
Court of Cassation, 15 September 2010, no. P101218F, published on: http://jure.juridat.just.
    fgov.be.
Court of Cassation, 20 October 2010, no. P090529F, published on: http://jure.juridat.just.fgov.be.
Court of Cassation, 24 October 2012, no. P121333F, published on: http://jure.juridat.just.fgov.be.

Labour Courts

Labour Court of Appeal of Liège, 13 February 1996, S.K. 1996, part 11, 568.
Labour Court of Verviers, 12 August, 1993, S.K. 1993, part 10, pp. 471-473.
Labour Court of Verviers, 22 March 1994, S.K. 1995, part 2, pp. 60-61.
Labour Court of Leuven, 12 December 2001, published on http://jure.juridat.just.fgov.be.
Labour Court of Brugge, 28 January 2002 published on: http://jure.juridat.just.fgov.be.
Labour Court of Appeal of Liège, 5 March 2002, published on: http://jure.juridat.just.fgov.be.
Labour Court of Appeal of Brussels, 17 April 2002, published on: http://jure.juridat.just.fgov.be.
Labour Court of Appeal of Brussels, 16 December 2004, S.K. 2005(pp. 264-265).

Domestic lower legislation

Dutch lower legislation

Adaption on the regulation on the provision of certain categories of aliens: Regulation of the
    Minister of Justice of 22 December 2006 no. 5458886/06/DVB. Original text in Dutch:
    Wijziging Regeling verstrekking bepaalde categorieën vreemdelingen; Regeling van de
    Minister van Justitie van 22 december 2006, no. 5458886/06/DVB.
Regulation of Internal Service of the Supreme Court of the Netherlands pursuant to Article
    75 Section 4 of the Law on the judicial organisation. Original title in Dutch: Reglement
    van Inwendige Dienst van de Hoge Raad der Nederlanden ex Artikel 75 lid 4 Wet op de
    Rechterlijke Organisatie.

Belgian lower legislation

Circular on the application of Article 9, third Section, of the Act of 15 December 1980 on the
    access to the territory, residence, establishment and removal of aliens, 9 October 1997,
    B.S. 14 November 1997. Original title in Dutch: Omzendbrief betreffende de toepassing
    van artikel 9, derde lid, van de wet van 15 december 1980 betreffende de toegang tot het
    grondgebied, het verblijf, de vestiging en de verwijdering van vreemdelingen.

Circular on the application of Article 9, third Section, of the Act of 15 December 1980 on the access to the territory, residence, establishment and removal of aliens, 15 December 1998, B.S. 19 December 1998. Original title in Dutch: Omzendbrief betreffende de toepassing van artikel 9, derde lid, van de wet van 15 december 1980 betreffende de toegang tot het grondgebied, het verblijf, de vestiging en de verwijdering van vreemdelingen.

Decree of the Flemish Community of 15 May 1991.

Decree of the German-language Community of 9 August 1991.

Decree of the French Community of 25 November 1991.

Decree of the Flemish Community of 15 July 1997, establishing the child impact report and review of the Government's policy to respect the rights of the child original text in Dutch: Decreet van de Vlaamse Gemeenschap van 15 July 1997, houdende instelling van het kindeffectrapport en de toetsing van het regeringsbeleid aan de naleving van de rechten van het kind.

Organic Law of 8 July 1976 on the Public Centres for Social Welfare. Original title in Dutch: Organieke Wet van 8 juli 1976 betreffende de openbare centra voor maatschappelijk welzijn, B.S. 5.VIII.1976 – err. B.S. 26.XI.1976).

Programme Act 22 December 2003, B.S. 31 December 2003.

Programme Act of 24 December 2004, B.S. 31 December 2004.

Programme Act 9 January 2006, B.S. 30 December 2005.

Royal Decree of 8 October 1981 on access to the territory, residence, establishment and the removal of aliens. Original title in Dutch: Koninklijk Besluit van 8 oktober 1981, betreffende de toegang tot het grondgebied, het verblijf, de vestiging en de verwijdering van vreemdelingen.

Royal Decree of 12 December 1996 on emergency medical assistance provided to aliens unlawfully residing in the State through Public Centres for Social Welfare. Original title in Dutch: Koninklijk besluit van 12 december 1996, betreffende de dringende medische hulp die door openbare centra voor maatschappelijk welzijn wordt verstrekt aan de vreemdelingen die onwettig in het Rijk verblijven.

Royal Decree of 13 July 2006, establishing the standards the care for children must meet to be recognised and amending the Royal Decree of 25 November 1997 establishing the standards the 'surgical day treatment' function must meet to be recognised. Original title in Dutch: Koninklijk Besluit van 13 juli 2006, houdende vaststelling van de normen waaraan het zorgprogramma voor kinderen moet voldoen om erkend te worden en tot wijziging van het koninklijk besluit van 25 november 1997 houdende vaststelling van de normen waaraan de functie 'chirurgische daghospitalisatie' moet voldoen om te worden erkend.

Statute of 20 July 1971 on the establishment of guaranteed family benefits. Original title in Dutch: Wet van 20 juli 1971 tot de instelling van gewaarborgde gezinsbijslag.

Statute of 31 March 1987 amending certain provisions relating to the origin. Original title in Dutch: Wet van 31 maart 1987 tot wijziging van een aantal bepalingen betreffende de afstamming.

Statute of 15 July 1996 amending Article 57 § 2 Organic Law of 8 July 1976 on public centres for social welfare.

Statute of 25 May 1999 on the Belgian international cooperation. Original title in Dutch: Wet van 25 mei 1999 betreffende de Belgische internationale samenwerking, B.S. 1 July 1999.

Bibliography

Statute of 22 December 1999 on the regularisation of the residence of certain categories of Aliens residing on Belgian Territory. Original title in Dutch: Wet van 22 december 1999 betreffende de regularisatie van het verblijf van bepaalde categorieën van vreemdelingen verblijvend op het grondgebied van het Rijk.

Statute of 4 September 2002, on the assignment to the public centres on social welfare of the guidance and financial social assistance to the most deprived on the supply of energy. Original title in Dutch: Wet van 4 September 2002, houdende toewijzing van een opdracht aan de openbare centra voor maatschappelijk welzijn inzake de begeleiding en de financiële maatschappelijke steunverlening aan de meest hulpbehoevenden inzake energielevering, B.S. 28 September 2002.

Statute of 21 February 2003, establishing a service for maintenance claims at the Finance Ministry. Original title in Dutch: Wet van 21 februari 2003, tot oprichting van een Dienst voor alimentatievorderingen.

Parliamentary Documents

Dutch Parliamentary Documents

Commitee-Van Eysinga, 9 July 1951, Eindrapport van de commissie nopens de samenwerking tussen Regering en Staten-Generaal inzake het buitenlandse beleid, NL-HaNA, BuZa/Code-Archief 45-54, 25.117, inv. no. 27297.

Constitutional reform of 1953, first reading

House of Representatives, 60th meeting, 13 March 1952.
House of Representatives, 61st meeting, 14 March 1952.
House of Representatives, 62nd meeting, 18 March 1952.
House of Representatives, 63rd meeting, 19 March 1952.
Senate, 40th meeting, 6 May 1952.
Parliamentary Documents, I 1951-1952, 2374, no. 113.
Parliamentary Documents, I 1951-1952, 2374, no. 113a.
Parliamentary Documents, II 1951-1952, 2374, no. 2.
Parliamentary Documents, II 1951-1952, 2374, no. 3.
Parliamentary Documents, II 1951-1952, 2374, no. 7.
Parliamentary Documents, II 1951-1952, 2374, no. 8.
Parliamentary Documents, II 1951-1952, 2374, no. 9.
Parliamentary Documents, II 1951-1952, 2374, no. 10.
Parliamentary Documents, II 1951-1952, 2374, no. 17.
Parliamentary Documents, II 1951-1952, 2374, no. 26.
Parliamentary Documents, II 1951-1952, 2374, no. 32.

Constitutional reform of 1953, second reading

Parliamentary Documents, I 1952-1953, 2700, no. 63.

Parliamentary Documents, I 1952-1953, 2700, no. 63a.

Ratification Bill of the ECHR

House of Representatives, 48th meeting, 3 March 1954.
House of Representatives, 49th meeting, 4 March 1954.
Senate, 50th meeting, 27 July 1954.
Parliamentary Documents, I 1952-1953 3043, no. 162.
Parliamentary Documents, II 1952-1953 3043, no. 3.
Parliamentary Documents, II 1952-1953 3043, no. 5.
Parliamentary Documents, II 1952-1953 3043, no. 6.

Constitutional reform of 1956

Parliamentary Documents, II 1955-1956, 4133 (R 19), no. 3.
Parliamentary Documents, II 1955-1956, 4133 (R 19), no. 4.
Parliamentary Documents, II 1955-1956, 4133 (R 19), no. 6.
Parliamentary Documents, II 1955-1956, 4133 (R 19), no. 7.

Raification Bill of the ESC (earliest version)

Senate, 3rd meeting, 31 October 1978.
House of Representatives, 37th meeting, 4 April 1978.
House of Representatives, 38th meeting, 5 April 1979.
Parliamentary Documents, II 1965-1966, 8606 (R 533), no. 2.
Parliamentary Documents, II 1965-1966, 8606 (R 533), no. 3.
Parliamentary Documents, II 1966-1967, 8606 (R 533), no. 5.
Parliamentary Documents, II 1966-1967, 8606 (R 533), no. 6.
Parliamentary Documents, II 1966-1967, 8606 (R 533), no. 7.
Parliamentary Documents, II 1970-1971, 8606 (R 533), no. 11.
Parliamentary Documents, II 1976-1977, 8606 (R 533), no. 12.
Parliamentary Documents, II 1976-1977, 8606 (R 533), no. 14.
Parliamentary Documents, II 1977-1978, 8606 (R 533), no. 15.
Parliamentary Documents, II 1977-1978, 8606 (R 533), no. 18.
Parliamentary Documents, II 1977-1978, 8606 (R 533), no. 21.

Ratification Bill of the ICCPR and the ICESCR

Parliamentary Documents, II 1975-1976 13932 (R 1037), no. 2.
Parliamentary Documents, II 1975-1976 13932 (R 1037), no. 3.
Parliamentary Documents, II 1975-1976 13932 (R 1037), no. 7.
Parliamentary Documents, II 1975-1976 13932 (R 1037), no. 8.
House of Representatives, 3rd meeting, 21 September 1978.
Senate, 5th meeting, 21 November 1978.

Bibliography

Ratification Bill of the CEDAW

Parliamentary Documents, II 1984-1985, 18950 (R 1281), no. 3.
Parliamentary Documents, II 1984-1985, 18950 (R 1281), no. 4.
Parliamentary Documents, II 1984-1985, 18950 (R 1281), no. 6.
Parliamentary Documents, II 1984-1985, 18950 (R 1281), no. 8.
Parliamentary Documents, II 1984-1985, 18950 (R 1281), no. 9.

Constitutional reform of 1983

House of Representatives, 62nd meeting, 18 March 1980.
House of Representatives, 66th meeting, 25 March 1980.
House of Representatives, 73rd meeting, 23 April 1980.
Parliamentary Documents, II 1977-1978 (R 1100), no. 2.
Parliamentary Documents, II 1977-1978 (R 1100), no. 3.
Parliamentary Documents, II 1977-1978 (R 1100), no. 4.
Parliamentary Documents, II 1977-1978 (R 1100), no. 6.
Parliamentary Documents, II 1977-1978 (R 1100), no. 7.
Parliamentary Documents, II 1977-1978 (R 1100), no. 10.
Parliamentary Documents, II 1977-1978 (R 1100), no. 14.
Parliamentary Documents, II 1977-1978 (R 1100), no. 19.

Ratification of the ICRC

House of Representatives, 81st meeting, 23 June 1994.
House of Representatives, 84th meeting, 30 June 1994.
Senate, 6th meeting, 22 November 1994.
Parliamentary Documents, II 1992-1993, 22855 (R1451), no. 2.
Parliamentary Documents, II 1992-1993, 22855 (R1451), no. 3.
Parliamentary Documents, II 1992-1993, 22855 (R1451), no. 4.
Parliamentary Documents, II 1993-1994, 22855 (R1451), no. 6.
Parliamentary Documents, II 1993-1994, 22855 (R1451), no. 10.
Parliamentary Documents, I 1994-1995, 22855 (R1451), no. 22.
Parliamentary Documents, I 1994-1995, 22855 (R1451), no. 22a.

Ratification Bill of the ESC (revised version)

Parliamentary Documents, II 2004-2005, 29941, no. 3.
Parliamentary Documents, II 2004-2005, 29941, no. 4.

Documents related to draft-Bill 10111

House of Representatives, 51st meeting, 10 February 1977 (Interpellation-Bakker).
Parliamentary Documents, II 1979-1980, 10111 no. 5.

Belgian Parliamentary Documents

Parliamentary Documents, House of Representatives, 1999-2000, doc. 50 0234/001.

Parliamentary Documents, House of Representatives, 1999-2000, doc. 50 0234/005.

Parliamentary Documents, Senate, 1999-2000, no. 2-202/3.

House of Representatives, 1999-2000, meeting of 24 November 1999, HA 50 plen. 017.

Periodic country reports

National periodic reports submitted by the Netherlands

European Social Charter

RAP/RCha/Netherlands/1(2008), 13 February 2008, 1st report on the implementation of the European Social Charter.

RAP/RCha/NE/1(2009), 14 April 2009, 2nd national report on the implementation of the European Social Charter.

RAP/RCha/NE/111(2010), 15 February 2010, 3rd national report on the implementation of the European Social Charter.

RAP/RCha/NE/V(2011), 4 November 2011, 4th national report on the implementation of the European Social Charter.

RAP/RCha/NL/VI(2013), 2 November 2012, 5th national report on the implementation of the European Social Charter.

International Covenant on Economic, Social and Cultural Rights

E/1980/6/Add.33, 8 November 1983.

E/1982/3/Add.35, 17 January 1986.

E/1984/6/Add.14, 20 February 1986.

E/1982/3/Add.44, 19 January 1988.

E/1984/6/Add. 20, 19 August 1988.

E/1990/6/Add.11, 5 August 1996.

E/1990/6/Add.12, 28 August 1996 (Netherlands Antilles).

E/1990/6/Add.13, 28 August 1996 (Aruba).

E/C.12/1998/SR.13, 7 May 1998.

E/C.12/1998/SR.14, 15 September 1998.

E/C.12/1/Add.25, 16 June 1998.

E/C.12/1998/26, 27 April; 15 May 1998; 16 November; 4 December 1998; CESCR report on the eighteenth and nineteenth sessions.

E/1994/104/Add.30, 23 August 2005.

E/C.12/ANT/3, 13 February 2006 (Netherlands Antilles).

E/C.12/NLD/Q/3, 2 March 2006.

E/C.12/NLD/Q/3/Add.1, 23 August 2006.

E.C.12/2006/SR.33, 22 November 2006.

Bibliography

E.C.12/2006/SR.34, 10 January 2007.
E/C.12/NLD/CO/3, 24 November 2006.
E/C.12/2007/SR.9, 10 May 2007.
E/C.12/2006/SR.34, 10 January 2007.
E/C.12/NLD/CO/3/Add.1, 31 January 2008.
E/C.12/NLD/4-5, 17 July 2009.
E/C.12/NLD/4/Add.1, 13 July 2009 (Netherlands Antilles).
E/C.12/NLD/Q/4-5, 22 December 2009.
E/C.12/NLD/Q/4-5/Add.1, 14 October 2010.
E/C.12/NLD/4/Add.2, 19 October 2010 (Aruba).
E/C.12/2010/SR.30, 8 November 2010.
E/C.12/2010/SR.43, 18 November 2010.
E/C.12/NDL/CO/4-5, 19 November 2010.
E/C.12/2010/SR.45, 24 November 2010.
E/C.12/NLD/Q/4-5, 22 December 2009.
E/C.12/NLD/Q/4-5/Add.1, 14 October 2010.
E/C.12/2010/SR.43, 18 November 2010.
E/C.12/NDL/CO/4-5, 19 November 2010.
E/C.12/2010/SR.44, 25 March 2011.

International Convention on the Rights of the Child

CRC/C/51/Add.1, 24 July 1997.
CRC/C/Q/NET.1, 30 June 1999.
CRC/C/SR.578, 16 November 1999.
CRC/C/SR.579, 8 October 1999.
CRC/C/SR.580, 18 November 1999.
CRC/C/90, 7 December 1999.
CRC/C/61/Add.4, 4 October 2001.
CRC/C/Q/NET-ANT/1, 8 February 2002.
CRC/C/118, 3 September 2002.
CRC/C/117/Add.1, 5 June 2003.
CRC/C/117/Add.2, 17 June 2003.
CRC/C/RESP/48, 16 December 2003.
CRC/C/Q/NLD/2, 27 October 2003.
CRC/C/SR.928, 23 January 2004.
CRC/C/SR.929, 23 January 2004.
CRC/C/15/Add.226, 26 February 2004.
CRC/C/NLD/3, 23 July 2008.
CRC/C/NLD/3, 23 July 2008 (Part 2).
CRC/C/NLD/3, 23 July 2008 (Part 3).
CRC/C/SR.1376, 20 January 2010.
CRC/C/SR.1377, 23 January 2009.
CRC/C/NLD/CO/3, 27 March 2009.

CEDAW

CEDAW/C/NET/l, 7 April 1993.
CEDAW/C/NET/1/Add.1, 17 September 1993.
CEDAW/C/NET/1/Add.2, 30 June 1993.
CEDAW/C/NET/1/Add.3, 18 October 1993.
CEDAW/C/SR.234, 25 January 1994.
CEDAW/C/SR.239, 4 February 1994.
A/49/38, 12 April 1994.
CEDAW/C/NET/2, 15 March 1999.
CEDAW/C/NET/2/Add.1, 15 March 1999.
CEDAW/C/NET/2/Add.2, 15 March 2000.
CEDAW/C/NET/3/Add.1, 27 October 2000.
CEDAW/C/NET/3, 22 November 2000.
CEDAW/C/NET/3/Add.2, 22 November 2000.
CEDAW/C/SR.512, 5 September 2001.
CEDAW/C/SR.513, 7 September 2001.
A/56/38, 31 October 2001.
CEDAW/C/NLD/4, 10 February 2005.
CEDAW/C/NLD/4/Add.1, 7 June 2005.
CEDAW/C/NLD/Q/4, 4 August 2006.
CEDAW/C/NLD/Q/4/Add.1, 27 October 2006.
CEDAW/C/SR.768 (B), 2 February 2007.
CEDAW/C/NLD/CO/4, 2 February 2007.
CEDAW/C/SR.767 (B), 5 March 2007.
CEDAW/C/NLD/5, 24 November 2008.
CEDAW/C/NLD/Q/5, 13 March 2009.
CEDAW/C/NLD/4/Add.2, 19 May 2009.
CEDAW/C/NLD/5/Add.2, 19 May 2009.
CEDAW/C/NLD/5/Add.1, 3 September 2009.
CEDAW/C/NLD/Q/5/Add.1, 19 October 2009.
CEDAW/C/SR.916, 3 March 2010.
CEDAW/C/SR.917, 18 March 2010.
CEDAW/C/NLD/CO/5, 5 February 2010.

Universal Periodic Review

A/HRC/WG/6/1/NLD/1, 7 March 2008.
A/HRC/WG.6/1/NLD/2, 13 March 2008.
A/HRC/8/31, 13 May 2008.
A/HRC/8/31/Add.1, 25 August 2008.
A/HRC/WG.6/13/NLD/1, 8 March 2012.
A/HRC/8/31, 13 May 2008.
A/HRC/8/31/add.1, 25 August 2008.

Bibliography

A/HRC/21/15, 9 July 2012.
A/HRC/21/15/Add.1/Rev.1, 12 October 2012.

NGO shadow-reports

Contribution to the Dutch Section of the International Commission of Jurists (NJCM) and the Johannes Wier Foundation (JWS) to the Committee on Economic, Social and Cultural Rights, provisional reaction to the Third periodic report submitted by the Netherlands under Articles 16 and 17 of the Covenant (E/1994/104/Add.30).
Joint Parallel Report to the Combined Fourth and Fifth Periodic Report of the Netherlands on the International Covenant on Economic, Social and Cultural Rights, 28 October 2009.
Addendum to the Joint Parallel Report to the Combined Fourth and Fifth Periodic Report of the Netherlands on the International Covenant on Economic, Social and Cultural Rights (ICESCR) (28 October 2009), 16 September 2010.
Shadow-report on the right to food in the Netherlands, 17 November 2009, European Institute for Food Law.

National periodic reports submitted by Belgium

ICESCR

E/1990/5/Add.15, 13 May 1993.
E/1990/5/Add.15, 13 May 1993.
E/C.12/1994/SR.15, 17 May 1994.
E/C.12/1994/SR.16/Ad.1, 18 May 1994.
E/1995/22, 20 May 1994.
E/1990/6/Add.18, 5 March 1998.
E/C/12/Q/BELG/1, 13 December 1999.
E/C.12/2000/SR.64, 27 November 2000.
E/C.12/2000/SR.66, 14 March 2002.
E/C.12/1/Add.54, 1 December 2000.
E/C.12/BEL/3, 21 September 2006.
E/C.12/BEL/Q/3, 10 April 2007.
E/C.12/BEL/NGO/3, 23 October 2007.
E/C.12/BEL/Q/3/Add.1, 1 November 2007.
E/C.12/2007/SR.41, 29 February 2008.
E/C.12/2007/SR.42, 10 December 2007.
E/C.12/BEL/CO/3, 4 January 2008.
E/C.12/BEL/4, 18 June 2012.

ICRC

CRC/C/11/Add.4, 6 September 1994.
CRC/C.9/WP.4, 17 February 1995.

Belg/1, 04 May 1995.

CRC/C/SR.222, 6 June 1995.

CRC/C/15/Add.38, 20 June 1995.

CRC/C/83/Add.2, 25 October 2000.

CRC/C/Q/BELG/2, 8 February 2002.

CRC/C/RESP/7, received on 3 May 2002.

CRC/C/15/Add.178, 13 June 2002.

CRC/C/SR.782, 24 July 2002.

CRC/C/SR.783, 29 May 2002.

CRC/C/BEL/3-4, 4 December 2009.

CRC/C/SR.1521, 10 June 2010.

CRC/C/SR.1523, 11 June 2011.

CEDAW

CEDAW/C/5/Add.53, 25 September 1987.

CEDAW/C/BEL/2, 8 April 1993.

CEDAW/C/SR.300, 1 February 1996.

CEDAW/C/SR.301, 6 February 1996.

A/51/38, 9 May 1996.

CEDAW/C/BEL/3-4, 29 September 1998.

CEDAW/C/SR.559, 25 June 2005.

CEDAW/C/SR.560, 20 December 2002.

A/57/38 part II, 8 October 2002.

CEDAW/C/BEL/6, 22 June 2007.

CEDAW/C/BEL/Q/6, 6 March 2008.

CEDAW/C/BEL/Q/6/Add.1.

CEDAW/C/SR.852, 7 January 2009.

CEDAW/C/SR.853, 26 December 2008.

CEDAW/C/BEL/CO/6, 7 November 2008.

NGO shadow reports

Shadow-report of the NGOs on the implementation of the International Convention on the Rights of the Child. Original title in Dutch: Alternatief Rapport van de NGOs over de toepassing van het Internationaal Verdrag inzake de Rechten van het Kind in België, 2010, available at: www.kinderrechtencoalitie.be.

Universal Periodic Review

A/HRC/WG.6/11/BEL/1, 16 February 2011.

Bibliography

Reporting formats

Universal Periodic Review, formats

A/HRC/DEC/6/102, 27 September 2007.
A/HRC/DEC/17/119, 19 July 2011.

ICRC formats

CRC/C/58/Rev.2, 23 November 2010.

CEDAW formats

HRI/GEN/2/Rev.1/Add.2, 5 May 2003.

Common Core Documents formats

HRI/GEN/2/Rev.4, 21 May 2007.

Websites

www.arableagueonline.org; website of the League of Arab States, includes a link to the activities of the Arab Human Rights Committee.
www.achpr.org; website of the African Commission on Human and Peoples' Rights.
http://www.const-Court.be; official website of the Belgian Constitutional Court.
www.african-Court.org; official website of the African Court on Human and Peoples' Rights.
http://www.denederlandsegrondwet.nl; website related to the Dutch Ministry of Internal Affairs and Kingdom Relations.
http://diplomatie.belgium.be/en/policy/development_cooperation/partnerships/special_programmes/belgian_fund_food_security; official website of the Belgian Fund for Food Security.
https://dofi.ibz.be; official website of the Belgian Immigration Department.
www.fao.org; Food and Agricultural Organisation website.
www.ifrc.org; official website of the International Red Cross.
http://jure.juridat.just.fgov.be; official website for Belgian case law.
www.kinderrechtencoalitie.be; website of Belgian NGO in the field of the rights of the child.
http://www.kruispuntmi.be; a Flemish expertise centre on migration, integration and cultural diversity.
http://www2.ohchr.org/english/bodies/crc/workingmethods.htm; official working methods of the Committee on the Rights of the Child.
www.raadvst-consetat.be; official website of the Belgian Council of State.
www.rechtspraak.nl; official website of the Dutch Judiciary.
www.righttofood.org; website of Special Rapporteur on the right to food Jean Ziegler.
www.rvv-cce.be; website of the Council for Aliens Disputes (Belgium).

www.srfood.org; website of Special Rapporteur on the right to food Olivier de Schutter.

http:/treaties.un.org/Pages/Treaties.aspx?id = 4&subid = A&lang = en; overview of signatories, accessions and ratifications of all UN human rights treaties.

http://treaties.un.org; official UN treaty database.

www.voedingscentrum.nl; website of the Dutch Food and Nutrition Information Office.

www.vwa.nl; website of the Dutch Food and Consumer Safety Authority.

www.wfp.org/hunger; the Section of the WFP website on which the concepts of hunger and malnutrition are further defined.

Speeches

H.E. Carlo Azeglio Ciampi, President of the Italian Republic, 10 June 2002, inaugural ceremony of the World Food Summit, five years later. Available at: http://www.fao.org/DOCREP/ MEETING/005/Y7106E/Y7106E02.htm.

Summary

Part I. Introduction, methodology and *tertium comparationis*

Chapter 1. Introduction

While the right to adequate food is often discussed in the context of developing countries, especially in situations where access to adequate food is a problem on a larger scale, this book focusses on the right to food in two western countries in which theoretically the circumstances allow this right to be enjoyed by each individual: the Netherlands and Belgium. This book addresses the question whether in those countries an individual, that falls through the cracks of all national safety nets and finds her/himself deprived of basic sustenance to lead a life in dignity, can invoke the right to adequate food, as embedded in international human rights treaties, in the domestic Courts.

The main research objectives of this thesis are:
1. to gain knowledge about the enforceability of the right to food as embedded in UN human right instruments in the Netherlands and Belgium through law comparison; and
2. where necessary critically evaluate both approaches in light of the UN human right doctrine regarding the enforceability of the right to food.

The main research question is:
What are the legal factors that explain the differences and similarities regarding the response of the Dutch and Belgium Judiciaries and Governments to the enforceability of the right to adequate food in view of the UN human rights system?

This research thus implies a triple comparison:
I.   a comparison between the legal practice in Belgium and the Netherlands (what the countries really *do*);
II. a comparison of those legal practices with the reporting behavior of both countries (what the countries *say* they do); and
III. a comparison between the legal practice and the interpretation on the enforceability of the right to food within the UN human rights system (what the countries *should* do).

To be able to make a comparison, it is necessary to first collect and analyze the relevant data. Therefore, three sub-questions need to be answered:
1. To what extent is the right to adequate food perceived to be an enforceable right within the UN human rights system?

2. A. What is the response of the national Judiciaries of the Netherlands and Belgium when the right to food as stipulated in the UN human rights system is invoked by individuals?
   B. And how can this response be explained?
3. What do the Governments of the Netherlands and Belgium communicate in their reports in the United Nations arena regarding the enforceability of the right to food in their domestic legal order?

Chapter 2. Methods

This study applies legal comparative methods. However, in comparative law, no coherent methodology seems to have been adopted. Most methodological considerations refer back to a short passage on the functional method, as proposed by Zweigert and Kötz. While it has often been criticized, the functional method was not successfully replaced by a better alternative. Therefore, this study applies functionalism, but in a modern interpretation introduced by Ralf Michaels. It can be qualified as a functional method based on equivalence functionalism and on an epistemology of constructive functionalism. The constructive move that comes from this approach to functionalism will be used to determine the research direction regarding the sources that need to be examined and compared. These sources are necessary to answer the threefold sub-questions on how Member States should respond, do respond and say they respond to the need of enforceability of the right to food. This constructive move is not as flexible as Michaels proposed, but has a rigid core that is determined by the UN human rights system. In the comparative process, the extent to which the right to adequate food is perceived to be an enforceable right within the UN human rights system will then serve as the *tertium comparationis*. This will be determined by inventorying the international, regional and national provisions that either stipulate the right to food, or are related to the right to food, and the viewpoints of the relevant UN or UN related institutions that respond to the need to further clarify its meaning (sub-question 1). In this light, the countries' social responses will be analyzed (answering sub-questions 2 and 3). In order to determine and explain the response of the national Judiciary of the Netherlands and Belgium when the right to food as stipulated in the UN human rights system is invoked by individuals, the reasoning patterns of the Courts are leading in establishing the research direction (sub-question 2). To determine the communication of the Governments of the Netherlands and Belgium in their reports in the United Nations arena regarding the enforceability of the right to food in their domestic legal order, the reporting cycles on the implementation of the ICESCR, ICRC, CEDAW, and the UPR will be analyzed (sub-question 3). Next to answering the sub-questions to find and explain the relevant data, the three comparisons are necessary to answer the main research question, and will be performed accordingly. Finally, in line with the second research objective, the Dutch and Belgian approach towards the enforceability of the right to food are critically evaluated in light of the UN human rights doctrine.

Chapter 3. The enforceability of the international human right to adequate food

Since World War II, considerable but also difficult progress was made to develop the right to food in the international human rights arena. This involved many actors who contributed from their different perspectives and out of different expertise. The work that has been done to further clarify the right to food, especially in light of Article 11 ICESCR, can be subdivided into three pillars. The first pillar is the work done by the treaty bodies, such as the CESCR. The second pillar is the work done within FAO context, which includes the World Food Summits and the adoption of the Voluntary Guidelines. The third pillar is the work done in the context of the Human Rights Council, in particular the works of the special rapporteurs.

The right to food is most discussed in the context of Article 11 ICESCR. Limiting the right to food to the interpretation of only this Article, however, would do no justice to its full meaning. A survey through international human rights instruments shows that the right to food is also recognized directly in other documents that mostly aim at the protection of a particular group of individuals. Furthermore, the right is inextricably linked to other human rights or human rights related issues. Also, the right to food is recognized outside the UN context on a regional level. Finally, on a domestic level, it is sometimes recognized directly or indirectly in constitutions or used as a directive principle.

As it appears, the relevant articles and explanatory documents written in UN context seem to offer sufficient guidance to at least come to a minimum specification of what 'adequate food' means as a substantive right. While the realization of human rights is a responsibility for all, the international human rights system specifically addresses States.

The traditional distinction made between civil and political rights on the one hand, implying negative state obligations, and economic, social and cultural rights on the other, implying positive state obligations has been criticized for decades now. There are many good arguments to oppose such a distinction both from a practical point of view and from a legal theoretical perspective. Instead, a typology of duties applicable to all human rights, consisting of duties to respect, protect and fulfill seems to do more justice to the meaning of human rights. It demonstrates that ECOSOC rights do not leave States an undefined margin of discretion in the realization of these rights.

In general, according to the CESCR, Article 2 ICESCR implies that there is a minimum core obligation regarding ECOSOC rights, that requires immediate realization. In addition, there are immediate obligations of conduct to move as expeditiously and effectively as possible towards the full realization of the right. Ultimately, this is an obligation of result.

It is difficult to establish whether in the UN context the right to food is unanimously considered to be an enforceable right, and if so, what it exactly means. The UN can hardly be considered to be one single entity speaking with one voice. A *tertium comparationis* in this light therefore, is not unambiguous. In general, no evidence could be found that immediate obligations of conduct are enforceable. The discussion on enforceability focusses more on the States' core obligations. The specialized bodies of the UN, basically the CESCR and the Special Rapporteurs, consider, with sound arguments, that the right to food and ECOSOC rights in general can be understood to imply immediate state obligations that should be enforceable through domestic Courts. There seems to be at least a core content, consisting of a minimum substance without whom the right would be stripped of its *raison d'être*, and a non-discrimination principle, that should be justiciable, or even self-executing. On the other hand, in the UN fora in which intergovernmental decision making procedures are used, the enforceability seems to be hardly supported. An approach that resembles the behavior of Member States towards the human rights treaties stipulating ECOSOC rights, in particular when considering the rich use of reservations and the reluctance to adopt complaint mechanisms.

Part 2. The Netherlands

Chapter 4. Dutch case law on enforceability

The international right to food is not directly applicable in the Netherlands. Article 11 ICESCR seems to be completely non-enforceable. However, case law seems to emerge in which ICRC standards, including Article 27, are used as an interpretative standard that is used in the treaty conform interpretation of national legislation. As a result, the ICRC provisions had a significant effect on the verdicts. The claimants are usually minors seeking asylum who lawfully reside in the Netherlands without a residence permit, and are in distressing humanitarian circumstances. Furthermore, the Supreme Court used Article 27 for a treaty conform interpretation in a landmark ruling. The Court decided that the COA could not remove illegally residing children from their facilities, and thus had to provide for the child's basic needs. However, such cases are still exceptions to the rule that Article 27 is not applied. More than once, in rejecting the direct applicability of ECOSOC standards, reference was made to the Parliamentary Documents in which the Government, in their function of co-legislator, expressed that in their view, direct applicability of the ICESCR treaty would not be possible. It seems that the Courts follow this lead of politics in determining whether these rights can be directly invoked.

Chapter 5. Dutch monism and constitutional reforms

Since 1953, the Netherlands have quite an obscure constitutional system regulating the position and effect of provisions under international law. This so called 'qualified monistic system' embedded in Articles 93 and 94 of the Dutch Constitution, enables the Courts to apply international Provisions directly, but only when they are binding on all persons and have been published. Article 93 Constitutional Act (CA) determines what international legal standards are binding on individuals. Article 94 Constitutional Act regulates the relation between these international directly applicable standards and national legislation.

Especially the prerequisite that a provision must be binding on all persons has led to much confusion, in particular with regard to the question whether the Government or the Courts have a final say in the matter. Officially, the Government and the Courts agree that it is up to the Judiciary to finally decide whether or not a provision is binding on all persons. It was however the intention of the Constitutional Legislator to establish a legal practice in which the legislator would voice his opinion on direct applicability of international legal standards in their Parliamentary Documents. This is indeed done ever since by the Legislature when adopting a Bill on the ratification of human rights treaties.

Chapter 6-8. The direct applicability of economic, social and cultural rights in the Netherlands

The Dutch Judiciary seems to follow the lead politics gave in the ratification Bill when a decision must be made on the direct applicability of ECOSOC rights. According to the legislator, civil and political rights in general are suitable for direct applicability, and ECOSOC rights are not, due to the fact that for the realization of ECOSOC rights policy decisions must be made, which is the responsibility of politics instead of the Courts. This point of view is persistently expressed by the Government in their various Explanatory Memoranda on the Bills on the ratification of the international human rights treaties, and during the various reporting procedures. The Government expresses similar opinions in case of international treaties for specific groups that address at the same time civil and political as well as economic, social and cultural rights, albeit not always as specific as for instance the Explanatory Memorandum to the ratification Bill of the ICESCR. It appears that when the Government did express itself clearly on the direct applicability of human rights provisions, normally, the Judiciary doesn't hesitate to apply this view in their rulings. However, when the Government is not so specific, the Courts appear to take the liberty in reaching their own conclusions with regard to the matter. Their considerations are then of a more practical nature and closer related to the case. This is clearly demonstrated with regard to the right to strike (Article 6 ESC), which is often referred to in literature when scholars address the issue of direct applicability of ECOSOC rights.

Four criteria can be distinguished to determine whether an international provision containing a human right is directly applicable: (1) the intentions of the state-parties of the international document expressed during the negotiations preceding its adoption; (2) the nature and content of the provision, as well as its wording; (3) the intentions of the national legislator expressed in its explanatory memorandum; and (4) the existence of legislation that has the purpose of fulfilling the obligations coming forth from the provision in question. The fourth criterion was only mentioned in the Parliamentary Documents of the Act of approval of the ICRC. It was hardly explained and has not been mentioned elsewhere since then. Furthermore, due to the various constitutional systems worldwide it seems unlikely that state parties will express their vision on direct applicability during the negotiations preceding and international human rights document. This leaves criteria (2) and (3) as the criteria that truly matter. There seems to be a difference between the hierarchy as explained by both the Dutch Legislature and the Judiciary on the one hand, and the hierarchy that is actually used by the Courts with regard to direct applicability ECOSOC rights. Officially, the Courts should look at the nature and content of the provisions, as well as its wording in the first place, and use the expressed intention of the legislator as a source of inspiration only. However, this research demonstrates that in situations in which the Legislature clearly expressed their view on the possibility of direct applicability of international ECOSOC standards, the Courts duly follow this interpretation, without giving due consideration to the nature, content or wording of the provision. Only when the legislator is unclear in their vision, the Courts feel free to base their verdict on the nature, content and wording of the article.

Chapter 9. The Dutch periodic country reports

At the time of ratification of the right to food in several treaty provisions, the Dutch Government underlined that the Netherlands already fulfilled the obligations that follow from the ratification of this right. This practice seems to be standard procedure in case of the ratification of any economic, social and cultural right. A treaty provision will only be ratified if the Dutch legislator is of the opinion that the Netherlands already meet the obligations resulting from the human right in question. If there is any doubt, the provision will not be ratified, or ratified with reservations or interpretative declarations.

Also during the process of the diverse report procedures that form part of the obligations coming forth from the ratification of international human rights treaties, the Netherlands seem to have no ambition to improve the recognition of human rights within its borders, which seems to be contradicting with the obligation of progressive realization that can be found in various human rights treaties. Especially in the context of the ICESCR, this attitude led to increasing differences of opinion between the Dutch Government and the Committee on Economic, Social and Cultural Rights. In addition, the Dutch Government does not seem to be willing to seriously answer the questions of the treaty based bodies that consider these reports, while the quality of the reports is more than once considered to be doubtful.

The Dutch Government reported that within the Netherlands, there is sufficient food of good quality and affordable for everyone, basically due to a sound system of social security. However, most cases in which the right to food is invoked concern foreigners, who do not always have access to this system of social security. This is often discussed with and criticized by the Committees during the reporting cycles.

In general, it can be concluded that the Netherlands, also in their reports, make a sharp distinction between on the one hand, civil and political rights, and on the other hand, economic, social and cultural rights, especially when it concerns the possibility of direct applicability. It is unambiguously clear that the Netherlands does not recognize direct applicability of ECOSOC rights.

Chapter 10. Evaluation and comparison

As it appears, the Dutch reports are window-dressing when painting a positive image of the Dutch implementation of internationally recognized ECOSOC rights, including the substantive right to food. Simultaneously, the Dutch do not seem to be eager to answer critical questions, or to reflect on points for improvement. This has led to some remarkable statements of the Dutch Government on the domestic case law. For example that the Dutch judges are familiar with the General Comments of the CESCR, and also apply them in their verdicts, or that Article 11 ICESCR has some effect in Dutch case law. It appears from the discussions with the various Committees that these treaty bodies have a rather accurate impression regarding the status of Dutch case law. However, it cannot be said that the Dutch reporting behaviour is in full conformity with legal reality.

Part 3. Belgium

Chapter 11-12. The legal practice of Belgium regarding enforceability

Due to differences in language, culture and economy, Belgium has, since their profound State reforms in 1970, a rather complex *trias politica*. The Legislature consists of three legislative powers, equally and exclusively competent. This had an effect on the organisation of the Judiciary, culminating in the instalment of the Constitutional Court. The Belgian Judiciary seems organized, based on purpose and functionality, rather than pragmatic considerations. The organisation is complex. On the one hand there is a subjective contentieux, with two Courts of last instance (the Court of Cassation and the Council of State), and on the other hand an objective contentieux, with two highest tribunals (the Council of State and the Constitutional Court).

The difference between both contentieux increasingly becomes less important for the parties involved in a court case. The legal effects of the verdicts seems to 'merge' rather than remain of a fundamentally different nature. Nevertheless, the Judicial organisation has a strong effect on the legal reasoning of the Courts. Belgium is a monistic State. This does not follow directly from the Constitution, but from case law of the Court of Cassation. This means that in principle, an international standard is superior to a contradicting (anterior and posterior) national rule, as long as the international rule has direct effect. For an international standard to have direct effect, some formal requirements have to be fulfilled. The main criterion, however, is that the State Parties involved have expressed the intention that the standard should have direct effect. Such expression can be recognized when the standard addresses citizens and not to States (the subjective criterion) and/or the standard is self-executing (objective criterion). Neither the Court of Cassation nor the Council of State appear to have recognized direct effect of the right to adequate food. As a rule, with some minor, seemingly random, exceptions, they consider that ICESC and ICRC provisions do not have direct effect. There is hardly any case law on (the direct effect of) CEDAW provisions.

The Constitutional Court is not bound by the principle of direct effect due to its function to review national legislation against the Constitution. As a result, it applies international standards only indirectly, initially via Articles 10 and 11 and 24 CA, and later, also via all other rights stipulated in chapter II CA. The 'school-pact' prompted the Legislature to broaden the reviewing competencies of the Constitutional Court. It is now authorized not only to settle conflicts between the three legislative powers, but also to review national legislation against fundamental rights. This is also reflected in the case law of the Court. The first rulings in which the Court indirectly reviewed against internationally embedded ECOSOC rights concerned Article 13 ICESCR. Only later, the Court also reviewed (mostly indirectly) against other ECOSOC provisions, including articles stipulating the right to adequate food.

It is mostly in cases concerning illegally residing aliens that right to food provisions, and other related ECOSOC rights embedded in UN treaties, play a role. In several cases, both in annulment procedures and in preliminary rulings, the Court reviewed Article 57 § 2 of the '*Organic Law of 8 July 1976 on public centers for social welfare*'. This provision restricts social benefits entitlements of illegally residing asylum seekers to urgent medical care only. The Court assessed whether this restriction, in light of the invoked fundamental rights, was objectively and reasonably justified, and proportionate towards a legitimate goal. The basic principle seems to be that it is the legitimate purpose of Belgium to restrict immigration, and that therefore it is not unreasonable to accept different obligations regarding the needs of legally residing persons in Belgium on the one hand and illegally residing persons on the other hand. That is to say that Article 11 ICESCR cannot reasonably be understood without limitations, and therefore, does not apply to persons Belgium is not responsible for, such as illegally residing aliens. In a series of rulings, the Court has further specified and nuanced this notion. Occasionally the Court allows for exceptions from this rule. In particular, the rights of the child usually seem to have more effect than rights stipulated in the ICESCR. In general, in line with its function, the Court shows little hesitation in nullifying legislation, or advising in preliminary proceedings that a Constitutional standard is violated, when it considers the disputed national legislation not objectively and reasonably justified, and/or a disproportionate means to reach a justified goal. Therefore, verdicts of the Constitutional Court often resulted in amendments to legislation, or the adoption of new regulations. However, while international provisions in this case law are often reviewed against indirectly, no evidence could be found that those provisions made any difference of significance towards the outcome of the rulings on top of the – mostly equivalent – invoked national provisions. This is due to the 'inward' focus of the case law, involving mainly an assessment of national legislation and policies.

Chapter 13. The Belgian periodic country reports

In its reports, Belgium seems to prefer to communicate initiatives, policies and legal measures that positively contribute to the implementation of human rights. The reports show little self-evaluation or self-critique. The right to food is seldom addressed as a right in itself. Instead the Government seems to focus on related rights such as the right to healthcare, the right to social security, and more in particular, the position of foreigners in relation with entitlements to social benefits. Regarding the latter, the Committees often express concern. In particular they do not seem to be convinced that withholding rights based on the status of illegally residing foreigners, is compatible with the relevant human rights treaties, especially not in case of accompanied and unaccompanied children.

Furthermore, the complexity of the Belgian federal structure seems to be a reason for concern. The Committees regularly question whether Belgium is capable of a uniform and consistent approach in implementing human rights. It also leads to misunderstandings, especially regarding the issue of direct effect in combination with the function of the Constitutional Court. The delegations underline the direct effect of international ECOSOC provisions by referring to case law of the Constitutional Court. However, they fail to point out that this is at odds with the case law of the Court of Cassation and the Council of State. It appears that more than once, the impression is created that international human rights treaties, such as the ICRC and the CEDAW, have direct effect in the national legal order, even though this is contrary to the legal practice. The Committees seem to be poorly informed on the Belgian constitutional system. They do not really address the issue in-depth. The Belgian Government on the other hand is reluctant in correcting those misunderstandings.

Chapter 14. Evaluation and comparison

It must be concluded that the Belgian reports on the enforceability of the ICESCR, the ICRC and the CEDAW hardly match legal reality. Indeed, the complex federal structure, especially regarding the organisation of the Judiciary, in combination with the Committees' lack of knowledge of Belgian constitutional law, seems to lead to misunderstandings and inaccurate assumptions. Legal practice shows that ECOSOC rights in general, with some rare exceptions, do not have direct effect in the subjective contentieux. Nevertheless, in several reporting cycles it was suggested or assumed that the both ICRC and CEDAW do in their entirety have direct effect. Interestingly, the case law of the Constitutional Court is often invoked by Belgium to demonstrate the direct effect of international human rights provisions, or to paint a positive picture on the implementation of treaty provisions, while such examples seem inaccurate, for there is at most a little indirect effect for the international provisions. In addition, its added value *in casu* should not be overestimated.

Part 4. Conclusion and recommendations

Chapter 15. Conclusions

This research found its origin and inspiration in the analysis of two Court rulings, one Dutch and one Belgian, in which entitlements based on the right to food were denied to a claimant. This sparked the question which added value such a human right would have in these two countries, which have the means to ensure adequate food to all within their jurisdiction. In line with the nature of human rights, one would expect that such rights basically would serve as an ultimate safety net, to catch those individuals that somehow fall through the domestic safety nets. The more so, as both countries communicate in their reports on the implementation of various treaties to fulfill all their duties the right to food implies. In short, the question was raised what the two countries actually do, what they say they do, and what they should do regarding the enforceability of the right to adequate food.

In light of the question what the countries should do, it is difficult to exactly establish whether the right to food is considered an enforceable right in the UN arena, when considering the viewpoints of all bodies that are related to the matter, including those in which intergovernmental decision-making procedures are followed, based on the principle of consensus. However, the bodies that are installed specifically to contribute to the realisation of ECOSOC rights in general, such as the Committees and the Special Rapporteurs, have frequently addressed the urgency of recognizing enforceability of these rights. They certainly consider a core content of the right to food justiciable, or even self-executing.

Regarding the question what the countries actually do, this research has shown that unfortunately it is the main rule both in the Netherlands and Belgium, that the right to food does not function as a safety net, and that it is hardly possible to successfully invoke the right to food in Court of law. I say 'hardly', because there are some developments in which the Courts do show some compassion for those people who are in most distressing situations, and allow international provisions that stipulate the right to food to have some (indirect) effect in their verdicts. It is those exceptions to the rule that we could be hopeful for. In the Netherlands that is mostly the increasing use of the ICRC as an interpretative standard, whereas in Belgium it is the increasing indirect review against international standards through Constitutional Provisions. The general impression however is that it are the coincidental Constitutional circumstances that mainly determine the enforceability of the right to food, rather than the actual content of the right in itself. In the Netherlands, that is mainly the rather obscure construction of Articles 93 and 94 CA, while in Belgium it is the complex organisation of *trias politica*.

As it seems, most cases in which the right to food is invoked, concern foreigners, illegally residing on the Dutch or Belgian territory. It is remarkable that in both Countries, any possible effect of an international Provision stipulating the right to food seems to emerge from a balancing of interests, in which the domestic policies to restrict immigration are the counterbalance of the individual interest of the foreigner. This leads to the impression that the Courts through their case law not only respond to the shared social problem that the right to food should be an enforceable right (in this research the *tertium comparationis*), but also respond to another problem, that is the restriction of immigration.

Regarding the question what countries say they do, this research has shown that there is a considerable discrepancy between the legal practice in Belgium and the Netherlands on the one hand, and what is reported on this practice on the other hand. Both countries paint a positive picture of their performances in human rights implementation, but devote little space to self-evaluation or self-criticism regarding the implementation of ECOSOC rights in general, and the right to food in particular. Neither country evidences much ambition to improve the implementation of the rights enshrined in the treaties they ratified. They certainly do not consider this to be part of their State's duties. In the international arena thus, both countries want to emphasize their accurate implementation of human rights, while this contradicts the domestic legal practices.

Considering the above, it can be concluded that the countries do not do what they should do, and do not say what they really do concerning the enforceability of the right to adequate food.

Chapter 16. Recommendations

Countries should act in accordance to the obligations they have subscribed to when ratifying human rights treaties and they should truthfully report what they do. To this end, the study concludes with recommendations to the Netherlands, Belgium, the UN Committees and legal comparists.

About the author

B.F.W. (Bart) Wernaart teaches law and ethics at Fontys University of Applied Sciences in Eindhoven and 's-Hertogenbosch, the Netherlands. He is also a professional musician (drums, mallets and percussion), composer and conductor. Bart Wernaart was born in Eindhoven, the Netherlands, in 1983. After graduating from high school, he took courses at Tilburg University and the music academy of Brabant. He holds a master degree of international law, with a specialism in human rights, and two bachelor degrees of arts, with specialisms in drums, vibraphone and conducting. Since 2007, he worked on his PhD-thesis under the supervision of Professor Bernd van der Meulen, at the Law and Governance group of Wageningen University and Research Centre, Wageningen, the Netherlands, of which this book is the final result. Bart Wernaart was raised in a true teachers family, has two younger brothers, and lives together with his fiancée Sylvia in Dommelen, the Netherlands.

List of publications

Wernaart, B.F.W., 2009. Veiled justice. The courts'compassionate case law regarding hunger. In: Hospes, O. and Van der Meulen, B.M.J. (eds.) Fed up with the right to food? Wageningen Academic Publishers, Wageningen, the Netherlands.

Wernaart, B.F.W., 2010. The plural wells of the right to food. In: Hospes, O. and Hadiprayitno, I. (eds.) Governing food security, law, policies and the right to food. Wageningen Academic Publishers, Wageningen, the Netherlands.

Wernaart, B.F.W., 2010. The right to water in the Netherlands. European Food and Feed Law Review 5: 361-364.

The research described in this thesis was financially supported by Fontys University of Applied Sciences.

Financial support from the Law and Governance Group at Wageningen University for editing this thesis is gratefully acknowledged.

Printed in the United States
by Baker & Taylor Publisher Services